C0-AVL-263

Aquatic Weeds

The Rawa Manumbo, a reservoir for the storage of irrigation water in West Java (Indonesia), covered by a dense vegetation of aquatic weeds (mainly water hyacinth, *Eichhornia crassipes*). (Photo: A. H. Pieterse)

Aquatic Weeds

The Ecology and Management of Nuisance Aquatic Vegetation

Edited by

ARNOLD H. PIETERSE
Royal Tropical Institute
Amsterdam

and

KEVIN J. MURPHY
University of Glasgow

OXFORD UNIVERSITY PRESS
1990

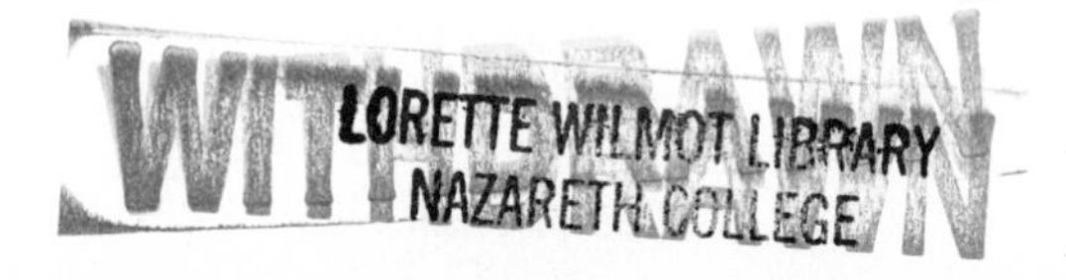

Oxford University Press, Walton Street, Oxford OX2 6DP
Oxford New York Toronto
Delhi Bombay Calcutta Madras Karachi
Petaling Jaya Singapore Hong Kong Tokyo
Nairobi Dar es Salaam Cape Town
Melbourne Auckland
and associated companies in
Berlin Ibadan

Oxford is a trade mark of Oxford University Press

Published in the United States
by Oxford University Press, New York

British Library Cataloguing in Publication Data
Aquatic weeds
1. Aquatic weeds. Control measures
I. Pieterse, Arnold H. II. Murphy, Kevin J.
632′.58
ISBN 0–19–854181–3

Library of Congress Cataloging in Publication Data
Aquatic weeds: the ecology and management of nuisance aquatic vegetation/edited by Arnold H. Pieterse and Kevin J. Murphy.
p. cm. Bibliography: p. Includes index.
1. Aquatic weeds—Control 2. Aquatic weeds—Ecology.
I. Pieterse, A. H. II. Murphy, Kevin J.
SB614.A73 1989 628.9′7—dc 19—89–3146 CIP
ISBN 0–19–854181–3

Printed and bound by
Courier International,
Tiptree, Essex

'Although the desire to control nature may be censured by the purist as an attitude of mind conceived in arrogance, it unquestionably reflects a contemporary economic and agricultural necessity. If resources of land and water are to be efficiently exploited in the provision of food, other natural products, power and communications, an effort must clearly be made to resist and, if possible, eliminate biological factors which would otherwise quickly thwart this aim. To acknowledge this need is not to condone such clumsy and short-sighted attempts to control weeds and pests as have in recent years evoked proper public outrage at the use of certain herbicides and insecticides.'

C. Duncan Sculthorpe
The biology of aquatic vascular plants
(1967)

'Human progress depends in part upon man's ability to regulate and control water supplies to suit his requirements. As he has acquired and developed the technological skill to manage resources by controlling floods, conserving water in reservoirs, tapping its underground resources, pumping or piping it from wet to arid regions, or using it as a medium for transport, so have the problems caused by aquatic plants increased. Unfortunately man's skill at, and knowledge of, engineering have not been matched by his understanding of the intricacies of biological systems and his ability to maintain them in a stable condition. Man's history shows him, for the most part, to be a despoiler of ecosystems and, from bitter experience, he has learnt that there is a limit to nature's bounty.'

David S. Mitchell
Aquatic vegetation and its use and control
(1974)

PREFACE

In 1984 the EWRS (European Weed Research Society) Working Group on Aquatic Weeds decided to take the initiative for publishing a new textbook on aquatic weeds. It was generally felt that such a book was urgently needed. The only textbook which specifically dealt with aquatic weeds and their management on a global basis was the UNESCO publication *Aquatic vegetation and its use and control*, edited by D. S. Mitchell, published in 1974. This included chapters on the ecology of aquatic weeds, the effects of excessive aquatic plant populations, means of control, and utilization. Although this valuable work is still very useful, there have been many new developments since 1974. However, the more recent publications which give general information on aquatic weeds are either connected with specific regions or are relatively brief texts in books which are not primarily concerned with aquatic weeds, such as books on weed control in general, or aquatic ecology.

As well as the members of the EWRS Working Group on Aquatic Weeds, scientists from all over the world were invited to contribute to this new book. The book is divided into three main parts. The first part is concerned with concepts, ecology, and characteristics of aquatic weeds and also includes chapters on flow resistance and the relation between aquatic weeds and public health. The second part covers the management of aquatic weeds, with chapters on various control methods, surveying and modelling of aquatic weed vegetation, utilization of aquatic weeds, and the relation between plant survival strategies and control measures. The third part deals with the present status of aquatic weed problems in the various continents.

As well as the editors, the other members of the EWRS Working Group on Aquatic Weeds: Maria Arsenović, Pip Barrett, Dale Robson, and Max Wade, have greatly assisted in the preparation of the book, from its conception. We wish to thank them, as well as the other contributors: L. W. J. Anderson, E. P. H. Best, G. Bowes, K. H. Bowmer, N. F. Cardarelli, R. Charudattan, C. D. K. Cook, F. H. Dawson, O. A. Fernández, I. W. Forno, E. O. Gangstad, B. Gopal, K. L. S. Harley, J. H. Irigoyen, J. C. Joyce, V. H. Lallana, D. S. Mitchell, R. H. Pitlo, M. R. Sabbatini, W. Spencer, D. L. Sutton, K. K. Steward, W. van Vierssen, and W. van der Zweerde. We also thank the Oxford University Press for support and encouragement throughout the production of the book.

A.H.P.

September 1988 K.J.M.

CONTENTS

EWRS WORKING GROUP ON AQUATIC WEEDS

The European Weed Research Society (EWRS) is an organization which promotes and encourages weed research and weed control technology in Europe for the benefit of the community as a whole. Within the EWRS, in the framework of main subject areas, working groups have been set up with a specific scientific and technical task.

The Working Group on Aquatic Weeds, which falls under the main subject area 'Weed Control in Non-Crop Areas', is the oldest working group within the EWRS. It started as an acknowledged working group within the European Weed Research Council (EWRC), an organization which was replaced by the EWRS in 1975. Its formation was the result of the activities of a preliminary committee which was set up at a meeting of the EWRC in Oxford in 1960. This committee, which consisted of W. Holz (Federal Republic of Germany), H. Johannes (FRG), J. Lhoste (France), J. V. Spalding (UK), J. A. Timmermans (Belgium), and H. G. van der Weij (The Netherlands), organized in 1964 an International Symposium (the first) on Aquatic Weeds in La Rochelle, France.

In 1966 the preliminary committee became officially acknowledged as a working group within the EWRC and in 1967 the second International Symposium on Aquatic Weeds was organized in Oldenburg, FRG. During this symposium T. O. Robson, from the Weed Research Organization in Oxford, UK, was selected as chairman of the working group, a function which he was to fulfil until his retirement in 1983.

In 1971 the third International Symposium was organized in Oxford, UK, and in 1974 the fourth in Vienna, Austria. Subsequently, when the EWRC was replaced by the EWRS, it was agreed that the group would continue. The main objectives of the group within the EWRS were:

(1) to promote the interchange of information on aquatic weed problems between members of the Society by organizing symposia at regular intervals;
(2) to encourage co-operation between research scientists working in similar fields on common problems, by forming groups to consider specific topics;
(3) to develop and maintain contact with other international organizations with similar interests; and
(4) to stimulate and encourage members to contribute items of information on aquatic weeds and news of individuals working on them to the EWRS Newsletter.

As in the previous period a main activity of the group remained the organization of international symposia, which included the production and publication of symposium proceedings. The fifth International Symposium on Aquatic Weeds was organized in Amsterdam, The Netherlands in 1978, and the sixth in Novi Sad, Yugoslavia in 1982. In 1983, A. H. Pieterse (the Netherlands), who together with J. C. J. van Zon (The Netherlands) had joined the group in 1976, took over the chairmanship. Subsequently, in 1986 the seventh International Symposium was held in Loughborough, UK.

In 1984 the Working Group on Aquatic Weeds, which then consisted of M. Arsenović (Yugoslavia), P. R. F. Barrett (UK), K. J. Murphy (UK), A. H. Pieterse (The Netherlands), T. O. Robson (UK), and P. M. Wade (UK), decided to take the initiative for the publication of the present book.

The work of the group continues, with new members joining, further symposia planned for future years, and new collaborative ventures also at the planning stage.

CONTRIBUTORS

Anderson, L. W. J.

United States Department of Agriculture, Agricultural Research Service, Aquatic Weed Research Laboratory, University of California, Botany Department, Davis, California 95616, USA.

Arsenović, M.,

Institute for Plant Protection, Veljka Vlahovica 2, 21000 Novi Sad, Yugoslavia.

Barrett, P. R. F.

AFRC Institute of Arable Crops Research, Aquatic Weeds Research Unit, Sonning Farm, Sonning-on-Thames, Reading RG4 0TH, UK.

Best, E. P. H.

Centre for Agrobiological Research, Bornsesteeg 65, 6700 AA Wageningen, The Netherlands.

Bowes, G.

University of Florida, Department of Botany, 220 Bartram Hall East, Gainesville, Florida 32611, USA.

Bowmer, K. H.

CSIRO, Division of Water Resources, Private Mail Bag, Griffith, New South Wales 2680, Australia.

Cardarelli, N. F.

Division of Engineering and Science Technology, University of Akron, Akron, Ohio, USA.

Charudattan, R.

Plant Pathology Department, University of Florida, Gainesville, Florida 32611, USA.

Cook, C. D. K.

Institute for Systematic Botany and Botanical Garden, University of Zürich, Zollikerstrasse 107, Zürich 8008, Switzerland.

Dawson, F. H.

Institute of Freshwater Geology, River Laboratory, East Stoke, Wareham, Dorset BH20 6BB, UK.

Fernández, O. A.

Departamento de Agronomía and Centro de Recursos Naturales Renovables de la Zona Semiárida (CERZOS), Universidad Nacional del Sur, 8000 Bahía Blanca, Argentina.

Forno, I. W.

CSIRO, Division of Entomology, Long Pocket Laboratories, Private Bag 3, Indooroopilly, Queensland 4068, Australia.

Gangstad, E. O.

US Army Corps of Engineers, Washington DC, USA.

Gopal, B.

Jawaharlal Nehru University, School of Environmental Sciences, New Mehrauli Road, New Delhi 110067, India.

Harley, K. L. S.

CSIRO, Division of Entomology, Long Pocket Laboratories, Private Bag 3, Indooroopilly, Queensland 4068, Australia.

Irigoyen, J. H.

Departamento de Agronomía and Centro de Recursos Naturales Renovables de la Zona Semiarida (CERZOS), Universidad Nacional del Sur, 8000 Bahía Blanca, Argentina.

Joyce, J. C.

University of Florida, Institute of Food and Agricultural Sciences, Center for Aquatic Plants, 7922 N.W. 71st Street, Gainesville, Florida 32606, USA.

Lallana, V. H.

Facultad de Ciencias Agropecuarias, Universidad Nacional de Entre Rios, 3100 Paraná, Entre Ríos, Argentina.

Mitchell, D. S.

The Murray-Darling Freshwater Research Centre, P.O. Box 921, Albury 2640, New South Wales, Australia.

Murphy, K. J.

University of Glasgow, Department of Botany, Glasgow G12 8QQ, UK.

Pieterse, A. H.

Royal Tropical Institute, Rural Development Programme, Mauritskade 63, 1092 AD Amsterdam, The Netherlands.

Pitlo, R. H.

Ministry of Agriculture and Fisheries, Advisory Group Vegetation Management, Bornsesteeg 69, 6708 PD Wageningen, The Netherlands.

Robson, T. O.

18 Kidsley Close, Chesterfield S40 4XA, UK.

Sabbatini, M. R.

Departamento de Agronomía and Centro de Recursos Naturales Renovables de la Zona Semiárida (CERZOS), Universidad Nacional del Sur, 8000 Bahía Blanca, Argentina.

Spencer, W.

University of Florida, Department of Botany, 220 Bartram Hall East, Gainesville, Florida 32611, USA. (*Current address*: Department of Biology, Natural Science Building, University of Michigan, Ann Arbor, Michigan 48109–1048, USA.)

Steward, K. K.

United States Department of Agriculture, South Atlantic Area, Aquatic Weed Research, 3205 College Avenue, Fort Lauderdale, Florida 33314, USA.

Sutton, D. L.

Institute of Food and Agricultural Sciences, University of Florida, 3205 College Avenue, Fort Lauderdale, Florida 33314, USA.

van Vierssen, W.

International Institute for Hydraulic and Environmental Engineering, Oude Delft 95, P.O.B. 3015, 2601 DA Delft, The Netherlands.

Wade, P. M.

Department of Geography, Loughborough University of Technology, Loughborough, Leicestershire LE11 3TU, UK.

van der Zweerde, W.

Centre for Agrobiological Research, Bornsesteeg 65, 6700 AA Wageningen, The Netherlands.

PART I

CONCEPTS, ECOLOGY, AND CHARACTERISTICS OF AQUATIC WEEDS

1

Introduction

A. H. Pieterse

DEFINITIONS

A WEED can be defined as 'a plant (or group of plants) which is not desired at its place of occurrence' (Coordinatiecommissie Onkruidonderzoek, 1984). Although 'not desired' may in general be referred to 'not desired by man', the definition could be narrowed to 'not desired by the manager(s) of the area where the particular plant(s) occurs'.

This definition can also be applied to an aquatic weed. It should be taken into consideration, however, that as a rule weed problems in aquatic habitats are of a somewhat different nature than those occurring in terrestrial habitats. In terrestrial habitats weeds are unwanted mainly because they compete with crop plants or ornamentals, i.e. they have a negative effect on the growth of plants which are wanted by man. Exceptions are plants which are unwanted on non-cultivated strips of land, like railways and road verges. In the aquatic habitat weed problems are generally caused by the growth of dense vegetation which hampers the (often multi-functional) use of water bodies, for example by interfering with irrigation and hydroelectric schemes, fisheries, or navigation. On the other hand, when aquatic plants occur in low densities they are usually very beneficial to the aquatic ecosystem because they produce oxygen, purify the water by trapping silt particles, take up toxic compounds (or support associated populations of micro-organisms which do this) and provide habitat for fish fry and other organisms.

Therefore, a more appropriate definition of an aquatic weed is 'an aquatic plant which, when growing in abundance, is not desired by the manager of its place of occurrence'. This definition, however, does not completely hold for aquatic weeds which interfere with the growth of crop plants cultivated in an aquatic habitat (e.g. weeds in rice fields). The role of aquatic weeds under such circumstances is more comparable to that of weeds in most terrestrial habitats. Thus, if we want to take into consideration aquatic weeds which compete with crops the definition could be 'an aquatic plant (or group of plants) which is not desired by the manager(s) of the water body where it occurs, either when growing in abundance or when interfering with the growth of crop plants or ornamentals'.

In this context it should be noted that most water bodies have various functions. An aquatic weed vegetation may hamper one function and favour another one. Therefore, the manager has to consider all the various aspects, and balance the pros and cons, in order to decide whether the vegetation is troublesome (or potentially so) to the functioning of the water body.

In connection with the definition of an aquatic weed it is also of interest to define the difference between terrestrial and aquatic plants. Often there is a gradual transition from the terrestrial to the aquatic habitat which makes it difficult to make a sharp distinction. Den Hartog and Segal (1964) defined aquatic plants as plants which are able to achieve their generative cycle when all vegetative parts are submerged or are supported by the water, or which occur normally submerged but are induced to reproduce sexually when their vegetative parts are exposed due to emersion. Sculthorpe (1967), Cook *et al.* (1974) and Denny (1985) also included plants with partial aerial foliage which inhabit the aquatic environment with their basal parts. In the present book the latter definition will be adopted, which implies that all plants growing in an aquatic habitat will be considered to be aquatic weeds when they interfere with man's use of water bodies by forming dense vegetation, or when they interfere with the growth of plants which are wanted by man.

FIG. 1.1. *Eichhornia crassipes* (water hyacinth).

CATEGORIES

In order to describe aquatic plant communities, aquatic plants are commonly divided into categories according to their growth form. Various growth forms may be distinguished (den Hartog and Segal 1964; Bloemendaal and Roelofs 1986). However, largely following Denny (1985), a simple classification of aquatic weeds will be used in the present book which implies a division into five main groups:

Group 1. *Free-floating weeds* (Free-floating with most of the leaf and stem tissue at or above the water surface). Examples are *Eichhornia crassipes* (Fig. 1.1), *Salvinia* spp. (Fig. 1.2(a), (b)), *Pistia stratiotes* (Fig. 1.3(a), (b)), and *Azolla* spp.

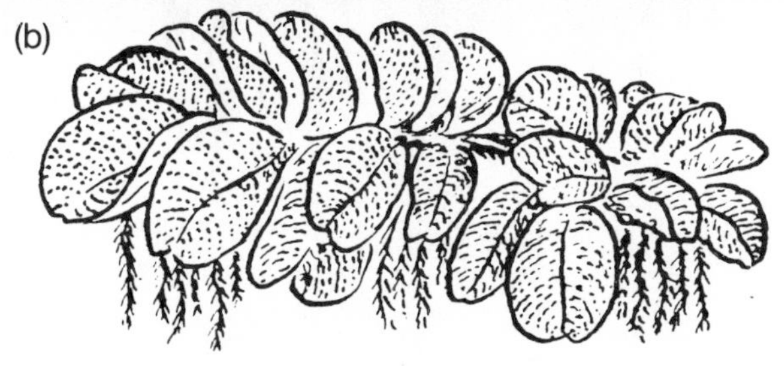

FIG. 1.2. (a) A dense mat of *Salvinia molesta* (Photo: A. H. Pieterse). (b) Habit of *Salvinia molesta*.

FIG. 1.3. (a) A vegetation of *Pistia stratiotes* (water lettuce) (Photo: A. H. Pieterse). (b) Habit of *Pistia stratiotes* (water lettuce).

FIG. 1.4. *Phragmites australis*.

FIG. 1.5. *Typha* spp.

Group 2. *Emergent weeds* (Rooted plants with most of their leaf and stem tissue above the water surface). Examples are *Phragmites* spp. (Fig. 1.4), *Typha* spp. (Fig. 1.5), *Carex* spp., and *Alternanthera philoxeroides* (Fig. 1.6).

Group 3. *Rooted weeds with floating leaves* (Rooted plants with most of the leaf tissue at the water surface). Examples are *Nymphaea* spp. (Fig. 1.7), *Nuphar* spp., and *Nymphoides* spp.

FIG. 1.6. *Alternanthera philoxeroides* (alligator weed).

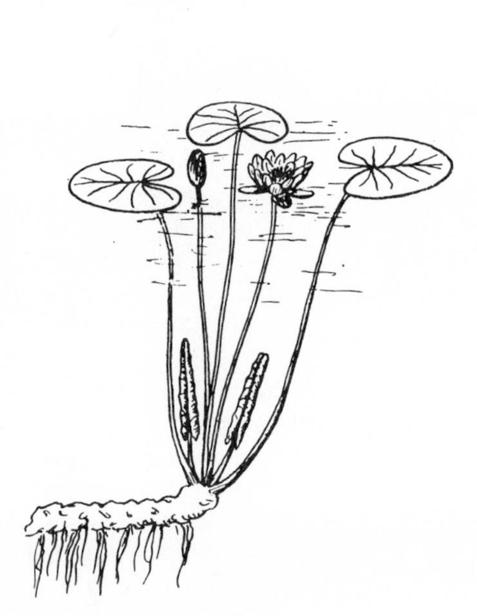

FIG. 1.7. *Nymphaea alba*.

FIG. 1.8. *Hydrilla verticillata*.

Group 4. *Submerged weeds* (Most of the vegetative tissue beneath the water surface; they are rooted or attached to the bottom of a water body by root-like organs). Examples are *Hydrilla verticillata* (Fig. 1.8), *Potamogeton* spp. (Fig. 1.9, 1.10), *Elodea* spp. (Fig. 1.11), and *Myriophyllum* spp. (Fig. 1.12).

Group 5. *Algae* (Unicellular or filamentous lower plants without differentiated tissues which grow at, or below, the water surface). Examples are *Microcystis* spp., *Spirogyra* spp., and *Hydrodictyon* spp.

The first four groups comprise spermatophytes (seed-bearing plants), pteridophytes (ferns and fern allies), bryophytes (mosses and liverworts), and also the charophytes (stoneworts), which, despite the fact that they are representatives of a special group of algae, are categorized in Group 4 because they are attached to the bottom and, as far as control is concerned, are more comparable to submerged vascular plants.

This classification is superficial and adopted only for practical reasons. Certain

FIG. 1.9. *Potamogeton pectinatus.*

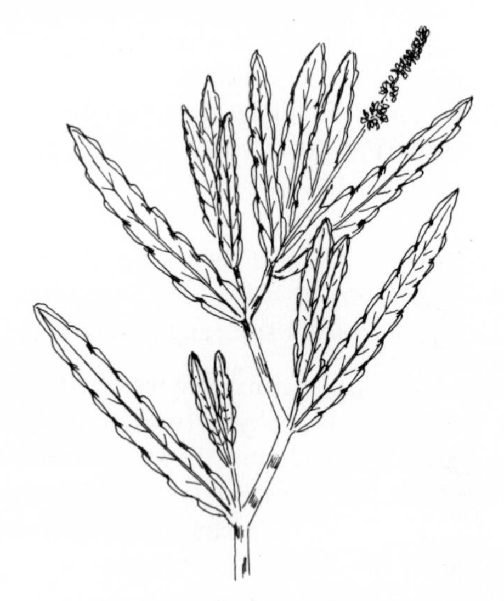

FIG. 1.10. *Potamogeton crispus.*

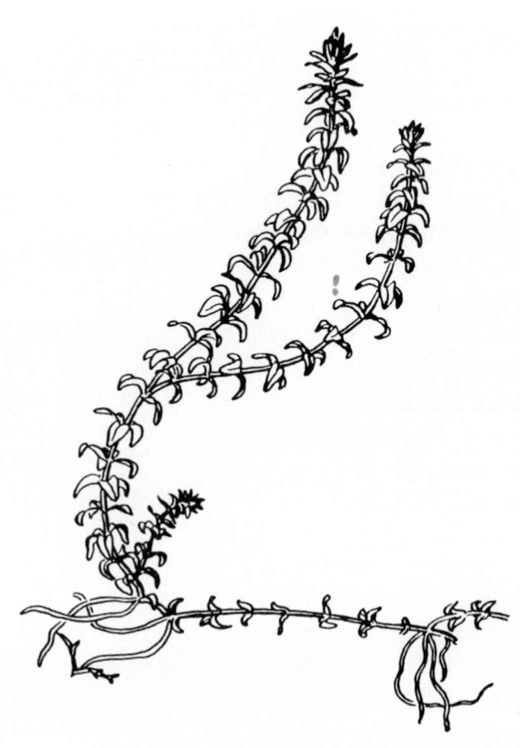

FIG. 1.11. *Elodea* spp.

FIG. 1.12. *Myriophyllum* spp.

plants may exhibit different life forms, depending upon the depth and the flow of the water. For example, in deep, flowing water, various species of *Sagittaria* and *Sparganium* remain submerged, whereas they normally emerge above the water surface. In addition, there are plants which generally remain submerged but under certain conditions become emergent, such as *Myriophyllum verticillatum*. Some plants are not satisfactorily categorized by this classification. This applies in particular to plants which float below the water surface, like *Lemna trisulca* and, under certain conditions *Ceratophyllum* spp. (Fig. 1.13), although the latter are commonly anchored in the bottom with root-like organs and may be included in Group 4. According to their growth form various algae would also belong to this type if they were not categorized as a separate group (mainly because they pose somewhat different problems from vascular plants). Plants with both floating and submerged leaves are here included in Group 3. The proposal of Denny (1985) to combine submerged plants, plants which are anchored to the substratum with floating leaves, and plants with both floating and submerged leaves, in a group called 'euhydrophytes', is not followed.

The habitats of the different life forms are often primarily determined by water depth. However, other factors may also play a role. Examples are turbidity of the water (and consequent light attenuation rate below the water surface), and water movement. A typical zonation, i.e. the occurrence of different zones (characterized by different plant species and growth forms) from shore to deeper water, com-

FIG. 1.13. *Ceratophyllum demersum*.

prises a community of emergent plants, followed by rooted plants with floating leaves, and then by submerged plants. Free-floating plants are commonly found in shallow as well as deep water; in general, their growth is strongly affected by water movement.

Water flow and wave action can also bring about an effect on the soil composition of the bottom of a water body, which may directly influence zonation. In relatively stagnant water smaller soil particles will settle out more readily than in turbulent water and, as a consequence, the zonation at the windward side of a lake can be different from the leeward side in response to the differing soil types present.

Apart from these horizontal differences in aquatic plant vegetation there are also vertical differences, i.e. the water column can be occupied by different growth forms. This stratification of aquatic plant vegetation is strongly influenced by the underwater light climate which is, in turn, influenced by water turbidity (Spence 1982).

It should also be noted that, with the exception of the free-floating species, the aquatic weeds which occur in water bodies used for the cultivation of crops such as rice (*Oryza sativa*), water cress (*Nasturtium officinale*), and water morning glory (*Ipomoea aquatica*), are generally different from the weeds which are troublesome in deeper waters (Soerjani, Kostermans, and Tjitrosoepomo 1987). These water bodies represent a different type of ecosystem: in most cases they are very shallow and temporarily created by flooding a piece of land, with a soil which becomes waterlogged.

CHARACTERISTICS

Aquatic weeds are generally characterized by a rapid vegetative growth and the ability to regenerate by asexual means, i.e. via fragmentation (production of new plants from small plant segments) or vegetative hibernating organs (tubers and turions). The factors, however, which contribute to their rapid development are often connected with the normal pattern of succession. Some of the most troublesome aquatic weeds are essentially primary colonizers of aquatic ecosystems. If control measures are not carried out and factors like nutrient content of the water, light, temperature, or water movement do not bring about a limiting effect on plant growth, shallow waters will eventually be filled up with vegetation. This will lead to swamp formation and, ultimately, transformation into a terrestrial habitat.

In addition to the natural succession of aquatic vegetations in shallow waters the development of aquatic weeds is enhanced by eutrophication, i.e. an increase in the nutrient content of the water as a consequence of human activities.

Another human influence on the environment which has caused severe aquatic weed problems is the spread of species out of their natural habitat. Notorious examples are *Eichhornia crassipes*, which originally occurred in South America and is now a pest in most tropical areas, *Hydrilla verticillata*, which was recently introduced to the USA with disastrous results, and *Elodea canadensis*, which was spread into Europe during the latter half of the nineteenth century. These plants were, and are, not particularly troublesome in their native areas, as they were in balance with natural enemies (see Chapter 3).

In general, it may be concluded that the characteristic explosive developments of aquatic weeds seen in many waters worldwide are a result of man's interference with the environment, either because succession is being prevented or because certain components of the ecosystem have been changed, or a combination of both.

HARMFUL EFFECTS

The negative effects which are brought about by aquatic weeds can be divided into direct and indirect effects. The direct effects include:

(1) impeding transport of irrigation and drainage water in canals and ditches (Fig. 1.14);
(2) hindering navigation (Fig. 1.15);

FIG. 1.14. An irrigation canal in the Nile Valley in Egypt blocked by an aquatic weed vegetation which mainly consists of *Eichhornia crassipes* (water hyacinth). (Photo: A. H. Pieterse.)

FIG. 1.15. A dense vegetation of water hyacinth hindering navigation in the Damietta Branch of the Nile in Egypt. (Photo: A. H. Pieterse.)

(3) interfering with hydroelectric schemes;
(4) increasing sedimentation by trapping silt particles;
(5) decreasing human food production in aquatic habitats (fisheries, crops);
(6) decreasing the possibilities for washing and bathing;
(7) adversely affecting recreation (swimming, waterskiing, angling).

The flow resistance of aquatic vegetation, which is especially important in irrigation and drainage canals, is discussed in detail in Chapter 5. As far as interference with boat movements is concerned this is mainly a problem in tropical areas where weed growth is not periodically retarded during a winter season, but can be a serious problem also in temperate, seasonally used, recreational boat-traffic-bearing waterways, such as the canal system of Great Britain (Murphy and Eaton 1983). In the tropics it can be of great economic importance in areas where the local transport mainly occurs on water. In particular, floating mats of aquatic weeds may cause serious problems as they can cut off remote settlements from the outside world completely.

Interference by aquatic weeds with hydroelectric schemes has generally not been a severe problem, although when the water intakes are close to the surface they can be blocked by drifting masses of aquatic weeds (see Chapter 15). It has been reported (van Zon 1982) that various algae may cause corrosion in concrete and steel.

When access to the water is hampered by vegetation along the shore by a water body this can be troublesome for fishing, swimming, and boating. Especially in areas where recreation is a major economic activity this may lead to considerable financial losses.

Indirect negative effects of aquatic weeds include:

(1) water loss by means of evapo-transpiration (transpiration via plants);
(2) increase of health hazards by the formation of habitats which are favourable for the development of vectors of human diseases, such as malaria and schistosomiasis (bilharzia).

It has been stated repeatedly in the literature that evapo-transpiration, especially in arid zones, can be an important cause of water losses (Little 1967; Timmer and Weldon 1967; van der Weert and Kamerling 1974; Mitchell 1974; Benton, James, and Rouse 1978). However, the data vary considerably. Estimates of the evapo-transpiration to open water transpiration ratio (E/E_0) were often above 1.4, and some were above 3.0. The discrepancies may be related to differences in methodology, but climatic conditions and growth form of the vegetation may also play an important role (DeBusk, Ryther, and Williams 1983).

Recently Boyd (1987) and Snyder and Boyd (1987) determined E/E_0 ratios for several species of aquatic plants over a six-month period, in Alabama (USA). The ratios varied from 1.17 to 1.75. Moreover, there are reports that E/E_0 ratios in swamps would be around 1.0 (Rijks 1969; Linacre *et al.* 1970). It may be concluded that aquatic plants, when they completely cover the surface of a water body, may increase water losses appreciably. However, in all probability the losses are not as great as previously suggested.

Health aspects are discussed in detail in Chapter 6. Mosquito larvae and aquatic snails which are intermediate hosts of human diseases thrive in dense aquatic weed vegetations in certain parts of the world. Therefore, aquatic weeds may play an indirect role in affecting the health of local human populations.

2

General biology and ecology of aquatic weeds

P. M. WADE

TO UNDERSTAND and manage aquatic weed problems effectively, it is important that the large body of knowledge which has been built up about the biology and ecology of aquatic plants is used as the foundation of management approaches. This chapter highlights those aspects of aquatic weed biology and ecology important in this context. Ecophysiological aspects are covered in more detail in Chapter 4.

The first requirement is an ability to recognize the various groups and species of aquatic plants which cause weed problems. The second need is to understand the ecological and biological mechanisms which regulate aquatic plant growth and species and community development. Armed with this knowledge it becomes possible to manipulate the development or structure of the macrophyte community, or to interfere at vulnerable stages of the growth or development of the target weeds in order to achieve management aims (see also Chapter 12).

PLANT RECOGNITION

Fundamental to the success of aquatic plant management is the correct taxonomic identification of the plant(s) to be managed. This information permits the selection of an appropriate control agent for use against the weed(s) (e.g. Ministry of Agriculture, Fisheries and Food 1986; Gangstad 1976; Westerdahl and Getsinger 1988). Correct knowledge of the other species occurring within a system will give an indication of the habitat conditions (Melzer 1976; Hellawell 1978; Standing Committee of Analysts 1987) and consequently the likely pattern of recolonization. Also of importance is the identification of important food plants for wildfowl (Mason 1957) and the occurrence of protected species (Perring and Farrell 1983).

Cook *et al.* (1974) provided a comprehensive catalogue of the genera of aquatic macrophytes of the world. None of the continents, however, enjoys a full description within the confines of a single publication though Hotchkiss (1972) comes close to this for the USA and Canada and Aston (1973) for Australia. Wade (1987) reviewed the provision made for the identification of aquatic plants around the world and listed over 100 manuals and other publications to aid the recognition of the aquatic flora of the six continents. The geographical coverage of the world is uneven and the provision for Africa, parts of Asia and South America is poor compared to Europe, Australia, and North America. Certain plant groups, notably bryophytes and macrophytic algae, are poorly covered or not dealt with at all within these texts.

Specialist handbooks are available to aid the recognition of some of these groups. For example, the charophytes are covered by Moore (1986) for the UK; Corillion (1957) for France; Zanveld (1940) for Malaysia; Wood and Mason (1977) for New Zealand; and Wood and Imahori (1965) for the world.

Other relevant publications include the first of a series of manuals for the aquatic and amphibious plants of Costa Rica and Central America (Gomez 1984), a text on the freshwater plants of Papua New Guinea (Leach and Osborne 1985) and a guide to the recognition of the aquatic plants of the British Isles (Spencer-Jones and Wade 1986).

A useful start to the recognition of an aquatic weed is made by determining the life and growth form adopted by the plant in question (see Chapter 1). All aquatic weeds can be described within the five basic categories: free-floating, emergent, floating-leaved rooted, submerged macrophytes, and algae.

GENERAL BIOLOGY

Aquatic plants have evolved a range of adaptations to survive in the freshwater environment. An understanding of these biological adaptations is important for successful management of aquatic weeds. For example, a knowledge of the pattern of rooting and root production is fundamental to the mechanical control of *Phragmites australis* (Haslam 1968; see also Chapter 8). Shading effects will be different depending upon the photosynthetic characteristics of the species concerned (Dawson and Hallows 1983), and the physiological state of a plant can influence the efficacy of herbicidal treatment (Barrett and Robson 1971; see also Chapter 7).

One of the earliest textbooks on the biology of water plants was by Arber (1920). Gessner (1955, 1959) made the physiology of aquatic plants (both freshwater and marine) the emphasis of his treatise on aquatic botany and both these texts are a valuable source of information on the importance of light, temperature, water movement, and metabolic processes. Subramanyan (1962) prepared a general treatise on aquatic angiosperms. Sculthorpe (1967) wrote an important and comprehensive text on the comparative biology of freshwater and marine vascular plants. Mitchell (1974) edited the first textbook on aquatic weeds, and Hutchinson (1975), although focusing on the lacustrine freshwater habitat, dealt with a substantial amount of information on the biology and ecology of aquatic species. His aim was to concentrate on those subjects not treated at length in such works as Gessner (1955, 1959) and Sculthorpe (1967). In addition to the higher plants, his text deals with charophytic and benthic algae, and bryophytes.

A number of other publications are useful sources of information. These include the following:

(i) the *Proceedings of the European Weed Research Council* and the *European Weed Research Society Symposia on Aquatic Weeds* (1964, 1967, 1971, 1974, 1978, 1982, 1986) available from EWRS Secretariat, CABO, Wageningen, The Netherlands;
(ii) the *Proceedings of the International Symposia on Aquatic Macrophytes* (e.g. Symoens, Hooper, and Compère 1982; Faculteit der Wiskunde en Natuurwetenschappen 1983; Sand-Jensen and Sondergaard 1986);
(iii) books and proceedings of other relevant symposia (e.g. Crawford 1987; Good, Whigham, and Simson 1978; Pokorny *et al.* 1987);
(iv) the journals *Aquatic Botany* (published by Elsevier, Amsterdam) and the *Journal of Aquatic Plant Management* (prior to 1976: *Hyacinth Control Journal*) (published by the Aquatic Plant Management Society, Vicksburg, Mississippi, USA);
(v) literature on aquarium plants (e.g. de Wit 1971).

ECOLOGY

The way in which a plant species or community interacts with its environment will ultimately determine the role of that species or community within a given habitat. This will depend on the biology of the constituent species and also upon conditions prevailing in the habitat, and changes in these conditions.

Autecological studies

The study of the individual species in relation to its environment (autecology) can provide valuable information about the way a species will react to management. Autecological studies were initially closely associated with taxonomic and distribution studies of aquatic plant taxa, for example *Alisma* (Bjorkqvist 1967, 1968) and *Eleocharis* (Svenson 1929, 1939). More recently studies have been undertaken, wholly or partly, because species or genera are aquatic weeds. Gopal (1987) for instance, provided a wealth of information about the distribution, morphology, anatomy, phenology, reproductive biology, and ecology of *Eichhornia crassipes*. Those autecological studies which have been undertaken have usually dealt with the emergent species, such as the genera *Carex* (Bernard 1988), *Glyceria* (Lambert 1946, 1947, 1949) *Phragmites* (Haslam 1969, 1972; Rezk and Edany 1981) and *Typha* (Morton, 1975; Smith 1987). Notable exceptions are: *Salvinia* (Mitchell 1970); water-lilies (Heslop-Harrison 1955*a, b*; Masters 1974; Swindells 1983), *Callitriche* (Schotsman 1967, 1977), *Ruppia* (Weber 1979); Lemnaceae (Landolt 1986, 1987), *Myriophyllum* (Anderson 1985), *Eichhornia* (Gopal 1987), and the macroalgal taxon Characeae (Imahori 1954; Olsen 1944; Pereyra-Ramos 1982; Forsberg 1956*b*; Krause 1981; Wood 1972). Intraspecific variation within macrophyte species also falls under this broad heading: for example studies on *Hydrilla verticillata* (e.g. van Vierssen, Breukelaar, and Peppelenbos 1986; Cook and Lüönd 1982; Verkleij and Pieterse 1986). Chapter 4 in this book deals in detail with the ecophysiology of the world's most troublesome aquatic weeds, including aspects of their photosynthesis and growth, nutrition and reproduction.

Although a knowledge of the autecology of a species can provide useful information for the management of a species of water weed, a sounder basis is provided by an understanding of the ecology of the community to which the species belongs.

Community ecology

Aquatic macrophyte community interactions have proved to be a challenging area for investigation. Many studies have simply described the assemblages of plants found in different locations and attempted to relate them to the conditions recorded there. Such studies tended to rely on extensive descriptions of plant communities with the emphasis on the individual species (Samuelsson 1934; Luther 1951*a, b*; Swindale and Curtis 1957; Hejny 1960; Koumpli-Sovantzi 1983).

Phytosociologists have attempted to classify and describe the various types of macrophyte communities according to the schools of European phytosociology (du Roetz *et al.* 1939; den Hartog and Segal 1964; Meriaux 1978; Spence 1964; Gehu 1983) although, as den Hartog (1983) pointed out, such classification of aquatic plant communities has always lagged behind that of terrestrial plant communities. Other classification techniques, based on association analysis (Williams and Lambert 1966) and ordination techniques (Orloci 1966), have been applied to freshwater macrophyte communities. These techniques have been used to generate working hypotheses of relationships between community composition and the factors in the environment which might regulate that composition.

Classification and ordination techniques have been applied successfully to plant communities in a variety of freshwater habitats, for example, lakes (Seddon 1965; Collins, Sheldon, and Boylen 1987), canals (Murphy and Eaton 1983; Murphy, Fox, and Hanbury 1987), and drainage channels (Doarks 1984).

Aquatic plant communities are dynamic entities and it is important to appreciate how and why they change. A programme of weed control does not usually simply remove one species from within a community, but will alter the relationships between the constituent species by disturbing the environment, altering competitive interactions, or creating stress, so disrupting the natural pattern of developmental change occurring in the community. Weed control measures can, therefore, play a major role in the community dynamics of managed freshwater vegetation. Some of these dynamic community processes have been investigated though there are many gaps in our knowledge.

A newly created water body offers considerable scope for colonization by aquatic plants. The seeds and spores of some species may be dispersed to the site by the wind (e.g. *Typha* spp. and *Phragmites australis*) but the most important agents are animals and particularly waterfowl (Ridley 1930; Samuelsson 1934; Proctor 1968). The seeds of many species can float for days or even months and if the new habitat is occasionally flooded, or an existing habitat is extended or altered, these seeds will be dispersed by water (Praeger 1913; Ridley 1930). Fragments of plants, winter buds, turions, and other propagules can also be transported via water, aided by the wind and currents.

Animals have also been shown to transport such material. Frankland, Bartley, and Spence (1987) and Cook (1987) considered dispersion in aquatic plants and concluded that the spread of many aquatic plants is largely dependent on vegetative propagules. These are often complex in structure and sometimes show dormancy and resistance to drying (see also Chapter 12). The factors which control the induction of germination of such propagules were discussed by Frankland, Bartley, and Spence (1987), with an emphasis on environmental signals which may be important under water, including light and oxygen concentration. Haag (1983) described the germination of macrophyte seedlings from lake hydrosoils, but Hutchinson (1975) and Cook (1987) drew attention to the limited knowledge about aquatic plant propagules and, in particular, their mobility.

Combining the data collected by Godwin (1923) on the aquatic flora of a series of small isolated water bodies, with data from contemporary surveys of the same sites and applying biogeographical theory based on MacArthur and Wilson (1967), the colonization rate by vascular plants in such waters is 0.20–0.22 species per year and the extinction rate 0.09–0.60 species per year (P. M. Wade, unpublished data). Hutchinson (1975) used data from Harris and Silvey (1940), collected from four artificial lakes in Texas, to calculate 2.5–6.1 introductions and extinctions per year at equilibrium. Further research is necessary to understand these processes more fully, and in particular the invasive behaviour of some aquatic plants, e.g. *Elodea canadensis* and *Eichhornia crassipes* (see Chapter 4). More attention should be paid to the control of the dispersal and germination of propagules as a part of aquatic plant management. The transfer of boats from one piece of water to another has been established as an important mechanism of infecting a water body with potential problem weed species (Johnstone, Coffey, and Howard-Williams 1985).

Ecophysiological and morphological adaptation

Certain species have the ability to establish themselves rapidly in a newly created or recently disturbed freshwater habitat. Macroalgae such as charophytes and filamentous algae are very successful in this respect. Another example is *Callitriche truncata* (Wade, Vanhecke, and Barry 1986). These opportunist species have evolved to exploit severely disturbed but often potentially productive environments, typically relying on high propagule production over a short period of time. These plants correspond to the ruderals, or disturbance-tolerators, (R-strategists) as described by Grime (1979), who emphasized the importance, as a regulatory factor for such plants, of partial or total removal of the species by disturbance to the habitat. Such disturbance can occur in aquatic habitats, for example, by increased water velocity in rivers (spates), or due to annual herbicide treatment, but displacement by other immigrant species within a few seasons of initial colonization is also important in influencing the survival of R-strategist aquatic macrophytes. Although seeds are of paramount importance in the reproductive phase of terrestrial species, aquatic plants have adapted to rely on vegetative propagules, such as turions and rhizome fragments, as well as seeds. R-strategist macrophyte species are likely to become aquatic weeds when the habitat is subject to regular severe disturbance (e.g. annual herbicide treatment of drainage channels). If the habitat is allowed to develop naturally such species will cease to be a problem.

As aquatic plants establish themselves in a freshwater habitat, communities become established which can be arranged along a gradient from those which are 'biologically accommodated' to those which are 'physically controlled' (Sanders 1968). The former refers to communities where physiological stress arises predominantly from biological interactions (e.g. competition). The latter refers to communities where physiological stress results primarily from physical factors (e.g. fluctuations in temperature) and thus where biological interactions are less important.

Aquatic plant species all utilize light, the ions of mineral nutrients, and space, and compete with each other for these resources. In contrast to terrestrial plants, water is not a resource in limiting supply. However, the supply of inorganic carbon for use in photosynthesis can be limiting, and aquatic plants can experience low oxygen levels (see also Chapter 4).

Emergent aquatic plants compete for light in similar ways to terrestrial plants and the radiation climate has been described for various emergent plant communities (Ondok 1978). A remarkably diverse group of floating plants has exploited the water surface of freshwater habitats. In many instances these plants have, thereby, secured for themselves a favourable radiation climate (Sculthorpe 1967), despite the limitations which can be experienced in this narrow niche. It is in the submerged plant that significant physiological and structural adaptations have occurred to enable growth at reduced light levels. Submerged plants invariably can be categorized as shade plants, because photosynthesis is saturated at an irradiance of less than half full sunlight, even for those species that inhabit shallow waters (Salvucci and Bowes 1982). The strategies adopted by aquatic weed species for resource capture are dealt with more fully in Chapter 4 and only the broad outlines are considered here.

Aquatic plants have undergone adaptations to enhance carbon gain and overcome the low oxygen levels. These adaptations involve two basic strategies. The first is structural and relies upon a network of enlarged intercellular airspaces or lacunae which can account for up to 60 per cent of some tissues. In emergent plants the lacunae can be continuous from the shoots, where they are in contact with the atmosphere, down to the rhizomes and roots, hence providing a pathway for gas movement within the plant (Drew 1983). Rhizomes exhibit a greater range of tolerance to anoxia than roots and their role in relation to survival mechanisms and habitat specificity has been highlighted (Braendle and Crawford 1987). Some submerged species such as *Isoetes lacustris* possess extensive air channels in their roots and leaves for the transport of gases from the sediments to the leaves (Wium-Andersen and Andersen 1972), whereas other submerged plants without extensive air-filled roots rely on the supply of oxygen and carbon from the water. They usually have thin leaves, sometimes finely dissected, with a thin cuticle to optimize exposure to the water. The second strategy relies on biochemical mechanisms, for example, tolerance of ethanol, the usual product of anaerobic metabolism, and the production of respiratory cytochromes which have unusually high affinity for oxygen (Henshaw, Coult, and Boulter 1961).

Aquatic plants usually enjoy a rich nutrient supply in the surrounding water and in the sediments (Denny 1972) though nutrient limitations may occur especially in tropical and subtropical freshwaters, particularly after large biomass levels have been attained. Other implications of nutrient availability and uptake from water and sediments are considered in Chapter 4. A third source of nutrients is exploited by aquatic carnivorous plants belonging to the genera *Aldrovanda* and *Utricularia* (Lollar, Coleman, and Boyd 1971). In oligotrophic waters, animals caught in the traps of such plants are a signficant source of phosphorus, nitrogen, and perhaps some minor elements.

Those aquatic plants which have adapted to achieve a high rate of acquisition of resources in a particular freshwater habitat come under the heading of competitor species (C-strategists), as defined by Grime (1979). They are potential weed species especially in productive waters. Typically such species achieve dense, productive stands, with extensive root or rhizome systems and are capable of rapid growth during favourable periods.

Special mention should be made of invasive plant species which have achieved spectacular success in invading a wide range of freshwater habitats outside their native range. The spread of such species as *Hydrilla verticillata, Elodea canadensis, Myriophyllum spicatum, Salvinia molesta,* and *Eichhornia crassipes* has been clearly described by Sculthorpe (1967), Hutchinson (1975), and Gopal (1987) (see also Chapter 3). The invasion of a body of water by such species is usually characterized by a very high rate of growth and spread, followed by a period of habitat stabilization and attainment of an equilibrium situation (Rørslett, Berge, and Johansen 1986). There are also indications that after a number of years the invader assumes a less significant role in the region into which it moved or was introduced and a species such as *Elodea canadensis*, which was a significant aquatic weed in the UK, although still widespread has declined to a lower problem status (Simpson 1984). Johnstone (1986) drew attention to the paucity of studies of the process of invasion which is usually associated with mechanisms of succession. He described the invasion of Lake Ohakuri, New Zealand by *Nymphaea mexicana* and introduced the concept of plant 'invasion windows' as a means of classifying the invasion potential of aquatic plant species.

In contrast to successful competitors, there are other species which have adapted to tolerate stress (S-strategists), for example, the fluctuations in the water level of a lake or the shading effect of epiphytic algae and bacteria (Hutchinson 1975; Phillips, Eminson, and Moss 1978). Other potential stress factors such as temperature, hydrogen ion concentration, and edaphic factors are described in Chapter 4. Such species are comparatively long-lived, able to endure conditions of limited productivity, and often exhibit a wide variety of growth forms. Keddy (1983, 1984) and Wilson and Keddy (1985) investigated the effects of exposure on zonation patterns and the physiological response of the component species in the macrophyte communities of a Canadian lake. They found that a changing environment allows species to coexist by shifting the competitive advantage from one species to another, or by creating gaps into which poor competitors can continuously migrate. Within Gillfillan Lake, Canada, there is obviously some specialization as different species grow in different water depths and at different degrees of exposure (Keddy 1984). Stress-tolerant plants do not typically produce large stands and rarely cause aquatic weed problems, except in unusual circumstances (see Chapter 15).

The consideration of environmental pressures such as disturbance, competition, and stress may be useful in understanding the ecology of species and plant communities with a view to deciding appropriate management (Kvet and Hejny 1986). In reality, species and plant communities will be controlled by more than one such pressure. Although the individual physiological responses of a species to changing physical factors explains much of the regulation of its position in the community (e.g. in a depth zonation), interactions with other plants may also be important in determining its distribution. This has been demonstrated for *Potamogeton pectinatus* and *Isoetes lacustris* (Misra 1938), *Glyceria maxima* and *Phragmites australis* (Buttery and Lambert 1965), and species of *Typha* (Grace and Wetzel 1981*c*).

Hydroseral succession and zonation

During the colonization process the constituent macrophyte species of a freshwater habitat will adopt a community structure which shows changes both in time and space. Those temporal changes which begin in the aquatic environment are commonly known as hydroseral succession. The spatial changes are typically manifested in some form of zonation of the plants. The concept of hydroseral succession is based to a large extent on the studies of temperate freshwater habitats by Pearsall (1917, 1920, 1921) and Tansley (1949). The process is dependent upon a raising of the bed of the water body toward the water surface either by accumulation of plant debris or by silting, or usually by both together. Those submerged and floating plants which had originally colonized the aquatic habitat would be succeeded by emergent species establishing themselves, for example, as reedswamp. With a further rise in the level of the substratum, shrubs and trees would be able to colonize with the eventual development of woodland. The nature and rate of this process is affected by a number of environmental factors especially sedimentation rates, the pH of the water, and nutrient status of the system. In temperate acid freshwaters the succession is based upon the accumulation of bryophytes, typically *Sphagnum* spp. leading to mire or bog formation; in temperate alkaline waters vascular plants are the basis of the development with the formation of fen and subsequently carr woodland habitats.

A similar process has been observed in the formation of sudd in tropical areas (Denny 1985) (the term sudd is used to describe any floating vegetation mat). Two mechanisms have been described for sudd formation. In one, the floating fringes of permanent swamps, for example, the rapidly growing *Cyperus papyrus*, break off and drift away with the wind or water currents. Such sudd is bound together by rhizomes (e.g. *Cyperus*) or floating stems (e.g. *Vossia cuspidata*) and the resulting raft provides a suitable habitat for non-mat-forming species (Denny 1985). Alternatively, floating vegetation such as *Salvinia molesta* and *Pistia stratiotes* can form mats of sufficient substance for other plants such as *Paspalum* spp., *Scirpus* spp. and *Cyperus* spp. to colonize and build up floating islands (Soerjani 1976).

Successional processes do not always develop toward the formation of terrestrial systems and retrogression is a natural phenomenon. A reversal in the direction of the succession towards open water can be brought about through a number of factors such as a cooling of the climate or a change in hydrology especially when the water level is raised. Particular interest has been shown in the die-back of *Phragmites australis* in temperate freshwater habitats (see also Chapter 15). A number of environmental factors may be associated with die-back. These include feeding by noctuid moths (Mook and van der Toorn 1982; van der Toorn and Mook 1982), grazing by *Myocaster coypus* (coypu) and wildfowl on the reed vegetation, and eutrophication (Boorman and Fuller 1981).

It is important for two reasons to take into account the stage of development of the target communities when aquatic plant management is under consideration. Firstly, such information enables predictions to be made about undesirable vegetation types. For example, the build-up of floating mats of *Salvinia* in tropical water indicates the potential for sudd formation, which is likely to be troublesome. Management, to break up or halt mat formation, would be much easier to effect than the control of substantial sudd. Secondly, any form of management will interfere with the normal community development processes. Understanding the implications of management is important in predicting the likely effects, not only on the vegetation, but also on associated animal communities. For example, the dredging of a drainage channel dominated by *Carex riparia* and *Phragmites australis* returns the habitat to an earlier stage in the hydroseral succession dominated by submerged and floating species (Wade 1978). This would be advantageous for fish but would present problems for nesting birds such as species of *Acrocephalus* (warblers). The importance of taking into account succession in management was emphasized by Dutartre and Gross (1982).

The development and succession of vegetation is a significant component in the ecology of the larger and/or slower flowing rivers. In those rivers in which the flow of water inhibits such processes of hydroseral succession, other factors will be of greater significance.

The spatial arrangement of aquatic plants commonly takes the form of a depth zonation (Spence 1982). The formation and establishment of such zonation is of particular importance in habitats with a vegetated littoral area and zonation patterns have been recognized for lakes in all parts of the world, for example, New Zealand lakes (Howard-Williams, Davies, and Vincent 1986). Knowledge of macrophyte zonation can be important in management, for example, in assessing the impact of drawdown (Hestand and Carter 1975), or recreation (Tivy 1980) on the plant communities and the importance of such communities in the stabilization of the substrate (Spence 1982).

Spence (1982) considered that the depth of the macrophyte zone in Scottish lakes (lochs) is governed by the depth of the wave-mixed zone which, in turn, is determined by the shape and size of the lake, together with factors controlling the underwater light climate. The extent to which wave action or light determines zonation depends on the relative amount of the macrophyte zone lying within or below the wave-mixed zone. Where the macrophyte zone lies wholly or mainly in the wave-mixed zone, wave action and the type of sediment exert the primary control of zonation, growth form and biomass of the vegetation. Species diversity decreases between sheltered and exposed shores. If more than half of the macrophyte zone lies below the wave-mixed zone, i.e. in the depositional zone, underwater light quality and quantity will exert the primary control on zonation. Species diversity decreases as water deepens. In a lake in which the macrophyte zone is not more than twice the depth of the wave mixed zone, both wave action and light determine zonation. Rørslett and Agami (1987) related observations on the distribution of *Isoetes lacustris* and *Najas marina* and drew attention to limitations in Spence's hypothesis. They showed that the distribution of these contrasting macrophytes provided support for a transient niche hypothesis (Rørslett 1987).

Not all freshwater habitats have plant communities which exhibit zonation. In some rivers the pattern of species distribution is governed by such factors as channel depth, flow, substrate, and shading (Haslam 1978), a pattern which can change with the season and from year to year. In habitats which exhibit a zonation of vegetation, such patterns are also seen within a particular zone of vegetation. Macan (1977) described a mosaic of submerged plant species for a small upland lake in the UK and noted the changes in the pattern over twenty-one years. The factors regulating such patterns of vegetation change in lentic systems are poorly understood.

PRIMARY PRODUCTIVITY

A combination of the luxuriant growth of aquatic vegetation and the rapid increase in areal cover of certain species of floating weeds suggests that aquatic weeds are unusually productive. A doubling time of 24 h has been reported for *Lemna aequinoctialis* (Landolt 1957) and *Wolffia microscopica* (Chang, Yang, and Sung 1977). This impression can be erroneous as primary productivity will vary from species to species, season to season, and from one part of the world to another.

Generally, submerged plants are less productive than emergent species with floating plants being intermediate between the two (Fig. 2.1). Submerged and floating species require less structural matter than emergent species, in which dry matter can account for 25–40 per cent of the plant (Sculthorpe 1967), though rooted submerged plants with access to nutrients in the sediments are more productive than phytoplankton (Fig. 2.1). Per unit area, some emergent aquatic plant communities are clearly among the most productive of all the world's vegetation types due to ample provision of water and nutrients, two factors which commonly limit growth on land (Moss 1982) (Fig. 2.1). Net average production of *Cyperus papyrus* may reach 48–143 t $ha^{-1}yr^{-1}$ (Thompson, Shewry, and Woolhouse 1979).

Denny (1985) summarized primary productivity in a range of aquatic plants and demonstrated that even within a single genus different species have different annual production estimates; for example, *Potamogeton* from different climates

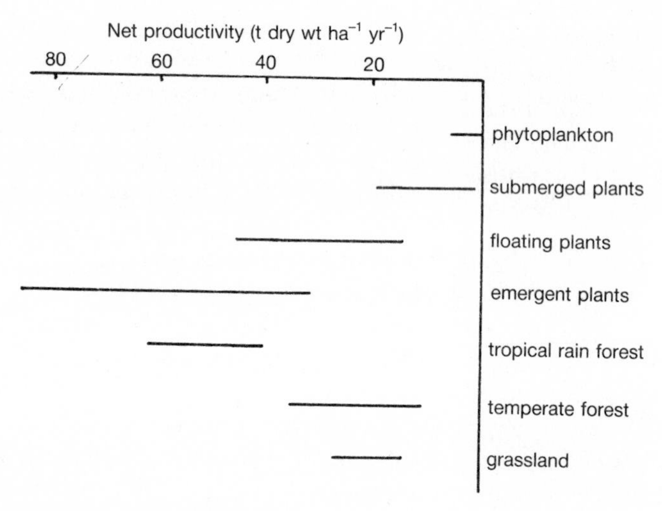

FIG. 2.1. Net primary productivity of aquatic plants in comparison with those of phytoplankton and terrestrial vegetation (Moss (1982) with additions from Denny (1985) and Landolt (1987)).

(dry weight estimates): *P. pectinatus* 25 t $ha^{-1}yr^{-1}$: south temperate (Howard-Williams 1978); *P. crispus* 0.5 t $ha^{-1}yr^{-1}$: subtropical (Rogers and Breen 1980) and *P. schweinfurthii* 2–40 t $ha^{-1}yr^{-1}$: tropical (P. Denny, unpublished). Other valuable data for emergent, floating, and submerged communities are summarized by Hejny, Květ, and Dykyjova (1981). Landolt (1987) reviewed productivity in the Lemnaceae in different regions of the world and demonstrated a range from 7–23.3 t $ha^{-1}yr^{-1}$ measured as dry weight. There may not be large differences between mean crop growth rates of tropical and temperate water weeds during their active growing seasons but, in tropical and subtropical regions, where rapid growth can proceed for a large part of, if not all, the year, the annual production values are significantly higher (Westlake 1975; Denny 1985). Growth can, however, be slowed down due to lack of nutrients during rainy seasons.

REGIONAL ASPECTS

The bulk of research on freshwater plants has taken place in North America and Europe (see Chapters 15 and 19) and only recently have texts been published reviewing aspects of aquatic plant ecology in other parts of the world. Denny (1985) provided a fascinating account of African wetland plants in relation to vegetation dynamics. The structure and functioning of aquatic vegetation was considered with special reference to zonation, distribution, primary production, decomposition, nutrition, and mineral cycling. The text also dealt with waterweed problems and other aspects of management and conservation. African aquatic weed problems are considered in more detail in Chapter 17.

Similar treatment of the South American and Asian regions has hitherto been sadly lacking. Pancho and Soerjani (1979) drew attention to the fact that the 'aquatic weeds of south-east Asia are poorly understood and have been much neglected' although Varshney and Rzoska (1976) brought together a wide range of investigations into aquatic weeds in Asia, many of which dealt with aspects of their biology and ecology (see also Chapter 16). Junk and Howard-Williams (1984) characterized the macrophyte communities of some of the principal wetland habitats of Amazonia, giving consideration to the main ecological factors which affect the aquatic plants and the importance of these plants to their habitats (see also Chapter 20).

The attention of botanists and ecologists in Australia and New Zealand has been focused on aquatic weed problems (Graham 1976; Sainty and Jacobs 1981; also Chapter 18) and there is a relative lack of accounts of the biology and ecology of freshwater macrophytes in general.

CONCLUSIONS

The growing knowledge of the biology and ecology of aquatic weeds, coupled with need for management practices which are both economical and environmentally acceptable, has led in recent years to general adoption of a more holistic approach to aquatic weed problems worldwide. Those responsible for management are aware of the need to be able to recognize different weed species and to relate this information to various management options. There are a variety of projects underway which are investigating manipulating environmental factors to control unwanted weed growth. These projects include examination of the potential of such options as nutrient-stripping, shading watercourses using bankside vegetation, and physically damaging perennating organs at critical periods of the weed life cycles. The potential for such management strategies is considerable—the propagules of a weed species alone provide considerable scope for manipulation (see Chapter 12). With time, such approaches will become integrated with the more traditional forms of management such as harvesting and the use of herbicides.

Aquatic plant managers have not necessarily received any biology or ecology training. Often they are more likely to have been trained as engineers. It will become increasingly important to provide in-service courses and other forms of training, not only to improve basic skills in plant identification, but to explain and demonstrate the value of understanding the autecology of a weed species and to appreciate the significance of a particular management approach to the community ecology of a given habitat. In the end all aquatic plant management comes down to manipulating the ecology of the target aquatic habitat, with the express purpose of altering the community ecology of that habitat, with minimum disruption, particularly when the water is used for more than one purpose. The more that is known and understood about such management, the more successful it is likely to be.

3

Origin, autecology, and spread of some of the world's most troublesome aquatic weeds

C. D. K. Cook

ORIGIN

Aquatic plants are only troublesome when they grow where they are not wanted. There is no aquatic plant that is *ipso facto* harmful and can be considered a weed throughout its geographical range. There are even cases of plants that are protected in one region and 'hunted' in another, for example, *Heteranthera reniformis* (mud plantain) is on the list of endangered plants in Connecticut, USA, and is perhaps the worst weed today in the northern Italian rice fields. *Trapa natans* (water chestnut) is extinct or endangered in many parts of Europe, an important crop plant in India, a noxious weed in eastern North America and a very serious threat to the sturgeon fisheries (caviar industry) in the southern part of the Caspian Sea.

Sometimes the decision of determining which plants are unwanted and which are wanted is capricious and the motives are often political or chauvinistic. Research money is easier to obtain for work on 'important' rather than 'unimportant' plants. For example, in North America the introduced species, *Myriophyllum spicatum* (Eurasian milfoil), is considered a weed and is attacked while the native species are to be protected. It requires expert knowledge to distinguish *M. spicatum* from a native species of milfoil (*M. sibiricum*) and it is an absurd situation when the water management authorities have to call in a taxonomic botanist to know if they have a weed problem or not. I myself have seen *Myriophyllum laxum* being sprayed in the south-eastern USA where it is a rare and endangered species. The reasons for the increased growth of *Myriophyllum*, native or introduced, in North America is usually the result of changes in land and water usage. The excessive and usually undesirable growth of the plants are a symptom of change and not its cause.

The concept of change is very important in deciding if a plant is a weed or not. For example, in Florida, USA, sawgrass (*Cladium jamaicense*) occupies about 700 000 ha (Steward and Ornes 1975) while *Hydrilla verticillata* occupied about 17 000 ha in 1981 (Schardt and Nall 1982). The sawgrass is native and receding while hydrilla is introduced and spreading. One has learnt to live with the former, it, therefore, does not hinder man's activities; the hydrilla one tries to eradicate because it causes change and hinders man's activities. Several aquatic weeds occupy very large areas without being considered weeds. For example, *Phragmites australis* (common reed) occupies about 5 500 000 ha in the USSR. In the Danube Delta it occupies about 200 000 ha of which about 150 000 ha are pure *Phragmites* stands. *Cyperus papyrus* (papyrus), a protected plant in Egypt, occupies about 1 000 000 ha in Sudan, the neighbouring country. Other aquatics such as *Typha* spp., *Scirpus californicus* (totora), *Nelumbo lutea* (American lotus), and the water-lilies (*Nuphar* spp. and *Nymphaea* spp.) often occupy very large areas. As long as these plants do not increase (change) they are not considered to be weeds. However, a visitor arriving between ten or a hundred thousand years ago may have had other views.

The origin of some weeds, as pointed out, is occasionally somewhat arbitrary. However, most aquatic plant infestations in recent years have presented serious threats. Usually it is very clear which species are undesirable. Nevertheless, the situation is often more important than the plant species. For example, *Typha* may be highly desirable in a nature reserve but in the same water system, a few metres away, it may be a serious pest in places used for transport or amenity (eg, boating, swimming, fishing).

AUTECOLOGY

Various classifications of aquatic plants according to their life forms, growth forms, and ecological tolerances have been suggested but none have found general acceptance. However, almost all divide the hydrophytes into three main categories: submerged, emergent, and floating. The more complex the subdivision of these categories becomes, the greater is the need to qualify the definitions to accept transitional examples. The enormous plasticity which so many aquatics show also tends to obscure distinctions and thus make the detailed classification unwieldy. Taking the aquatic plants known to be weeds somewhere in the world and comparing them with aquatics that have never been reported to be weedy it is not possible to say that any particular growth form is more likely to become weedy than another. In this book (see Chapter 1) floating weeds are divided into free-floating weeds and rooted weeds with floating leaves; the algae (except the Charophytes) are considered as a separate category. One growth form, however, the haptophytes—plants which are attached to, but do not penetrate, the substrate (lichens, bryophytes, Hydrostachydaceae, Podostemaceae, and some Lythraceae) – has never developed weedy species. When one divides the aquatic plants according to their ecological tolerances then the plants specially adapted to very swiftly flowing water, mostly haptophytes, and the species specialized for very oligotrophic conditions are the least likely candidates for weeds. However, most aquatic plants show relatively wide ecological tolerances and most are potentially, at least, capable of becoming weeds.

Floating weeds

Taking the three major categories – submerged, emergent, and floating – all have weedy representatives. However, it is among the floating plants that the most spectacular and alarming weeds are found. This is because firstly, the vegetation spreads over the surface of the water and is thus clearly seen with immediate visible consequences: navigation, fisheries, and amenity are interfered with or blocked. A biological and, perhaps, more serious consequence is that dense surface cover reduces the quantity of light penetrating the water which reduces or eliminates the submerged photosynthetic organisms which, in turn, reduces the oxygen level leading to death of animal life. In developing countries this may lead to famine.

Another serious consequence of floating vegetation is that the dead and living plant material may be transported by water currents or wind and eventually pile up and cause blockages. In southern India I have seen *Salvinia molesta* blown by monsoon winds into piles up to 3 m high and I have been told that the piles may reach up to 5 m high. Such masses of vegetation may block waterways and cause flooding which can directly threaten human life.

On a world-wide scale the floating weeds are the most important in terms of distress to human life. The 'world's worst aquatic weed' is generally agreed to be *Eichhornia crassipes* (water hyacinth) which is a free-floating weed; in some parts of the Palaeotropics another floating aquatic *Salvinia molesta* is, today, a more serious threat. There are other floating aquatics that are locally important weeds: for example, *Pistia stratiotes* (water lettuce), *Azolla* spp., *Hydrocharis morsus-ranae*, *Trapa* spp. and *Ipomoea aquatica*. Some essentially rooted emergent aquatics may develop floating mats which can cause serious problems: such as, *Alternanthera philoxeroides* (alligator weed), *Ludwigia* spp., and several grass species from the following genera: *Echinochloa* (barnyard grass), *Panicum* (torpedo grass) and *Paspalum*.

From an ecological point of view these floating aquatics are pioneer plants because they occupy the water surface and spread essentially in two dimensions. The growth rates are usually high (summarized by Westlake 1981): for example, *Eichhornia crassipes* has a doubling time of about 13 days and in *Salvinia molesta* the doubling time is between 7 and 17 days. In cultivation a doubling time of

1.54 days has been reported by Blackman (1960). In Lake Kariba (between Zimbabwe and Zambia) *Salvinia* was first reported in 1959, less than one year after closure of the dam; 13 months later it occupied 39 000 ha. and by 1962 it occupied about 100 000 ha.

These large growth rates are obtained through clonal growth. The clone itself is usually long-lived and the individual ramets are mostly short-lived. The spread is of the phalanx-type: from one or more inoculation points it spreads with a closed front; the occupied area remains closed and may be colonized by other species. Sustained development can take place only in equitable climates in permanent water. In climatically temperate regions or in areas of temporary water floating aquatic plants are less commonly considered to be serious weeds.

Most terrestrial weeds are competing with crop plants and, therefore, rarely make such spectacular conquests. Also crops are usually maintained in a more or less permanent pioneer phase so that following invasion of a weed the damage remains constant and predictable. Invasions of floating aquatics are usually worse, not only because the habitat is disturbed but also because an autogenic succession usually takes place shortly after the initial colonization. At first, small herbs, grasses and sedges grow on the vegetation binding both living and dead material together. Later larger plants come and eventually woody shrubs and small trees become established. A typical succession on floating mats of *Salvinia* is described by Cook and Gut (1971). Vegetational succession simply leads to loss of water carrying capacity which may have different effects in different situations, the most important being lack of ability to control flood water, loss of energy (hydroelectricity), reduction in fisheries, or disruption of transport.

In aquatic crops, such as rice, floating aquatics usually play a minor role; the weeds are relatively easily removed before the rice is set and as an emergent aquatic it can out-compete floating plants.

Submerged weeds

Few species of submerged aquatics have become troublesome on a continental scale like some floating aquatics. Locally, however, they may be very serious. It is, therefore, very difficult to make a list of the world's worst submerged weeds. A list made by a New Zealander would be rather different from one made by a north American. Another important aspect of submerged aquatic weeds is that they tend to persist for a relatively short time and then become replaced by other species. In Europe, for example, it seems that *Elodea canadensis* (Canadian waterweed) is being replaced by *E. nuttallii* (another North American species) and, in parts, by *Lagarosiphon major* (a species from South Africa) while in New Zealand *E. canadensis* has been largely replaced by *L. major* and *Egeria densa*, and in the USA *Myriophyllum* is being replaced in some regions by the more harmful plant *Hydrilla verticillata*.

Many submerged aquatic species have very large distributional areas and, consequently, wide ecological tolerances. In disturbed water it is quite common for completely native species to increase their growth and become a threat to man's usage. Frequently, the disturbance is an increase in trophic level of the water or the substrate and the following submerged plants have often been seen to become troublesome within their native range: *Ceratophyllum demersum*, *Najas* spp., *Potamogeton* spp. (particularly *P. crispus* and *P. pectinatus*), *Ranunculus fluitans*, and *Vallisneria* spp. Some species of *Najas* and *Potamogeton* have also become troublesome beyond their native range.

Many submerged weeds, like most floating ones, spread by vegetative growth, even to the extreme of precluding sexual reproduction, for example, *Egeria densa* is only known as male plants outside its native range while *Lagarosiphon major* is represented only by female plants. *Elodea canadensis* is exclusively female in Europe and exclusively male in Australia, while in New Zealand from about 1872 to about 1964 females only were known but now apparently males are taking over.

Emergent weeds

The majority of aquatic plant species are emergent and it is perhaps surprising that relatively few have become serious pests. Emergent aquatics may occupy very large areas, but usually inhabit shallow or temporary water and are thus often easily controlled unless they grow in floating mats. Emergent aquatics can be serious pests on a smaller scale when they grow with crop plants. The more they ecologically resemble the crop plant the worse they are. Perhaps the worst weeds of rice are wild species of rice that shed their seeds before the crop is ripe and have seeds with dormancy. This kind of problem is found in all crop plants and has nothing to do with aquatics specifically. Among emergent aquatics, like submerged aquatics, no precise ecological diagnosis can separate weedy from non-weedy species. All species given the best conditions can become weedy.

SPREAD

Cook (1985) attempted to review the range extensions of aquatic vascular plants. Altogether 198 species were considered. Twenty-six species are very widespread or disjunct, it is unlikely that all of them are native throughout their ranges but the alien and native areas are unknown or disputed. Several species in this category are sometimes weedy: for example, *Ceratophyllum demersum* (coontail, hornwort), *Echinochloa crus-galli* (barnyard grass), *Eclipta prostrata*, *Eleocharis dulcis* ('Chinese' water chestnut), *Ipomoea aquatica*, *Oryza rufipogon*, *Phragmites australis*, and *Pistia stratiotes*.

The remaining 172 species have all become established outside their native range. This does not mean that they are all weeds, in fact, many are useful and desirable but the majority are considered to be insignificant. Potentially, however, virtually all could become weeds if given the right conditions for luxuriant growth.

Of the 172 species discussed by Cook (1985) about 30 per cent are common to the Old and New World, 32 per cent are exclusively native to the Old World and have become established in the New World while 38 per cent have made the journey from the New to the Old World. The migration of aquatic plant species between east and west is fairly well balanced.

From the point of view of aquatic plant species, India is floristically very rich, probably as a result of its migration from the southern to the northern hemisphere during the Upper Cretaceous. Thirty-two species native to India have become established elsewhere, while 21 species have been introduced in recent times. Such give and receive calculations are rarely so well balanced. The most extreme unbalanced case is probably that found in New Zealand. It has 'received' four species but has 'given' no more than one (*Crassula helmsii*) which might have reached Europe from Australia where it is also native. The reason for this imbalance is probably due to the extraordinarily poor native aquatic flora which obviously left many ecological niches unfilled.

In a review of the aquatic species endemic to Europe and the Mediterranean (Cook 1983), sixty-one taxa were considered to be endemic and of these forty-six were considered to be neo-endemics, which originated during or after the Pleistocene. The only European endemic that has established itself outside is *Ranunculus hederaceus*, which has become naturalized in eastern North America, it is one of the few palaeoendemics. It was first recorded in 1821 and is believed to have come with ships' ballast. It has remained near its original sites of introduction along the Atlantic coast.

It is unlikely that all the remaining 60 European endemic aquatics have been denied transport to other regions. The problems of establishment in new lands are obviously great and it is perhaps rather naive to say that the aquatic environment is uniform and allows plants a wide geographical range without further qualification. Some aquatics such as *Brasenia* and *Dulichium* became extinct before or during the ice age in Europe, but persisted in North America where they are common and very widespread. These plants have not become re-established in Europe in spite of the fact that they are in cultivation in Europe. This observation also lends weight to the argument that ecological niches in the aquatic environment are rather more complex than we commonly accept today.

Floating weeds

The spread of the most noxious aquatic weeds is relatively well documented. The worst aquatic weed today is probably still *Eichhornia crassipes* (Pieterse 1978; Gopal 1987; Gopal and Sharma 1981). It is native in the northern neotropics; it spread to southern N. America about 1860; in Africa it arrived first in Egypt about 1879; in Asia around 1888 in India, 1894 in Java and about 1900 in Japan; in Australia it arrived about 1890. In spite of the fact that it is native in the neotropics it has recently become a pest in man-made reservoirs in Surinam (e.g. Lake Brokopondo), El Salvador (Lake Rio Lempa), Nicaragua (Lake Apanas), and other central and south American states (see Chapter 20). The water hyacinth has very attractive flowers and originally was probably spread deliberately by horticulturalists.

Salvinia molesta is another neotropical plant that has invaded the Old World. Its spread in Sri Lanka and Africa is well documented by Sculthorpe (1967) and in India by Cook and Gut (1971)—in both these publications it is called *Salvinia auriculata*. In 1972 Mitchell first described this invading *Salvinia* under the name of *S. molesta*; he suggested that it is a horticultural hybrid involving *S. auriculata* and *S. biloba* or *S. biloba* and *S. herzogii*, all species from South America. *Salvinia molesta* has 45 chromosomes and is considered to be pentaploid; it has a disturbed meiosis and has not yet been found to reproduce by spores. Schneller (1980) found *S. herzogii* has 63 chromosomes and is a heptaploid and, therefore, unlikely to be parent of *S. molesta*. In 1981 Schneller reported *S. auriculata* to be hexaploid ($2n=54$). *Salvinia biloba* reproduces by spores in Zürich (Schneller pers. comm.) and on morphological grounds is likely to be involved in *S. molesta* but the question of *S. auriculata* as the other parent remains open.

Pistia stratiotes is probably the next most important floating weed but it had spread throughout the warmer regions of the world before botanists documented the distribution of plants. This does not prevent it becoming a pest.

Because floating aquatic weeds are essentially pioneer plants they may be expected to recede naturally unless the habitat is maintained in a pioneer state. *Salvinia molesta* in Kariba Lake occupied 100 000 ha in 1962, but had reduced to 50 000 ha by 1966. The management of *Salvinia* was discussed by Mitchell (1976) (see also Chapter 17).

Submerged weeds

The spread of the more spectacular submerged weeds is documented in Sculthorpe (1967) with more recent information on *Egeria* (Cook and Urmi-König 1984), *Elodea* (Cook and Urmi-König 1985), *Hydrilla* (Cook and Lüönd 1982; Verkleij *et al.* 1983*a*, *b*; Pieterse 1981; Pieterse, Verkleij, and Staphorst 1985), and *Myriophyllum* (Aiken, Newroth, and Wile 1979). Like floating weeds the submerged species are mostly spread by specialized or unspecialized vegetative fragments; sexual reproduction is often lacking. Also, following the initial spread there is a tendency for them to retract before becoming stabilized and integrated into the local flora.

Emergent weeds

The spread of emergent weeds is reviewed by Sculthorpe (1967) and Cook (1987*a*, *b*). Unlike floating and submerged weeds most emergents rely on sexual reproduction and long-range dispersal is usually by means of diaspores. Of the three principle agents of dispersal—wind, water, and animals (including man)—wind plays a minor role. Within a particular body of water the spread may well be by vegetative means.

CONCLUSIONS

There often seems to be an almost blind belief that really troublesome aquatic weeds have been introduced from elsewhere and that the main reason for their aggressiveness is that they have been relieved of the ecological restraints within their native range. Also it is often believed that if one can discover the 'ecological restraint'—often considered to be a disease, parasite, or herbivore—then the introduction of this 'restraint' to the new range will control the weed in its new area. Such thinking is, unfortunately, an oversimplification. The aquatic environment is complex. Even if it is possible to eliminate a troublesome weed, one can almost be sure that the niche it occupied will be filled by another. At least 172 species of aquatic plants have become established outside their native range (Cook 1985). Of these, only a handful have become seriously troublesome but all are potential weeds. Very frequently the disturbance or change that man has brought about in the habitat is the obvious cause of the 'trouble'. The weed, whether native or introduced, cannot help growing when man offers it the best conditions for growth. Much more effort should be taken to manage the water body and its environment. Aquatic weeds are symptoms of trouble and very rarely their cause.

4

Ecophysiology of the world's most troublesome aquatic weeds

W. SPENCER AND G. BOWES

GENERAL INTRODUCTION

A variety of aquatic plant species, under the right conditions, can become nuisance infestations in a particular location. However, only about a dozen or so higher plant species have a global reputation as major aquatic weeds. In this chapter we will restrict our discussion to those that cause the most problems on a global scale. These include six submerged species: *Ceratophyllum demersum, Elodea canadensis, Hydrilla verticillata, Myriophyllum spicatum, Potamogeton crispus,* and *P. pectinatus*; three free-floating species: *Eichhornia crassipes, Pistia stratiotes,* and *Salvinia molesta*; and five emergent species: *Alternanthera philoxeroides, Echinochloa crus-galli, Phragmites australis, Panicum repens,* and *Typha* spp. These species, by virtue of their prolific growth and reproduction, may interfere with human utilization of freshwater resources, become aesthetically displeasing, or displace desirable indigenous vegetation. Other species of more local importance as weeds are not included, because of the general nature of this review.

The ecology and applied aspects of aquatic weeds are covered elsewhere in this volume. This chapter will emphasize the physiology of aquatic weeds in relation to primary production. Particular attention will be paid to photosynthesis, as it is intimately linked to growth. We also discuss how photosynthetic and growth responses to environmental parameters are expressed as weed characteristics.

Most terrestrial species, including weeds, with regard to photosynthetic carbon metabolism fall into two categories: C_3 or C_4 plants. In C_3 plants, ribulose-bisphosphate carboxylase-oxygenase (rubisco) is the initial carboxylation enzyme that fixes atmospheric CO_2 directly into the Calvin or photosynthetic carbon reduction (PCR) cycle. Due to the oxygenase activity of rubisco, the photorespiratory carbon oxidation (PCO) cycle operates in C_3 plants, so they show inhibition of photosynthesis by atmospheric O_2 levels, and photorespiratory CO_2

loss. In C_4 plants, phosphoenolpyruvate carboxylase (PEPC) initiates CO_2 fixation into CO_4 acids which, after decarboxylation, provide CO_2 for rubisco and the PCR cycle. The benefit of the C_4 system is that it concentrates CO_2 at the site of fixation by rubisco, in specialized bundle sheath cells, and thus overcomes the inhibitory effects of O_2 on rubisco and eliminates photorespiratory CO_2 losses. The C_4 system is most advantageous under high temperature and light regimes. Consequently, in terms of distribution, C_3 species make up the larger proportion of temperate floras, whereas C_4 species are more abundant in subtropical/tropical regions. Because of high photosynthetic rates, high growth rates, and high productivity, many of the world's most aggressive terrestrial weeds, such as crabgrass (*Digitaria sanguinalis*) and pigweed (*Amaranthus retroflexus*), fall into the C_4 photosynthetic category (Holm *et al.* 1977; Bowes and Beer 1987).

As will be discussed in later sections, although the terrestrial C_3 and C_4 photosynthetic categories can be applied to emergent and floating aquatic plants which photosynthesize in air, they are not valid for submerged species that are totally bathed in water.

SUBMERGED AQUATIC WEEDS

Introduction

The submerged habitat is occasionally portrayed as one where the extremes of terrestrial environments are moderated. This can be an over-simplification of the constraints imposed upon freshwater plants. Although water acts to buffer air temperature fluctuations, substantial diel and seasonal extremes can be observed (Bowes 1987). One of the major distinguishing characteristics between aqueous and aerial environments, is the high resistance exerted by water to solute diffusion. This results in potentially wide fluctuations in O_2 concentration and the type of dissolved inorganic carbon (DIC) available for photosynthesis, especially in densely vegetated areas. Also, in contrast to aerial leaves, the photosynthetic organs of submerged plants are bathed by a dense, polar, nutrient-containing solvent, the pH of which can vary widely even on a diel basis. Light is selectively attenuated in its passage through water, which produces changes in light quality as well as quantity at increasing depth (Spence 1981); yet the irradiance in surface water can be as high as in terrestrial habitats.

However, both non-weed and weed species may encounter the same environmental parameters. Thus the difference lies in biological and physiological traits that submerged weeds have acquired which enable them to exploit conditions in an opportunistic and competitive manner, often to the detriment of desirable submerged plants. Submerged weeds often show competitive or competitive-ruderal plant strategies (Grime 1979), similar to those of terrestrial weeds with rapid canopy formation and prolific vegetative reproduction. Submerged weeds are successful because they can obtain adequate resources for growth and reproduction, even under suboptimal conditions and in the face of competition, and can interfere with the resource acquisition of other non-weed species. In the following sections we examine the environmental constraints and the adaptations to these constraints that result in certain submerged species becoming major weed problems.

Growth habit and weed expression

From a shoreline perspective, the biomass in a dense 'mat' of submerged weeds appears to be enormous. However, measurements of total biomass and productivity indicate they are small in comparison with many terrestrial plant communities, although higher than non-weed submerged species (Westlake 1963, 1967; Bowes, Holaday, and Haller 1979; Spencer and Bowes 1985).

Several factors contribute to this apparent anomaly. The biomass of submerged weeds is usually unevenly distributed in the water column, more so than non-weeds (Ikusima 1965; Yeo 1965; Adams, Titus, and McCracken 1974; Haller and Sutton 1975; Titus and Adams 1979; Baudo *et al.* 1981; Howard-Williams 1981; Sand-Jensen 1983; Getsinger and Dillon 1984). Thus, the submerged weeds under consideration have similar growth habits in terms of the vertical placement of biomass, with potentially over 60 per cent located in the upper third of the water column. The growth of *M. spicatum, Egeria densa,* and *H. verticillata* to the surface develops with season, light and/or nutrient availability (Adams, Titus, and McCracken 1974; Haller and Sutton 1975; Getsinger and Dillon 1984). During early spring, growth begins at the hydrosoil where light is often limiting. As temperature and daylength increase, the plants elongate to the surface. This growth, and profuse branching, result in characteristic surface mats of dense vegetation that overlay and shade other submerged species. However, in locations where light is not limiting, such as Florida springs, *H. verticillata* may remain close to the hydrosoil (W. Spencer, unpublished observation). Thus the growth habit is contingent upon environmental conditions.

The surface placement of biomass, where it is easily seen and interferes most with human activities, is a primary reason that some submerged plants are regarded as weeds. In contrast, non-weed or desirable species usually possess only a benthic, prostrate, or rosette growth form with little or no apical branching (Ikusima 1965; Haller and Sutton 1975; Titus and Adams 1979). Most submerged weeds exhibit terminal growth via numerous actively growing apical buds, which increases both the invasive potential throughout the water column, and the placement of photosynthetic tissue in less light-limited locations.

A further factor in the relatively low biomass of submerged plants is their diminutive fresh to dry weight ratios in comparison with floating, emergent, or terrestrial plants (Van, Haller, and Bowes 1976; Rogers and Breen 1980; Jana and Choudhuri 1982; Pokorny *et al.* 1984; Spencer and Bowes 1985). This difference is related to the physical medium in which the plant grows and reflects the reduced need for support tissue to present the leaves to light. Freshwater is 775 times as dense as air, and provides a buoyancy almost 1000 times greater, which renders extensive structural tissue superfluous for submerged weeds (Sculthorpe 1967).

Coinciding with the low dry weight, the metabolic machinery of submerged plants, including weeds, is much less than that of emergent or terrestrial species, as demonstrated by low chlorophyll and rubisco activities (Van, Haller and Bowes 1976; Holaday, Salvucci, and Bowes 1983). Furthermore, photosynthetic rates of submerged plants are far lower than those of aerial species (Bowes 1987).

Environmental constraints on photosynthesis and growth

Light

Physical and chemical water parameters in a weed mat differ substantially from those of adjacent uninfested water (Van, Haller, and Bowes 1976). Attenuation of light with water depth is a common characteristic (Wetzel 1975), which is greatly accentuated by a mat of weeds (Haller and Sutton 1975; Fair and Meeke 1983). Since light attenuation depends upon wavelength (Wetzel 1975), both the quality and quantity of light varies with depth. Submerged plants, and weeds in particular, have morphological and physiological traits that maximise exploitation of the lower light environment and minimize deleterious effects of the high surface irradiance.

Morphologically, submerged weeds have some similarities with terrestrial shade plants. The leaves of submerged weeds are often only a few cells thick, some are finely dissected, and most have chloroplasts in the outer cell layers, including the epidermis (Bowes 1987). These characteristics may reduce shading, and present the maximum amount of chlorophyll to light, though the benefits in a dense canopy are unclear. In contrast, the chloroplasts of terrestrial sun leaves are generally restricted to the mesophyll and are not commonly found in the epidermis, except in guard cells.

Similarly, growth at the surface increases the interception of solar energy and also deprives competing species of access to a major resource – light. For *H. verticillata*, shoot elongation and branching are stimulated by different wavelengths. Shoot elongation is promoted by green (500–540 nm) and inhibited by red ($>$580 nm) light (Van *et al.* 1977). Shoot branching, however, shows the opposite effect, as red light stimulates and green light reduces branching. As shorter wavelengths are less attentuated in passing through water (Spence 1981), these characteristics would favour elongation of the plant to the surface and reduce energy expenditure on branch and leaf production at depths where the irradiance was insufficient to sustain photosynthesis.

Besides wavelength, the irradiance influences the growth and morphology of submerged weeds. *Hydrilla verticillata, E. densa,* and *M. spicatum* grown under 50 per cent or greater shade all showed substantial increases in shoot length, while *H. verticillata* and *M. spicatum* also exhibited a 20 to 30 per cent reduction in shoot number (Barko and Smart 1981*a*). This response is similar to that due to wavelength and enhances the tendency for canopy formation at the surface. Leaf production also decreased under low irradiance, which would reduce respiratory losses from non-photosynthesizing tissue. At increased depth, chlorophyll content and chlorophyll *a* : *b* ratios change, though whether these chromatic adaptations have any ecological relevance is disputed (Van *et al.* 1977). Chlorophyll *a* : *b* ratios of submerged plants are less than those of terrestrial C_3 and C_4 species (Black 1973; Van *et al.* 1977).

Light penetration may influence the distribution of weeds and any increase potentially increases the hydrosoil area covered by submerged weeds. Naïve use of water level management as a tool for submerged weed control may allow further infestation, particularly if the basin morphology is characterized by a shallow slope (Gophen 1982).

Physiologically, all submerged species are shade plants, in that leaf photosynthesis is saturated by a fraction of full sun irradiance. Thus, for photosynthesis, the light saturation points (the irradiance at which net photosynthesis is maximal) and compensation points (the irradiance at which photosynthetic CO_2 uptake is balanced by photorespiratory and respiratory CO_2 loss) are low in comparison with terrestrial sun species (Bowes 1985).

It is important to delineate between leaf and canopy photosynthesis in this regard. Since most gas exchange measurements on submerged weeds are for detached leaves or stems, an irradiance above the saturation point is 'wasted' from an individual leaf perspective since no increase in photosynthesis occurs. However, in a canopy, the reflected or transmitted quanta could be captured by shaded lower leaves that are not light saturated. Thus canopy photosynthesis may not show saturation at full irradiance, even though leaves at the surface are saturated. Light saturation points for individual leaf photosynthesis of submerged weeds range from 10 to 50 per cent of full sun (i.e. 2000 μmol quanta $m^{-2}s^{-1}$) irradiance (Brown *et al.* 1974; Van *et al.* 1977; Bowes *et al.* 1977; Hough 1979; Baudo *et al.* 1981). This situation for submerged weeds is unusual, as the most aggressive terrestrial weeds, and also floating and emergent weeds, show sun plant characteristics with leaves capable of utilizing full sun irradiance (Gloser 1978; Matsunaka 1983; Rani and Bhambie 1983; Sale *et al.* 1985; Spencer and Bowes 1986; Bowes and Beer 1987). It is clear that the leaves of submerged plants are adapted to the lower radiant energy fluxes of their environment, rather than the surface sunlight (Bowes 1987). This shade nature of all submerged leaves, including those of weed species, may represent a compromise with the massive constraint on photosynthesis imposed by the resistance of water to DIC diffusion (Bowes 1987). Whether differing leaf light saturation points play a role in the weed expression of a submerged plant has not been determined.

The elongation of submerged weeds to the surface poses a problem in that parts of these shade plants then are exposed to high irradiance and O_2, that could cause photo-inhibition and photo-oxidation. Whether they possess biochemical adaptations to deal with this problem, such as more carotenoids or superoxide dismutase activity, has not been investigated. Cytoplasmic streaming and aggregation of chloroplasts into a narrow band in *H. verticillata* cells at high irradiance (G. Bowes, unpublished observations) could result in chloroplast self-shading and delay the onset of photo-inhibition (Raven 1984). Alternatively, altruistic shading may be provided by bleached upper parts for the canopy below.

Irradiance is a factor in species dominance of submerged weeds (Brown *et al.* 1974). In a subtropical river (Waikato River, New Zealand), turbidity and light attenuation increased towards the sea. *Elodea canadensis* was the dominant species at the headwater, but as turbidity increased it was replaced by *Lagarosiphon major*, and near the river mouth where turbidity was greatest, *E. densa* replaced *L. major*. A correlation was found between the light compensation points of these species and turbidity. The light compensation points for *E. canadensis, L. major,* and *E. densa* were 36, 22, and 10 μmol quanta $m^{-2} s^{-1}$ respectively (Brown *et al.* 1974). It could not be determined if the species distribution was due to the variation in light compensation point, or other environmental parameters, with the compensation point simply acclimatizing to the growth conditions.

The importance of the light compensation point is that for net carbon gain to occur, the incident irradiance must be above this value. The light compensation points of several submerged weed species have been shown to be as low as 0.5 per cent of full sun (Brown *et al.* 1974; Van, Haller, and Bowes 1976; Bowes *et al.* 1977); whereas light compensation points for non-weedy, submerged plants can be as high as 3.0 per cent of full sun (Van, Haller, and Bowes 1976). Although this difference is small, it may have important ecological implications since solar irradiance not only varies spatially within a body of water, but also chronologically (Van, Haller, and Bowes 1976). Irradiance increases exponentially after sunrise, but for the first hour or two a submerged plant community is exposed to an irradiance below the light compensation point of some of its members. During the day when adequate light is available, DIC can become a major limiting resource due to depletion by photosynthesis. A weed species with a low light compensation point growing from the hydrosoil would achieve net photosynthesis earlier in the day and successfully compete for the limited DIC. Repetition of this event over several days could result in a greater photosynthesis and growth of the weed species in comparison with other submerged plants. This scenario has been implicated in the success of *H. verticillata* in Florida lakes (Van, Haller, and Bowes 1976, 1978).

The photosynthesis of *H. verticillata* acclimates to the irradiance received during growth (Bowes *et al.* 1977). Light compensation points of 7, 10, 15, and 20 μmol quanta $m^{-2}s^{-1}$ were measured for plants grown for three weeks at 6, 30, 120, and 300 μmol quanta $m^{-2}s^{-1}$ respectively. Furthermore, the $K_{1/2}$ (irradiance) for photosynthesis was reduced by growth at lower irradiance, indicating that the photosynthetic apparatus was able to maintain higher net photosynthetic rates at the lower light levels than non-adapted plants. However, the maximum rate of photosynthesis was reduced in the low-light grown plants. These results suggest that the photosynthetic machinery of submerged weeds rapidly acclimates to maintain net carbon gain at decreased irradiance. Dark respiration decreased in the low-light grown plants, suggesting there is less energy expenditure for maintenance and synthesis when the energy input from light is greatly limited. These acclimations could enhance the survival and success of weed species growing under low irradiance, as compared to non-weed species.

Temperature

Water tends to buffer the temperature to which submerged plants are exposed, but extremes from near zero to as much as 40 °C can be found (Van, Haller, and Bowes 1976; Bowes *et al.* 1979). Although some submerged plants can photosynthesize and grow at temperatures as low as 2 °C (Boylen and Sheldon 1976), it is generally at higher temperatures (20 to 35 °C) that weed problems become most severe. Different metabolic processes show differing responses to temperature and growth represents an integration of all these processes and responses. Consequently, caution is required when extrapolating from the temperature effect on any one process to overall growth.

In short-term measurements, *H. verticillata, M. spicatum, C. demersum,* and *E. densa* exhibit net photosynthesis between 10 and 44 °C, and temperature optima that range from 28 to 37 °C, with *H. verticillata* having the highest optimum (Saitoh, Narita, and Isikawa 1970; Stanley and Naylor 1972; Van, Haller, and Bowes 1976). Similarly, the photosynthesis to dark respiration ratio of *H. verticillata* is greater than that of *E. densa* above 20 °C, but this is reversed at lower temperatures (Barko and Smart 1981*a*).

These effects of temperature on photosynthesis correlate with growth responses. Thus, *H. verticillata* shows greater growth responses to increased temperature (32 °C) than either *M. spicatum* or *E. densa* in terms of shoot biomass, shoot length, and number of roots produced (Barko and Smart 1981*a*). For *E. densa*, terminal growth ceased during the summer and the plants began to deteriorate when the temperature reached 30 °C (Getsinger and Dillon 1984). Growth resumed in autumn and surprisingly the peak biomass occurred when the water temperature declined to 10 °C. *Elodea canadensis* also acclimates to lower temperatures. Plants under ice over-winter in a photosynthetically active state, and possess 50 per cent greater net photosynthetic rates than summer-grown plants measured at similar low temperatures (Boylen and Sheldon 1976). The nature of this winter acclimation is undetermined.

It appears that *H. verticillata,* at least the dioecious form from the USA which was designated 'USA hydrilla I' by Verkleij *et al.* (1983b), has a greater capacity for growth at elevated temperatures than most other submerged weeds. The higher temperatures stimulate increases in shoot length and biomass and thus in mat formation, which may explain why this plant has its major weed expression in tropical and subtropical waters. In contrast, *M. spicatum* and *E. canadensis* are more of a problem in temperate areas. The evidence suggests that species differences in the response of photosynthesis and growth to temperature may be important in the latitudinal distribution of submerged weeds.

Both *H. verticillata* and *E. densa* show an inverse relationship between growth temperature and CO_2 compensation point. For *H. verticillata,* low values were associated with plants grown at 32 °C, while high CO_2 compensation points were produced by growth at 16 °C (Barko and Smart 1981*a*). The CO_2 compensation point is a parameter easily measured by lowering the CO_2 concentration around a plant to the point where photosynthesis and photorespiration are equivalent, and *net* CO_2 exchange becomes zero. Consequently, it indicates the degree to which photorespiration is occurring; with high values denoting substantial photorespiration and low values the reverse. Thus, these two submerged weeds, when grown at high temperature, restrict the DIC loss that normally occurs through photorespiration, thereby improving their carbon economy in an environment with limited DIC availability. This plasticity in photosynthetic metabolism, under differing growth conditions, is a major difference between terrestrial and submerged plants (Bowes 1985).

Hydrogen ion concentration

The pH of freshwater varies from acid to alkaline (pH 2–12), though for the majority of open lakes it is between pH 6 and 9 (Wetzel 1975). It is modulated by H^+ and OH^- ions, most often formed by the dissociation of carbonic acid and hydrolysis of bicarbonate, respectively. In some well-buffered freshwaters the pH may be a very stable component, whereas in relatively static water bodies that are heavily vegetated, photosynthetic and respiratory activity during the day and night, respectively, can result in large diel pH changes at a single location (Brown *et al.* 1974; Bowes, Holaday, and Haller 1979). Furthermore, pH differences of 4 units (10 000-fold) within distances of less than a metre can be found where dense patches of vegetation are interspersed with open water. Consequently, submerged plants and especially weed species, have to compensate rapidly for pH extremes.

The pH of water has direct and indirect effects on photosynthesis and growth of submerged weeds. The transport mechanisms that move essential nutrient solutes across the plasmalemma and into the cell, include active processes that co-transport H^+ ions (Raven 1984). Proton co-transport cannot support the active transport of other ions at extremes of pH and, as a consequence, the external pH has a direct effect on nutrient uptake abilities. Under some nutrient regimes, aquatic plants may be net producers of H^+ ions, and thus must export them to the medium as part of their internal pH regulation (Raven 1984). The more acid the medium, the greater the energy expenditure to regulate the internal pH. Some submerged weeds, including *E. canadensis* and *P. lucens*, secrete H^+ and OH^- ions from the abaxial and adaxial leaf surfaces, respectively, thereby regulating the pH of the aqueous boundary layers surrounding the leaf (Prins *et al.* 1982). Such large micro-environmental pH effects make gross measurements of water pH at best a poor indicator of the true pH to which the plant surface is exposed.

The pH of freshwater has a number of critical indirect effects on submerged plants. It influences the complexing and chelation of phosphate ions with metal cations which, in turn, affect phosphate and, to a lesser extent, metal ion, availability (Wetzel 1975). Thus, the solubility of ferric phosphate, is minimal below pH 6.0, making access to it difficult in acidic waters. The presence of calcium leads to the formation of hydroxylapatite, $Ca_5(OH)(PO_4)_3$, and at alkaline pH the phosphate precipitates so the concentration available for uptake declines. The supply and form of nitrogen is also influenced by pH (Wetzel 1975). At acid pH, nitrification of ammonium to nitrate ions is severely curtailed, so in acid bog lakes, nitrate, the nitrogen form preferred by many aquatic species, may be virtually undetectable. At increasingly high pH, the proportion of ammonium ions in the toxic ammonium hydroxide form rises exponentially.

For submerged plants, one of the most important indirect effects of pH is on the equilibrium among the forms of DIC (free CO_2, H_2CO_3, HCO^-_3, and CO_3^{2-}) in the water. In low pH (<7.0) or soft water most of the DIC exists as free CO_2 and concentrations above air equilibrium values are encountered only infrequently. In high pH (>7.0) or hard water, large amounts of DIC may exist as HCO^-_3 and/or CO_3^{2-} anions in addition to air equilibration levels of CO_2. For a closed aqueous system the pH determines the distribution of DIC among its various forms. However, most natural water bodies are open, to varying extents, to equilibrium with CO_2 in air and, thus, one cannot infer from the pH alone the amount of DIC available for photosynthesis. If the water is truly an open or closed system in relation to the atmosphere the quantity of the various forms of DIC, given the total DIC concentration and pH, can be calculated (Wetzel 1975; Beer and Eshel 1983). Since natural systems are seldom entirely open or closed, caution should be exercised when making interpolations from pH to DIC.

Short-term photosynthesis by submerged weeds can occur across a wide pH range. For *H. verticillata* at subsaturating DIC levels, net photosynthesis is greatest from pH 3.0 to 6.0, while above pH 6.0 the rate declines along with the concentration of free CO_2, which indicates a photosynthetic preference for free CO_2 rather than HCO_3^- (Van, Haller, and Bowes 1976). Similarly, the photosynthesis of *C. demersum* is greatest between pH 5.8 and 6.2 (Shiyan and Merezhko 1972; Ondok and Pokorny 1982), and less than 20 per cent of maximum at pH values higher than 9.0. The decline at high pH reflects more on the form of DIC available than on pH *per se*. When sufficient DIC is present, the maximum photosynthetic rate at pH 8.0 is similar to that at pH 4.0 for *H. verticillata, M. spicatum,* and *C. demersum* (Van, Haller, and Bowes 1976). Thus, although freshwater species prefer free CO_2 most can use HCO_3^- to some degree, but the affinity for HCO_3^- is lower and, consequently, more is required to achieve a given photosynthetic rate.

At very high pH, declines in photosynthesis have been attributed to either a CO_3^{2-} inhibition of the active uptake of HCO_3, or increased buffering which neutralizes the efflux of H^+ ions from the leaf. This latter effect would reduce the production of CO_2 from HCO_3^- at the localized acid regions next to the leaf surface (Sand-Jensen 1983).

Although measurements of net photosynthesis show it is often greater below pH 7.0, aquatic plants do not always grow most profusely under such conditions (Kadono 1982a). This anomaly is partially explained by the fact that natural waters of low pH are usually low in DIC (Wetzel 1975). Waters of high pH have larger DIC pools, which include CO_3^{2-} ions and DIC used in photosynthesis can be replenished. Thus the growth of *H. verticillata* is 10-fold greater at pH 9.0 than at pH 7.0; though appreciable growth occurs across the entire pH range from 5.0 to 9.0 (Spencer and Bowes 1985). Both *Limnophila sessiliflora* and *Hygrophila polysperma* show a different response in that their growth is greatest at pH 5.0 and 7.0, while at pH 9.0 it is negligible. Thus for growth their tolerance to pH is more limited than that of *H. verticillata* (Spencer and Bowes 1985), possibly because they have less ability to access HCO_3^- ions for photosynthesis (Bowes 1985, 1987), though other pH effects cannot be ruled out. These data suggest that *L. sessiliflora* and *H. polysperma* are unlikely to become major weed problems in very alkaline waters. In contrast, the broad pH tolerance of *H. verticillata* allows it to express its weed potential in many disparate bodies of water, and this is probably one of the reasons why it is among the world's worst submerged weeds (Spencer and Bowes 1985).

Most submerged plants can tolerate highly alkaline conditions (pH 10–11), at least for short periods (Brown *et al.* 1974; Van, Haller, and Bowes 1976; Bowes, Holaday, and Haller 1979; Keeley 1983); however, very few (an example is *Juncus bulbosus*) can survive exposure to highly acidic conditions (below pH 4.5). Thus, under very acid conditions, submerged macrophytes do not usually cause weed problems.

Dissolved inorganic carbon and oxygen concentrations

Freshwater in equilibrium with air contains at 25 °C about 10 μM free CO_2 and 250 μM O_2. Unlike the situation in air, these values are variable. The large resistance to diffusion of dissolved gases can lead to free CO_2 values that range from zero to over 350 μM, and O_2 levels from zero to over 500 μM, or 200 per cent air-saturation (Brown *et al.* 1974; Bowes, Holaday, and Haller 1979; Keeley 1983). In unproductive, oligotrophic waters CO_2 and O_2 levels are largely regulated by physical means. However, under eutrophic conditions, biological activities such as photosynthesis, respiration, and decomposition play an important role in determining the CO_2 and O_2 concentrations.

In softwater lakes of low DIC the CO_2 invasion rate is about 700 μmol $CO_2\,m^{-2}h^{-1}$ (Schindler *et al.* 1972; Emerson and Broecker 1973), but in alkaline waters the hydration of HCO_3^- and CO_3^{2-} may produce a net movement of CO_2 to the atmosphere. In low productivity systems, the CO_2 invasion rate alone may support phytoplankton photosynthesis (Schindler *et al.* 1972). However, a dense weed mat of *H. verticillata* containing about 10 kg fresh weight m^{-2} (Bowes, Holaday, and Haller 1979), with a chlorophyll to fresh weight ratio of 1.0 mg g^{-1} (Holaday, Salvucci, and Bowes 1983), and a net photosynthetic rate of 10 μmol $CO_2\,mg^{-1}$ Chl h^{-1} at natural DIC levels, could potentially consume 10 000 μmol $CO_2\,m^{-2}h^{-1}$. This is 100 times greater than the estimated CO_2 invasion rate, which suggests there must be auxiliary sources of inorganic carbon for photosynthesis, other than just free CO_2 invading the mat from the atmosphere. The sediment and carbonate-containing mineral substrates, can contribute significant amounts of DIC to photosynthesis. However, the depletion of free CO_2 and HCO_3^- in dense weed mats after only a few hours of photosynthesis (Van, Haller and Bowes 1976; Bowes, Holaday, and Haller 1979; Beer, Spencer, and Bowes 1986) points to a substantial DIC limitation on the growth of submerged weeds.

The reduced photosynthetic rate as a result of DIC depletion is further aggravated by the high pH, O_2 levels, and temperatures that can accompany DIC depletion in dense submerged weed vegetation. These water conditions are very conducive to O_2 inhibition of photosynthesis, and photorespiratory CO_2 loss, both of which further lower net photosynthesis.

Adaptations which enhance carbon gain

For aerial leaves, the biochemical events associated with fixation limit photosynthesis. In contrast, submerged leaves are faced with a CO_2 diffusion rate that is 10^4-times slower than in air and this, together with a large aqueous boundary layer, makes DIC diffusion the major limiting factor to their photosynthesis (Bowes 1985, 1987). The limitation imposed by DIC accessibility is evident from measurements which show that at pH 8.0, 30–50 mM DIC is needed to saturate photosynthesis and at more natural DIC levels photosynthesis operates at only about 10 per cent of maximum (Van, Haller, and Bowes 1976; Browse, Dromgoole, and Brown 1977). As a consequence, the photosynthetic rates of all submerged plants, including weed species, are far lower than rates measured on terrestrial plants, or even aerial leaves of an amphibious species (Van, Haller, and Bowes 1976; Salvucci and Bowes 1982). Similarly, the apparent $K_m(CO_2)$ for photosynthesis, which is an overall measure of the affinity of the process for CO_2, is greatly improved, and photosynthesis increases when leaves are measured in air, as opposed to under water (Lloyd, Canvin, and Bristow 1977; Salvucci and Bowes 1982; Bowes 1987). The conclusions from these short-term physiological measurements have been confirmed by studies which indicate DIC can greatly limit the growth of *M. spicatum* (Barko 1983).

Probably as a result of constraints from the underwater existence, submerged species cannot be characterized by terrestrial photosynthetic categories (Bowes 1985). A terrestrial species shows very little variation in the photosynthesis to photorespiration ratio, as evidenced by a relatively constant CO_2 compensation point (Ogren 1984). Thus the photosynthesis to photorespiration ratio and CO_2 compensation point are high in C_3, but low in C_4 plants (Ogren 1984). In contrast, for a submerged freshwater species the CO_2 compensation point is variable and dependent upon the growth conditions (Holaday, Salvucci, and Bowes 1983). A high CO_2 compensation point is associated with low net photosynthesis, but also with increased photorespiration and O_2 inhibition of photosynthesis, which are C_3-like characteristics. The reverse is true when the plant has a low CO_2 compensation point, which makes it appear C_4-like (Bowes 1985). Thus, a submerged species can exist in a continuum of photorespiration (PR) states, from high-PR to low-PR. Winter-like growth conditions induce the high-PR state, whereas summer-like conditions with low DIC induce low-PR plants (Holaday, Salvucci, and Bowes 1983). This is the reason why contradictory reports concerning the photosynthetic and photorespiratory capacity of submerged weeds have appeared in the literature (Winter 1978; Howard-Williams 1981; Meulemans and Best 1981; Sand-Jensen 1983; Getsinger and Dillon 1984). Such biochemical plasticity is not found in terrestrial C_3 and C_4 plants, but is a feature common to submerged freshwater species and thus they have been referred to by the acronym 'SAM' (submerged aquatic macrophyte) to indicate a different photosynthetic category (Bowes *et al.* 1978; Bowes, Holaday, and Haller 1979). Plasticity in the photosynthetic mechanism of SAM may reflect the variability in DIC and O_2 levels which they, but not terrestrial plants, encounter.

Given that submerged weeds encounter conditions that limit DIC availability, including some that stimulate photorespiration, it is not surprising that carbon conservation adaptations are found in these plants. The adaptations range from anatomical and morphological to physiological and biochemical. They function variously to reduce the DIC diffusion resistance, to reduce CO_2 losses from respiratory processes, or to elevate the CO_2 concentration at the site of the primary carboxylating enzyme, rubisco.

Anatomical and morphological

The thin membranous, and sometimes dissected, leaf morphology of most elodeid submerged weeds serves to increase the surface area to volume ratio, and hence diffusion. Since the leaves are usually only two to a few cells thick, with a chloroplastic epidermis, the distance from the environment to the site of CO_2 fixation is kept to a minimum. Furthermore, the leaves have no, or non-functional, stomata, and little cuticle.

Lacunal gas spaces in the stems and leaves of some species also improve the movement of gases, especially O_2, though whether they facilitate CO_2 fixation by submerged weeds is questionable. There are some isoetid non-weedy species which possess lacunal systems that play a role in obtaining DIC (Wium-Anderson 1971; Søndergaard and Sand-Jensen 1979). These plants typically occur in nutrient and DIC-poor waters and derive CO_2 for photosynthesis from the hydrosoil via the roots. The CO_2 is piped to the leaves in the lacunal gas channels. The diffusion resistance of water and a thick cuticle, in some cases, inhibits the escape of CO_2, while the low-growing nature of the plant, close to the hydrosoil, minimizes the CO_2 diffusion pathway.

Use of hydrosoil CO_2 is not a common attribute of submerged weed species. The placement of biomass at the surface, and use of hydrosoil CO_2 seem mutually exclusive, as the pathlength from roots to leaves is generally too long for transfer of sufficient CO_2 to support photosynthesis. The surface canopy growth habit of many submerged weeds permits the converse situation, as it places the photosynthetic surface close to the atmosphere and minimizes the distance CO_2 has to travel in water before reaching the plant. Wind and wave action at the surface increases the dissolution of CO_2 (Kanwisher 1963) and, concomitantly, the supply of DIC, though dense surface vegetation may reduce the effectiveness of this strategy because it moderates wave action.

Utilization of bicarbonate ions

Although the diffusion resistance of water is a substantial limitation to CO_2 fixation, many submerged plants partially compensate by using an inorganic carbon source that is unavailable to aerial species, namely HCO_3^- ions, though apparently no submerged plant makes direct use of CO_3^{2-} ions (Raven 1984). Whether submerged species directly transport HCO_3^- ions, or convert them to free CO_2 prior to uptake, has been an on-going debate since the turn of the century (Steemann-Nielsen 1947); though possibly both mechanisms operate. Even though a freshwater species may use HCO_3^-, based on affinity, free CO_2 is always the preferred DIC form (Steemann-Nielsen 1947; Van, Haller, and Bowes

1976; Browse, Dromgoole, and Brown 1979; Raven 1984). Consequently, much higher HCO_3^- levels, as compared to CO_2, are needed for photosynthesis.

Freshwater submerged plants have been broadly categorized as HCO_3^- 'users' and 'non-users' (Steemann-Nielsen 1947; Hutchinson 1970), but this is simplistic. Certainly, some non-weed species, such as mosses and isoetids, show little ability to use HCO_3^- ions (Bain and Proctor 1980; Boston and Adams 1983), but most species can, to a greater or lesser degree (Bowes 1985; Prins *et al.* 1982). Furthermore, recent findings indicate that for a species, or even an individual plant, the capacity for HCO_3^- use varies with the growth conditions and the PR-state of the plant (Salvucci and Bowes 1983*a*, *b*; Bowes 1987; Sand-Jensen and Gordon 1984).

Laboratory-grown unicellular green algae and, particularly, cyanobacteria (blue-green algae), actively use HCO_3^- ions to raise the internal DIC to levels many times the external concentration; in effect a CO_2 concentrating mechanism (Badger, Kaplan, and Berry 1978; Beardall and Raven 1981). Growth at low CO_2 induces the system, whereas high CO_2 suppresses it. The CO_2 concentrating mechanism overcomes the effects of O_2 on rubisco; thus the algae do not exhibit photorespiration and it also improves the photosynthetic affinity for DIC. Consequently, these organisms are very effective at scavenging DIC from low-DIC environments. There is very little research on the role of the CO_2 concentrating mechanism in naturally-occurring algae and cyanobacteria. However, the increasing weed threat in Florida from the filamentous cyanobacterium *Lyngbya birgei*, is largely attributable to a HCO_3^- use system that raises its internal CO_2 and enables it to obtain DIC from high pH water (Beer, Spencer, and Bowes 1986). Then dense surface growth makes it difficult for submerged macrophytes to compete with *L. birgei*.

There is growing evidence that submerged macrophytes, including some angiosperms, also use HCO_3^- ions to produce high internal CO_2 pools, though not as well as cyanobacteria (Salvucci and Bowes, 1983*a*, *b*; Bowes and Reiskind 1987). Internal CO_2 well above the external levels has been measured in *H. verticillata*, especially in the low-PR state (Bowes and Reiskind 1987), and in *E. nuttallii* (Eighmy, Jahnke, and Fagerberg 1987). Furthermore, *H. verticillata* uses HCO_3^- effectively. For *M. spicatum* the use of HCO_3^- ions, in conjunction with the enzyme carbonic anhydrase, to equilibrate HCO_3^- and CO_2 rapidly, seems to be the mechanism producing the low-PR state (Salvucci and Bowes 1983*a*). In the low-PR state, *M. spicatum* uses HCO_3^- ions more effectively (Bowes 1987). Also the presence of ethoxyzolamide, an inhibitor of carbonic anhydrase, increases the CO_2 compensation point, O_2 inhibition, and K_m(DIC) of photosynthesis; thus the low-PR plants behave as if in the high-PR state (Salvucci and Bowes 1981, 1983*a*).

The distribution of freshwater macrophytes in hard and soft waters has been linked to their ability to use HCO_3^- ions for photosynthesis (Hutchinson 1970;

Bain and Proctor 1980). The ability to use HCO_3^- cannot be accurately quantified (Spence and Maberly 1985), which makes comparisons uncertain. However, species with minimal HCO_3^--use capacity are often slower-growing, benthic plants, that are restricted to low pH waters and generally do not become weed problems. In contrast, the major submerged weed *M. spicatum* appears to have a substantial capacity for photosynthetic HCO_3^- use, even in the high-PR state; more so than *L. sessiliflora* which is much less of a weed threat (Spencer and Bowes 1985; Bowes 1987). It is probably no coincidence that the submerged plants listed as global weeds seem in the main to use HCO_3^- ions for photosynthesis. However, further research is needed before a causal relationship between HCO_3^- use and aquatic weed potential or distribution can be unequivocally established.

C_4 acid metabolism

During photosynthesis, a number of submerged plants show evidence of substantial production of the C_4 acids malate and aspartate. They include the weed species *E. densa, L. major* (Brown *et al.* 1974), *E. canadensis* (Degroote and Kennedy 1977), *P. pectinatus* (Winter 1978) and *H. verticillata* (Bose 1924; Bowes *et al.* 1978; Holaday and Bowes 1980). Initial pulse-chase experiments with $^{14}CO_2$, to resolve the fate of the first products of photosynthesis, indicated that although up to 50 per cent of the ^{14}C entered malate, it was not further metabolized (Degroote and Kennedy 1977; Browse, Dromgoole, and Brown 1979; Browse, Brown, and Dromgoole 1980). Consequently, malate was suggested to function as a pH stat, or an ion balance to compensate for excess cation uptake, or a source of organic carbon for respiration, rather than having a direct photosynthetic role (Degroote and Kennedy 1977; Browse, Dromgoole and Brown 1979; Browse, Brown, and Dromgoole 1980).

Although malate does not turnover in high-PR *H. verticillata*, when stress induces the low-PR state, malate is decarboxylated and the CO_2 passes into the PCR cycle, in a manner reminiscent of terrestrial C_4 photosynthesis (Holaday and Bowes 1980; Salvucci and Bowes 1983*b*). Furthermore, all the enzymes necessary for C_4-like photosynthesis increase in activity as the low-PR state is induced (Salvucci and Bowes 1981). Thus, in some submerged weeds, PEPC activity may be greater than that of rubisco, with a PEPC/rubisco ratio over 14 (Degroote and Kennedy 1977; Browse, Brown, and Dromgoole 1980; Salvucci and Bowes 1981; Helder and van Harmelen 1982; Salvucci and Bowes 1983; Sand-Jensen 1983). Increases in activity of NADP malic enzyme, which decarboxylates malate in the chloroplast, and pyruvate orthophosphate dikinase, which regenerates PEP, are circumstantial evidence for a complete C_4 acid cycle in low-PR plants (Salvucci and Bowes 1981).

Why, under non-stress conditions so much DIC is fixed into malate by plants in the high-PR state, and then apparently not metabolized further in photosynthesis, is still uncertain. The situation for *H. verticillata* in the low-PR state is much clearer. The malate serves as an internal source of CO_2, to concentrate CO_2 around rubisco in the chloroplasts and thereby reduce the effects of O_2 on photosynthesis and photorespiration and improve the affinity of photosynthesis for DIC. Thus in *H. verticillata*, this C_4-like system, possibly in conjunction with HCO_3^- ion use, is responsible for the low-PR state. A similar situation may apply to some other submerged weeds, such as *E. canadensis, E. densa*, and *L. major*, when stress conditions induce them to shift to the low-PR state.

Even though some submerged species form C_4 acids during photosynthesis, they cannot be classified as C_4 plants. Aquatic plants when submerged do not possess the characteristic Kranz leaf anatomy (concentric rings of mesophyll and bundle sheath cells to which the chloroplasts are restricted) of terrestrial C_4 plants (Bowes 1985). As a consequence, unlike terrestrial C_4 plants, the PEPC and rubisco CO_2 fixation events are not separated in mesophyll and bundle sheath cells (Edwards and Huber 1981). Instead, there exists *intra*cellular compartmentation, with PEPC and rubisco located in the cytosol and chloroplast, respectively, of the same cell (Bowes and Salvucci 1984). Also, where it occurs in submerged species, the C_4-like system is induced by stress, especially low-DIC conditions; but the system in terrestrial C_4 plants is not environmentally induced.

To summarize, it now seems that to concentrate DIC internally two inducible systems, both biochemical, occur in submerged plants: the HCO_3^- utilization and the C_4 acid system. *Myriophyllum spicatum* exhibits only the HCO_3^- utilization system, whereas *H. verticillata* possesses both. More research is needed before other submerged weeds can be shown unequivocally to operate one or both of these systems. Either system would appear to be advantageous for a submerged weed that potentially functions in dense vegetation, where O_2 is high and DIC low. It is interesting that under most field conditions submerged weeds exist in the high-PR state (Bowes, Holaday, and Haller 1979; Holaday, Salvucci, and Bowes 1983; Maberley and Spence 1983). Possibly induction of the low-PR state only becomes cost-effective when the plants are stressed. In contrast, unicellular and filamentous algae and cyanobacteria seem normally to exist in the low-PR state (Bowes 1985; Beer, Spencer, and Bowes 1986).

Carbon fixation at night

Dark CO_2 fixation rates as high as 30 per cent of photosynthesis have been reported for a number of submerged plants, including *H. verticillata, P. crispus*, and *Vallisneria spiralis* (Holaday and Bowes 1980; Baudo *et al.* 1981; Helder and van Harmelen 1982). This phenomenon is like dark CO_2 fixation in terrestrial Crassulacean acid metabolism (CAM) plants (Kluge and Ting 1978). However, unlike CAM plants *net* carbon gain at night does not occur with submerged weeds, unless DIC levels are very high (Holaday, Salvucci, and Bowes 1983). Instead dark fixation conserves carbon by fixing and recycling respired CO_2.

Dark fixation of CO_2 by PEPC leads to the production of malate, which accumulates at night, and is decarboxylated during the day, to provide CO_2 for the PCR cycle when external DIC is scarce. Thus diel fluctuations in malic acid and, concomitantly, titratable acidity, occur in submerged weeds which exhibit substantial PEPC activity and C_4 acid metabolism (Bose 1924; Holaday and Bowes 1980). For *H. verticillata*, it is the low-PR state which has the greatest capacity for dark fixation (Holaday, Salvucci, and Bowes 1983), and thus it may be a strategy primarily invoked when daytime DIC levels are low.

Several submerged isoetids, which are not weeds, are capable of net CO_2 uptake at night from low DIC levels, much like terrestrial CAM species (Keeley 1981; Keeley and Bowes 1982; Boston and Adams 1983). Estimates of the contribution that dark fixation makes to the carbon economy of these plants during growth range up to 50 per cent (Boston and Adams 1986). However, these are relatively slow-growing species. No submerged weed relies to this degree on dark fixation; possibly their more rapid growth potential precludes it.

Reduction of dark respiration

For submerged weeds, little is known about the interaction between respiration and primary production, even though the ratio of dark respiration to photosynthesis is a factor in the seasonal decline of submerged weeds (Jana and Choudhuri 1979; Barko and Smart 1981*a*). Dark respiration may profoundly influence primary production (Pearcy *et al.* 1987), but in this regard, submerged plants have not received the attention accorded agronomic crops (Lambers 1985).

Dark respiration for terrestrial plant leaves is about 10 per cent of the photosynthetic rate (Zelitch 1971), but in submerged plants it can be more than 50 per cent of net photosynthesis at ambient CO_2 levels (Van, Haller, and Bowes 1976; Spencer and Bowes 1985). Thus, for the two weed species *H. verticillata* and *P. pectinatus* the dark respiration to net photosynthesis ratio was about 0.3 (Jana and Choudhuri 1979). This is a function of both low, DIC-limited photosynthesis in submerged leaves and high dark respiration. The relatively high dark respiration may be due to an additional alternative oxidase system (Raven 1984), though its function in submerged plants is unknown. From the high rates, dark respiration might be anticipated to have a more pronounced effect on the primary production of submerged, than of terrestrial weeds.

The dark respiration of submerged plants is influenced by a number of factors, including species, temperature, nutrient-status, time of day and PR-state of the plant. In terms of their response to temperature, submerged species vary considerably with Q_{10} values reportedly ranging from 1.0 to 4.4 (McDonnell

1971). Plants of *P. crispus* and *E. canadensis* grown in nutrient-rich waters had higher respiration rates than those in more nutrient-poor waters (McDonnell 1971). Community dark respiration may fluctuate on a diel basis, with maximum rates occurring just after sunset and lowest rates just before dawn (Beyers 1966). Consequently, late in the day respiration may exceed photosynthesis. The ratio between dark respiration and photosynthesis shows greatest variation between plants of different PR-states; high-PR plants have higher ratios that low-PR plants (Salvucci and Bowes 1981; Holaday, Sulvucci, and Bowes 1983; Spencer and Bowes 1985). For those weeds capable of substantial C_4 acid metabolism, some of the reduction in dark respiration probably results from recycling of the respired CO_2, as discussed in the previous section. The decreased dark respiration of the low-PR state undoubtedly conserves carbon under adverse conditions, but its contribution to dry matter production or weediness has not been quantified.

Nutrient and edaphic effects

Inorganic nutrition is integral to both photosynthesis and growth and a major part of the submerged plant literature deals with this subject. Unfortunately much is descriptive and focused on a few species and elements which makes causal relationships difficult to determine. Also, until recently inorganic carbon was not widely recognized as a nutrient with limiting or co-limiting effects (Raven 1984; Bowes 1987).

Because submerged weeds are bathed by a nutrient medium they differ from most aerial plants in having access to mineral ions by foliar absorption. The rootless submerged species *C. demersum* must absorb its nutrients through the shoots. Early work assigned all nutrient absorption to the leaf tissue, with roots serving only as organs of attachment (King 1943; Sutcliffe 1962; Den Hartog and Segal 1963).

Subsequent research has shown that absorption of nutrients from the hydrosoil via the roots can occur (Mitra 1960; Bristow and Whitcombe 1971; Best and Mantai 1978; Bole and Allan 1978; Barko and Smart 1980; Carignan and Kalff 1980; Barko and Smart 1981*b*; Barko 1982). This is to be expected as over 95 per cent of aquatic species have extensive root and root hair systems, and sediment nutrient levels may exceed by several orders of magnitude those in the overlying water (Wetzel 1975). However, problems with anoxia and long distance transport are inadequately resolved. Roots in the low-O_2 environment of the hydrosoil may have limited respiration and hence energy to drive the active uptake of ions (Raven 1984), though lacunal systems can compensate to some extent by allowing O_2 to diffuse in the gas phase to the roots, especially if the path length between shoot and root is short and uninterrupted (Wetzel 1975). The mechanism of long distance ion transport in submerged plants is still unknown. Xylem tissues are much reduced and a transpiration stream driven by solar evaporation, as found in aerial plants, obviously cannot occur. Root pressure, to force liquid from root to shoot, has been reported, but evidence that this drives any substantial movement of water and nutrients to the shoots is lacking (Raven 1984). The transport problems of root oxygenation and their provision of nutrients to the foliage, are compounded in submerged weeds which place foliage at the water surface and connect it to the roots by long (several metres) stems of narrow diameter. How these are overcome has yet to be demonstrated.

Micronutrients include Fe, C1, Mn, Zn, Cu, B, Co, Mo, and V; in some cases they can limit algal photosynthesis and growth (Goldman 1972), but apart from work with micronutrient effects on *H. verticillata* in Florida, little is known about their effects on submerged weed growth (Martin and Reid 1976; Martin, Victor, and Dooris 1976). In general, micronutrient concentrations in natural waters seem sufficient to sustain plant growth (Wetzel 1975). This does not preclude micronutrient limitations, expecially Fe and Mn, under specific environmental conditions (Wetzel 1975).

Macronutrients include N, P, K, Ca, Mg, S, and also C, O, H, though these latter three are not often regarded as nutrients. Comparisons of tissue and freshwater nutrient concentrations suggest that N, P, and possibly C, are most likely to limit photosynthesis and growth under natural conditions (Raven 1984); but high levels of an element do not necessarily mean it is in a form available to the plant (Wetzel 1975). The elements N, P, and S are generally absorbed as NH^{+}_{4} or NO^{-}_{3}, PO^{-}_{4}, and SO_4^{2-}, respectively, using active transport processes and thus ultimately relying on photosynthesis-derived energy for their uptake (Raven 1984). Of the macronutrients, N, P, K, S, and Mg are relatively mobile within the plant, which suggests that Ca has to be absorbed by roots and leaves. Similarly, N, P, and K may be both root and foliar absorbed (Toetz 1974; Nichols and Keeny 1976; Best and Mantai 1978; Barko and Smart 1981*a*; Barko 1982; Huebert and Gorham 1983; Steward 1984). In some instances, the N and P mobilized from the sediment is sufficient to support growth (Best and Mantai 1978; Barko 1982). For *E. canadensis* and *M. spicatum*, from 60 to 90 per cent of the P requirement may be supplied by the roots (Wetzel 1975). In fact, submerged weed mobilization of nutrients from the hydrosoil may hasten eutrophication by leakage of nutrients to the water from plant injury or senescence (McRoy, Barsdate, and Nebert 1972; Reimold 1972; Toetz 1974; Bole and Allan 1978; Carignan and Kalff 1980; Barko and Smart 1980; Huebert and Gorham 1983).

Species differences in nutrient uptake exist. Thus for growth of *H. verticillata*, the hydrosoil apparently only supplies a fraction of the tissue content of K, with most coming from shoot absorption (Barko and Smart 1981*b*; Barko 1982). In contrast, root absorption of K by *P. pectinatus* is sufficient to support the biomass accumulation (Huebert and Gorham 1983). At most locations, K and Mg levels do not limit aquatic weed growth (Wetzel 1975). Although Ca levels are usually adequate for growth, concentrations in productive hard water can fluctuate markedly, down to undetectable levels (Wetzel 1975). This element may be involved in the distribution of submerged weeds, including *M. spicatum* (Grace and Wetzel 1978).

Submerged weeds may have lower nutrient requirements than other species. Low critical concentrations for N and P have been implicated in the distributional and competitive success of *M. spicatum* (Grace and Wetzel 1978). Also, submerged macrophytes contain lower tissue concentrations of N and P, relative to C, than unicellular autotrophs (Raven 1984). Submerged weeds have minimal needs for structural and metabolic components, as evidenced by low dry to fresh weight ratios, and enzyme activities. Thus in comparison with floating or emergent species they probably need far less nutrient inputs to place a nuisance cover of plants over an area of water.

The quantity and quality of organic matter influences the growth and distribution of submerged aquatic weeds. Several submerged weeds disappeared from two Florida lakes after a sudden inflow of particulate organic material (Dooris and Martin 1980; Barko 1982). This same phenomenon has been observed in English lakes (Pearsall 1920; Misra 1938). The growth of both *H. verticillata* and *P. nodosus* is inhibited by the addition of 5 per cent organic material to the substrate (Barko and Smart 1983). The inhibition due to labile organic material decreased with time, but that from refractory organic material lasted over 14 weeks. The inhibition was caused by toxic, soluble organic compounds from anaerobic decomposition of the organic substrate.

Organic material in the sediment or water also can influence the growth of submerged weeds indirectly, by altering, through chelation, the availability of nutrients, especially cation ratios. In some cases, growth is enhanced by improving the solubility of ions such as Fe and Mn (Wetzel 1975).

The salinity of coastal freshwaters has a significant effect on the growth and distribution of submerged weeds. Only a few submerged species tolerate brackish water. For example, *H. verticillata* and *M. spicatum* grow together in Crystal River, Florida, but where the river enters the Gulf of Mexico, and salinity increases, *H. verticillata* is replaced by *M. spicatum* (Haller, Sutton, and Barlow 1974). The growth of *M. spicatum* is unaffected by up to 10 per cent salinity, but that of *H. verticillata* is significantly reduced at 6 per cent salinity. Similarly, the photosynthesis to dark respiration ratio of *M. spicatum* did not decline until a salinity value of 8 per cent was reached (McGahee and Davis 1971), which suggests that salinity influences growth partly through effects on photosynthesis.

Reproduction

For terrestrial and aquatic plants alike, reproductive ability is one of the foremost factors determining whether a plant is regarded as a major weed. The capacity to invade new territory successfully, especially after some form of disturbance, to maintain existing populations or re-infest an area after control and to propagate rapidly and easily, are important weed characteristics. The submerged weeds under consideration reproduce both sexually and asexually. Sexual reproduction is by seed/fruit formation. Asexual reproduction is primarily by vegetative means, including regrowth from stem fragments, stolons, rhizomes, or green rootcrowns, and the formation of underground tubers or turions, or aboveground turions (Sculthorpe 1967; Haller 1976).

It is a philosophical issue whether sexual or asexual reproduction is more important to weed expression. Sexual reproduction produces offspring with variability, and the potential for survival when conditions change, or the species invades a different environment. But for a plant that is well-adapted to prevailing conditions, vegetative reproduction with minimal variety in the offspring, ensures they will be similarly suited, and successful.

Sexual reproduction

Seed production is an important means to aid survival during periods of adverse environmental conditions and to improve the chances for successful invasion of a new area. There are constraints on seed production by submerged plants, which is not surprising considering that pollen and seeds are structures adapted for aerial, not aquatic, reproduction. Flowers are typically borne above the water surface, even for otherwise totally submerged weeds and pollen floats across the water surface to the stigma. Consequently, fertilization of angiosperm weeds occurs at the water surface, not within the water, because of difficulties associated with submerged pollination (Sculthorpe 1967). The production of reproductive structures at the surface can increase the likelihood of herbivory and mechanical damage.

Seeds produced by submerged weeds are usually small and contain little carbohydrate reserve, but they may remain viable for several years (Sculthorpe 1967). When mature, the seeds sink to the hydrosoil and remain dormant until conditions are favourable for germination. Seed germination under only marginally-survivable conditions, which could result in the loss of new individuals, is usually minimal (Sculthorpe 1967).

Prolific seed production is not a sure indicator of success. Germination provides an entrance into the environment, but does not guarantee success (Haag 1983). Just as important is the subsequent ability to acquire light, nutrients, and possibly moisture, in competition with other individuals. Certainly, plantlets from tubers or underground turions of submerged weeds possess aggressive growth habits (Bowes *et al.* 1977; Van, Haller, and Bowes 1978), and seedlings might be expected to perform similarly.

There is considerable work on sexual reproduction in *Potamogeton*. The flowers of *P. pectinatus* are monoecious and pollination, fertilization, and fruit development all occur at the water surface (Yeo 1965). Seed production can be abundant. Over 10 000 seeds m^{-2} can result from the growth of one tuber in a growing season. There is an inverse relationship between seed yield and tuber size, which suggests sexual and vegetative reproduction are inversely related. However, the regulation of photosynthate allocation to seeds versus vegetative propagules, or environmental stimuli that may trigger sexual as opposed to vegetative reproduction, are unknown.

The germination of *P. pectinatus* seed does not occur under conditions that normally prevail during growth (Yeo 1965). Seeds subject to a cold (2°C) storage period, especially if kept dry, germinate better than those stored at 25 °C. Light also is required for the germination of some *Potamogeton* species (Spence *et al.* 1971). Furthermore, the impermeable integument has to be ruptured; seeds with the integument completely ruptured showed 100 per cent germination, while for those just abraded it was 14 per cent (Yeo 1965). It is evident from these studies that germination is less likely for seeds that remain underwater and especially under a weed cover. Events that could destroy the population, such as drying out or low temperatures, promote seed germination and thus favour reinfestation.

The infestation of the USA by *H. verticillata* is an interesting study in weed reproduction. This plant exists in monoecious and dioecious forms (Cook and Lüönd 1982; Steward *et al.* 1984). In the Gulf States and California only female plants of the dioecious form occur. Seeds are not produced, and reproduction is solely vegetative. In the mid-Atlantic States the monoecious plant is found. It produces large numbers of flowers (Langeland and Schiller 1983), and seed production has been recorded in North Carolina lakes (Langeland and Smith 1984). Whether the ranges will eventually overlap, and crossing of the monoecious and dioecious forms occur (which would greatly increase the potential for adaptation to different habitats) is still conjectural. In this context it should be noted that the two forms, designated 'USA hydrilla I', and 'USA hydrilla II' by Verkleij *et al.* (1983*b*), represents two markedly different genotypes and probably represent separate introductions to the USA. The chromosome number is 24 in 'USA hydrilla I', and 16 in 'USA hydrilla II' and there are also substantial differences in isoenzyme patterns between the two forms of *Hydrilla verticillata* in US populations (Verkleij *et al.* 1983*a*, *b*; Verkleij and Pieterse 1986).

To summarize, seed/fruit production by submerged weeds enhances their survival during adverse conditions, such as drought, cold, or salinity. It also aids dispersal and colonization of new locations. Seeds usually do not function in seasonal regeneration (Haag 1983), but together with vegetative propagules buried in the hydrosoil they are the aquatic equivalent of terrestrial seed banks, and serve to re-populate an area following any event which destroys the established population (van der Valk and Davis 1978).

Vegetative reproduction

Vegetative reproduction is the primary means by which most submerged weeds both regrow and infest new areas. The most prevalent method of vegetative dispersal is fragmentation, in which small fragments of two to several nodes of allochthonous or autochthonous origin are carried by wind, water, animals, or humans into an uninfested area (Langeland and Sutton 1980; Kimble 1982; Bowmer, Mitchell, and Short 1984; Kunii 1984; Spencer and Bowes 1985). If the conditions and substrate are appropriate, the fragments produce roots and become established in the new area. There is substantial variation in the growth potential of fragments, both among species and among fragment sizes within a species. For *H. verticillata* and *L. sessiliflora* few single node fragments produce new roots and shoots (Langeland and Sutton 1980; Spencer and Bowes 1985), but over 50 per cent of these fragments from *H. polysperma* are capable of initiating new plants (Spencer and Bowes 1985). With *H. verticillata* three-node fragments are needed to achieve this percent regrowth, whereas with *L. sessiliflora* even six-node fragments only achieve 25 per cent regrowth. Given the importance of fragmentation in the spread of submerged weeds, there is surprisingly little work on the factors influencing their production and subsequent regrowth.

Fragments of *E. nuttallii* produced in outdoor pools in Japan show a relationship between irradiance and location in the water column (Kunii 1984). Detached fragments are negatively buoyant and remain submerged as long as they receive at least 17 per cent of full sun irradiance, which allows photosynthetic starch production to increase their density. In contrast, fragments growing in less than 4.5 per cent of full sun irradiance float, because of low starch levels. This suggests that fragments will float until they reach a site with sufficient light for growth, whereupon they will tend to sink and establish a new attached plant. Whether other submerged weeds show a similar fragment dispersal strategy is not known.

There is a relationship between starch content and survival of the allofragments of *M. spicatum* (Kimble 1982). Successful over-wintering of the allofragments requires high starch levels in the meristematic tissue. Also the following spring, productivity of the high starch is twice that of the low starch fragments.

Over-wintering of submerged weeds in temperate climates is essential if the plants are to re-establish the following spring. The major submerged weeds are very successful in this regard. The various over-wintering strategies include fragmentation, tuber and/or turion production, green root-crowns, regrowth from stolons and rhizomes, and even survival of whole plants growing close to the hydrosoil.

To survive cold winters, *E. densa* relies upon green root crowns to regrow during the spring (Getsinger and Dillon 1984). Spring warming of the water stimulates root crowns to develop numerous shoots, and roots. Some submerged weeds, such as *E. canadensis*, can actually remain in an active state and photosynthesize under ice (Boylen and Sheldon 1976).

The production of tubers or turions, or regrowth from rhizomes and stolons serves to maintain a population of plants through unfavourable growth conditions rather than to disperse them. During tuber or subterranean turion formation, starch allocation to the rhizomes from the photosynthetic tissue causes the meristematic tips to enlarge and form a tuber or turion. When mature the tubers/turions abscise and remain dormant until conditions are favourable for germination (Haller 1976). In the hydrosoil they are protected from herbivory, exposure, and herbicide treatment. In fact, species that produce tubers/subterranean turions in significant numbers are amongst the most difficult to control. The entire biomass above the hydrosoil may be removed and the weed infestation can return the following season via germinating tubers (see also Chapter 12).

Most of the work on tuber/subterranean turion formation concerns *H. verticillata*, *P. pectinatus*, and *P. crispus* (Yeo 1965; Van, Haller, and Bowes 1978; Rogers and Breen 1980; Sutton, Littell, and Langeland 1980; Klaine and Ward 1984). In *H. verticillata* subterranean turion formation is initiated by short photoperiods, relatively independent of temperature (Klaine and Ward 1984). Under long-day conditions, treatment with the plant hormone abscisic acid (ABA) at 10 μM stimulates subterranean turion production to the short-day level. Light during the night inhibits subterranean turion formation, as does ethylene or gibberellin applications (Van, Haller, and Bowes 1978; Klaine and Ward 1984). It is possible that short days are perceived by the phytochrome system and that ABA transmits the stimulus and induces tuberization.

In north Florida, subterranean turion production usually occurs in autumn and winter, but is dependent upon growth during the previous spring and summer (Haller 1976). When plantings were made in February or March, less than 50 subterranean turions m^{-2} were produced after 16 weeks, but in late winter densities in Florida lakes can reach 4 000 turions m^{-2} (Sutton, Littell, and Langeland 1980). There is little information concerning the seasonality of subterranean turion production in more temperate or tropical latitudes, even though this weed has a wide geographic range. Similarly, the germination requirements for *H. verticillata* subterranean turions are inadequately known. Light, and temperatures between 18 and 32 °C, favour germination; whereas CO_2, but not low O_2, may inhibit the process (Miller, Garrard, and Haller 1976).

Tuber production by *P. pectinatus* begins in late summer but declines rapidly in early winter (Yeo 1965). Over 1 500 tubers m^{-2} may occur in 0.5 m deep water. Mechanical harvesting of vegetation in August, can reduce tuber production by 91 per cent (Yeo 1965), because over 70 per cent of the annual biomass is present by this date and its removal severely reduces the accumulation of carbohydrate reserves for tuber production. The management implications are obvious, though as with *H. verticillata*, several seasons of timed control may be needed to deplete the hydrosoil of tubers.

Above ground turions develop from apical or leaf axial meristematic tissue and eventually abscise and sink to the hydrosoil. Although for *H. verticillata* and *P. pectinatus* they are not usually produced as prolifically as subterranean turions or tubers, in *P. crispus* they are an important means of propagation (Rogers and Breen 1980; Yeo 1965). In *P. crispus*, more seeds were produced than turions, but only 0.001 per cent of the seeds germinated compared with 60 per cent of the turions (Rogers and Breen 1980). Turion germination is greater at water temperatures below 25 °C, and can occur at an irradiance insufficient for net photosynthesis. The plants which emerge at low light lack chlorophyll. They rely upon turion carbohydrate reserves to elongate in the water column until they reach an irradiance that is above their light compensation point and allows them to achieve independent growth. Although in *H. verticillata* and *P. pectinatus* subterranean turions/tubers are the most important propagule for new seasonal growth, in *P. crispus* this role is filled by turions.

EMERGENT AND FLOATING WEEDS

Introduction

The emergent aquatic plants which demonstrate the greatest weed propensity include: *Alternanthera philoxeroides, Echinochloa crus-galli, Panicum repens, Phragmites australis,* and *Typha* spp. Among floating plants, *Eichhornia crassipes, Pistia stratiotes,* and *Salvinia molesta* (Mitchell (1972) has separated *S. molesta* as a species distinct from *S. auriculata*) are among the worst causes of weed problems. These are the species to be given foremost consideration in this section.

The emergent weeds are rooted in water-saturated hydrosoil, but produce aerial leaves which are not supported by water. The floating weeds are generally not attached to the hydrosoil, but produce roots within, and leaves that float on or are held above, the water column. Like submerged weeds, emergent and floating species seldom encounter water stress and in this major respect they differ from true terrestrial weeds. Unlike the submerged weeds, all of the emergent and floating weeds listed above photosynthesize in an aerial environment and thus rely on free CO_2 as their inorganic carbon source. Furthermore, the leaves are not faced by the potentially wide fluctuations in DIC, O_2, and pH that submerged species encounter and, in most instances, being out of the water, the leaves are not bathed by a nutrient solution. Consequently, for emergent and free-floating weeds, nutrient uptake is mainly by roots; though even in this respect they differ, as nutrients are obtained from the hydrosoil and water column, respectively. The emergent weeds especially, with their roots in the sediment, may experience root anoxia which potentially affects nutrient uptake. Thus, the environmental conditions faced by emergent, floating, and submerged plants differ considerably; it is erroneous to assume, just because they may grow in close proximity, that plants with these different growth habits are experiencing the same limitations.

The aerial environment does not attenuate light nor affect its spectral quality to the degree that the submerged environment does; therefore, emergent and floating weeds are less likely to experience light limitation of growth, apart from self-shading. Consequently, emergent and floating weeds often have the advantages of terrestrial growth conditions, in terms of rapid CO_2 diffusion and/or active transport into the leaf, and high irradiance, without the disadvantage of limited water availability. Therefore, it is not surprising that they are among the most productive plants in the world (Westlake 1963).

Among the most important environmental factors that constrain the photosynthesis and growth of emergent and floating weeds are light, temperature, and nutrient availability. The following section describes the physiology of photosynthesis and growth and characterizes the effects of environmental factors on them.

Photosynthesis and growth

Gas exchange and enzymatic data indicate that the floating weeds exhibit C_3 photosynthesis (Tjitrosemito, Soerjani, and Mercado 1977; Sale and Orr 1981; Spencer and Bowes 1986). For example, *E. crassipes* fixes CO_2 predominantly by rubisco, has a rubisco to PEPC ratio greater than 16, its photosynthesis is inhibited by 21 per cent O_2, it has a high CO_2 compensation point, and it shows photorespiratory CO_2 release; all of which are characteristics of C_3 photosynthesis (Patterson and Duke 1979; Larigauderie, Roy, and Berger 1986; Spencer and Bowes 1986). There is no evidence that any floating plants possess C_4 photosynthesis. Although most emergent plants also exhibit C_3 photosynthesis, including *A. philoxeroides, P. australis,* and *Typha* spp. (McNaughton and Fullem 1970; Pearcy, Berry, and Bartholomew, 1974; Gloser 1977, 1978; Reddy, Sutton, and Bowes 1983; Longstreth, Balanos, and Smith 1984; Bowes and Beer 1987), there are some notable exceptions that have C_4 characteristics, such as the two emergent weeds *E. crus-galli* and *P. repens* (Gutierrez, Gracen, and Edwards 1974; Downton 1975; Hattersley and Watson 1976; Raghavendra and Das 1978; Garrard and Van 1982; Simon 1987).

Because they are not inhibited by atmospheric O_2, C_4 terrestrial plants have the potential for greater photosynthetic and growth rates than C_3 species (Bowes and Beer 1987). However, the high productivity of emergent and floating weeds is not just predicated on whether they are C_4 or C_3 plants, with greater or lesser maximum photosynthetic rates. An important factor is the absence of water stress, which allows stomata to remain open and photosynthesis to proceed unhindered throughout the daylight hours (Westlake 1963). Thus the C_3 aquatic plant *T. latifolia* can achieve an above-ground productivity of 45 tonnes $ha^{-1}yr^{-1}$.

In nutrient-rich cultivated systems *E. crassipes* can produce biomass at rates equivalent to over 200 tonnes $ha^{-1}yr^{-1}$ (Reddy, Sutton, and Bowes 1983), and high rates of growth are also seen in other free-floating species such as *Salvinia* (Gaudet 1973). These values are comparable to those of the most productive terrestrial C_4 plants, for example, sugar cane. Another component of the high productivity is that in eutrophic water with an abundant supply of N and P, C_3 emergent and floating weeds can afford substantial nutrient investments in rubisco and the PCR cycle, thereby maximizing CO_2 fixation. Although not as elegant as C_4 photosynthesis, it can be an effective strategy to improve the photosynthetic and growth potential. In contrast, submerged weeds such as *H. verticillata* have relatively low biomass production, ranging up to only 10 tonnes $ha^{-1}yr^{-1}$, even in tropical areas (Bowes and Beer 1987). This is attributable to low photosynthetic rates that are greatly limited by the diffusion of CO_2 in water (Bowes 1987).

Although emergent and floating weeds photosynthesize in air and thus are not subject to the very low diffusion rates faced by submerged leaves, this should not be interpreted as an absence of CO_2 limitation, since for C_3 plants short-term photosynthesis is increased by increasing CO_2 levels (Flock, Klug, and Canvin 1979). Thus, when grown under elevated CO_2, biomass production by *E. crassipes* increases (Spencer and Bowes 1986). The CO_2 enrichment produces an initial increase in net photosythesis, presumably by reducing the inhibitory effects of O_2 on C_3 photosynthesis. This leads to greater leaf area and leaf number, but not more plants. After the initial increase, the plants acclimatize to the higher CO_2 and the photosynthetic rate declines, but the increased leaf area is then sufficient to sustain the higher CO_2 fixation and growth.

The emergent and floating weeds are sun plants, whose photosynthetic light saturation and compensation points occur between 75 and 100 per cent, and 3 to 12 per cent of full sun irradiance, respectively (Dykyjova, Veber, and Priban 1967; Slamet and Sukowati 1975; Gloser 1978; Longstreth and Mason 1984; Longstreth, Bolanos, and Goddard 1985; Sale *et al.* 1985; Center and Spencer 1981; Spencer and Bowes 1986). Thus, each leaf can potentially utilize for growth all the solar energy it receives. These values are similar to those of terrestrial sun species (Beadle *et al.* 1985), but almost an order of magnitude greater than those for submerged weeds, which are all shade species (Bowes 1985). Not all emergent or floating plants are sun species; for example *Lemna* spp. seem to be shade plants, and this coincides with lower productivity (Bowes and Beer 1987).

Despite being sun plants, *E. crassipes* and *P. repens*, particularly the younger leaves, can acclimatize to low irradiance (Siregar and Soemarwoto 1976; Patterson and Duke 1979). Under low light, *E. crassipes* has a reduced specific leaf weight, protein content, stomatal and mesophyll conductance. The relationship between photosynthetic acclimatization to growth at low light and competition or community development is not known.

Terrestrial C_3 and C_4 plants typically differ in their temperature requirements for photosynthesis and growth (Edwards and Huber 1981). However, the temperature requirements for C_3 and C_4 emergent and floating weeds are not as narrowly delineated as terrestrial C_3 and C_4 species. For example, the annual emergent weed *E. crus-galli*, thrives in both northern and southern latitudes in North America, while the temperature optimum for photosynthesis of the C_3 emergent weed *P. australis* can range from 40 °C to 18 °C depending upon location.

The temperature optimum for photosynthesis of *P. australis*, varies with geographical location. In soils of high moisture content in Death Valley, California, *P. australis* has a photosynthetic leaf temperature optimum of 30 °C (Pearcy, Berry, and Bartholomew 1974) at an air temperature of 40 °C. The leaf temperature is maintained at 5 to 10 °C lower than the air temperature, unlike other desert plants in which air and leaf temperatures are more equivalent. At leaf temperatures equivalent to air temperature, the photosynthesis of *P. australis* drops by 50 per cent, suggesting that substantial transpirational cooling is needed to sustain high photosynthetic rates. For *P. australis* growing in Czechoslovakia, the photosynthetic temperature optimum is considerably lower than that of the California plants and varies with time of day. It is 13 °C during the early morning and increases to 18 °C at midday, when the irradiance is greatest (Ondok and Gloser, 1978*b*). These values more typically reflect the C_3 nature of the plant.

In temperate climates, biomass production of *P. australis* is greatest in July and August, coinciding with high photosynthesis and low dark respiration (Dykyjova, Veber, and Priban 1967; Ondok and Gloser 1978*b*, *c*). For *T. latifolia* in greenhouse and growth-chamber studies, an increase in temperature from 10 to 25 °C increases the shoot biomass by 275 per cent (Reddy and Portier 1987). Similarly, the shoot : root biomass ratios increase by 300 per cent at higher temperatures. However, a two-year growth analysis of *T. angustifolia* shows that above-ground biomass accrual is not only directly related to the cumulative degree days, but also to the length of the growing season and cumulative precipitation (Hill 1987).

At locations where elevated air temperatures become a limiting factor to C_3 photosynthesis, the transpiration of emergent and floating weeds may cool the leaf sufficiently to allow high photosynthetic rates to be sustained. Aquatic plants can afford more luxurious water-use to cool the leaf, since water is in plentiful supply. It would be interesting to examine the photosynthetic and growth responses of water-stressed aquatic plants at higher temperatures.

Ecotypes of the C_4 emergent weed *E. crus-galli* (which is common in rice cultivation) occur in both northern and southern latitudes, unlike most terrestrial C_4 species which are more typical of the tropics (Potvin 1986). Reciprocal transplant experiments show that the ecotypes do not produce as much biomass or reproductive material when transplanted outside their native geographical range. Furthermore, activities of the C_4 enzymes PEPC, $NADP^+$ malate dehydrogenase, pyruvate Pi dikinase, and $NADP^+$ malic enzyme from southern, but not northern, ecotypes are inhibited by exposure to temperatures between 4 and 7 °C (Simon 1987).

For floating weeds, both roots and leaves are susceptible to freezing damage, therefore, in temperate climates vegetative tissue seldom survives the winter. Neither *E. crassipes* nor *P. stratiotes* produce vegetative reproductive structures, such as tubers or turions, capable of long-term survival under adverse conditions of drought or freezing; also, they have only limited reproduction from seed. Consequently, these reproductive factors, combined with the freezing damage to the plants, limit the distribution of the major floating weeds to climates characterized by mild winters (Holm *et al.* 1977).

In contrast, *Lemna* spp., though causing more localized weed problems, survive freezing temperatures by various strategies, including over-wintering as turions or seed, 'hibernation' of vegetative tissue at depths where the temperature remains above freezing, and even by survival of leaf tissue within the ice (Landolt and Kandeler 1987). The distribution of *Lemna* is, therefore, pan-latitudinal, unlike the other major floating weeds.

The emergent weeds produce substantial amounts of tissue beneath the hydrosoil where it is protected from freezing. Even after a killing frost, new plants can arise from the meristematic tissue of buried rhizomes the following spring (Sculthorpe 1967). Thus all the emergent weeds can survive severe winters and like *Lemna*, possess a pan-latitudinal distribution.

Nutrient and edaphic factors

Because of their substantially greater capacity for producing biomass, emergent and floating weeds might be expected to have a larger requirement for nutrients than their submerged counterparts and, concomitantly, show a greater response to eutrophic conditions (Reddy, Sutton, and Bowes 1983). In many situations the key nutrients are N and P. Thus N fertilization not only increases the net photosynthetic rate of *P. australis*, but also its growth, and that of *T. latifolia* and *E. crassipes* (Wahlquist 1972; Overdieck 1978; Reddy and Portier 1987). In contrast, K fertilization had no effect on *P. australis* growth (Ulrich and Burton 1985). For *E. crassipes* in culture solution, N and P concentrations below 42 and 7.8 mg l^{-1}, respectively, may be limiting for growth (Gossett and Norris 1971). At present, data are not available on the critical levels of nutrients needed to achieve maximum growth for most emergent and floating weeds.

The nutrient uptake ability of *E. crassipes* in particular, has been exploited for the biological removal of nutrients from wastewater. One hectare of *E. crassipes* growing under optimal conditions can absorb the N and P waste production of over 800 people (Rogers and Davis 1972). Similarly, there are a number of reports concerning the nutrient uptake of emergent weeds as related to their growth and capacity for nutrient removal from water (Sheffield 1967; Boyd 1970a; Rogers and Davis 1972; Slamet and Sukowati 1975; Reddy 1984; Reddy and Debusk 1984; Joglekar and Sonar 1987). In general, the greater potential for biomass production of emergent and floating weeds makes them more effective than the submerged weeds in nutrient removal from water (Steward 1970) (see also Chapter 14).

Nutrient deprivation experiments have shown that the root biomass of emergent and floating weeds increases when nutrient supply, especially N and P is limited (Haller and Sutton 1973*b*; Barko and Smart 1979; Lieffers 1983). Whether low levels of other nutrients cause similar effects, or if other emergent and floating weeds respond similarly is unknown. Despite a report of successful foliar uptake of nutrients by *E. crassipes* (Shiralipour, Haller, and Garrard 1981), nutrient uptake by floating and emergent weeds under natural conditions occurs almost exclusively through the roots. The rate at which nutrients are root absorbed appears to be related to the metabolic activity of the plants. Nutrient absorption was shown to increase during the regrowth of *P. repens* from subterranean rhizomes following herbicidal destruction of the above-ground tissues (Peng and Twu 1979).

Especially for emergent plants, the underground biomass often exists in an anaerobic environment. As documented earlier for the submerged weeds, a lack of O_2 can hinder nutrient acquisition. The emergent weeds ameliorate the potentially anoxic conditions by production of aerenchyma, which allows the diffusion, and even mass flow, of O_2 from the photosynthetic leaves to the submerged organs (Sale and Wetzel 1983; Sebacher, Harriss, and Bartlett 1985). In fact, some control of both *T. latifolia* and *T. angustifolia* has been obtained by sequential sub-surface cutting (Sale and Wetzel 1983), which depletes the supply of O_2 to the roots and rhizomes (see also Chapters 7 and 16).

Unlike the situation for submerged weeds, the photosynthetic tissue of emergent and floating weeds is aerial and so the pH of the water has little effect on the supply of DIC for photosynthesis. However, there are reports that *E. crassipes* is able to take up HCO_3^- by the roots and incorporate it into organic acids (Ultsch and Anthony 1974; Asensio 1985). The degree to which this supplements aerial CO_2 for photosynthesis is probably small and it is not known if HCO_3^- absorbed by the roots is subsequently incorporated into the PCR cycle. In some emergent weeds, the aerenchyma tissue may not only serve to aerate the roots, but also provide a pathway for diffusion of free CO_2 from the sediment, where it is often in very high concentrations, to the leaves (Sebacher, Harriss, and Bartlett 1985). Elevated internal CO_2 levels in the leaves of C_3 emergent weeds would suppress photorespiration and allow for higher net photosynthesis. The extent to which this is a factor in their high productivity needs further assessment.

As described for the submerged weeds, water pH influences the nutrient supply by regulating the prevalent ionic species, especially of N and P and also nutrient solubility. Growth experiments have demonstrated considerable variation for growth in the pH range that emergent and floating weeds can tolerate. For instance, *S. molesta* produces maximum biomass at pH 6.0, while for *E. crassipes* the range is much broader, being between pH 4.0 and 8.0 (Chadwick and Obeid 1966; Haller and Sutton 1973*b*; Cary and Weerts 1984*a*). *Eichhornia crassipes* also possesses a broad range of pH, tolerance being between 6.0 and 12 (Haller and Sutton 1973*b*). *Panicum repens* is most often found in soils with a pH range of 5 to 7 (Holm *et al.* 1977). Generally a pH outside the range of 4 to 8 is restrictive to growth of emergent and floating weeds. It is interesting that *E. crassipes* can alter the pH of the medium in which it grows (Haller and Sutton 1973*b*; Tamil *et al.* 1985). Whether the above phenomenon enhances growth is unknown.

Reproduction

Vegetative reproduction is the principle method by which emergent and floating aquatic weeds over-winter and colonize new locations. Particularly adapted to this method of reproduction are the floating weeds *E. crassipes, P. stratiotes,* and *S. molesta* which produce many ramets, or daughter plants, from meristematic areas on the rhizomes of parent plants (Penfound and Earle 1948; Holm *et al.* 1977; Room 1983). The new ramets in turn produce additional ramets throughout the growing season. These floating weeds can double in number of ramets every three to ten days, under optimal growth conditions (Penfound and Earle 1948; Room 1983). Thus, within three months *E. crassipes* produced 1 610 plants from an original ten (Penfound and Earle 1948). This propensity for such rapid, vegetative reproduction over the water surface, is probably the foremost reason these plants are regarded as major weed problems.

The emergent weeds also reproduce vegetatively, though usually from the nodes of stolons or rhizomes. Thus, *Typha* spp. and *P. australis* produce new plants from subterranean rhizomes (Haslam 1972; Grace and Wetzel 1981*a, b*). Populations of *T. latifolia* in the temperate USA typically consist of new ramets, because frost kills the ramets produced the previous year (Grace and Wetzel 1982). Each new ramet is capable of producing an average of 2.2 ramets during the growing season. However, *T. angustifolia* produces fewer ramets and regrowth often results from the original rhizome. In northern communities of *Typha*, there appears to be an inverse relationship between vegetative and sexual reproduction, such that plants which flower and produce seeds have a marked reduction in vegetative reproduction (Grace and Wetzel 1981*a, b*).

Mature stands of *P. australis* can develop from rhizome fragments, or seed, within three years (Haslam 1973*b*). Rhizome fragments are transported by water, animals, or man, and will sprout if they lodge in an area of negligible current, which is partially exposed (Haslam 1973*b*). Once established, up to 120 shoots m^{-2} can be produced by a mature *P. australis* population (Haslam 1972), which then advances by either rhizome and/or legehalme production. A mature plant can spread 2 m per year by rhizome growth. A legehalme is produced when an aerial shoot lodges to the substrate and begins to elongate. Although much rarer than rhizome production, legehalme development can cause the stand to advance by as much as 5 m per year (Haslam 1973*b*).

Similar to *P. australis*, the dissemination of *P. repens, A. philoxeroides,* and to some extent *E. crus-galli*, depends upon vegetative reproduction (Penfound 1940; Siregar and Soemarwoto 1976; Holm *et al.* 1977; Peng and Twu 1979; Yabuno 1983). Over 20 per cent of the latent auxiliary buds on the rhizomes of *P. repens* germinate when planted as either single nodes or larger rhizome fragments (Peng and Twu, 1979). Germination of aerial stem latent buds was much less because of the influence of apical dominance. Rhizomes of *P. repens* penetrate the soil to a depth of 20 cm, with the greatest density occurring at 0 to 10 cm depth (Siregar and Soemarwoto 1976). The depth and density of rhizome production makes control of this species difficult.

Alternanthera philoxeroides has even greater potential for vegetative regrowth than *P. repens*. Each node on its floating stolons has the potential to germinate and produce a new plant. The 'germinated' nodes break off from the stolon and remain floating until they root in shallow water (Penfound 1940). The stolon tissue of *A. philoxeroides* has a low dry weight to stem length ratio which enables it to cover a substantial surface area with very little dry weight. The prolific ability to reproduce from stolon fragments and the low stem dry weight to length ratio are major factors in the weediness of *A. philoxeroides* (Penfound, 1940). The sprouting of submerged stolon nodes of *A. philoxeroides* requires light, but will occur under low O_2. It has been suggested that O_2 production from photosystem II activity in the light may account for nodes sprouting under these conditions (Quimby, Potter, and Duke 1978).

Although both floating and emergent weeds reproduce vegetatively, the floating weeds generally appear more prolific in this regard. This disparity may be related to the site and method of colonization. The emergent weeds usually colonize littoral areas of relatively shallow water. Although the spread of the population is slower than that of floating weeds, the area once infested is not easily acquired by other invading species. Additionally, the production of reproductive tissue in the hydrosoil ensures that new individuals will survive the winter and again inhabit the site. The floating weeds, however, have no such hold upon the habitat. Strong winds, killing frosts, or other floating weeds may displace the entire population within a growing season. Regrowth after a frost often depends upon the production of new ramets from few surviving plants. Although the high rate of vegetative reproduction possessed by the floating weeds allows them to dominate a given habitat quickly, this domination, unlike that of the emergent weeds, is often ephemeral.

It is generally understood that, as for the submerged weeds, sexual reproduction by the emergent and floating weeds is less important to their weed potential than vegetative reproduction. Among the floating weeds, *E. crassipes* and *P. stratiotes* produce flowers and viable seed (Das 1969; Datta 1969; Pieterse 1977b; Pieterse, de Lange, and Verhagen 1981) and the fern *S. molesta* produces spores (Holm *et al.* 1977), but their viability and role in the spread of these plants is of questionable importance.

Eichhornia crassipes plants growing in 10 cm of water produce more flowers and viable seed than either floating or rooted plants (Ueki and Oki 1979). Since seed germination is also greater in shallow water, there appears to be some relationship between site of seed production and germination. Seeds may remain dormant for up to five years, but there is disagreement concerning the need for or length of the quiescent period (Hitchcock *et al.* 1949; Obeid and Tag el Seed 1976; Ueki and Oki 1979). Seeds of *E. crassipes* apparently require scarification of the seed coat and light for germination to occur (Penfound and Earle 1948; Das 1969).

In the case of *P. stratiotes*, Datta (1969) reported that field-collected seed did not germinate when planted in mud under water. It was concluded that germination was inhibited by either low O_2 and/or high CO_2 concentrations. Pieterse, de Lange, and Verhagen (1981) have subsequently shown that submerged seeds do germinate if they are exposed to high light intensity. When released from the plants, seeds of *P. stratiotes* may float for up to two days, before sinking to the hydrosoil. Similarly, upon germination the presence of an air-chamber at the hilum provides the buoyancy necessary to bring the seedling to the water surface (Holm *et al.* 1977; Mercado-Noriel and Mercado 1978). Seed viability is reduced by drying at 30 °C but some seeds survive, which suggests that *P. stratiotes* has the potential to endure drought via its seeds (Pieterse, de Lange, and Verhagen 1981).

In regard to sexual reproduction in the emergent weeds, *Typha* spp. are among the most studied. *Typha* produces many air-borne seeds whose role in regeneration seems to be restricted to germination in uninfested, unshaded sites (McNaughton 1968, 1975; Bonnewell, Koukkari, and Pratt 1983; Smith and Kadlec 1983). Germination of *T. latifolia* seed requires exposure to light, low O_2 levels, and high temperatures (Morinaga 1926*a*, *b*; Sifton 1959; McNaughton 1966; Hutchinson 1975; Bonnewell, Koukkari, and Pratt 1983). Greatest seed germination occurred between 2.3 and 4.3 mg $O_2\ l^{-1}$ (i.e. 28 to 40 per cent of air equilibration values), but not under complete anoxia (Bonnewell, Koukhari and Pratt 1983). Red, as opposed to far-red, light stimulated germination with twelve uninterrupted hours of exposure needed for maximum effect. The optimum temperature for germination was 35 °C. These germination parameters are characteristic of shallow water, bare mud flats where *Typha* seedlings commonly occur (Bonnewell, Koukhari, and Pratt 1983). Sharma and Gopal (1979*b*) reported that even after germination, seedling survival requires a minimum of 7 per cent of full sun irradiance. Therefore, light is critical for both seed germination and seedling survival.

Autotoxic inhibition of *Typha* spp. seed germination has been suggested to occur (McNaughton 1968; van der Valk and Davis 1978), but the data of Grace (1983) do not bear this out. The absence of *Typha* seedlings in *Typha* stands probably represent a lack of the necessary germination conditions, rather than autotoxic effects. The germination of *Typha* seed and seedling survival is restricted to open areas not colonized by other plants. It does appear that seeds, rather than vegetative reproductive structures, are the primary vehicle whereby *Typha* colonizes new locations.

Unlike *Typha* spp., *P. repens* and *A. philoxeroides* seldom produce seed. In the case of *A. philoxeroides*, ovaries with immature seeds have been observed (Penfound 1940; Siregar and Soemarwoto 1976; Holm *et al.* 1977), but the biology of seed germination in *A. philoxeroides* is unknown.

Similar to *E. crassipes* and *P. stratiotes* the optimum habitat conditions for germination of *P. australis* seeds are much narrower than for growth of the adult plant. Therefore, the spread of an existing stand is sustained largely by vegetative means. However, like *Typha* spp., introduction to uninfested areas usually occurs by seed, which may remain viable for several years. The germination of *P. australis* seed requires shallow water, light, fluctuating temperatures between 10 and 30 °C, and sufficient aeration to prevent anoxia (Hürlimann 1951; Haslam 1972, 1973*b*; Spence 1964). Thus the germination of *P. australis* seed may be largely restricted to exposed mudflats, where the conditions are appropriate for germination and seedling survival.

Unlike *P. australis, E. crus-galli* seed germination commonly occurs under anaerobic conditions (Rumpho and Kennedy 1983). Successful seed germination and seedling development under anoxia is supported by ATP production via glycolysis and the anaerobic reduction of pyruvate to ethanol. Also, an increase in the oxidative pentose phosphate pathway provides reducing power in the form of NADPH for the biosynthesis of lipids, nucleotides, and RNA required by the developing seedling (Rumpho and Kennedy 1983). This plant can produce over 20 000 seeds m^{-2} and up to 90 per cent may germinate 16 days after flowering (Yabuno 1983). It seems likely that *E. crus-galli* is less restricted in terms of germination requirements than *P. australis* or *Typha* spp., and thus possesses a potentially greater range of habitats in which it can become established via seed.

CONCLUSIONS

Fewer than twenty of approximately 700 aquatic plant species are considered to be major weeds. These weeds, by virtue of their prolific growth and reproduction, often interfere with human utilization of freshwater resources and displace indigenous vegetation. The aquatic weeds include plants from three distinct groups: submerged, floating, and emergent; each of which encounter certain environmental conditions which limit photosynthesis and/or growth. The environmental conditions faced by emergent, floating, and submerged plants differ considerably; therefore, it is erroneous to assume that they experience the same limitations.

Low levels of both light and dissolved inorganic carbon often limit the growth of submerged weeds in the aqueous environment. Low light compensation points, the ability to vary photorespiratory CO_2 losses, use of HCO_3^- ions, reduction of dark respiration, and in some cases C_4-like metabolism; and surface growth are physiological and morphological adaptations respectively, to the above environmental constraints. Because of their lower dry weight biomass, submerged plants tend to have lower nutrient requirements than emergent and floating plants. Also, prolific vegetative reproductive strategies ensure the survival and spread of these weeds. Basic research concerning biochemical management of growth habit and/or vegetative reproduction is likely to result in novel and environmentally acceptable control methods.

Since the emergent and floating weeds seldom experience water, light or major CO_2 limitations of photosynthesis they comprise some of the most productive plants on earth. The most important environmental factors that constrain the growth of these plants are nutrient availability, temperature and root anoxia. Increased root growth under nutrient limitation, transpiration-driven leaf cooling and extensive aerenchymatous tissue are among the adaptations the floating and emergent weeds have acquired. Vegetative reproduction is the principle method of population growth. However, unlike most floating weeds the emergent weeds colonize new areas primarily by seed dissemination.

ACKNOWLEDGEMENTS

We are indebted to The Aquatic Plant Information and Retrieval System, Center for Aquatic Plants, IFAS, University of Florida for providing electronic literature searching and access to their reprint library. Portions of this work were supported in part by subventions from Florida Department of Natural Resources contract #3635, and from grant 82-CRCR-1-1147 from the Competitive Research Grants Office, Science and Education Administration, United States Department of Agriculture.

5

Flow-resistance of aquatic weeds

R. H. PITLO AND F. H. DAWSON

INTRODUCTION

THE MOVEMENT of water flowing in natural and artificial watercourses is influenced mainly by the following factors:

(1) the dimensions of the water course (width, depth, and cross-sectional shape);

(2) the slope or hydraulic gradient of the water course;

(3) the resistance to flow (bed and bank roughness, aquatic vegetation, longitudinal changes in both cross-section and alignment).

For an existing watercourse with low sediment transport, wetted perimeter and gradient can be measured and the hydraulic roughness can be estimated by reference to tables relating channel boundary conditions and roughness (Chow 1981). However, vegetation within the watercourse and along its margins can cause major variations in the resistance to flow, a problem most frequently experienced in lowland watercourses. This increased vegetative resistance depends both upon the type and species and on changes in the biomass of vegetation during the seasonal cycle of growth and decay. The interaction between the flowing water and the vegetation is complex, depending on many factors including the degree of complexity of the plant stand (e.g. from single cylindrical stems to many branched stems of irregular cross-section), the form, or shape, which is presented to water flow, the flexibility, the cross-sectional area and spacing of the stems, and the ratio of the depth of water to the height or length of vegetation.

The presence of vegetation may affect flow in a channel in the following ways:

(1) reduce water velocities thus raising water levels in the channel and also in the water table in adjacent land causing it to become waterlogged;

(2) increase the flooding or overbank spill (i.e. increases flood risk);

(3) encourage the deposition of suspended sediment (i.e. aggregation) which may have to be removed to maintain the waterway depending upon the rate of accretion, which may vary seasonally;

(4) change the magnitude and direction of currents within a channel thus causing local erosion or reducing bank erosion, depending on the location, extent and density of the vegetation, and

(5) interfere with other water uses, e.g. navigation and recreation.

It is the responsibility of the river, irrigation or drainage engineer to predict these effects and to ensure that the waterway conveys the flow efficiently throughout the year with the minimum of inconvenience to water users and people and property in the vicinity, while having regard for conservation interests. A better understanding of the effect of vegetation on the flow of water in channels, by engineers, botanists, conservationists, fishermen, and all those who are concerned with natural and artificial waterways, should lead ultimately to the more efficient management of the waterways with ecological and financial benefit to the whole community.

THE DETERMINATION OF FLOW RESISTANCE

Although much has been written, no single method of estimating the effect on flow is suitable for, or applicable to, all possible situations because many variables affect channel flow when plants are present (Henderson 1966; Ackers *et al.* 1978; Kinori and Mevorach 1984; see also British Standards Institution BS3680). It is imperative, therefore, that when examining a particular watercourse the conditions affecting flow are appreciated and the most appropriate method of estimating hydraulic roughness is chosen.

Calculations of flow resistance are often based upon the work of Prandtl (1904), von Kármàn (1932), and Chow (1981) which assumes that the vertical distribution of velocity follows a logarithmic relationship:

$$\bar{V} = 5.75 V^* \log 12.3 \frac{R}{k} \tag{5.1}$$

where $\bar{V}$ = mean velocity of flow ($\mathrm{m\,s^{-1}}$); V^* = shear velocity ($\mathrm{m\,s^{-1}}$) ($\equiv (gRS)^{0.5}$) in which g = acceleration due to gravity ($\mathrm{m\,s^{-2}}$); S = hydraulic gradient; R = hydraulic radius or hydraulic mean depth (m) (i.e. the ratio of the cross-sectional area of the water to the wetted perimeter) and k = equivalent grain roughness (m).

This equation was derived from pipe flow and its application to open channel flow is not exact due to variation in boundary roughness, the free water surface and secondary currents; the latter are perpendicular in direction to the main flow and impart a spiral action to flowing water. It is thus impractical to determine a friction factor theoretically because the velocity distribution over the cross-section is unknown. Nevertheless, the use of this equation has been recommended in preference to the more widely used empirical equations because they do not discriminate between hydraulic rough conduits and conduits in the intermediate zone between rough and smooth. Here the conduit resistance is also influenced by the viscosity in the laminar sub-layer (Task Force on Friction Factors in Open Channels 1963).

Another equation in common use was derived empirically by Manning (1891). When computing Manning's '*n*', knowing discharge and channel dimensions, it is essential that the difference between uniform and non-uniform flow, and steady and unsteady flow is understood. Broadly the definitions are as follows.

1. Steady flow: At any point in a channel, discharge, depth, and water surface slope remain constant or steady with time.
2. Unsteady flow: It follows from above that discharge, depth, and slope change from instant to instant, as, for example, when a flood is rising or falling or a wave passes down a channel.
3. Uniform flow: Under conditions of steady flow in a channel, the water surface slope is parallel to the bed (i.e. the depth remains constant at any instant at points along the channel).
4. Non-uniform flow: Under conditions of steady flow in a channel, the water surface slope is not parallel to the bed (i.e. in going upstream or downstream the depth may increase or decrease).

In most situations the aim is to develop steady uniform flow; this then allows the use of Manning's equation which relates velocity, roughness, hydraulic depth, and gradient. Using metre per second units this is:

$$\bar{V} = \frac{1.000}{n} R^{2/3} S^{1/2} \tag{5.2}$$

where $\bar{V}$ = mean velocity of flow (m s^{-1}); n = resistance coefficient: this is dimensionless, hence the need to introduce a dimensional constant depending upon units: 1.000 in metres per second units (above) or 1.486 in foot per second units; the value of n is the same irrespective of system of units used [see Chow 1981: pp. 98–9]; R = hydraulic 'radius' or hydraulic mean depth (m) = A/P, A = area of cross-section of water course (m^2); P = wetted perimeter of water course (m); S = hydraulic gradient or hydraulic energy gradient. In uniform flow $S = S_b$; S_b = bed gradient of watercourse. When flow is not uniform, however, $S \neq S_b$; and S must be determined using a complex computation. This is the reason for trying to develop uniform flow.

Manning's equation was derived for use in large unvegetated rivers where friction acts over the wetted perimeter. For channels without vegetation the average velocity of flow ($\bar{V}$) is proportional to the square root of the hydraulic gradient or slope (S). It is possible for design purposes to use eqn (5.2) to calculate the dimensions for a channel since 'n' is a function of the equivalent grain roughness (k). In addition, Manning's 'n' is related to the water depth (h), since the boundary roughness decreases relatively as the wetted cross sectional area increases:

$$\frac{1}{n} = v \,.\, h^{\delta} \tag{5.3}$$

where: v = 'vegetative roughness factor', which changes from 34 during the winter period to 23 during the summer and $\delta = 1/3$ (Werkgroep Afvoerberekeningen 1979).

Attempts have been made using eqn (5.3) to estimate (for temperate zones) the seasonal effects of the growth of vegetation on 'n' using a range of values for v. However, in practice it was evident that the summer value of 'n' (eqn (5.3)) not only changes during the season, but the relation between $\bar{V}$ and S (eqn (5.2)) depends on the density, type, and stiffness of the vegetation.

In the determination of the hydraulic roughness of dense stand-forming species (e.g. submerged plants such as *Ranunculus* spp. and *Callitriche* spp., and emergent plants such as *Rorippa* spp.), it is important to account for the change in wetted surface, because water velocity through stands can be less than a tenth of that in the channel. This has led to the concept of a 'blockage factor', that is, the ratio of effective to actual cross-section of watercourse and the separation of the main flow from that flowing through the plants (Hydraulics Research Ltd. 1985). A similar factor: 'the percentage of "open" water' (V_o) was introduced by Pitlo (1986). The determination of the actual cross-sectional area occupied by plants which severely restrict flow is not always easy to define but such a concept may prove useful. Complications may, in addition, arise from the decision to include or exclude, fine sediments accumulated as a direct result of such dense growth of plants (Dawson 1978*a*). This leads to the requirement for fuller descriptions of study sites which include the type, extent, and stage of growth of vegetation, if the study is to be of use elsewhere.

For rivers and unlined canals with vegetation, a plot of roughness (n) against the product of velocity and hydraulic radius (VR) has been found useful (Kouwen and Li 1980). This assumes a unique relation between n and VR for a particular type of vegetation regardless of the season and the values of V and R. Whilst this is a valuable method for determining the roughness in large rivers of fairly uniform cross-section, studies of such plots have shown that large errors can occur in other channels. There is concern that the method is often used outside its intended limits (Ree and Crow 1977; Kouwen and Li 1980).

VEGETATIVE FLOW RESISTANCE

The growth of both submerged and emergent aquatic plants, can progressively increase the hydraulic resistance to water flow (Hillebrand 1950). The mechanisms by which the plants restrict flow are not well understood and an empirical approach has been used, as is often expedient in complex flow problems in hydraulics. Hydraulic resistance or roughness values are available both from tables (Chow 1981: pp. 109–13) and from annotated photographs (Fasken 1963; Barnes 1967). It is frequently difficult to relate the broad, often subjective, descriptions to field sites. Vegetated sites are normally referred to using descriptive phrases ranging from 'some weeds' to 'very weedy'. Problems with the above arise primarily because, for a particular species, the ratio of cover (the main term used for plant abundance) to plant biomass (the current best gauge of vegetative resistance) can vary significantly during the growing season. The seasonal variation of resistance has, in the past, been disguised within broad descriptions of the amount of vegetation as seen above, but more recent work has sought to be more precise by relating resistance to biomass during the plants' seasonal growth cycle, or for particular months (Powell 1972; Brooker, Morris, and Wilson 1978; Pitlo 1978, 1982; Dawson 1978*a*; Dawson and Robinson 1984).

Flexible flow resistance

The reaction of vegetation to water flow, and hence the flow resistance offered by aquatic vegetation, varies with water velocity mainly due to the flexible nature of both plant stems and their leaves, which are bent in the direction of flow when velocity increases (Fig. 5.1). Work on flexibility was probably first undertaken at the Stillwater Outdoor Hydraulic Station, Oklahoma, USA in the late 1920's, but this was related to the suitability of various terrestrial grass types for use in intermittent low-velocity wide agricultural spillways and in erosion control. The tables of flow resistances obtained under different velocities continue to be used but have been expanded to include the majority of vegetated watercourses.

The reaction to flow can be separated into different stages as velocity increases. The importance of each stage varies considerably with species and is additionally influenced by the preferred habitat range of each species but, in general, the stages follow this progressive pattern:

(1) downstream orientation or slight displacement of stems: leaves similarly orientated;

(2) vibration of vertical stems or sinuous movement of oblique or elongated horizontal stems: strong vortices downstream of plant stands;

(3) stems inclined, submerged leaves strongly orientated, surface leaves submerging or tipping: progressive loss of moribund or dead parts, turbulent water flow over or below stands;

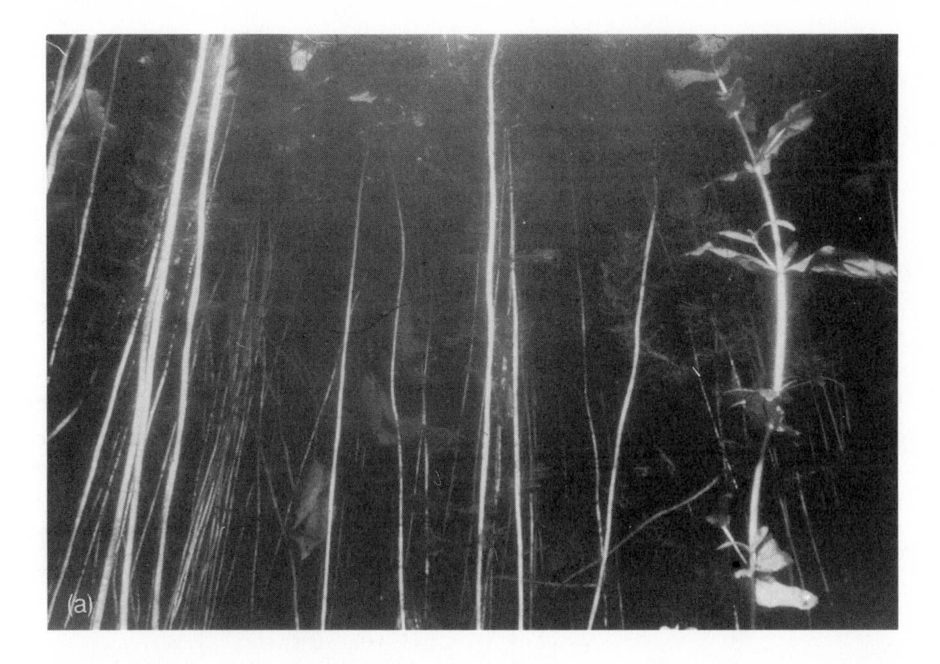

FIG. 5.1. Effects of water velocity upon vegetation (a) at low velocity, with vegetation erect and (b) at high velocity, with vegetation nearly prone and surface leaves pulled under water. (Photo: R. H. Pitlo.)

(4) stems prone or densely compacted: surface leaves submerged (filamentous submerged leaves compacted), and

(5) damage and loss of leaves and parts of, or whole, stems; washout of some whole plants.

Flow-resistance normally rises during the initial two or three stages before declining; plants can recover fully with subsequent reductions in velocity unless there has been loss of stems or leaves producing irreversible damage. Variation in biomass of a particular plant species between sites may well be linked to the minor floods which remove moribund materials; major floods often cause catastrophic loss although the effect may be less severe if overbank can occur. One study in a lowland river dominated by *Ranunculus penicillatus* var. *calcareus*, a submerged macrophyte, showed a progressive series of changes in flow resistance as the seasonal cycle of growth and decline proceeded. This sequence was only interrupted by plant washout during autumn floods (Fig. 5.2; Dawson and Robinson 1984).

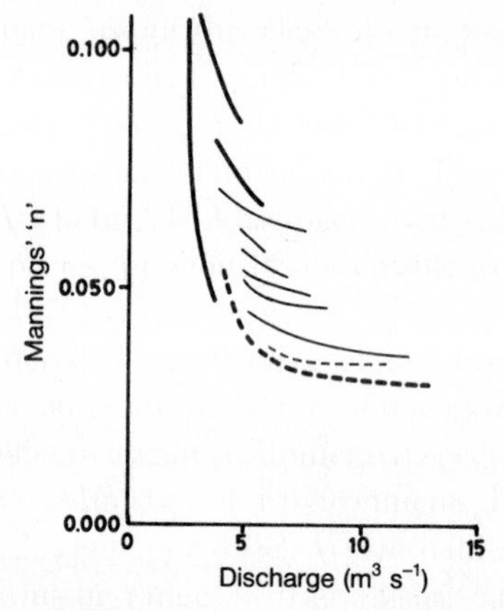

FIG. 5.2. Seasonal variation of low-resistance with discharge for periods between major discharge changes from summer 1981 to summer 1982 in a vegetated lowland river in UK (Solid thick line = summer, plant biomass rising to 2.5 kg m^{-2} fresh weight; thin line = winter, biomass of 0.2; 0.4 kg m^{-2}; thick dashed line = spring, biomass rising to 1.5–2.0 kg m^{-2} before removal reduced it to 0.7 kg m^{-2} and regrowth started (from Dawson and Robinson 1984).

PRACTICAL APPLICATIONS AND FURTHER STUDIES

Low velocity study

Calculated roughness factors (eqn (5.3)) and measured Manning hydraulic resistance coefficients (eqn (5.2)) of three species of plant were compared at a range

Table 5.1. A comparison of calculated and measured values of Manning '*n*' flow-resistance at an average velocity of 0.06 m s^{-1} for three vegetated experimental channels in The Netherlands (after Pitlo 1982)

Plant species	Water depth (m)	Mannings coefficient '*n*'					
		Calculated, Eq. 5.3		Measured, Eq. 5.2			
		Winter	Summer	Feb	Mar	May	June
Potamogeton natans	0.6	0.035	0.052	0.033[a]	0.084	0.14	0.22
Nymphaea alba	0.75	0.032	0.048	0.036[a]	0.035	0.10	0.12
Nymphoides peltata	1.2	0.028	0.041	0.039	0.036	0.063	0.22

[a] Only one measurement, after cleaning in February.

of velocities at intervals during the growing season (Table 5.1; Pitlo 1978). The study was carried out using a channel of 800 m in length, trapezoidal in section with a bed width of 12 m and 1.2 m depth, and a three-channel experimental ditch system, 65 m in length, trapezoidal in section with either 0.55 m or 1.5 m wide base, with pumped recirculation (0.02–0.20 m s^{-1}) and a manometric system of water slope measurement. For an average water velocity of 0.06 m s^{-1}, and three different water depths, it was concluded that the calculated value of '*n*' for winter could only be obtained for channels free of all vegetation. If small quantities of dead vegetation remained, then '*n*' values of *c*. 0.04 or higher were obtained. Values of *n* of 0.05 or even greater were measured by April. It was concluded that the use of the roughness factor equation (eqn (5.3)) for design purposes, leads to cross-sections which make no allowance for the presence of vegetation. These results also illustrated the variation of flow resistance with species, with that of *Nymphaea alba* being only half that of *Potamogeton natans* or *Nymphoides peltata* by June. Small changes in water velocity and depth, with populations of *P. natans* and *N. peltata* cause little change in resistance but roughness decreases with further increases in velocity as surface leaves are submerged (Fig. 5.1; Pitlo 1978, 1982).

Regular short-term experimental increases in water velocity during the growing season of a population of *Potamogeton trichoides* showed that '*n*' decreased with increasing values of velocity when the hydraulic radius was maintained at a constant; no irrevocable damage occurred to the submerged plants at the higher velocities (Fig. 5.3). Analysis using the Manning equation (eqn (5.2)), showed that if the power to which the hydraulic gradient term was raised changed during the growing season then the balance of the equation could be maintained using 0.5 in March to 1.0 in June and then back towards 0.5 by September (Fig. 5.4). Such problems are, however, best resolved by investigating the relationship between flow resistance and the Reynolds number in order to characterize the species of vegetation in 'hydraulic' terms (van Ieperen and Herfst 1986).

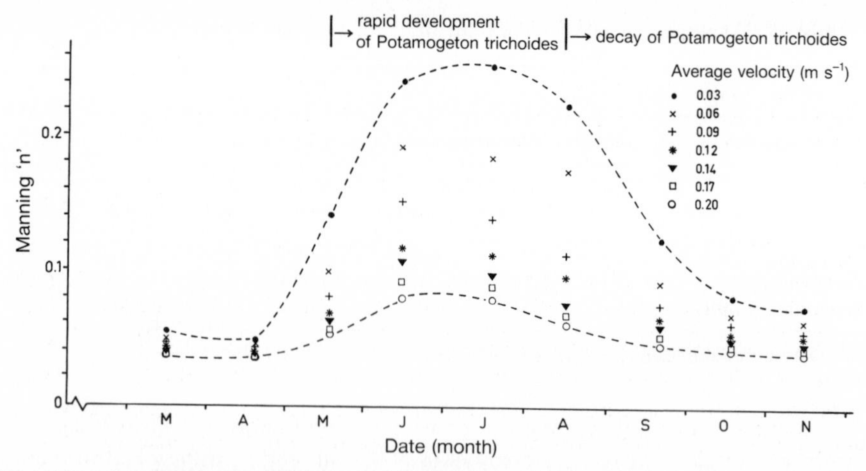

FIG. 5.3. The variation in Mannings '*n*' at different water velocities during the growing season for *Potamogeton trichoides*. The hydraulic 'radius' (R) was kept constant during these tests (from Pitlo 1979, unpublished data).

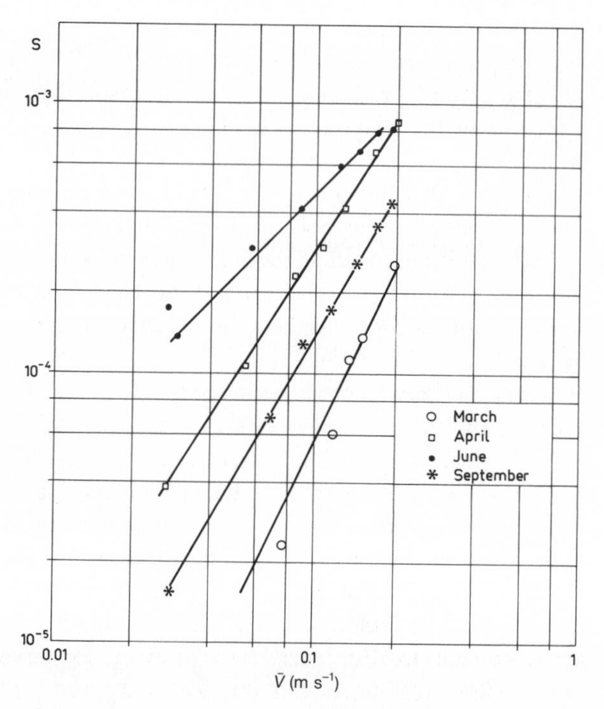

FIG. 5.4. The relationship of average water velocity ($\bar{V}$) and slope during the growing season for *Ranunculus circinatus* (from Pitlo 1978).

Medium velocity study

During measurements of the seasonal change in flow resistance in a typical densely-vegetated lowland river, the forces acting on a selection of submerged plants were also determined, *in situ*, using a single tensiometer (strain gauge) (Dawson and Robinson 1985). These forces, taking account of the number of plants, were shown to account directly for about half the flow resistance of the section of river. Subsequent removal of the majority of the above-bed plants, for flood control, allowed an assessment of the channel resistance in late spring. The value of '*n*' was reduced by 0.02–0.04 from *c*. 0.11 when 1.0–1.5 $kg\,m^{-2}$ fresh weight of plants were removed; water levels were reduced by 0.2–0.3 m from 1–1.2 m or about one-third of the reduction achieved by removing the maximum biomass in summer. More detailed analysis of the forces acting upon shallow and deep water plant stands was attempted by considering the typical structure of plant stands, plant width, stem length, stem buoyancy, and biomass (Fig. 5.5). The seasonal interrelations between these main factors were not separated and although plant width was used as a surrogate for the other factors, it did not fully explain the 'approach area' presented to flow ('drag form') or more fundamental aspects of flow-resistance by plants. This study did, however, introduce several new concepts, firstly that of minimum drag forms for plants, which were consistent with either maintaining their presence or for optimal growth; secondly, the distortion of velocity profile in shallow water by flow between plants, leading to development of within-channel 'pseudo-braiding'; and, thirdly, maximum velocities at significantly lower levels in the water column than normal, because of the development by some species of dense surface mats; water flow between their anchoring stems was, in addition, sinuous and this complex situation could be considered as a type of 'pipe or tube' flow.

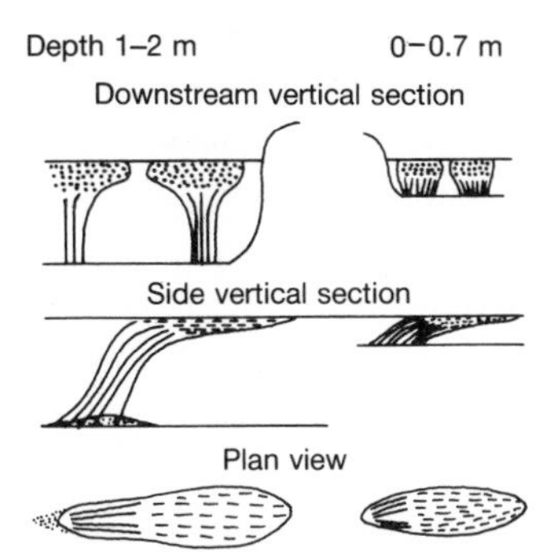

FIG. 5.5. Diagramatic representations of the shape of a typical mature plant of *Ranunculus penicillatus* var. *calcareus* growing amongst others in shallow and deep water. (The lines represent the density of anchoring stems, the large dots and dashes, the surface stems, and the dotted areas near the streambed, silt (from Dawson and Robinson 1985).

Fundamental and theoretical studies

Analysis of flow resistance by vegetation is complex; apart from the great diversity of form of water plants, each plant species undergoes its own series of transitions of varying duration with increasing water velocity. Additionally, the physical habitat of plants may affect the flow pattern, although the range of habitats is not unrestricted, being limited for each species. Plant form varies enormously and some attempts have been made to analyse, both physically and theoretically, the simpler types, e.g. vertical stems, and the flow-transition stages through which they pass (Li and Shen 1973; Thompson and Roberson 1975; Petryk and Bosmajian 1975). These and other models frequently include the effects of water velocity upon the length, height, width, shape, stiffness, deflection and areal distribution, and layout of elements. The principles of flow resistance for such models have applications in management, for example, by the bias which could be introduced in favour of a species with less flow resistance. Use of these models is not restricted to aquatic vegetation for they have also been used in the optimisation of the patterns of tree planting for flood wave retardation in a flood plain (Bonham 1978).

CONCLUSION

In conclusion, it can be seen that studies on vegetative resistance have started to produce data for application to the design and maintenance of watercourses, but progress is often hampered by incomplete descriptions of the method or type and amount of vegetation at a site. The use of the logarithmic type of equation (eqn (5.1)) is preferable to the older Manning equation. The nature of the flexible flow-resistance has been adequately demonstrated but together with the fundamental nature of vegetation, it is complex in its interrelations between the biotic and physical environments (Dawson and Charlton, 1988). In consequence, several studies have tried to simplify the problem by observing the effect of flow resistance using physical models or constrictions to represent the parts or physical elements of water plants and their characteristics at different velocities under chosen controlled conditions. These simplified studies, whilst not providing a general solution, give useful information which assists both the practical problems and the fundamental explanations. It is apparent, however, that true progress will not be made in isolation by either specialists in fluid mechanics or biology but in combination.

ACKNOWLEDGEMENT

The authors wish to thank F.G. Charlton for constructive criticism of the draft manuscript.

6

The relation between aquatic weeds and public health

E. O. Gangstad and N. F. Cardarelli

INTRODUCTION

Dense growths of aquatic plants may have an adverse effect on public health, mainly because they provide a habitat which favours the development of vectors of human diseases. The plants give shelter to these vector organisms, provide a food source (organisms and/or detritus associated with the plants) and can also be a source of oxygen (oxygen which is present in the plant tissues). The most important vectors of human diseases which, at least during part of their life cycle, depend on the environmental conditions prevailing in dense aquatic weed vegetation are certain types of mosquitoes and aquatic snails. The disease-carrying mosquitoes which thrive under these conditions can be divided into mosquitoes with larvae which breathe at the water surface, i.e. the anopheline mosquitoes, and mosquitoes which are able to obtain oxygen from plant tissues, i.e. mosquitoes of the genus *Mansonia*.

In a few cases aquatic plants may bring about a direct toxic effect to man due to the release of poisonous compounds. These effects are mainly restricted to blooms of blue-green algae and, in marine waters, blooms of Protista (flagellates in which a photosynthetic pigment occurs). The Protista are eaten by molluscs which accumulate the toxin and, subsequently, when man eats these molluscs this may lead to acute paralytic poisoning.

VECTOR DISEASES

Anopheline mosquitoes

These species of mosquitoes have larvae which inhabit stagnant waters, feeding and breathing at the surface. The larvae are protected and sheltered by aquatic weeds; weed control measures may, therefore, to a certain extent, control the diseases which they transmit (Angerelli and Beirne 1982; Brown and Deom 1973; Mulrennan 1962; Ferguson 1971, 1980; MacKenthun and Ingram 1967).

Malaria, the most serious vector-transmitted disease, is common throughout tropical and subtropical regions of the world. The malaria parasite responsible for human infections is a protozoon belonging to the genus *Plasmodium*. The malaria parasite is transmitted solely by the mosquito genus *Anopheles*. Dengue, which is caused by an arbovirus, is also prevalent throughout the same areas and is transmitted by the genus *Aedes*. This disease, which is normally of low endemicity, exhibits occasional outbreaks, such as in 1941 in Panama, in 1954 in Yemen and in 1963 in Puerto Rico. Although dengue is not often a fatal disease, it may be complicated with Dengue Haemorrhagic Fever (DHF) which does have a high mortality rate (Ferguson 1971, 1980; Gholson 1982).

Yellow fever, another virus disease, largely confined to Africa and to Central and South America but also reported from tropical areas of Asia, is mainly transmitted by the genus *Aedes*. However, the disease is sometimes transmitted by the genus *Hoenagagus*. Filariasis (caused by the roundworm *Wuchereria bancrofti*) can be transmitted by the genera *Aedes*, *Anopheles*, and *Culex*, while encephalitis of several different types, can be transmitted by various species of *Culex* and *Aedes* (US Department of the Navy, Army and the Air Force 1971).

Virus encephalitis has been a problem in parts of Latin America and the USA for decades. The larvae of *Culex tarsalis*, the feared vector of St. Louis and Western encephalitis in the midwestern and western USA, thrive in vegetation-clogged irrigation and drainage ditches, seeps and roadside impoundments.

Mansonia mosquitoes

Mansonia mosquitoes, which are responsible for transmitting rural filariasis and encephalitis, are unique in that both larvae and pupae derive their oxygen by puncturing air chambers in plant stems and roots by means of a specialized air tube. Two species of *Mansonia* are reported to be totally dependent on *Pistia stratiotes*.

As a consequence *Mansonia* mosquito breeding can only be controlled by the elimination of all aquatic vegetation. In this context it should be noted that a heavy production of *Mansonia* is directly proportional to a uniform and constant depth of the water. In nature, the water level generally fluctuates and when the water body periodically dries up the required breeding conditions are eliminated. Since *Mansonia* mosquitoes have only one to two reproductive cycles per year, it takes some considerable time for the mosquito density to build up again after a water body has dried out and refilled.

Certain aquatic plants produce natural substances toxic to mosquitoes which may be used in mosquito control programmes in the future (Angerelli and Beirne

1982; Ferguson 1971, 1980; Gangstad 1980; Gass *et al.* 1983; MacKenthun and Ingram 1967; US Department of the Navy, Army and the Air Force 1971).

Aquatic snails

Some aquatic molluscs, especially certain fresh water snails, are in many areas of the world medically important because of their role in the transmission of disease organisms. In various regions they affect the health of entire human populations. The disease organisms are, in general, trematode worms (flukes) that are parasitic in man and the snails are obligate intermediate hosts. The life cycle of these flukes consists of an asexual phase in the snail and a sexual phase in man. The eggs hatch upon reaching water and a minute larva, the miracidium, escapes and seeks its particular intermediate host. Once penetration into the snail has been effected the miracidium loses its cilia and transforms into a sporocyst. The sporocyst buds off daughter sporocysts which, in turn, produce another type of larva, the cercarium. The cercaria are discharged by the snail and attack man. Usually these worms are specific for a single or a very few species of snails.

The most important diseases which are caused by these parasites are schistosomiasis (bilharziasis), paragonimiasis, clonorchiasis and fasciolopsiasis. Schistosomiasis is one of the most important public health problems of the tropics and subtropics. *Schistosoma haematobium*, *S. mansoni*, and *S. japonicum*, the three major species of schistosomes which infect man, are widely distributed; *S. haematobium* occurs in Africa and south-west Asia; *S. mansoni* is distributed in parts of Africa and the western hemisphere and *S. japonicum* is an oriental species. In the human body the adults of *S. mansoni* and *S. japonicum* commonly inhabit the mesenteric veins, although occasionally they are found in aberrant locations. On the other hand, the adults of *S. haematobium* usually occur in the veins of the pelvic plexuses. The eggs leave the body via the faeces or urine. The cercaria infect the human body via the skin.

The main endemic areas of paragonimiasis are in the Far East but the disease is also present in Africa and northern South America. It is an infection caused by the fluke *Paragonimus westermani*, or related species. The worms may enter various parts of the human body but usually occur in the lungs, the bronchi and the liver. Apart from a snail it needs a second intermediate host, i.e. a crab or a crayfish. Infection is usually caused by eating raw crabs and crayfish, which is a habit in many parts of the Far East.

Clonorchiasis is caused by trematode infections of the liver. It is a serious health problem in certain regions of the Far East. A second intermediate host is a fish and man is infected by eating raw fish. Fasciolopsiasis, which is caused by the giant intestinal fluke *Fasciolopsis buski*, is an infection of the intestines. It is influenced by the Asian custom of eating raw aquatic plants, such as water chestnuts (*Trapa* spp.), on which cercaria are encysted (Brown and Deom 1973; Cardarelli 1980; Ferguson 1971, 1980; Sculthorpe 1967).

The habit preferences of the various aquatic snails are quite different, ranging from temporary woodland pools through mudflats, stagnant pools and large lakes to rapidly flowing streams. Some snails are amphibious rather than strictly aquatic and are found along the banks of streams or canals just above the water surface. Briefly outlined here are two widely different types of life histories, representing the two most important snail genera in this context. The genus *Oncomelania* (species of this genus are intermediate hosts of *Schistosoma japonicum*) occurs in Japan, Taiwan, China and the Philippines and the species *Australorbis glabratus*, which is the intermediate host of *S. mansoni* in the West Indies and South America. The genus *Biomphalaria*, which transmits *S. mansoni* in Africa (*S. haematobium* is mainly transmitted by the genus *Bulinus*), is probably taxonomically identical to *Australorbis*.

1. *Oncomelania*. This snail genus is common in the Far East. There are distinct species with slightly different life cycles. Mating activities are typically at a peak in early winter and egg-laying is most common in late spring and summer. However, where cold weather does not force hibernation, both mating and egg-laying may continue all year. After the eggs hatch, full development of adult snails may require 3 to 6 months. Hibernation takes place in the top 10 cm of mud beneath shallow water. These snails are most numerous in dense vegetation at the edge of a water body, especially in small irrigation ditches.
2. *Australorbis glabratus*. This species is common throughout the Caribbean and South America. The snails are, in contrast to *Oncomelania*, hermaphroditic and, as a consequence, each specimen can mate with each other member of the species. They mate at any season and their reproduction capacity is extremely high. A single snail may produce 20 000 eggs in its lifetime. The eggs, which are highly resistant to drying, are laid in translucent albumoid masses on any convenient, firm surface in the water (including plants). After 6 to 10 days the young snails hatch from the eggs and begin a period of rapid growth. After 4 to 5 weeks of growth the snails are sexually mature and may begin reproduction (Brown and Deom 1973; Cardarelli 1980; Gangstad 1980; US Department of the Navy, Army and the Air Force 1971).

Blackflies

The larvae of the blackfly (various species of the genus *Simulium*), vector of onchocercosis, are usually attached to rocks in clean, rapid flowing mountain streams. However, they are sometimes plant-associated in slow flowing waters.

The disease, which is due to the filarial worm *Onchocerca volvulus*, occurs in parts of Africa, particularly the western part of the equatorial belt and Central America. The microfilariae (in contrast to the adult worms which are seldom found in the eye) infiltrate the anterior portion of the globe of the eye. The disease ('river blindness') is of great opthalmological importance as many cases, if untreated, lead to early blindness.

SNAIL–AQUATIC WEED INTERACTIONS

Studies in Puerto Rico have shown that of 38 aquatic plant species 14 species were distinctly associated with the snail *Biomphalaria glabrata* (the intermediate host of *Schistosoma mansoni*). The largest numbers of these snails were found on *Colocasia esculentum*, *Sagittaria latifolia*, and *Eichhornia crassipes* (Ferguson 1971, 1980; Smith 1982; US Department of the Navy, Army and the Air Force 1971). Dawood *et al.* (1965) have shown that, in Egypt, the bilharzial snails prefer *Potamogeton crispus*, followed by *Eichhornia crassipes* and *Panicum repens*.

Investigations in Egypt have shown that there is a positive relationship between the bilharzia-bearing snails *Biomphalaria alexandrina* and *Bulinus truncatus*, and aquatic vegetation (El Gindy 1962; Dazo, Hairston, and Dawood 1966; van Schayck 1986). Laboratory experiments in aquaria indicated that the reproduction of these snails is reduced significantly when there are fewer aquatic plants (van Schayck 1985). When four methods of aquatic weed control in Egypt were compared, i.e. manual control, chemical control by means of acrolein, mechanical control with a dragline, and biological control by means of grass carp at a stocking rate of 200 kg ha^{-1}, it appeared that in the biologically controlled waterways the number of bilharzia-bearing snails was significantly lower than in waterways with one of the other methods of control (van Schayck 1986). This effect of grass carp was explained by the fact that the fish graze continuously on the vegetation and that the feeding preferences of the two snail species and grass carp largely coincide. In this context it should also be noted that predation on snails by omnivorous fish is increased when the aquatic vegetation is maintained continuously at a low level (Coates and Redding-Coates 1981).

DIRECT TOXIC EFFECTS

Various blue-green algae produce toxins which are harmful to man and animals (Steyn 1945; Schwimmer and Schwimmer 1968; Gentile 1971; Carmichael and Bent 1980; Kappers *et al.* 1981). Water which contains large quantities of these algae ('algal blooms'), may be poisonous. In general, the adverse effects of blooms of blue-green algae to man have been limited to forms of dermatitis, irritation of the mucous membranes and dysentery-like troubles of the intestines. On the other hand, there are various cases known of animals which died after drinking water containing a dense mass of algae (Prescott 1948; Schwimmer and Schwimmer 1968). It may be assumed that the unattractiveness of water in which an algal bloom occurs (the algal blooms usually cause a foul stench) has usually prevented people from drinking the water.

Poisonous algal blooms are mainly caused by the genera *Anabaena*, *Aphanizomenon* and *Microcystis*. However, there are also reports of intoxication by the genera *Coelosphaerium*, *Gloeotrichia*, *Nodularia*, *Nostoc*, and *Oscillatoria* (Kappers 1973). It has been estimated that for an adult man with a body weight of 68 kg, drinking of 1.1 to 3.4 litre of a toxic blue-green algal suspension is lethal. However, the degree of toxicity varies between algal species, and may be influenced also by environmental conditions (Kappers 1973).

Blooms caused by Protista in marine waters may cause a very serious form of human intoxication. These organisms produce an extremely poisonous toxin, saxitoxine (the minimum lethal oral dose for man is thought to be between 1 and 4 mg kg^{-1} of the body weight), which is bio-accumulated in molluscs feeding on these organisms. When the molluscs are subsequently eaten by man this may lead to a respiratory paralysis, which is generally called 'shellfish poisoning'. Death may occur within a few hours. The clinical effects are not completely understood, nor is the mechanism by which the poison accumulates in the molluscs. The Protista blooms, which attract public attention as 'red tides' or 'red water', also frequently cause massive fish kills.

Clinical symptoms which resemble those of shellfish poisoning may also be the result of consumption of certain marine fish (Brown and Deom 1973; Cardarelli 1980; Ferguson 1971, 1980; Gangstad 1980). This fish poisoning may also be associated with organisms lower in the food chain. Current evidence favours certain blue-green algae, as the likely causative factor.

PART II

MANAGEMENT OF AQUATIC WEEDS

7

Physical control of aquatic weeds

P. M. WADE

THE PHYSICAL control of those aquatic plants inimical to man's activities goes back to the first time that man used his hands and sticks to clear away weeds blocking the passage of boats or denying access to good fishing spots. As irrigation and drainage channels were constructed, so the need to control aquatic plants increased. At present, the physical control option continues to play an important role in aquatic plant management together with chemical and biological control. To assess the role of the physical approach in the armoury of techniques available to manage aquatic weed infestations, it is necessary to examine the different forms which physical control can take: manual, mechanical, and other. The more important techniques need to be compared with other forms of control, both from the economic standpoint and in terms of the ecological implications of their use.

MANUAL TECHNIQUES

The time-honoured method of aquatic weed control is that of manual labour: hand-pulling, raking, cutting, and other techniques. Until recently these were the only means of weed control and today they are still important in many parts of the world. An example is the manual weeding of *Echinochloa stagnina* and *Phragmites australis* in Egypt (Fig. 7.1).

A range of hand-held tools is used in Europe for weed control. Their application and mode of operation are summarized by Robson (1974) and Brooks (1981). They include scythes, sickles, grass hooks, rakes, forks, hoes, chain scythes, and chain knives, most of which are used in rivers, canals, and drainage channels. All have evolved from implements used originally for agricultural purposes. A typical example would be the use of a chain scythe, a series of interconnected serrated blades for drainage channel maintenance. Two men, one on each bank, pull the chain scythe with ropes through the vegetation. Four to five men follow on, pulling out the cut weed on to the banks using cromes, a type of fork with the tines bent over and a long handle (G. Cave pers. comm.; Brooks 1981). Implements used in tropical regions are similar but the diversity of techniques

FIG. 7.1. Manual clearance of *Echinochloa stagnina* and *Phragmites australis* in Egypt. (Photo: A. H. Pieterse.)

is greater, including devices for clearing lakes, such as floating booms, rope or nets used for removing free-floating weeds. Druijff (1979) described the use of chain scythes, clearing scythes, ditch-bank knives, digging forks, and rakes, emphasizing the preference for long-handled tools to prevent the worker from having to enter the water. Applications of these and other techniques were given by Philipose (1968).

Manual techniques are an important means of weed control in countries where labour is readily available and cheap (Philipose 1986; Pieterse 1978; Druijff 1979; Ramaprabhu, Ramachandran, and Reddy 1982), but success is variable due to such factors as the extent of weed removed. For example, in India 50 per cent of manual treatments of *Eichhornia crassipes* achieved partial success, with 25 per cent total success and 25 per cent failure in the remainder of the treatments (Varshney and Singh 1976). Shibayama and Miyahara (1977) effected short-term control of *E. crassipes* by hand picking seedlings and overwintering plants in Kyushu Island, Japan. Sixty men cleared 3000 m^2 of *Salvinia* from a reservoir in southern India over a three-month period (Cook 1976) and amelioration of *Salvinia* infestations in paddy fields has been achieved by flooding the fields before transplanting the rice and sieving off the sporophytes which float to the water surface (Thomas 1976). Follow-up manual control ensured lasting success in both of these programmes. Such follow-up treatment is vital and one-off attempts to achieve control have usually failed. In 1952 there was a '*Salvinia* Week' – an all out attempt by the Government of Ceylon (now Sri Lanka) to clear *Salvinia* from thousands of hectares of infested waters. Complete reinfestation occurred in a few months despite the removal of huge amounts of weed (Williams 1956). The failure of similar '*Salvinia* pick-up days' in Kenya was described by Gaudet (1976*b*). Varshney and Singh (1976) considered that, in India, manual treatment was partially successful in 65 to 90 per cent of treatments of *Nelumbo nucifera*, *Pistia stratiotes*, *Nymphaea stellata*, and *Hydrilla verticillata*. The problems and expense of such temporary control were summarized by Malhotra (1976).

Manual clearance may be inefficient as it can leave 10% or more of the weed untouched (Soulsby 1974; Kern-Hansen 1978). Data from Mehta and Sharma (1976) suggest that the improvement of flow in an Indian irrigation canal, following manual clearance of submerged weed, was consistently poorer than that produced by mechanical cutting, the herbicide fenac, or dredging. Flow reductions of 27 to 46 per cent occurred in the manually-cleared sections, though this was still an improvement on the 68 to 83 per cent flow reduction in uncleared, heavily infested stretches. Regrowth to nuisance density tends to be rapid and more than one cut per season is usually required. In European trout streams manual weed cutting may be needed every month during the summer (Robson 1974). Price (1981) reported that three manual cuts per year are the norm for drainage channels in the UK.

In such parts of the world as Europe, North America, Australia, and Japan, manual forking and raking, following mechanical cutting, and manual cutting of weeds in the littoral zones of small lakes and drains, are still widely used techniques. Robson (1970) estimated that pre-1970, 70 per cent of the 8050 km of rivers, which receive weed control in the UK, were managed using hand labour and that 60 per cent of drainage authorities' expenditure on weed control operations was spent on manual removal. The drainage authority for a network of 653 km of drainage channels in Lincolnshire, England, spent 55 per cent of its expenditure on hand labour, mainly in the smaller watercourses (Miles 1976). Dunn (1976) reported the removal of 610 tonnes of weed by three men from a reservoir in England over a six-week period. The use of labour for weed cutting in these regions has diminished. A second drainage authority in Lincolnshire, England, employed 709 men in 1950 for 'roding', a hand raking operation on drains. By 1980 the cost of employing such a labour force was prohibitive and only 25 men were engaged in such work (Price 1981). Miles (1976) considered the cost of manual weed cutting in 20 m of drain in 1975 to be more than three times as expensive as flail mowing from a tractor. Murphy, Eaton, and Hyde (1980) found that only 8.5 per cent of the British navigable canal system received manual cutting or raking.

Despite the expense and the slow and laborious nature of manual techniques, they do have special applications and should be used more frequently. For example, repeated cutting at two-week intervals during one summer was sufficient to eliminate fairly small stands of *Nuphar lutea* (Riemer and MacMillan 1967). Likewise, if mature *Typha* plants are cut off beneath the surface of the water and left underwater for at least two weeks, they will not grow back (Riemer 1984).

The importance of using hand control methods for clearing-up operations and to prolong the period of weed control after mechanical cutting was emphasized by Livermore and Wunderlich (1969) who identified the problem of cut vegetation re-rooting. Manual control *per se* in such a situation has changed to manual harvesting of mechanically cut weeds. True control can be achieved by hand cutting plants which remain after mechanical or chemical treatment, e.g. three or four manual cuts of the feeble growth of *Fimbristylis* and *Isachne* remaining after machine cutting gave complete eradication of these weeds (Philipose 1968). Esler (1966) described a follow-up operation in which hand-pulling was used to clear *Trapa natans* remaining after a mechanical cut. A novel approach, which combines manual and mechanical methods, is based on the early recognition of new exotic weed infestations and the selective removal of the shoot and root systems by diver-operated suction hoses (Clayton 1977). A filtration technique is used to collect and separate weeds from water and sediments. The efficiency of this approach was found to be inversely proportional to the degree of plant development, hence its suitability for containing exotic species.

Manual control also provides a highly selective approach to weed control which has been applied in management of wildfowl (Swift 1976; Hertzman 1985), nature reserves (Brooks 1981; Lewis and Williams 1984), and fisheries (Philipose 1968; George 1976) including highly valued salmonid fisheries such as UK chalk streams, which support dense growths of Batrachian *Ranunculus* (Ham, Wright, and Berrie 1982). Brooks (1981) considered the advantages and disadvantages of the digging-out of plants, cutting, and hoeing for wildlife conservation management, with practical advice on how to use such techniques in different situations.

Manual methods avoid access problems, particularly where farmers grow crops right up to the water's edge. They enable a selective approach to management which is of particular importance in nature reserves and other sites containing rare and diverse aquatic macrophyte communities. However, the increasing cost of labour in many countries has tended to make manual control increasingly expensive compared with other techniques. On the other hand, in Third World countries with cheap labour manual control remains the most economic method in many cases, though it is of limited use in large water bodies, e.g. lakes, and the chances of regeneration are great. In some parts of the world the man-power is exposed to serious health hazards, e.g. bilharzia (Pieterse 1978; Druijff 1979), malaria (Dawood *et al.* 1965), and snakes (Malhotra 1976). The lack of advice and information about the most efficient way to employ manual control is surprising. Likewise, there has been little research into the design and effectiveness of manual control methods. Robson (1975) reported 15 drainage authorities in the UK which had recently employed or modified hand tools, e.g. a V-shaped knife attached to a chain. Although most were minor modifications to traditional tools, no follow-up studies were undertaken and there is little known about the potential of manual control for different species and in different situations, e.g. paddy fields, drainage channels, and lakes. In Egypt small watercourses (<2 m bed width) are particularly suitable for manual control because there is no need for the labourer to go into the water, limiting the chance of contracting bilharzia (El Sayed, Tolba, and Druijff 1978).

MECHANICAL TECHNIQUES

The diversity of machines devised to cut, shred, crush, suck or roll aquatic weeds would be large enough to fill a museum. This assemblage of machines can be usefully divided into two groups:

(1) those aimed at cutting and/or otherwise removing solely the aquatic plants;

(2) machines which have other functions apart from weed cutting and/or removal, for example, dredgers.

Some of these machines are water-based on boats and barges, others work from the bank and shore, mounted on tractors or as purpose built machinery. Useful reviews of machines are provided by Livermore and Wunderlich (1969), Koegel, Livermore, and Bruhn (1975), Livermore and Koegel (1979), Bagnall (1981), Canellos (1981), Ramey (1982) and Gopal (1987).

Cutting, chopping, shredding, and harvesting

Floating machines

Floating machines are used mainly to manage floating and submerged weeds and early devices were simply rakes or other pieces of farm machinery, weighted to keep them from riding up over the weed beds whilst dragged behind boats. The scratching and scraping action dislodged and broke off the weeds. After the First World War, smaller weed-cutting boats were developed in Europe and North America (Robson, 1974) which used a V-shaped knife with either a serrated or a straight edge pulled along the channel bottom behind the boat. Robson describes the efficiency and limitations of these devices which dulled easily and required an even bed devoid of solid obstacles, such as tree branches and rocks. The design of these weed cutting boats soon advanced to make use of reciprocating cutter (or mower) bars, initially horizontal straight bars 1 to 2 m long, lowered and fixed to cut at a required depth. In some devices the bar could also work at an angle to and above the water surface cutting emergent vegetation along lower parts of the bank. These and a range of other cutting devices soon became available for the clearance of canals, irrigation and drainage channels and rivers in Europe, Australia (Philipose 1968), and North America (Livermore and Wunderlich 1969; Ramey 1982). U-, L-, and inverted T-shaped cutter bars were the next step forward in design, coupled with hydraulic control of the depth and angle of the cutter bar in the water (Fig. 7.2, 7.3). Further details of such weed cutting boats are given by Price (1981).

Some of the first, large, experimental floating machinery was designed for aquatic weed management in North America and, in particular, the solution of the specific problems created by *Eichhornia* in the Gulf States of the USA (Livermore and Wunderlich 1969; Ramey 1982). One device, built in 1900, consisted of a conveyor-belt and a sugar cane crusher mounted on a steamboat. It lasted only two years of operation and other machines were slow to be developed. In 1937 a motorized barge with conveyor-belt and crusher, the 'Kenny', was developed, again for *Eichhornia* control, the crushed plants being washed overboard to decompose in the water. Sawboats, boats with numerous circular-saw blades mounted approximately 2 to 3 cm apart on a horizontal shaft in front of the boat, were also effective against floating vegetation. The blades

FIG. 7.2. Rolba Gibeaux Water Weed Mower with 'T' shaped cutting mechanism. The propulsion system is based on helical steel propellers.

FIG. 7.3. Aqua-Scythe (John Wilder (Engineering) Ltd) with a 2 m wide 'U' shaped cutting mechanism. Propulsion is by a large diameter screw.

were lowered into the water, revolving at 900 to 1000 rev. min^{-1}, and shredding the plants. Although such machines had, after the Second World War, to compete with the new aquatic herbicides, e.g. 2,4-D, they were demonstrated to be effective against *Eichhornia* and *Alternanthera* infestations. A shredding device, tested on *Alternanthera*, produced an end product which, when returned to the water, had a regrowth rate of about 5 per cent (Livermore and Wunderlich 1969). Other approaches to weed removal have been experimented with, e.g. the use of a hydraulic jet to dislodge rooted plants (Taussig 1969), but the most successful line of development has been the use of reciprocating cutter bars on large boats.

The 1 to 2 m swath cutter bars of the smaller boats were increased to 3 to 5 m to fit onto barges, improving their capacity for cutting weed. A wide variety of models based on this design has been marketed. In addition to these machines (Fig. 7.4, 7.5), other designs are shown in Livermore and Wunderlich (1969), Robson (1974), Ramey (1982) and Riemer (1984). The basic design is a flat, self-propelled barge with a steel hull capable of working in very shallow water. Propulsion is typically by paddle wheels, which increase manoeuverability and

Fig. 7.4. Rolba-Aquamarine harvesting submerged weed (cutting width 240 cm and cutting depth 0–150 cm). The harvester is propelled by two hydraulically driven paddle wheels (reproduced by permission of Kent & Sussex Courier and Tunbridge Wells Advertiser).

give a shallower draft. Two hydraulically-controlled arms extend from the front of the boat and a U-shaped reciprocating cutter bar is fastened between them.

The problems of propulsion of weed cutting boats have led to the exploration of alternative methods designed to overcome the problems of fouling by weeds (Robson 1974). Different designs are needed for different types of water: independent hydraulically-driven paddles are ideal for small craft in lakes and sluggish water, with steel propellers being more appropriate in fast-flowing rivers. The type of hull also depends upon the situation in which the boat is to be used. Steel is usually the preferred material, though moulded fibreglass is better for boats which have to be moved from one water to another.

The early cutting, crushing, and shredding devices, both small and large, had a major drawback in that the treated plants remained in the water. Decomposition of the shredded material and a certain proportion of cut material caused undesirable effects by depressing concentrations of dissolved oxygen and producing unsightly heaps and obnoxious smells along the margins of water bodies. The breakdown of the organic material in the water also released inorganic nutrients which resulted in algal blooms and the increased growth of aquatic macrophytes. Cuttings of submerged plants could float in the water almost indefinitely and

FIG. 7.5. Rolba-Aquamarine unloading harvested weed on to the shore (reproduced by permission of Kent & Sussex Courier and Tunbridge Wells Advertiser).

fragments of many species have the ability to root and regrow. The free-floating nature of the material meant that plants were able to move around or along a water body with the potential for new infestations (Johnson and Bagwell 1979; Gangstad 1986). The fragment of plant from which regrowth can occur may be very small: in *Panicum repens*, for example a one-node cutting 5 cm long is all that is necessary (Siregar and Soemarwoto 1976). Problems also arise from cut plants blocking screens, spillways, and channels (British Waterways Board 1981).

Two solutions were explored to overcome these drawbacks: improving the effectiveness of the shredding and chopping, and harvesting the weeds. The former approach does not overcome the problems of deoxygenation and nutrient release. Harvesting, or removing the weed from the water body, has become an essential part of physical control (Koegel *et al.* 1973). Such harvesting is often time consuming and is usually the limiting factor in mechanical control (Culpepper and Decell 1978).

Techniques for harvesting the mass of cut weed range from manual raking using wind and current to concentrate the weed, through the application of dragline cranes, to sophisticated machinery with dewatering and baling facilities. Machines can also be fitted with fragment barriers. A typical system involves one or more porous conveyor belts which pick up the weed from the water and transfer it either to the deck or a hopper on board the cutting boat/barge or into another boat/barge which then transports the weed to the shore, although an attached barge substantially reduces manoeuverability (Koegel, Livermore, and Bruhn 1977). Further transportation of the cut weed is frequently necessary, to a site where the nuisance weed may be utilized and/or allowed to decompose. Smaller weed cutting boats now have the facility of changing the cutter bar for a rake (4 m in width), which collects the weeds together and using the hydraulic arms lifts and dumps loads of up to 300 kg of weed onto the bank. The nature of the bucket or rake depends on the species being harvested: for example, free-floating non-rooted plants require a fine mesh bucket (British Waterways Board 1981).

The large quantities of unwanted water associated with harvested weeds are a major problem and presses can be used on board the harvesting craft to reduce the weight of the load by 68 per cent and the volume by 16 per cent, although some organic matter is lost to the water (Bruhn, Livermore, and Aboaba 1971; Bagnall 1980*a*,*b*).

The enormity of the task of weed control in heavily infested waters of the world gives an immediate indication of the limitation of this method of weed control. Given a standing fresh weight crop of vegetation of 376 tonnes ha^{-1} and a modern weed harvesting operation which can remove ca. 1 ha of weed per hour, this means that a crew attempting to control such an infestation from a 160 ha lake would still be working four–five weeks later (Ramey 1982). The limitations are equally apparent with species capable of rapid regrowth. Culpepper and Decell (1978) calculated that harvesting systems with a disposal rate of 80 to 100 tonnes per hour were necessary for such species as *Eichhornia crassipes* and *Hydrilla verticillata*. Although some come close, few harvesters appear able to achieve such a performance consistently (McGhee, 1979). Even if one assumes an efficient operation, there are a number of drawbacks which need to be appreciated in the use of floating machinery, particularly in the management of large water bodies.

1. Much of the machinery in use today has been developed with specific, often local, needs in mind which has produced a proliferation of different types of machines (Bryant 1974). Most devices, which are suitable for channel and canal maintenance, are not large enough for lakes and big rivers (Wunderlich 1967). A system which cuts to a maximum depth of 1 m may not effectively control those plants which grow in deeper water (Ramey 1982).
2. The economic effectiveness of these machines is hard to estimate due to the complexity of the operation: cutting, harvesting, transportation, and dumping. The period during which the machine is out of action (down-time) also needs to be taken into consideration and the hidden advantages of nutrient removal are even more difficult to quantify. Comparisons with other treatments, e.g. herbicides, are therefore difficult (Keeves 1983).

3. Maintenance of machinery of this type is difficult, particularly for machines manufactured in one country and used in another and often spare parts are costly (Hall 1969).
4. Access to water bodies may be difficult, and shoreline development, piers, etc. present problems to the larger craft. Long distances may need to be travelled, not solely to launch a weed cutting boat but also for the transportation of harvested weed away from the site.
5. Shallow waters also present severe problems, not just in terms of the draught of the boat but also with respect to the distances which have to be travelled in a large shallow water body.
6. The high cost of such management, exacerbated by the need for repeated treatment and the fact that the harvested weed has little or no value in many countries, may make the operation prohibitive.

On the other hand, there are a number of significant advantages of floating machinery. They include:

(1) the degree of selectivity which may be applied. This is of particular importance when vegetation needs to be left for example to benefit and maximize fish production;
(2) the removal of nutrients from the water. This latter factor is considered in more detail below;
(3) the removal of pollutants from a waterway (Seidel 1971; McNabb 1976);
(4) a reduction in the dependence upon foreign currency as harvesting reduces the need to purchase herbicides;
(5) compatibility with terrestrial crops growing near the water body, not necessarily achievable with herbicides;
(6) the potential for quick and predictable removal of weeds from specified areas which can be achieved rapidly;
(7) the production of useful materials, e.g. green manure and animal feeds;

Machines operating from the bank

A range of cutting mechanisms has been developed which operate from the bank of a canal, drain or river. Early devices were pulled by a team of horses and later by traction engine (Dunk 1954). An example of such a technique from southern Africa was a heavy chain or rake with downward-projecting teeth dragged along, usually by two tractors, effective in controlling *Aponogeton*, *Nymphaea*, *Nymphoides*, and *Potamogeton* (Wild 1961).

Several devices have been developed for use with dragline excavators and the hydraulic attachments available on modern tractors and excavators. The reach of such machinery varies: a tractor-mounted flail mower has a reach of up to 7.24 m (Fig. 7.6); a weed cutting bucket mounted on a hydraulic excavator 11 m and a weed cutting bucket on a dragline 18 m (Cave 1981). Deighton (1984) described a long reach excavator, which combines hydraulics and a drag line, with a reach of 15 m. The machine may be fitted with a weed cutting basket or a dredging

FIG. 7.6. Herder MBK hydraulic arm with mowing bucket up to 4 m in width.

bucket (Fig. 7.7). The most widely used device is the weed cutting bucket, considered to be the most important development in recent years in drainage channel maintenance (Robson 1975; Arts and van Wijk 1978). The bucket is attached to the hydraulic jib of a tractor or excavator and, in operation, the lower edge has a cutter bar which may range from 2 to 4 m in length. The bucket is lowered parallel to the substrate surface and pulled towards the excavator by the jib, cutting the weeds on the way. The bucket is able to cut weeds on the banks and the bed of the watercourse and, given a sufficient reach, both banks can be cut in one sweep. The cut weeds collect in the bucket which does not retain the water and are lifted out and dumped on the bank or in a truck. Depending on the skill of the operator, the bucket can cut above or slightly below the sediment/silt. The main problems with this technique are trees and other similar obstructions which reduce accessibility of the weeds and watercourse from the bank. Additionally, there is a disruption of land use especially where regular maintenance is required. This area of land, termed the 'maintenance path' by

Hoogerkamp and Rozenboom (1978), is 1.7 to 2 m wide although some machinery is available requiring paths only 1.2 to 1.5 m wide.

A weed rake operated from a dragline excavator is also a popular device, more robust than the weed-cutting bucket. Robson (1975) gave details of such machinery and the development of a weed-cutting bucket for use with a dragline.

FIG. 7.7. Priestman VC15 long reach excavator combining hydraulics and a drag line.

A range of other equipment has been specifically designed for weed removal from the bank, for the removal of cut weed and especially filamentous algae, and for use on screens at pumping stations. One drainage authority in the UK used a mobile crane for weed removal, which was found to be more versatile than a dragline, and can be used or hired out for other work (Robson 1975). Some machines have been designed to throw the cut weed up on to the bank (Canellos 1981; Arts and van Wijk 1982). Weed cleared from ditches or canals by weed buckets mounted on a bankside excavator may also be dumped straight onto the bank or onto barges (Robson 1974; Price 1981; Eaton and Freeman 1982). In flowing waters cut weed is usually allowed to drift downstream, for collection by boom systems (Price 1981; Westlake and Dawson 1982).

Rotary, reciprocating, and flail cutters provide an important range of machines for cutting emergent and bankside vegetation (Robson, 1975). A wide variety of small self-propelled pedestrian and ride-on cutters can be used on slopes with gradient less than 2:1 although Allen motor scythes can work across steeper gradients under suitable conditions (Price 1981) (Fig. 7.8). Operation of this type of equipment is difficult and tiring and, where access is available, tractor-mounted cutters provide a useful alternative (Krinke and Dyckova, 1974) (Fig. 7.9). Such

FIG. 7.8. Bradshaw H-6710 tracked mower with three different types of flail cutters, designed to work along banks to an angle of 45°. The mower also operates in 30 cm of water using a reciprocating knife in place of a flail head.

FIG. 7.9. B77-77 (Bomford and Evershed Ltd.) swing-over dyke mower with a reach of 7.8 m.

devices are usually operated from a tractor or excavator attached to hydraulic arms. A choice exists between lightweight and heavy duty flails with cutting heads in the region of 1.5 m wide (Price 1981) and long reach arms. Hoogerkamp and Rozenboom (1978) drew attention to the difficulty of adjusting the cutting depth on rotary cutters and the possible damage to the sward. Damage to the machinery can result from stones and other hard objects such as wire and string which can add considerably to down-time.

The variety of equipment which can be operated from the bank is extensive. An exhibition in 1978 of machinery suitable for channel management in the Netherlands included 31 different types available at that time. In 1978, the National College of Agricultural Engineering (NCAE), England, undertook a

FIG. 7.10. Bicycle tractor (Inter-Drain (England) Ltd.). A hydraulically driven, three wheeled machine. Two wheels are fitted in line (as a bicycle) and the third wheel is attached to a hydraulic arm allowing it to run on the bottom, near or opposite side of the ditch. Two additional hydraulic arms are fitted to accommodate implements such as cutter bar, flail or weed cutting bucket.

review of existing machinery for the management of drainage channels (Murfitt and Haslam 1981) and indicated that reciprocating drum and disc mowers could be used in any section within the water channel. The rotary devices, although needing higher power, are very much more robust than the reciprocating cutters. Flail cutters/mowers are limited in use to the area above the water. The NCAE also defined the design objectives for an ideal machine, which should be robust,

FIG. 7.11. (a) and (b) 'Spider' (Inter-Drain (England) Ltd.). The machine straddles the ditch with driving wheels on each bank. It can be fitted with implements such as cutter bar, dredging bucket, rake, and weedcutting bucket (reproduced with permission of Kent & Sussex Courier and Tunbridge Wells Advertiser).

reliable, and able to operate in water. It should control rooted and non-rooted weeds at one pass without affecting the stability of the banks. It should have variable geometry to cater for a channel bed of 0.6 to 1.2 m and bank slopes of

30 to 45 degrees and remove the weeds to a stable position above 1.2 m up the bank (Price 1981), criteria very similar to those listed by Kemmerling (1978). Price (1981) presented data on the characteristics of eight weed control machines related to these criteria, giving advantages and disadvantages. Two of the machines which came the nearest to satisfying the criteria, allowing for normal down-time and obstructions (e.g. culverts, side dykes) had an estimated output of approximately 2.4 km d^{-1} and 4.16 km d^{-1}. Examples of recent developments in machines operating from the bank are illustrated in Figs. 7.10 and 7.11.

Dredging

A major disadvantage of cutting and harvesting aquatic macrophytes, as a means of direct control, is that the underground material is left behind. This is particularly relevant for the submerged and rooted floating plants. Hitchcock *et al.* (1950) reported an abundant regeneration within a month from chopped material of *Alternanthera philoxeroides* and *Eichhornia crassipes*. More thorough control is achieved by dredging which removes both plant material, including much of the stem and leaf growth, and accumulated sediments (Robson 1974*a*; Riemer 1984). Such operations are usually undertaken from the bank using either dragline or hydraulic excavators. Tractor-mounted mud scoops are also produced, usually for use by individual land owners (George 1976). The draglines have the advantage of a considerable reach whereas the hydraulic excavators may be used more easily to create a steep, uniform batter on the bank or banks. Recent engineering has combined both approaches (Deighton, 1984) (Fig. 7.7). The types of machinery and the costs of small-scale dredging operations in inland lakes, were reviewed by Pierce (1970). Dredging is also necessary in cases where the system has become terrestrialized through the accumulation of organic material and silt (Hejny 1978) and where other control measures would be ineffective.

The effectiveness of dredging depends upon a number of factors and, in particular, the depth of mud dredged from a water body and the depth of water after dredging. Dredging of shallow water can have serious limitations, e.g., in a pond dredged to a depth of 1 m, either vascular macrophytes or *Chara* were a continual problem beginning the season after treatment (Born *et al.* 1973). Fragments of vegetation, rhizomes, turions, and other propagules tend to remain after management, e.g., in drainage channels in the east of England, of *Elodea canadensis*, *Lemna* spp., *Potamogeton* spp. and *Sparganium* spp. (George 1976).

The intervals between dredging are much longer than intervals between weed cutting, for example, and the degree of control is usually sufficient to negate the need for other control, e.g. herbicide application or cutting, for at least one full season. Further advantages accrue in that sediment removal extracts plant nutrients (Livermore and Wunderlich 1969) and where the depth of water increases, the amount of light penetrating to the bottom may be reduced.

The cost and time involved in dredging are considerable and there is also a problem with the disposal of the spoil/sediment—such sediments are not as useful as one might expect them to be (Riemer 1984). In the case of drainage channels and rivers, this waste material is usually dumped on the adjoining land and levelled. In the case of lakes and similar large water bodies, careful, long-term planning of the shoreline land use is important (Livermore and Wunderlich 1969). Dredging, because it is expensive and slow, is commonly used only when a channel or lake has deteriorated severely and other forms of maintenance are no longer effective (Robson 1974), usually in conjunction with the removal of accumulated mud and other material.

Improving the efficiency of mechanical techniques

A number of factors should be considered in attempting to improve the efficiency of mechanical techniques.

Timing

The season in which the control is effected is likely to alter the success of the operation (Murphy, Fox, and Hanbury 1987). For example, the season in which *Ranunculus* beds are cut in the UK has a profound effect on the regrowth (Westlake and Dawson 1982, 1986) whereas cutting *Butomus* and *Sagittaria* to prevent winter flooding, was found to be a waste of effort. After cutting beds of *Callitriche stagnalis*, *Berula erecta*, and *Ranunculus penicillatus* var. *calcareus* in a tributary of the River Itchen, UK, the percentage cover of each species changed in comparison with the previous year, and the growth pattern of the most dominant species also changed. The cut was not effective in controlling *B. erecta* and may have stimulated early winter growth (Soulsby 1974).

In India, the pre-rainy season (April–May) is best for *Eichhornia crassipes* control when the weed is restricted to smaller water areas (Gopal 1987) and infestations should certainly be cleared before seed formation (Parija 1934). Cutting of *Zizaniopsis miliacea* works best when applied in the later summer and autumn and least in the winter and spring (Birch and Cooley 1983). Plants with marked seasons of flowering and fruiting (e.g. *Najas* and *Trapa*) or turion formation (e.g. *Hydrilla*), are best controlled before propagules are formed or shed (see Macrophyte species section below). On the other hand, weed clearance on a regular basis can deplete the carbohydrate stores in perennating organs, effecting more lasting control (Perkins and Sytsma 1981; Kimbel and Carpenter 1981; Wallsten 1983).

The prevailing weather conditions will also dictate to some extent the time when control is undertaken (Gangstad 1986), especially when floating machines are used.

Improving the efficiency of existing machines/processes

More effort is needed to improve the efficiency of existing machines/processes and to reduce the cost of mechanical weed control. Bryant (1974) described reductions in the cost of harvesting aquatic weeds in the early 1970s as a result of improved equipment design. Every effort should be made to operate the machinery continuously and a good maintenance service with attendant resources (Kemmerling 1974) is essential to this end. Particular attention should be focused on improving the efficiency of removing the plant material from the water body and the subsequent processing of harvested material. This may be reduced in volume and in weight by dewatering (Bagnall 1980) and through improvements in handling characteristics (Koegel *et al.* 1973). The potential for the use of the cut material should be exploited (Livermore *et al.* 1975).

A better understanding of the impact of returning the cut weed to the water is needed. This has been investigated for certain species, e.g. *Hydrilla verticillata* (Sabol 1987) but reliably predicting the effect on dissolved oxygen concentrations, phytoplankton, and other environmental components is not possible for many species.

The efficient use of any piece of machinery needs training and the acquisition of skill, a principle which extends to the maintenance of the machinery and reduction in down-time. More effort should be made to improve the training of personnel involved in such operations (Feichtinger 1974; El Sayed, Tolba, and Druijff 1978).

The development of new machines

The development of equipment for aquatic weed control lags behind the advances made in agricultural equipment, companies being inhibited by the restricted sales such machinery is likely to achieve. Nevertheless, new machines have been produced (Arts and van Wijk 1978) and interesting consideration has been given to the criteria which such plant should meet (Livermore *et al.* 1975; Murfitt and Haslam 1981). In England, a National College of Agricultural Engineering study raised some fundamental ideas about the use of machines for controlling weed growth in drainage channels from the banks (Murfitt and Haslam 1981). These were largely based on three premises.

1. The need to know the relationships between flow characteristics and the density, species composition, and physiological condition of the plants in the drainage channel. Mechanical control is seen not as a process of destruction but more as environmental management ('biotope shift'), a view emphasized by Kemmerling (1978) although such processes are very poorly understood.
2. The prediction of the reaction of the various species to cutting. This is only partially known and only for a few species.

3. Cutting is not necessarily the most effective method of controlling weed growth. It is a technique developed from harvesting forage crops such as grass, a practice which must ensure the survival and regeneration of the cut crop. A more reasonable objective for weed control is a device or mode of action which will discourage regeneration. Murfitt and Haslam (1981) listed a number of possibilities aimed at damaging nutrient reserves or otherwise making unfavourable changes to the habitat. Included within these ideas are crushing/bruising underground parts, repeated cutting/chopping, drought, rolling, discing, and damage due to altering light regime.

Hertzman (1985, 1986) has demonstrated the effectiveness of rotavating reed beds in order to check *Phragmites* development and reverse the hydroseral succession as part of the restoration of Lakes Kvismaren and Hornborga in Sweden. Novel machinery has been produced through the co-operation of ecologists and engineers. One such machine is a six-wheel floating machine with a 135 hp diesel motor which drives a 3.2 m rotavator. The rotavator cuts reeds, sedge and small bushes reaching down 30 cm into the substrate. It can go deeper but this has not been found necessary. This management is similar to that of beushaning in the ricefields of India, a local practice of tight ploughing with a narrow wooden plough in standing water to control the weeds about a month after sowing (Misra, Patro, and Tosh 1976).

Hertzman (1986) also described a machine which has a trampling effect. This is an amphibious vehicle which is used to cross an area of reed and sedge several times. The effect is particularly good if the trampling takes place when the shoots for next year's growth are becoming established. The effectiveness of both rotavating and trampling is improved if preceded by burning.

Kemmerling (1978) placed emphasis on the need to develop a single machine with the potential for different functions using different component parts and the advent of the hydraulic capability has seen the development of machinery in this direction although there is still room for improvement.

The cost of mechanical control

Decisions on whether to use mechanical or chemical control are often based on cost, either relative to efficiency, or on the assumption that both will be equally effective. One of the early attractions of herbicidal control was its low cost as compared with manual and mechanical methods. Dassanayake and Chow (1954) reported that the clearance of *Pistia* in Sri Lanka cost 300 rupees ha^{-1} using manual clearance whereas herbicide treatment cost only 40 to 50 rupees ha^{-1}. Soerjani (1978) found in Indonesia that mechanical control was 125 to 175 per cent more expensive than chemical control and Guppy (1967), in a comparison of various methods for *Elodea* control, discarded mowing as an option early in the programme. This generalization must not be accepted for all situations and further exploration of the cost of physical methods is worthwhile.

Koegel, Livermore, and Bruhn (1977), in an excellent comparison of two aquatic plant harvesting machines, demonstrated the complexity of assessing the cost of any given operation. They explored the breakdown of such factors as the distribution of machine hours in operation and in down-time. This latter factor may be very significant; Blanchard (1970), for example, reported a 67 per cent efficiency for weed harvesting machines. To compare such an assessment effectively with an operation based upon chemical treatment is difficult. Mechanical control is based on a high initial capital investment whereas this is low in the case of herbicide treatment. The consequences of the two approaches are also hard to quantify financially—harvesting may remove significant amounts of nutrients, with attendant advantages, whereas chemical control might release nutrients, thereby significantly increasing the potential of future weed development.

In order to ensure a valid comparison, the biomass of plant material needs to be of the same order in the different situations, which can cause problems (Conyers and Cooke 1982). Likewise, consideration must also be given to the weed species involved. In India, the cost of a single manual clearance of thick infestations of *Eichhornia*, *Pistia* and *Lemna* from small water bodies was estimated at approximately 150, 75 and 52.5 rupees ha^{-1} respectively (Philipose 1963). An infestation of *Eichhornia* in a large water area cost 437 rupees ha^{-1} (Philipose 1968) (all these estimations are based on a rate of 1.5 rupees for six man-hours).

Other factors which need to be considered are the access for machinery, the skill of individual operators and the agency undertaking work, i.e. whether or not machinery is owned or hired (Cave 1981).

A number of attempts have been made to make realistic comparisons between physical and chemical operations. Mara (1976*a*, *b*), using a dynamic linear programming model to estimate minimum costs for different levels of *Eichhornia crassipes* infestation, concluded that mechanical control was very expensive (the model did not include the cost of the disposal of harvested plants). Costs could have been reduced by combining herbicides in the programme. A metropolitan county in Florida, USA assessed costs over a 4-year period (1964 to 1968) (Hall 1969) and found that chemical treatment for floating and submerged weeds in secondary canals cost \$746 km^{-1} yr^{-1}. However dragging the bottom, using an amphibious DUKW truck with a toothed metal A-frame, and removal with a rake mounted on the DUKW, cost \$249 km^{-1} yr^{-1}, although Hall (1969) considered that the chemical treatment was more effective.

Brooker and Baird (1974) compared the unit costs of weed clearance, by hand and chemical methods, of emergent vegetation in drainage channels in different areas in Essex, UK, for 1972. The unit cost per annum of treatment with dalapon and 2,4-D was some 50 to 75 per cent of the cost of hand clearance and disposal. A useful comparison of the cost of controlling a 2 m wide fringe of emergent

Table 7.1. Comparative costs (£ sterling) of controlling a 2 m wide fringe growth of emergent vegetation on one side of a large watercourse in the UK ('Typha' 1983)

Method of control	Remarks	Cost of single treatment (£km^{-1})
Hand cutting, weed raked to bank top	Maximum of two cuts for full control. No access problems	160
Weed cutting boat, with drag-line at intervals to remove cut weed	Maximum of two cuts for full control. No access problems	45
Weed cutting bucket mounted on hydraulic excavator	In arable areas one cut only after harvest unless compensation paid	120
Bicycle tractor with cutter bar and weed rake	Based on data from one UK user only	60
Tractor mounted sprayer using glyphosate or dalapon	Access problems in August where root crops are present	30
Hand spraying using glyphosate	No access problems but even application is difficult	35
Helicopter spray using dalapon	No access problems, 150 km can be completed per day	70

Table 7.2. Comparison of costs for controlling emergent weed growth in 100 m of channel, UK (£ sterling) (Cave 1981)

Cross-sectional area of channel	Diquat	Glyphosate	Dalapon (aerial spray)	Weed cutting boat	Weed cutting hydraulic excavator	Bucket dragline	Hand roding
2 m^2	3	4	5	—	9	12	50
7 m^2	3	4	5	6	15	18	50
40 m^2	3	4	10	21	—	40	—

Table 7.3. Comparison of costs for controlling submerged weed growth in 100 m of channel, UK (£ sterling) (Cave 1981)

Cross-sectional area of channel	Dichlobenil	Terbutryne	Diquat	Weed cutting boat	Weed cutting hydraulic excavator	Bucket dragline	Hand roding
2 m^2	7	7	9	—	9	12	16
7 m^2	20	20	27	6	15	18	100
40 m^2	20[1]	100	140	21	—	40	—

[1] = partial treatment of channel.

growth on one side of a large water course in the UK was given by 'Typha' (1983) (Table 7.1). Cave (1981) presented data supporting these comparisons (Table 7.2) for emergent vegetation in channels of different cross-sections. Herbicides are consistently cheaper and more effective than cutting, although the use of chemicals may be less beneficial than cutting from the point of view of stability and erosion prevention. Cave (1981) also provided some useful comparisons for submerged vegetation (Table 7.3). The comparisons are of particular interest when different-sized channels are considered. Although hand clearance is more expensive than mechanical and herbicidal techniques, weed cutting is cheaper than herbicide use in the 7 m channel.

These figures are supported by calculations presented by Miles (1976) for the control of vegetation in 11 km of a drainage channel (>20 m in width) in the UK. The costs for terbutryne, cyanatryn, and weed cutting/removal were UK £4 620, £7 700 and £2 576 respectively.

The operation of shore/bank based dredgers and dragline excavators is expensive (Speirs 1948; Hall 1969) but the effectiveness of the weed control aspect of the operation means that subsequent maintenance is delayed for one or two seasons.

There are no easy answers when a decision has to be made as to whether mechanical control is cheaper than chemical control. The cost of herbicidal treatment is easier to calculate and this could be a factor in deciding in favour of this method. More effort should be put into examining the costs of mechanical methods, particularly where lasting control is achieved and/or a significant amount of nutrient is removed. That such data are lacking has been recognized for a long time (Grinwald 1968).

Westlake and Dawson (1982, 1986) have demonstrated, through careful ecological investigation of the efficiency of cutting of *Ranunculus penicillatus* var. *calcareus* in chalk streams in the UK, that savings of more than 30 per cent may be produced compared with normal cutting practices. This significantly affects a comparison with herbicidal control. They also documented other advantages of an improved cutting regime including benefits of lowered water levels to riparian users, shorter periods of disruption for fishermen, and less disturbance to wildlife.

THE ECOLOGICAL EFFECTS OF CUTTING, HARVESTING, AND DREDGING

Macrophytes

All the management described so far is followed by the re-establishment of an aquatic macrophyte community and, usually, the necessity for further management in the same or the following season. The resilience of aquatic macrophytes to perturbation is surprising and occasionally remarkable (Wade and Edwards 1980; Nichols 1984*b*). Knowledge of the reaction of the vegetation to physical control is important for a number of reasons. Management will only be effective where the replacement community is either less of a nuisance than the preceding community or the recovery occurs over a long period. Weed control should not have a deleterious effect on the uses of the water body, e.g., fishing, nor damage its wildlife conservation value. The recovery of the vegetation may be looked at from two standpoints:

1. the rate of recovery of macrophyte communities and of individual species;
2 the nature/pattern of the recovery, i.e. re-establishment of the aquatic macrophyte community and of individual species; changes in other components of the aquatic ecosystem, e.g. algae, invertebrates, and fish.

Macrophyte communities

Re-vegetation is often rapid following clearance. This is particularly marked in cutting and harvesting management. For example, *Myriophyllum spicatum* regrew to pre-harvest levels within 30 d of spring and summer harvesting (Mikol, 1985), and communities in a dredged lake were re-established during the first summer (Nichols, 1984*b*), and 4 months after dredging (Collett *et al.* 1981). A year after dredging in certain areas, macrophyte biomass was about 60 per cent of the pre-dredging value. Narayanayya (1928) denuded a 12 m section of an irrigation canal in India of *Potamogeton indicus*, *Hydrilla verticillata*, and *Vallisneria spiralis*. Eight months after the clearance *Nitella* occupied almost the whole area and *V. spiralis* occurred throughout. Thirteen months after clearance *Nitella* and *V. spiralis* had still further multiplied. Twenty months after clearance *Nitella* was decreasing and *V. spiralis* increasing, whilst *Potamogeton indicus* and *P. perfoliatus* 'had made considerable growth', a similar situation to that present before clearance. Similar observations were made on another section of canal which was re-colonized by *Potamogeton pectinatus* four months after clearance. Although vegetation quickly recolonizes dredged drainage channels, the habitat can take two to four years to recover its former floristic diversity (George 1976). The rate of recovery in this latter situation was affected by a number of factors, such as the time of year treatment is effected (Westlake and Dawson 1982, 1986), whether or not livestock had access to the water's edge, whether or not the edges of the drainage channel were managed in any other way, and the number of adjoining water bodies. If the latter had also been dredged out, recovery would have been prolonged and the area would have experienced floristic impoverishment (George 1976).

The recovery of a mill-stream in southern England after dredging was a slow process (Crisp and Gledhill 1970). A pre-dredging community dominated by *Elodea canadensis* with some *Nuphar lutea* and *Potamogeton pectinatus* was succeeded by sparse growth of *E. canadensis* in the spring following management, extending through the summer to cover most of the bed despite a minor incident of pollution. The *E. canadensis* was removed by spate flow and immigrant *Ranunculus penicillatus* var. *calcareus* developed extensive growth in the second year after dredging, being replaced by *E. canadensis* and floating mats of *Apium* during the summer of that year. Plants in Lake Chemung, USA, harvested annually over a three year period, were slower in reaching the water surface in the fourth year than plants in control plots (Wile 1978).

Where dredging fundamentally alters the habitat, recovery can take a long time. The recolonization of the Boro River, Botswana, was only in its early stages of recovery eight years after dredging, with small colonies of *Rotala myriophylloides*, *Ottelia kunenensis*, *Najas pectinatus*, *Potamogeton thunbergii*, *P. trichoides*, and *Nymphaea caerulea* (Lubke, Raynham, and Reavell 1981; Lubke, Reavell, and Dyr, 1984). The dredging had resulted in the creation of a deep channel in the centre with steep sides. At greater depths reduced illumination was due to turbidity; the removal of natural meanders also increased water flow considerably. Likewise colonization of the steep margins was very slow.

Figure 7.12 provides a scheme summarizing the possible routes which re-establishment of the vegetation might take. A noticeable feature of this scheme is the cyclical nature of the recovery process. This has been observed by George (1976) and Wade (1978) in drainage channel habitats in Norfolk and Gwent, UK, respectively. Where the interval between management is very long and no comparison can be made with earlier communities, the development can only be described as a succession, usually following the classic hydroseral succession (Tansley 1949). The dredging of the Boro River, Botswana, initiated a succession in which submerged *Rotala myriophylloides* was the first aquatic plant to colonize the exposed silts. This community was replaced by *Limnophila ceratophylloides*, *Wiesneria schweinfurthii* and other species in shallower waters (Lubke, Reavell, and Dye 1984). In even shallower areas, rooted aquatic plants with floating leaves such as *Nymphaea caerulea*, *Nymphoides indica* and *Potamogeton* spp. appeared along with floating sedge islands of *Cyperus* spp., *Scirpus cubensis*, and other species.

Even where immediate follow-up control is undertaken after weed cutting/harvesting, certain species will escape control. Operations in which 20 to 50 per cent of the vegetation remains after treatment are normal (Eaton and Freeman 1982; Westlake and Dawson 1982). Some plant material will be missed in the cut and other cut material will not be harvested. In addition to intact growing plants and fragments of plants, there will be root and rhizome material plus a variety of plant propagules either laid down in the sediments or floating on the surface. Although dredging is a more thorough technique as regards removing plant

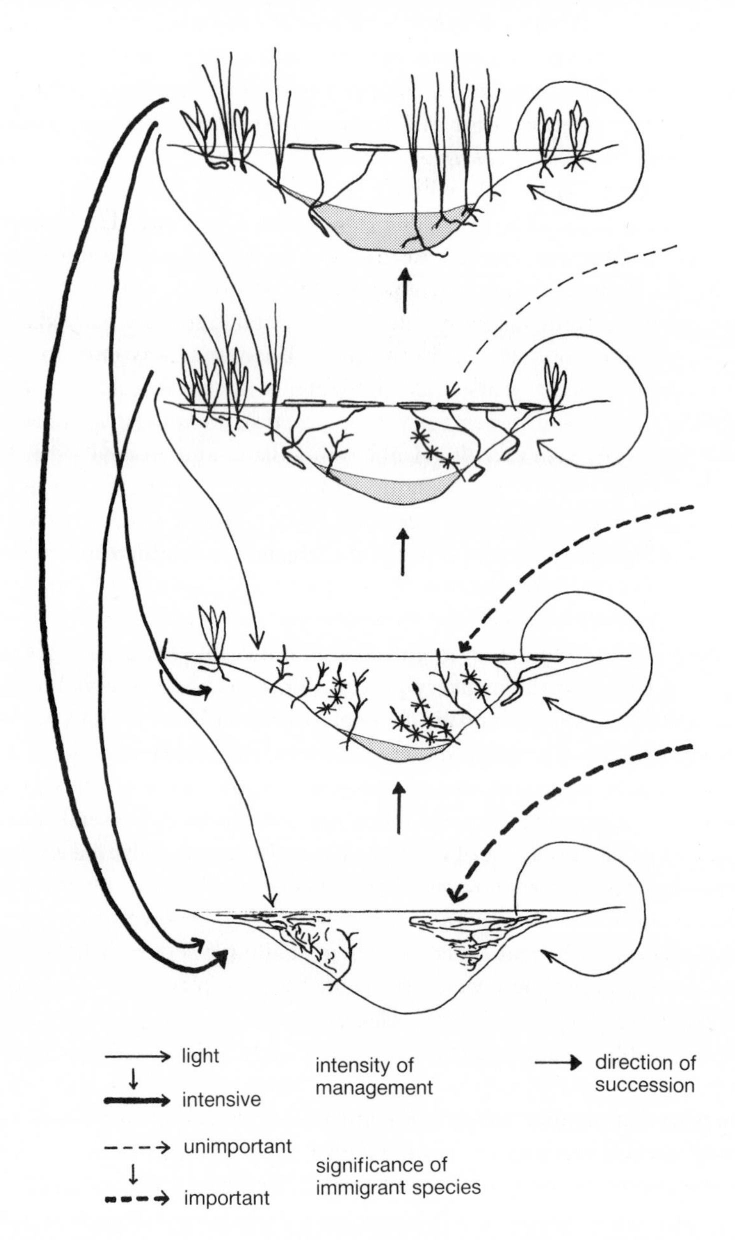

FIG. 7.12. A scheme for the re-establishment of aquatic vegetation after management in a small water body.

material, fragments of root and rhizome will be left and a significant proportion of plant propagules will remain after dredging. Evidence for the ability of propagules to survive for long periods in freshwater sediments and reappear after mechanical control comes from Wade and Edwards (1980).

The re-establishment of vegetation in a water body can follow different patterns:

(1) dominance by species not present immediately prior to management;

(2) dominance by those species which were dominant immediately prior to management;

(3) dominance by species which were present before management but not dominant.

The existence of species which are able to exploit open or disturbed freshwater habitats has been recognized by Eelman and van der Ploeg (1979), Wade and Edwards (1980) and Wade, Vanhecke, and Barry (1986). The Characeae provide good examples. Species of *Chara* and *Nitella* have a propensity for colonizing newly-created habitats and water bodies disturbed by weed control (Nichols 1984*b*). The genus *Myriophyllum* has a similar ability (Oglesby *et al.* 1976; Reed 1977; Nichols 1984*b*), and as a number of *Myriophyllum* species can be serious weeds, development of communities containing these species would be undesirable.

Thomas, Allen, and Grose (1981) found that removal of *Glyceria maxima* by dredging from drainage channels on the Ouse Washes, UK, produces a habitat suitable for a range of colonizing species. Propagules dormant in the mud can be brought to the surface, where suitable conditions for growth may be more likely. This might account for a higher floral diversity in channels one year after dredging compared with the diversity at any other time. Typically, emergent weed stands removed from watercourses are replaced by a community of submerged macrophytes (Wade 1978; Wade and Edwards 1980; Eaton, Murphy, and Hyde 1981; Murphy and Eaton 1981*a*).

In a dredged lake in Wisconsin USA, a different pattern was observed: all of the post-dredging species had been recorded in the two years prior to dredging but there was a major shift in dominance. *Chara* spp., *Najas flexilis* and *Myriophyllum* spp. became the dominant plants in place of *Potamogeton* spp., *Megalondonta beckii*, and *Elodea canadensis*, and two pre-dredging species, *Potamogeton amplifolius* and *M. beckii* were not found in the two years after dredging (Nichols 1984*b*).

Bernatowicz (1965) described the development of *Ceratophyllum demersum*, *Potamogeton perfoliatus* and *Ranunculus circinatus* after two years of mowing emergent macrophytes in Dgal Maly Lake, Poland. Other species appeared, not previously recorded from the lake: *Equisetum fluviatile*, *Scirpus tabernaemontani*,

Potamogeton lucens, and *Stratiotes aloides*. Other aspects of species succession in relation to management are discussed by Dawson, Castellano, and Ladle (1978).

Nichols (1984*b*) found that all communities were less diverse after the dredging of a 37 ha lake, with a substantial reduction in biomass and a 16.2 per cent decrease in the vegetated area. Driscoll (1975) and George (1976), considered that an increase in the frequency of dredging is associated with a decrease in the quality of the macroflora and it is known that some species, such as *Acorus calamus*, *Sium latifolium*, and *Ranunculus lingua* (George 1976), and *Alisma gramineum* (Gibbons 1975) cannot withstand frequent dredging. If the channel is too frequently dredged it cannot achieve all the successional stages which a developmental cycle would encompass. Doarks (1980) drew attention to the conservation value of recently dredged drainage channels which contain an improved quality of macroflora compared to channels overgrown with emergent species.

The practice of dredging all the channels in a given area in any one year also leads to a decline in the floral quality of the channels due to there being few intact channels from which aquatic macrophytes can re-invade (Driscoll 1975). Modern methods of dredging are too efficient at removing aquatic macrophytes and insufficient propagules remain within the drainage channel to enable the immediate recolonization of all the constituent species of the flora. Thomas, Allen, and Grose (1981) considered that the floral richness of the drainage channels of the Ouse Washes, UK, would be enhanced if some were dredged less frequently to cater for long-term colonists such as *Carex acuta*. The recovery after dredging is, in many ways, a simpler process to describe than the re-establishment after weed cutting. In most situations a new substrate is created for the vegetation to recolonize and the process is much slower. Scotter *et al.* (1978) and Wade (1978) have described the recovery of the communities in drainage channels.

The replacement vegetation might be more or less diverse than the pre-management community but it is likely that the standing biomass and the extent of the vegetation will be substantially reduced.

Macrophyte species

The examples presented below indicate the need to consider carefully the biology and autecology of the nuisance species before undertaking physical control measures. Studies have been undertaken on the effects of physical control on certain genera of aquatic macrophytes. Three example genera are chosen to illustrate this point.

Phragmites australis

Phragmites australis can be a pernicious weed in many parts of the world particularly in drainage and irrigation channels. Dredging has been used to control the plant and removal of vertical rhizomes slows down the recolonization process. If horizontal rhizomes are also removed re-invasion may be slow depending upon the inoculum potential. If the dredging leaves water over about 1.5 m deep, invasion will be postponed until sediment accretion has lessened the depth (Haslam 1968). Hoeing and ploughing which cut the rhizomes produce a replacement crop if the water is warm enough and *P. australis* has been known to persist as a weed under such conditions in damp areas for at least a century (Haslam 1972). Cutting is, however, a more common form of physical control.

In the Netherlands, cutting in spring and early autumn kills *P. australis*, and in Britain, July cutting, sometimes with a cut in early autumn, either depresses or kills *P. australis*, the result depending on the available competitors. In the Breckland fens, UK, a cut in July brought about a 40 per cent reduction of the crop in the following year (Haslam 1968). Shekhov (1974) reported that the control of *P. australis* in the littoral zone of some Polish fish ponds was most effective when the reeds were cut twice in the spring. *Phragmites australis* almost ceased growth for at least four or five years with only a very sparse stand of plants being produced. The first mowing must begin when the shoots appear above the water and the first cut should be completed before the plants reach full development (20–25 days). The second mowing should be 20–30 days after the first cut.

In Czechoslovakian ponds control of *P. australis* is carried out by cutting below the water surface in summer (Husak 1978). The truncated shoot base becomes flooded with water, dies and rots. This effect is emphasized when the cut is applied twice during one growing season. The ability of *P. australis* to grow back after being mown off under water, can depend on the nature of the substrate. For example, regrowth is considerably faster and stronger on a hard substrate than on soft sediments (Bernatowicz 1965).

A cut in April and/or May in the UK, during the early emergence period, gives a full replacement crop and a June cut, during the main emergence period, will give a good replacement crop (Haslam 1968). A September cut is less effective as a control measure and winter cutting when the reeds are hardened and the carbohydrates are in the rhizomes is not damaging. Nevertheless, all cutting exposes stands to frost (Haslam 1972). Hand cutting of *P. australis* in pond systems in Czechoslovakia has shown that winter cutting of *P. australis* culms has a favourable effect on the development of a reed stand in the subsequent growing season (Husak 1978). The removal of standing dead reeds improves and levels out the spring microclimate of the reed bed although the risk of late spring frosts is increased (Smid 1973). Winter cutting also removes fungi and insect larvae hibernating in dead reed culms, reducing subsequent infection by these agents. In the long-term the removal of litter reduces the rate of terrestrialization of the habitat (Husak 1978). Winter-cut *P. australis* stands acquire a more uniform vertical and horizontal structure and are more productive than uncut stands (Krotkevic 1959). Further studies by Husak (1973, 1978) demonstrate differences between a stand of *Phragmites* cut 180 cm above bottom level, a second stand cut 120 cm above bottom level, and an uncut stand.

Typha

Species of *Typha* are recognized as important weeds throughout the world (Morton 1975). Where heavy equipment is available control can be effected by dredging, otherwise cutting is a widely used practice. In mowing trials in Montana and Utah, USA, cutting off shoots below the water surface two or three times in one growing season during preheading or early heading stages reduced the stand by 95 to 99 per cent (Stodola 1967; Riemer 1984). Similar management is undertaken in the littoral zone of Czechoslovakian fishponds in which *T. angustifolia* is usually cut below the water surface twice in the summer (Husak 1973). Studies were undertaken of *T. angustifolia* by subjecting stands to either removal of inflorescences, a cut 180 cm above bottom level or a cut 80 cm above bottom level and comparing them to an uncut stand. During the season in which the treatments occurred, in the stand with inflorescences removed neither the aboveground biomass nor the leaf area index altered significantly but the underground biomass increased substantially in comparison with the control stand. In subsequent years *T. angustifolia* so treated did not differ from the untreated control stand. The cut at 180 cm did not induce any regeneration of the damaged shoots; the cut at 80 cm deprived the stand of nearly all its assimilating parts and provoked a large-scale regeneration of the leaves at the expense of a proportion of the reserves stored in the rhizomes; rhizome biomass fell to about 45 per cent of that in the control stand.

In the first year after the treatments both the stands cut at 80 cm and 180 cm showed that the shortage of rhizome reserves had resulted in both decreased density and shoot production, though the number of flowering shoots increased. In the second year after treatment the 180 cm cut and the control stand had produced the same amounts of biomass; the 80 cm cut stand was slightly denser than the other stands and the shoot biomass was higher. Other studies reported by Husak (1978) demonstrate that *T. latifolia* reacts to partial removal of its aboveground parts in a similar way to *T. angustifolia*.

Although *T. angustifolia* is sometimes cut in fishpond reed belts in early winter, there is no difference between winter-cut and uncut *Typha* stands (Husak 1978). Other studies have been undertaken of other species, for example, Sharma (1977) reported the effect of cutting on growth and flowering in *T. elephantina*.

Myriophyllum

The autecology of some species of *Myriophyllum* is now fairly well understood (Coffey and McNabb 1974; Aiken, Newroth, and Wile 1979), and a number of accounts have been made of the effects of harvesting on *Myriophyllum spicatum* (Mossier 1968; Cottam and Nichols 1970; Nichols 1973; Nichols and Cottam

1972; Neel, Peterson, and Smith 1973; Rawls 1975; Kimbel and Carpenter 1981).

In the short term two trends are evident. First, when single harvests are considered, recovery declines as the date of harvesting becomes progressively later. Substantial regrowth in the same year can follow a harvest early in the growing season. The most effective time for cutting is often considered to be prior to periods of peak biomass (Nichols 1975*a*). Secondly, multiple harvests are more effective than single harvests in reducing the amount of regrowth (Kimbel and Carpenter 1981; Perkins and Sytsma 1987). Considering the long-term response, biomass is commonly reduced in the season following harvesting. Thus, sustaining harvesting, even only once per year, reduces *M. spicatum* production the following year (Kimbel and Carpenter 1981). Similar conclusions were drawn from observations of the biomass of *M. spicatum* in areas of Chemung Lake, USA. The species harvested annually over a three year period was found to have increased at one-half the rate of adjacent unharvested areas (Wile 1978). Plots of *M. spicatum* were harvested either once, twice or three times during a season and changes in stem numbers and stem heights were compared to control plots (Wile 1978). Multiple harvests were most effective in reducing stem numbers. Single harvests resulted in slight increases in stem numbers. Similarly, with one exception, plant heights did not reach the levels attained by the control plots.

The effects of physical control on the physiology of most weed species, although poorly understood, should be considered more often. Investigations into the patterns of total non-structural carbohydrate (TNC) in *M. spicatum* have shown that this is an important factor in developing sound management strategies (Kimbel and Carpenter 1981). Biomass, TNC per plant and TNC per unit area were lower in the harvested plot eleven months after the harvest than in the unharvested plot. The low TNC levels in plants in the harvested area might have reduced their ability to regrow following a second harvest though Perkins and Sytsma (1987) found that depletion of TNC following harvesting was negated by overwinter accumulation. Kimbel and Carpenter (1981) reported a study which found that survival of *M. spicatum* vegetative propagules increased with TNC content of the propagules. Reduction in whole-plant TNC levels could lead to increased mortality of vegetative propagules and a decrease in colonization. The shoot and root phosphorus, nitrogen and carbon of *M. spicatum* were also altered by harvesting. This was particularly noticeable in the spring following an autumn cut (Painter and Waltho 1985). By the summer, the plants had returned to normal except for a significant reduction in TNC in the roots.

Potential control through inhibiting seed germination was explored by Coble and Vance (1987). Despite the high potential for sexual reproduction, the area of vertical distribution available for substantial rates of germination is restricted, i.e. less than 0.5 m of water under $9\,\mu E\,m^{-2}s^{-1}$ for 2–3 days for 50 per cent germination. Even then seedling survival is problematical. In relation to the study site, Coble and Vance (1987) considered that if anything, the energy which the stand expends on flowering and seed formation may simply be counted as wasted energy with respect to stand growth.

Since the plant can spread rapidly from vegetative parts, it is important to minimize disturbance of established plant beds. Motor boats travelling through beds can produce many fragments, all of which can take root and produce new plants (Nichols 1975*a*). Any measure to control the species mechanically should remove plant fragments from the lake. Regularly clearing small problem areas such as around harbours, docks or swimming beaches with a rake is effective (Nichols 1975*a*).

Phytoplankton

Many workers have reported an antagonistic relationship between aquatic macrophytes and phytoplankton (Hasler and Jones 1949; Goulder 1969) and the removal of macrophyte vegetation can result in increased algal production, for example, in Lake Sallie, USA (Peterson, Smith, and Malveg 1974). The reasons for increases in phytoplankton have been associated with nutrient availability, e.g. elevated chlorophyll *a* levels were observed after chopped *Hydrilla* was returned to the water (Sabol 1987), and the absence of chemicals excreted by the macrophytes which inhibit the growth of algae. Such increases in phytoplankton have not always been observed; e.g., in Chemung Lake, Canada, increases in algal biomass were not observed following commencement of harvesting operations and the phytoplankton community composition remained unchanged (Wile 1978). In British canals Murphy and Eaton (1981*a*) found little evidence to suggest that weed removal caused increased phytoplankton production. Nichols (1973) described a significant increase in algal biomass after two years of harvesting in the shallow water areas of Lake Mendota, USA, whereas in the deeper water, a drop in macrophyte biomass was accompanied by a drop in algal biomass. Productivity differences were also found between shallow and deep water areas. This aspect of mechanical weed control is poorly understood and needs further investigation.

Macro-invertebrates

Concern has been expressed that cutting, harvesting, and dredging remove most of the invertebrates attached to the vegetation and found in or on the mud of a drainage channel bottom. Free-swimming species, such as Coleoptera and Hemiptera are left behind although in reduced numbers, and recolonization by flying and swimming species does not usually take place until suitable habitat conditions have developed in the channel. For example, oviposition by Odonata will not occur unless the required type of vegetation is available (Corbet, Longfield, and Moore 1960). Recolonization by species with limited powers of movement, e.g. Gastropoda and Lamellibrancha, is an even slower process, and if dredging is carried out on an extensive scale, the status of such animals will decline and, in intensively managed areas, become eradicated altogether (George 1976).

Where dredging radically alters the habitat, the implications for the fauna are considerable. Toner, O'Riordan, and Twomey (1965), investigating a tributary of the River Moy, Ireland, observed that the recovery of the invertebrates after dredging was slow although large numbers of Chironomidae and Ephemeroptera were found two years after management. Trichoptera larvae were more seriously affected than other taxa. In other situations macro-invertebrate re-establishment has been rapid and recovery has occurred within twelve months. In a lake, dominated pre-dredging by *Zostera capricorni*, *Ruppia megacarpa* and the macroalga *Chaetomorpha linum*, recovery occurred within eight months of the management to the extent that the control and dredged sites were indistinguishable (Collett *et al.* 1981). Crisp and Gledhill (1970) considered that the recovery of the macrofauna of a mill stream in southern England was complete about one year after draining and dredging. Chironomidae, Oligochaeta, and Mollusca together formed 85 per cent of the benthos and all three groups were present from the autumn of the year of management, but whereas the Chironomidae were numerically important from the outset, the Oligochaeta and Mollusca took almost a year to build up their numbers. In a similar investigation on a canalized section of the River Hull, UK, no dramatic changes in community composition were shown. Most species were recovering their numbers six months after dredging and both variety and numbers were expected to have returned to normal by 8 to 9 months after dredging (Pearson and Jones 1975). As well as removal by the dredger, it was suggested that a large number of animals were lost in downstream drift.

The effects of weed cutting on the macro-invertebrate fauna of the River Hull (Pearson and Jones 1978) were similar to those made in the earlier dredging study (Pearson and Jones 1975). As the substrate was not disturbed the recovery was more rapid although large numbers of invertebrates were removed in the cut weed. The timing of weed cutting proved to be important. In this case, Chironomidae and *Caenis horaria* had emerged prior to weed cutting with no impact on the breeding population. The effect of weed cutting on invertebrate drift was investigated by Kern-Hansen (1978) in four small Danish lowland streams. Total macrophyte clearance increased the drift of *Gammarus pulex* while partial cutting programmes resulted in only small increases in drift intensity. Similar increases in invertebrate drift density have been recorded in a small stream in Germany (Statzner and Stechman 1977).

Szumiec (1963) investigated the difference in the macrofauna between cut and uncut plots within the littoral zone of fish ponds in Poland. There was little difference in the biomass per unit area, but the numbers of individuals per unit area were greater in the mowed plots and hence the average biomass of individuals, primarily Chironomidae and Oligochaeta, was less than in the uncut plots in which Hirudinea were predominant. Tabanidae and Coleoptera larvae were also more numerous in the vegetated plots.

Fish

Fish are not able to respond as rapidly to changes resulting from weed control as are invertebrates and macrophytes. Hence fishermen often regard mechanical weed control as deleterious to their fishery. On the Hampshire Avon, UK, anglers felt that the timing, extent, and severity of weed cutting had caused destruction of cyprinid spawn, removal of habitat for fry development, a decrease of invertebrate diet and loss of cover for adult fish (Frake 1979). This view can certainly be substantiated for rivers. Swales (1982) has shown how potentially serious mechanical weed control is, through disruption of feeding, reproduction, and behaviour. Cutting, harvesting, and dredging can result in the removal of large numbers of invertebrates with the weed and increases in the level of invertebrate drift (Kern-Hansen 1978; Pearson and Jones 1975, 1978; Statzner and Stechman, 1977). A 30 per cent decrease was observed in the fry of salmonid fish after dredging, though part of this was due to natural mortality and within two years population estimates compared favourably with pre-dredging estimates (Toner, O'Riordan, and Twomey 1965). A cut of *Callitriche*, *Berula*, and *Ranunculus* in a tributary of the River Itchen, UK, to improve fishing did not encourage increased use of the stretch for the spawning of *Salmo trutta* although it increased the area of exposed gravel. A count of redds showed no increase over the two previous years (Soulsby, 1974).

Some fish, such as *Leuciscus leuciscus* and *Tilapia* spp., utilize aquatic macrophytes as a direct food source (Swales 1982; Lubke, Raynham, and Reavell 1981). Removal of large amounts of vegetation can, therefore, have serious effects on fish feeding and growth. Macrophytes are an important spawning substrate for phytophilous fish (Mills 1981). Weed cutting and dredging may remove potential spawning areas and incubating eggs, and eliminate the slack water needed for fry development. Mortensen (1977) found that weed cutting and the clearing out of stream weed beds increased mortality of *Salmo trutta*. Reductions in water level through weed cutting may adversely affect spawning by stranding incubating eggs (Mills 1981).

A crucial factor determining the impact of weed cutting on fisheries and spawning in particular is the timing of the weed control (Pearson and Jones 1978). Swales (1982) attributed low population levels of *Rutilus rutilus* in the River Perry, UK, to the effects of annual weed cutting on the reproductive success of the species. He also showed fish distribution in uncut and partially cut sections of river to be strongly associated with weed cover and, particularly, cover provided by overhanging bankside vegetation. Increased fish movements were caused by the operation of the weed-cutting launch and manual cutting was suggested as a preferable form of control because it is more selective. Indeed selective weed-cutting can be used to improve species diversity and fishing, although this is difficult to achieve and is not normally attempted in the UK except in some of the more valuable trout fisheries (Barrett 1978*b*). Frake (1979) described a reduction of the weed cut on the Hampshire Avon, UK, based on a flexible programme fitted to the pattern of weed growth and land drainage requirement with marginal strips of weed left on one or both banks where possible and in some cases a strip down the centre.

The effect on fish in lakes and large ponds is less well understood. Cutting and harvesting *Myriophyllum spicatum* from Saratoga Lake, USA on four occasions over a two year period removed 2 to 8 per cent of the standing crop of fish, primarily *Lepomis macrochirus* (Mikol 1985). Wile (1978) reported that 8.9 kg ha^{-1} of fish, mostly *Perca flavescens*, were removed from Chemung Lake, USA, during harvesting, although this did not have any significant effect on the fish populations. Haller, Shireman, and Durant (1980) considered that such removal could be an underrated problem, whereas Wunderlich (1967) takes the opposite view that fish stocks quickly recover, a view supported by observations of the mowing of *Phragmites australis* in a Polish lake. The area occupied by submerged vegetation increased and the replacement of *P. australis* by fine-leaved vegetation improved the breeding and growth of phytophilous fish especially *Abramis brama* and *Alburnus alburnus*; the absence of *P. australis* made control of *Esox lucius* possible (Bernatowicz 1965).

Effects on other components of aquatic habitats including water quality and sediments

Cutting, harvesting, and dredging can have effects on other, abiotic, components of the aquatic habit. The removal of vegetation from the water will result in an initial reduction in community photosynthesis with possible implications for concentrations of oxygen, carbon dioxide, and bicarbonate ions. Deoxygenation rarely occurs and, given the normally rapid regrowth of the plants, any other effects on dissolved gas concentrations will be short-lived (Robson 1974; Carpenter and Gasith 1978; Wile 1978; Murphy and Eaton 1981). In the densely vegetated littoral zone of a shallow hardwater eutrophic lake, Lake Wingra, USA, the effects of cutting on concentrations of seston, dissolved organic carbon, biological oxygen demand of dissolved organic carbon, and particulate, dissolved unreactive, and dissolved reactive phosphorus were short-lived or insignificant (Carpenter and Gasith 1978). In shallow areas of Lake Wingra, community photosynthesis and respiration were decreased, reflecting the type of processes (for example, oxygen depletion and increase in algae and ammonia production) described by Ahmed, Ito, and Ueki (1982) in experiments on the decomposition of *Eichhornia crassipes* in enclosed containers.

The harvesting of large areas of submerged macrophytes may, rarely, contribute to erosion of littoral sediments by wave action (Carpenter and Gasith 1978). The control of emergent vegetation is more likely to have implications for exposing shorelines to erosion. Mowing of *Phragmites australis* in Dgal Maly Lake, Poland from 1961 to 1963 resulted in exposure of shallow areas of the shore to wave action, transforming the littoral zone. Stagnant water disappeared, the amount of slime and mud decreased and in some parts sand was exposed (Bernatowicz 1965).

An attractive feature of mechanical harvesting is the potential for removing nutrients from the habitat, so alleviating future weed problems (Björk 1968; Grinwald 1968; Carpenter and Gasith 1977; Musil and Breen 1977). Other measures such as herbicidal control can have the opposite effect (Carpenter and Adams 1978). A harvesting operation at Chemung Lake, Canada which removed 3.0×10^6 kg of plant material containing 560 kg of phosphorus effectively removed 47 per cent of the gross and 92 per cent of the net annual phosphorus loading to the southern half of the lake (Wile 1978). Carpenter and Adams (1978) reported the removal of 37 per cent of the annual net phosphorus load to Lake Wingra, USA by harvesting of macrophytes. In Lake Sallie, USA, however, only 1.4 per cent of the total phosphorus loading to the lake was removed (Peterson, Smith, and Malveg 1974). Lake Sallie is a deeper body of water with a macrophyte cover of 158 ha compared to 435 ha in Chemung Lake. The harvest from Lake Sallie was only 4.28×10^5 kg of plant material and the lake had a history of nutrient enrichment with an annual phosphorus input of 7 285 kg compared to 1 190 kg for Chemung Lake.

In order for harvesting to control nutrient levels effectively in the water and sediments, cutting and harvesting would need to be undertaken year after year. There are two important problems related to harvesting aquatic macrophytes as a means of nutrient removal. Firstly, the amount of weed harvested will affect the efficiency of nutrient removal; the low biomass of plants produced by regular cutting will not represent optimum removal. This optimum has been explored for certain plant species (Musil and Breen 1977) and floating species such as *Eichhornia* have a greater potential than submerged species because of their higher productivity (Steward 1970). Secondly, the available phosphorus in the sediment is continually replenished by a variety of processes and is available to the plants. The proportion of uptake from the sediments by different species and at different times of the year is poorly understood. Likewise, the contribution by plants to the sediments can be important. Doarks (1984) revealed that the total phosphorus content of the sediment of drainage channels increases with time since the last dredging operation, due to the build-up of organic detritus on the channel bottom and the storage of phosphorus in the surface sediment layers. When dredging occurs, the phosphorus-rich surface sediment is removed exposing a less enriched parent material.

Various changes have been observed in the nature of the sediments. Following dredging of lakes in New South Wales, Australia (Collett *et al.* 1981), deeper areas filled with soft mud with a high organic content composed mainly of macrophyte detritus. The nutrient and carbon content of sediments in dredged and control plots were generally similar. Narayanayya (1928) observed increasing amounts of fine silt accumulating in an Indian irrigation canal after clearance of vegetation. Eight months after clearance the silt was 3 cm deep, 13 months, 5 cm and after 20 months up to 7 cm deep. Narayanayya regarded this accumulation to be associated with the growth of *Nitella* beds and the increase in silt enabled *Potamogeton* species to become established and subsequently crowd out the *Nitella*.

ALTERATION OF WATER LEVEL

Exposure to air of part or all of the bed of a water body by lowering the water level has been used to manage aquatic vegetation for many centuries especially in shallow fish ponds and waterfowl habitats (Uhler 1944), irrigation systems (Malhotra 1976) and reservoirs (Richardson 1975; Manning and Sanders 1975; Beard 1969; Bond and Roberts 1978). Control can be achieved either through dehydration of the vegetation, its exposure to low temperatures, or by altering the characteristics of the sediment. The process is termed lake lowering, drawdown or dewatering (Leslie 1988). An increase in water level can be used to control emergent species by 'drowning' (Singh, Pahuja, and Moolani 1976), free-floating species by flooding and 'washout' (Ramaprabhu *et al.* 1987), and submerged species through increased shading caused by the deeper water column above the plants.

Partial, short-term drawdown of the Waikato hydroelectric lakes in New Zealand reduced the quantity of weed handled at the generating stations in the subsequent growing season (Coffey 1975). Clayton (1982) considered that this technique had slowed but not halted the competitive displacement of native species from submerged macrophyte communities, by the alien *Lagarosiphon major*. Substantial reductions in all the problem submerged species except for *Myriophyllum heterophyllum* in Lake Miccosukee, USA were achieved over an eight month drawdown (Tarver 1980). Winter drawdown in the Tennessee Valley Authority system killed all of the *Myriophyllum spicatum* plants on well-drained shorelines (Smith 1971). Hestand *et al.* (1973) described the effective reduction of *Eichhornia crassipes* following drawdown through 'root rot'. The effective reduction of *Glyceria declinata* by lowering the water level has also been observed (Hestand *et al.* 1973).

Infestations of *Myriophyllum spicatum* were reduced in Melton Hill Reservoir, USA by drawdowns of short duration during cold weather. Prolonged exposure sufficient to freeze the hydrosoil was found to be necessary. The management also incorporated the use of the herbicide 2,4-D (Goldsby, Bates and Stanley 1978: see also chapter 10). Palazzo *et al.* (1984) have explored the susceptibility of *Ceratophyllum demersum* to low temperatures and found that ice in the plant retards regrowth, and temperatures of −6°C will kill it. Similar studies have been undertaken for *Myriophyllum spicatum* (Stanley 1976; Stanley *et al.* 1976).

Consecutive drawdowns may be more effective than an individual drawdown, a practice which eradicated *Egeria densa* and achieved very good control of *Cabomba caroliniana* and *Ceratophyllum demersum* in Black Lake, USA (Goldsby and Sanders 1977). Nichols (1975*b*), however, reported that little additional control was gained by a second drawdown in the Mondeaux Flowage, USA except for *Nitella variegatum* which was probably controlled by the mechanical action of the ice. The first drawdown gave good control of a *Potamogeton robbinsii* dominated community, but Nichols considered that repeated water lowering would allow drawdown tolerant plants to replace susceptible species.

Drawdown used to control *Lagarosiphon major* in Aratiatia Lake, New Zealand, achieved an alteration to the submerged community, in which non-nuisance species such as *Elodea canadensis*, *Nitella hookeri* and *N. tasmanica* replaced *L. major* for a significant proportion of the following year (Coffey 1975). A similar effect on the macrophyte community was observed in Black Lake, USA, with an increase in *Chara* sp. and *Eleocharis acicularis* following drawdown (Manning and Sanders 1975; Goldsby and Sanders 1977).

Cooke (1980) reviewed the drawdown approach. He considered that results are variable and few submerged plants were consistently controlled in a range of lakes in the USA. The success of such a programme is limited by:

1. The formation of a 'skin' due to the drying of the canopy, which can afford dehydration protection to underlying plants and viable root stocks. A cover of filamentous algae is particularly effective. The effect of the protective skin can be reduced by mechanically turning the weed (Coffey 1975).
2. The nature of the substrate. Sandy sediments will dry out much more quickly than clay sediments.
3. The rate of desiccation as well as the degree of desiccation achieved.
4. The restrictions imposed by usage of the water, e.g. fishery or wildfowl refuge.
5. Susceptibility of nuisance species to dehydration. Nichols (1975*a*) found *Zizania aquatica* tolerant to overwinter drawdown, its dormant seeds being unaffected. *Najas flexilis* is also resistant to drawdown (Tazik, Kodrich, and Moore 1982) and can even be enhanced by it (Nichols 1975*b*).

6. Release of nutrients from decaying vegetation and exposed sediments after refilling, resulting in the development of algal blooms (Cooke 1980).
7. Growth of plants in any remaining water. Such growth became a problem in Rodman Reservoir, USA following drawdown (Hestand *et al.* 1973).
8. Dissolved oxygen reductions in the shallower water with the risk of fish kills during both summer and winter drawdowns.
9. Drastic changes in the benthic invertebrate communities of the littoral on reflooding with repercussions for waterfowl and fish (Cooke 1980).
10. Direct effects on fish spawning success and possible interference with migratory fish movements (Cooke 1980).

An important advantage of drawdown is its relatively low cost. The widespread adoption of a system for an irrigation network in Hissar, India based on an annual five day drawdown period incurred nominal initial expenditure and nil recurring costs (Malhotra 1976).

Drawdown is frequently used in conjunction with other weed control measures such as herbicides (Manning and Johnson 1975; Goldsby, Bates, and Stanley 1978), shade barriers (Cooke and Gorman 1980), and burning (Mohamed and Bebawi 1973*a*). Integrated into such control programmes the technique has significant potential (see chapter 10).

BURNING

Burning, in common with cutting and drawdown, removes aboveground parts of plants and exposes the top soil to increased temperature fluctuations. Burning is also a useful way of reducing the bulk of cut material, e.g. *Phragmites australis* and *Carex* (Hertzman 1985), and vegetation which has been killed by herbicide (Swift 1976). It can also reduce the viability of plants and propagules. For example, in order to prevent stranded and harvested *Eichhornia crassipes* plants and seeds being washed back into the rivers in times of flood they can be burnt. Mohamed and Bebawi, (1973*a*) recommend piling up the material and burning it in a light-to-moderate wind and in Sudan the dry conditions in March are best for undertaking this practice. There is a difference between the backburning and headburning of piles of *E. crassipes*, the former producing a more intense and thorough burn ensuring destruction of seeds (Mohamed and Bebawi, 1973*b*). Hertzman (1986) emphasizes the need for stability in both the direction and stength of the wind. This is important when the area to be burned is to be limited. The risk of undesired spread is limited by fire-breaks.

Stanley (1975) reported that the viability of *Myriophyllum spicatum* fragments drawn through the cooling systems of thermal electricity generating systems was severely reduced after 10 min at 45–50°C. Other applications of heat and burning are given in Pierce and Opoku (1971) and Junk (1977).

Damage to the fauna can be great in burning especially if done carelessly or if the basal litter is allowed to burn. Other problems are the risk in peaty areas of the peat catching fire (V. Holt pers. comm.) and the difficulty of burning green material. Propane flame-throwers have been used to reduce top growth but the process is slow and tedious. Burning can result in denser stands after a single growing period, for instance, in *P. australis* (Haslam 1986; Swift 1976; Shay, Thompson, and Shay 1987). Burning breaks internal dormancy and allows all preformed buds, and buds which are about to be formed, to emerge a month after the fire. Winter burning produces a very rapid early emergence of buds in spring, a competitive advantage for *P. australis*. A spring fire gives a replacement crop, perhaps denser than the original because of the stimulating effect on the dormant axillary buds on the rhizome, although the emergence may be late. A summer burn is less damaging than summer cutting since only the former produces a proper replacement crop. A fire sufficiently fierce to scorch the surface soil causes up to two months delay in emergence the next spring. These plants will not complete their growth before the autumn frosts and so the crop is diminished (Haslam 1968). The complete recovery of *P. australis* after burning can take at least four years (Shay 1983; Shay, Thompson, and Shay 1987).

Burning is often cheaper than cutting but summer burning may not be possible in wet years (Haslam 1973*a*). The fire can be started using a flame-thrower carried as a back-pack or on an amphibious vehicle (Hertzman 1986). Burning can be effectively integrated with other forms of control. It has been used in conjunction with drawdown and, when followed by herbicidal application with dalapon, is effective in controlling *P. australis*.

REDUCTION OF LIGHT

Light is essential for aquatic plant growth and limitation of photosynthetically active radiation (*PAR*) has been suggested as an ideal technique for aquatic weed control and has been used effectively in a variety of situations (Pitlo, 1978; Dawson and Kern-Hansen 1979; Dawson 1981; Jorga, Heym, and Weise 1982). There are currently two main approaches:

Limiting the quantity of light reaching the water

The quantity of light reaching the water can be limited by means of bankside vegetation (e.g. trees) or floating plant species (e.g. *Nuphar*) (Dawson 1978*b*; Pitlo 1978; Dawson and Kern-Hansen 1979; Dawson 1981; Dawson and Haslam 1983). The use of trees or other plants in this way is essentially a biological control technique (see Chapter 9(a)) and is not considered further in this chapter.

Limiting light penetration of the water

The limitation of light penetration can be achieved by the addition of *PAR*-absorbing dyes such as 'Aquashade' (Nichols 1974; White and Lembi 1976; Boyd and Noor 1982; Barltrop, Martin, and Martin 1982). Suitably prepared clays, e.g. bentonite and even slurry mix dispersed by outboard motor have also been considered (Dawson 1985).

Similar situations can be induced by stocking the water body with bottom feeding fish at an appropriate density such that they cause the suspension of lake sediments. Cyprinid fish have been successfully used in this way. Artificial suspension of sediments can be achieved by using boats to increase turbidity although direct mechanical effects on the plants are also likely (Murphy and Eaton 1983).

The appearance, as a result of using such techniques, is usually unsightly, which may be a problem in amenity waters, and the technique may sometimes be rejected for this reason alone. Other considerations are the rate of fade, settlement, or dilution of the agent; toxicity to man, fish, and other life forms (Dawson 1981) and the unsuitability of the technique in flowing waters. Studies should be undertaken to determine the effect of such techniques on non-target organisms.

Another method of reducing light penetration is to raise the water level (see Alteration of water level above). Robel (1962) described a decreased production of 35 per cent in the deepened parts of a piece of water near Bear River, USA resulting from raising the water level. Dredging has also been used to deepen lakes to increase the percentage of the bottom below the photic zone (Pierce 1970; Nichols 1974, 1984*b*). This latter practice can also remove nutrient rich sediment.

Inhibition of the penetration of light through the water column can also be achieved by using shade-barriers suspended over the water to control emergent or floating weeds, or lying on top of submerged beds of weed (Bernatowicz 1966; Lewis, Wile, and Painter 1983; Engel 1984; Bulthuis 1984). Continuous membrane screens such as black polythene, have the disadvantages of fragility, screen instability, and a tendency to trap gas bubbles which can lead to water quality problems (Pullman 1981). The polythene may be weighted down with a blanket of silt or sand (Born *et al.* 1973) but regrowth may rapidly occur in this layer, reducing the period of weed control (Dawson 1981; Engel 1984; Engel and Nichols 1984). The use of porous fibreglass screens largely overcomes these problems. Fibreglass barriers which reduce incident visible light by about 60 per cent have been shown to produce effective control of submerged weed growth (Mayer 1978; Perkins, Boston, and Current 1980). A recent development is the use of permeable polypropylene fabric as the barrier material (Cooke and Gorman 1980; Dawson and Hallows 1983). The cost of such material is quite high: 'Typar' (manufactured by Du Pont), for example, costs UK £500–1100 ha^{-1} (Dawson and Hallows 1983) though this was comparable with the cost of mechanical weed control for UK drainage ditch maintenance.

Cook and Gorman (1980) considered that such techniques might be most appropriately used on a small scale, for example, around harbours, or in swimming areas and boat moorings where weed growth had to be minimized continuously. Where screens were used in this way regrowth of filamentous algae (*Spirogyra*, *Oedogonium*, *Microspora* and *Oscillatoria*) was observed together with small stands of *Najas flexilis* above the screens, but the original problem of *Potamogeton* spp. was effectively overcome. In general, when using materials which reduce irradiance by 50 to 70 per cent, some two–three months shading is required to control established stands of aquatic plants (Mayer 1978; Perkins, Boston, and Current 1980; Dawson and Hallows 1983). Bernatowicz (1966) reported on the sensitivity of a number of species to shading including *Phragmites australis*, *Chara fragilis*, *Myriophyllum*, and *Ceratophyllum demersum*.

Mayer (1978) claimed that fibreglass screens produced no adverse ecosystem effects in Chautauqua Lake, New York, but gave few supporting data. Perkins, Boston, and Curren (1980) also found no significant variation in nine physical and chemical lake water variables, and phytoplankton chlorophyll *a*, between screened and control plots. Little or no regeneration of nitrate-nitrogen in the open water and only a moderate potential for phosphate-phosphorus regeneration occurred as a result of the death and decay of macrophytes beneath fibreglass screens (Boston and Perkins 1982). Dissolved oxygen concentrations could be reduced by plant decomposition but this may be minimized by placement of the screens in early spring whilst macrophyte biomass was still low. The use of 'Aquascreen' in Cox Hollow Lake, USA had adverse effects on the benthic fauna under the screen, which was almost eliminated, although chironomid larvae colonized the upper surface of the 'Aquascreen' suggesting that the material was not toxic to these organisms (Engel 1984).

Further work is desirable to assess the ecological impacts of screening techniques particularly if such methods achieve more widespread use.

MISCELLANEOUS TECHNIQUES

Booms

Barriers to aquatic weed movement such as booms are an important means of preventing plant material blocking screens and other outflow points from lakes and reservoirs. A typical example consists of netting hanging vertically from cable to a depth of about 4 m, the cable being suspended by a string of drum floats (Graham 1976).

Laser radiation and ultrasound

The use of laser radiation was advocated for aquatic weed control in 1969 (Gangstad 1971) and further tests were undertaken using *Lemna minor* (Couch and Gangstad 1974). The technique has subsequently been explored with *Eichhornia crassipes* (Long and Smith 1975). High-frequency sound is also being explored as a means of controlling *Myriophyllum spicatum* (Anon 1985*b*).

Although existing technology has not proved effective using these techniques, rapid development in these fields may provide a real alternative to physical cutting in the future. Whether they would be an economic means of control remains to be demonstrated.

Aeration

The aeration of model lake systems was found to control the growth of *Hydrilla verticillata* by 20 per cent fresh weight and 18 per cent dry weight on average after 21 d (Cooley, Dooris, and Martin 1980). This control was linked to a significant reduction in iron availability.

Flow manipulation

The movement of water can be effectively used to manipulate stands of nuisance plants. Spillett (1981) used wooden groynes as an effective long term measure in the River Thames, UK, to direct water flow on to downstream weed beds and silted sections of river bed. Water flow can also be manipulated to collect floating stands of plants such as *Salvinia* (Soerjani 1976).

CONCLUSIONS

For effective management of aquatic plants, it is necessary to consider the full range of available options. Physical control is one such option which, in turn, includes a wide variety of possibilities ranging from hand-pulling, cutting and the use of complex mechanical harvesters, to altering water levels and shading. Any one approach would need to be evaluated with respect to the management required, paying particular attention to such factors as adjacent land use and cost. Simple comparisons with other management options such as the use of herbicides are impossible and it is recommended that more effort is expended in helping managers make the correct decision as to which option to choose. This is true not only for cost but also with regard to environmental impact. Compared with our understanding of the effect of herbicides on the ecology of freshwater systems, we are surprisingly ignorant about the implications of many physical control measures and more research is needed to fill these gaps. Likewise, more attention should be paid to combining the use of physical, chemical and biological approaches.

8

Chemical control of aquatic weeds

K. J. MURPHY AND P. R. F. BARRETT

INTRODUCTION

HERBICIDES can offer a cheap, effective, and rapid method of aquatic weed control. However, they are powerful tools which require knowledge and understanding to be used safely and effectively. If misused, they can have side effects which may be harmful to aquatic organisms, wildlife in general and, ultimately, to man. This chapter describes the main herbicides used in or near water, discusses the factors which influence their performance and indicates how they can best be used.

Most of the herbicides used in water bodies were developed originally for terrestrial use, so their basic behaviour and properties were already known before they were tested and adapted for aquatic use. The subsequent testing procedure examined, in more detail, the toxicity to aquatic fauna, persistence and breakdown products in water and hydrosoil, effects on irrigated crops, and the efficacy of the product against target weeds. This information allows the manufacturer to provide detailed instructions on the product label concerning the timing, dose rate, susceptible weed species, and safety precautions required by the operator. Applications made without following these instructions can, at best, result in a poor level of weed control and, at worst, cause unnecessary damage to the target ecosystem. Even so, the user must decide the degree of weed control required in a particular body of water because overmanagement can be as harmful, in the long term, as undermanagement. The optimum level of control depends on the uses and priorities in each individual situation. A land drainage or irrigation channel may require total removal of aquatic weeds for the longest possible time, whereas in a fishery, 'trimming' of encroaching emergent or floating weeds may be all that is necessary (Barrett 1983). Both of these extremes, and intermediate levels of control can be achieved by herbicides (Barrett 1981*c*). The choice of the appropriate chemical and the correct application method require detailed knowledge of the capabilities and limitations of each herbicide. In many countries, government- or industry-sponsored training schemes are available which provide the user with both theoretical and practical experience of selecting and applying these herbicides.

Herbicides may have a direct toxic effect on non-target aquatic organisms or an indirect effect resulting, for example, from the removal of the target weeds. Laboratory-based toxicity tests often indicate greater toxicity than is found in the field. Thus they tend to err on the side of safety. Laboratory tests are followed by field experiments which can confirm laboratory results but may also show unpredicted toxic effects. By the time that the chemical receives official approval for aquatic use, the information available is such that direct toxic effects are unlikely to occur if the manufacturer's instructions are followed correctly. Some indirect effects are the inevitable result of the changes to the ecosystem caused by effective weed control. Thus, they are not limited to herbicides but can occur after any weed control operation. However, since herbicides can produce more thorough and, sometimes longer-lasting, control than other methods, the indirect effects can be more pronounced. The more common environmental side-effects associated with aquatic herbicide usage are described in this chapter.

Several of the terms used to describe the behaviour and properties of herbicides cannot be defined absolutely because these properties vary under different conditions of use. For example, a herbicide may be termed 'selective' if it controls only a limited range of plant species. However, it may become 'non-selective' at higher rates of application. The terms described here follow Robson and Spencer-Jones (in Roberts 1982).

The term 'active ingredient' (a.i.) refers to the concentration of herbicidally-active chemical within a formulation. It is expressed in terms of weight of active ingredient to volume (w/v: liquid formulations) or to weight (w/w: solid formulations), and may be shown either as grams per litre or percentage (e.g. the usual commercial formulation of glyphosate contains 360 g a.i. l^{-1} or 36 per cent w/v). Herbicides may be selective (e.g. dalapon which controls grasses but not broad-leaved weeds), or non-selective (e.g. glyphosate which controls almost all green plants). Contact herbicides (e.g. diquat) kill only those parts of the plant on which they fall (usually, the foliage), but if sufficient damage is caused, the whole plant may die. Translocated herbicides (e.g. dichlobenil) are absorbed by one part of the plant but move within the plant and act on other tissues or growing points. Persistent herbicides (e.g. fluridone) retain their activity in the soil or water for some time, usually measured in weeks or months. Non-persistent herbicides (e.g. glyphosate) act only when sprayed directly onto foliage and lose their phytotoxic activity very quickly on contact with soil or water. Some herbicides may show both characteristics; for example, diquat is non-persistent in an active form when sprayed onto terrestrial emergent plants. The droplets of chemical which miss the plant fall directly onto the soil and are rapidly and irreversibly absorbed onto soil particles, where they persist in non-phytotoxic form. In water, diquat molecules remain active in solution until they are absorbed by plant cells, or adsorbed onto sediments. The term 'availance' (Hartley and Graham-Bryce 1980) is defined as the combination of residue concentration and period of residue persistence in the aquatic environment, which produces a phytotoxic effect on the target plants.

The chemical names used in this chapter follow those given in the 7th edition of the *British Crop Protection Council Pesticide Manual* (Worthing and Walker 1983). Full chemical names are listed in Appendix 3.

HERBICIDE CHARACTERISTICS AND USAGE IN FRESHWATERS

Herbicide application

The objective of any herbicide application is to get the active ingredient to the target weeds in sufficient quantity, and maintain it there for sufficient time, for a phytotoxic dose to be absorbed into the plants. This objective is more difficult in aquatic than in terrestrial situations because of the diluting and dispersing effects of the water surrounding and, sometimes covering, the weeds (Brooker and Edwards 1975).

Perhaps the simplest herbicide formulations to use in aquatic situations are those applied as sprays directly onto emergent or floating foliage (Barrett 1974*a*; Lee and Furtado 1977; Anderson 1981). The application rate is calculated, as for terrestrial applications, on a unit area basis and neither the depth nor the velocity of the water need be considered. However, accurate application requires the spray nozzle, if not the operator, to move at a constant speed and height over the target. This can be difficult when working in or near water, where the operator may be working along a steep bank or spraying reed beds growing above head height. Boat-mounted equipment is also inaccurate because of the difficulty of moving the boat at constant speed through emergent or floating vegetation. The boat also tends to submerge the weeds over which it passes, washing off the herbicide before it can be absorbed. These problems are sometimes overcome by using a long-throw jet. Although less accurate, this can reach the target weeds from a firm platform away from the weedbed. In some instances, a second application is made once any remaining unaffected areas can be distinguished. Aerial application techniques may be used where the size of the problem warrants the expense involved (Little, Robson, and Johnstone 1964).

Conventional spraying techniques are ineffective against submerged weeds or algae. Also, some floating weeds are resistant to herbicides applied in this way. These weeds are controlled by applying the herbicide to the water so that it is absorbed by the submerged foliage or the roots. The application rate is usually calculated on the basis of the volume of water being treated rather than the area of water surface. Factors such as water velocity and dilution rates can have a highly significant effect on the performance of the chemical. The simplest form of submerged herbicide treatment is to apply an aqueous solution which disperses throughout the water body. In flowing water herbicide may be added at a constant rate to maintain a given concentration for the requisite period, so that a pulse of herbicide moves downstream, being absorbed by the target weeds as it passes (Bowmer and Smith 1984).

Special formulations have been developed to enhance, or take advantage of, the particular properties of individual chemicals (Steward, Van, and Jones 1982). Controlled release formulations of copper were an early development (e.g. Janes 1975). Dichlobenil, which is preferentially absorbed by roots, has been produced as a slow-release granule. This carries the chemical down to the the root region before releasing it (Spencer-Jones 1971). Diquat is rapidly absorbed by submerged foliage but is inactivated by mud and clay. It has been formulated as a liquid and as a viscous gel which sticks to weeds (Fig. 8.1), holding the herbicide in contact with the target foliage and reducing the amount lost to the sediment

FIG. 8.1. Viscous gel formulation of diquat-alginate: close-up of herbicide sticking to underwater foliage of *Ranunculus penicillatus*. (Photo: A. Fox.)

(Barrett 1978*a*; Clayton and Tanner 1982). This latter approach also improves the performance of diquat in flowing water (Barrett and Murphy 1982). Other controlled-release formulations for aquatic herbicides include impregnated rubber, polymer pellets and invert emulsions (Stovell 1966; Steward and Nelson 1972; Gates 1974; Osgerby 1975; Harris 1985; Harris and Talukder 1982; Van and Steward 1985). Westerdahl (1986) identified two controlled-release carriers, polyGMA polymer and polycaprolactone fibre as candidates with good potential for use in slow-release aquatic herbicide formulations. PolyGMA has side-chains which form a chemical bond with polar herbicides such as 2,4-D. The rate of hydrogen exchange, when submerged in water, governs the release of herbicide. Polycaprolactone forms a matrix within which herbicides can be entrapped, and then released by slow diffusion. In all cases, the aim is to transport the herbicide to the weed and release it in a way which enhances uptake and performance (Killgore 1984).

The application technique is dependent on the particular formulation and on the area involved. Granules and pellets can sometimes be applied by hand, but mechanical and motorized granule-spreaders are usually preferred. They give a more even application over the water surface and are safer for the user. Liquids may be applied directly from the original container, although it is more usual to dilute them first and apply with standard spray equipment. This tends to give a more even distribution of the chemical and improved environmental and user safety. They are also applied through metering devices (Bowmer and Smith 1984), high powered jets and submerged trailing hoses (McLintock, Frye, and Hogan 1974). Gel formulations cannot be applied through conventional sprayers but require specially-modified sprayers in which the spray nozzle is replaced by an orifice which produces a pencil jet. This gives a long range and allows the operator to spread the chemical over the water surface while working from the bank or in a boat (Fig. 8.2). The majority of application systems use machinery developed for terrestrial use, with modifications to improve their performance in aquatic situations. Frequently, these modifications are carried out by local water managers.

The choice of the correct application equipment is usually simplified by the herbicide manufacturer specifying, on the product label or in the accompanying technical leaflet, the type of equipment appropriate to the product. However, calibration of the equipment and calculation of dose is left to the operator. Mistakes in this, or in the timing of application, can result in poor levels of control, or cause fish mortalities, loss of irrigated crops, and contamination of water supplies (Bowmer *et al.* 1976; Bryan and Hellawell 1980). In some countries, operators are required to pass examinations in order to obtain a licence before using specified herbicides. Most countries have a legal requirement for consultation with an appropriate authority before a herbicide is used in water.

FIG. 8.2. Application of gel formulation herbicide (diquat-alginate) to flowing water (Crummock Beck, Cumbria, UK): note pencil jet application. (Photo: K. J. Murphy.)

Major groups of herbicidal compounds used in water

1. Inorganic compounds

Three simple inorganic compounds have been used to varying extent in different parts of the world for aquatic weed control: sodium arsenite, copper compounds, and hydrogen peroxide.

Sodium arsenite

Sodium arsenite was amongst the first inorganic aquatic herbicides to be used (Mackenthun 1950) and gained fairly widespread acceptance in the USA and elsewhere up to the 1960's. The compound is cheap and particularly effective at 5–8 mg a.i. l^{-1} against submerged vegetation (Mackenthun 1955; Johnson 1965; Hooper and Cook 1957). Subsequently, problems associated with the accumulation of toxic concentrations of arsenic, from the use of sodium arsenite or alkyl arsenical compounds in aquatic ecosystems were recognised (Mackenthun 1960; Woolson 1975; Isensee *et al.* 1973). The environmental consequences of continued usage of sodium arsenite as an aquatic herbicide are largely unacceptable and the herbicide is now relatively little used.

Copper

Copper was first used as an algicide in the nineteenth century and is still used for this purpose against a range of submerged weeds in some countries (Moore and Kellerman 1904; Pearlmutter and Lembi 1976; Khobot'ev *et al.* 1975; McGuire *et al.* 1984). Its toxicity derives from the ability of heavy metals to precipitate proteins in the cell. In its simplest form it is applied as crystals of hydrated copper sulphate. The crystal size helps to control the rate at which the chemical dissolves into the water and different sized crystals are used in some countries to achieve a form of controlled release (Sharif el Din and Jones 1954; FAO 1968). More recently, chelated copper complexes have been produced for use as aquatic herbicides (Rabe, Schuster, and Kohler 1982; Anon. 1980). The use of copper sulphate in fisheries waters was reviewed by Jackson (1974). Copper appears to be more active as a herbicide at higher water temperatures and in acid or neutral waters. Under alkaline conditions, copper reacts with dissolved carbonates and bicarbonates and is precipitated as insoluble copper carbonate. The chelated complexes are reported to be less susceptible to water hardness and less toxic to fish. Copper is normally applied to water at concentrations ranging from 0.5 to 2.0 mg a.i. l^{-1} depending on water temperature, water hardness, and susceptibility of weed species. In some irrigation channels, however, it is applied as a continuously metered supply, maintaining concentrations in the range of 0.005 to 0.02 mg a.i. l^{-1} for periods of days or weeks (Gangstad 1986; McKnight, Chisholm, and Morel 1981; Rabe, Schuster, and Kohler 1982).

Copper sulphate controls unicellular and filamentous green algae and, at higher doses, some Cyanophyta (Khobot'ev *et al.* 1975; Pearlmutter and Lembi 1976; Jackson 1974; McGuire *et al.* 1984).

Copper has long been used against submerged vascular plants in some countries (e.g. USA, Egypt: Banerjea and Mitra 1954; Sharif el Din and Jones 1954; Mitra 1977; Thakurta and Mitra 1977; Hodgson and Linda 1984; Lopinot 1963; Sutton and Blackburn 1971*a*,*b*; Brown and Rattigen 1979). Copper-tolerant strains of aquatic weeds may develop in waters treated for many years with copper herbicides (Muehlberger 1969).

Copper and diquat are synergistic when used for the control of *Hydrilla verticillata* and mixtures of the two compounds are used against this weed in the southern USA (Sutton *et al.* 1972; Anon. 1980). Combinations of copper and endothall are also synergistic against aquatic plants (Sutton, Blackburn, and Barlowe 1971).

Copper is rather similar to diquat in its persistence characteristics and although lost more slowly from water, copper is readily absorbed onto sediments, where it remains for prolonged periods (Frank 1972; Brown 1978). Copper is a trace element and, therefore, essential for the survival of many plant and animal species. However, the difference between beneficial and toxic concentrations is small and there is little difference between concentrations toxic to aquatic plants and those which kill fish and invertebrate animals. For this reason, many countries restrict the use of copper to those waters where fish are absent, or of secondary importance in the function of the watercourse.

Hydrogen peroxide

The biocidal properties of hydrogen peroxide have been known for many years, for example, as a bactericidal agent used as a wound dressing. Hydrogen peroxide is an effective sterilizing agent, permitted in drinking water within the European Community at concentrations up to 10 mg a.i. l^{-1}. The potential of hydrogen peroxide as a herbicide was discussed by Quimby (1981). Kay, Quimby, and Ouzts (1982, 1984) showed that the activity of hydrogen peroxide was related to light intensity and suggested that its activity might be very similar to diquat.

Relatively few field trials have been undertaken with this chemical. Kay, Quimby, and Ouzts (1983) reported a preliminary trial on *Ceratophyllum demersum* which gave satisfactory control. Fowler and Barrett (1986) achieved some control of *Elodea canadensis* and *Cladophora* sp. at concentrations of 10–20 mg a.i. l^{-1} under semi-natural conditions in large tanks. However, further work (P. R. F. Barrett, unpublished) has so far failed to show any control of *Vaucheria dichotoma* or *Fontinalis antipyretica* in ditches when used at concentrations up to 25 mg a.i. l^{-1}.

The short persistence and absence of organic breakdown products make hydrogen peroxide an attractive chemical for aquatic weed control, particularly in irrigation or potable supply waters. However, development is likely to be slow because it is not a proprietary chemical protected by the patent of a single company and may not, therefore, attract the development capital necessary for full and rapid testing and evaluation.

2. Dalapon

Dalapon, and its analogue TCA, are classed as halogenated-alkanoic acid derivatives (Roberts 1982). Dalapon–sodium became available in the mid-1950s (1957 in the UK: Chancellor 1960*b*). Dalapon, at 18–25 kg a.i. ha^{-1}, selectively controls grass weeds and other emergent monocots (Chancellor 1960*b*; Barrett and Robson 1971, 1974; Kramer, Schmaland, and Nanzke 1974; Barrett 1976; Agaronian, Aslanian, and Gevorkian 1980*a*; Comes, Marquis, and Kelly 1981; Eaton, Murphy, and Hyde 1981). Treatment is by simple spray application, sometimes from the air if extensive treatments are needed (Robson 1966*a*).

Dalapon is thought to damage protein (enzyme) structures at growing points (Worthing and Walker 1983). The herbicide is translocated within the plant and may be most effective against rhizomatous emergent monocots, such as *Phragmites australis*, at the time of year (mid–late summer) when carbohydrate food reserves are being laid down by translocation of sugars to the rhizome (Barrett and Robson 1974).

Dalapon and TCA are rapidly broken down by microbial action in both hydrosoils and water (Thiegs 1955; Magee and Colmar 1959; Agaronian *et al.* 1980*b*; Bowmer 1987*a*). Usually less than 1% of the applied dose persists, after only 10–20 days (Foy 1969; Niehuss and Borner 1971).

3. The (aryloxy) alkanoic herbicides

The herbicides of this group most commonly used in water are 2,4-D, fenac, 2,4,5-T, and silvex (Schultz and Harman 1974; Evans and Gallagher 1969; Sikka *et al.* 1982; Rosenberg 1984). They are often called the 'hormone' herbicides as their effects mimic those of natural growth-regulating plant hormones. Their mode of action is uncertain, but appears to involve interference with nucleic acid metabolism (Quee Hee and Sutherland 1981).

Studies of the use of 2,4-D against free-floating weeds were amongst the earliest published papers on the use of modern organic herbicides (Hildebrand 1946; Gerking 1948; Greco 1951; Hitchcock *et al.* 1950). The introduction of 2,4-D could fairly be said to have revolutionized approaches to aquatic weed control. Recently the aquatic use of this class of compounds (especially 2,4,5-T and silvex) has declined, mainly due to public fears over their environmental and toxicological safety (Gangstad 1982*a*, 1983), but 2,4-D remains in very widespread use worldwide some 40 years after its introduction. There is now some evidence that, in parts of the USA where 2,4-D has been used regularly over this period, tolerant strains of weed species, previously susceptible to 2,4-D, may have arisen (Haller and Tag el Seed 1979).

2,4-D is usually considered to be effective against broad-leaved weeds (Quee Hee and Sutherland 1981), but, in water, some monocots (e.g. *Eichhornia crassipes*) are also susceptible. It is effective against submerged species such as *Myriophyllum spicatum* (Taubayev 1958; Ministry of Environment, British Columbia 1981; Elliston and Steward 1972; Hall *et al.* 1982; Westerdahl and Hall 1983) at about 1.0 mg a.i. l^{-1}. At doses in the range 2–10 kg a.i. ha^{-1} it is active against free-floating and dicotyledonous emergent species, such as *Alisma plantago-aquatica*, (Hitchcock *et al.* 1950; Behl, Pahuja, and Moolani 1973; Hook 1977; Ransom, Oelke, and Wyse 1983; Bebawi and Mohamed 1984).

The efficacy of 2,4-D is strongly influenced by formulation. Addition of very low rates of gibberellic acid (a naturally-occurring plant hormone) increases 10-fold the *in vitro* potency of 2,4-D against *Eichhornia crassipes* (Pieterse, Roorda, and Verhagen 1980; Pieterse and Roorda 1982). However, under field conditions, this effect was hardly noticeable (Joyce and Haller 1985). The presence of surfactants also enhances 2,4-D activity against floating and emergent weeds, probably by enhancing herbicide uptake (Thayer and Haller 1982). Several other adjuvants are used in 2,4-D formulations (Killgore 1984). The butoxyethanol ester of 2,4-D (2,4-D BEE) has been incorporated in controlled-release formulations for use in submerged weed control (Gangstad 1977a; Getsinger and Westerdahl 1984). Westerdahl and Hall (1983) showed that the minimum sustained (threshold) 2,4-D concentrations (to control *Myriophyllum spicatum* and *Potamogeton pectinatus*), which should be achieved by controlled release from a suitable carrier, were in the range 0.05–0.1 mg a.i. l^{-1}. Combinations of 2,4-D with other herbicides, such as dalapon, are sometimes used to control mixed stands of emergent weeds (Brooker 1976*a*). Mixtures with paraquat, ametryne or 2,4-D can produce synergistic effects against *Eichhornia crassipes* (Seth, Fua, and Yusoff 1973; Diem and Davies 1974; Widyanto *et al.* 1977).

2,4-D is moderately persistent in aquatic systems (Westerdahl and Hoeppel 1982). Robson (1966*b*, 1968) found that phytotoxic concentrations of 2,4-D residues could persist for up to 9 weeks in a range of British waters. According to Gangstad (1982*b*), seasonal factors, water volume, and dosage rate were the most important variables influencing the dissipation of 2,4-D from the waters of large reservoirs in North America. Hoeppel and Westerdahl (1982) observed that, in 10 ha lake-plots treated with a controlled-release formulation of 2,4-D to control dense *Myriophyllum spicatum*, residue concentration dropped from a maximum of 3.6 mg l^{-1} to <0.03 mg l^{-1} within 13 d. In the lake sediments a maximum of 7.0 $\mu g\ g^{-1}$ (dry weight) 2,4-D was present 68 days after treatment (d.a.t.), dropping to $<0.2\ \mu g\ g^{-1}$ (dry weight) a.i. after a further 182 d.

2,4-D is only weakly adsorbed onto clays, silts, or organic particles (Grover and Smith 1974; Harris and Warren 1964). Microbial biodegradation, the rate of which is influenced by environmental factors such as pH, temperature, and nutrient availability, appears to be the principal mechanism of 2,4-D breakdown in the aquatic environment (Nesbitt and Watson 1980*a*, *b*; Gambrell *et al.* 1984).

4. Dichlobenil

The phytotoxic properties of dichlobenil, a non-selective, translocated herbicide, were described by Koopman and Daams (1960). It is used in water to control a wide range of submerged and floating-leaved species.

Although dichlobenil is preferentially absorbed by the roots of aquatic plants, foliar uptake also occurs in submerged weeds, as evidenced by its ability to control non-rooted species such as *Ceratophyllum demersum*. After absorption it is translocated to the growing points, where it interferes with cell division, causing the death of actively growing meristematic tissue (van Busschbach and Elings 1967; Verloop and Nimmo 1970; Newbold 1975). Treatment is most effective early in the growing season when the chemical is translocated rapidly to growing points, and when meristematic activity is highest.

Dichlobenil is applied directly to water to give a nominal concentration of 1–2 mg a.i. l^{-1}. The formulation recommended for aquatic use is a clay-based granule containing 20% active ingredient. The granules have a slow-release property which enables them to carry the chemical down to the root region and release it slowly into the hydrosoil/water interface region (Tooby and Spencer-Jones 1978). The chemical is adsorbed onto mud (Massini 1961) where it remains biologically active. Because of removal by plants and sediments the measurable concentration in the water seldom reaches the nominal applied dose. The slow-release formulation can be used to localize the area of control within a water body, by retaining the chemical within the hydrosoil in the treated area (Terry, Robson, and Hanley 1981). Another granular formulation, containing 6.75 per cent a.i., is less effective for localized control but can be used effectively when entire water bodies are to be treated.

The persistence of dichlobenil in aquatic systems is variable. For example, Ogg (1972) found that residues of a 1 mg a.i. l^{-1} treatment persisted for only 16 d whereas Cope, McRaren, and Eller (1969) found dichlobenil residues in treated water up to 189 d.a.t. In general, for normal doses of 0.5–1.5 mg a.i. l^{-1}, residues of about 0.05 mg a.i. l^{-1} remain in water up to 90 d.a.t., dropping to <0.01 mg a.i. l^{-1} after 120 d.a.t. (Pieters and de Boer 1971). These authors pointed out that *c.* 0.3 mg a.i. l^{-1} persisted in hydrosoils for up to 90 d.a.t. Volatilization, uptake, and re-release by plants and animals, and microbial breakdown (but not photodecomposition) are all implicated in the variability of dichlobenil persistence in freshwaters (van Busschbach and Elings 1967; Fowler and Robson 1974; Newbold 1975; Miyazake, Sikka, and Lynch 1975).

When applied to an entire water body, the dichlobenil concentration in the water is sufficient to control most species of submerged weed, whether rooted or not, and a number of floating species. Resistant species, including many monocot emergents, appear to be able to metabolize dichlobenil rapidly to non-phytotoxic breakdown products (Sikka, Lynch, and Lindenberger 1974) but may be susceptible to high doses (Kliemand 1974; Newbold 1975). When applied to localized areas, the concentration in the water (as opposed to the mud) is lower and control is limited more specifically to the rooted plants within the treated area. Dichlobenil is not generally regarded as an algicide, although it does control *Chara* spp. (Spencer-Jones 1971).

5. Diuron

Diuron was described by Bucha and Todd (1951). It is a substituted urea herbicide quite widely used for aquatic weed control (Mitchell 1957; Hambric 1969; Robinson and Leeming 1969; Dalrymple 1971; Skender, Filicić, and Steuk 1974; Hiltibran 1980). Several analogues, including the more recent sulfonylureas such as the methyl ester of sulfometuron (DPX 5648), have also been tested or used in freshwaters (McHendry 1963; Walker 1965; Liu and Cedeno-Maldonado 1976; Dechoretz and Anderson 1983; Langeland, Haller, and Thayer 1983).

The ureas are mainly photosynthesis inhibitors (although the sulphonylureas appear to be inhibitors of cell division at growing points), and are readily translocated (Cedeno-Maldonado and Liu 1974; Worthing *et al.* in Roberts 1982; Anderson 1983). Diuron is sometimes described as a total herbicide in freshwater usage, active against algae as well as submerged, floating, and emergent macrophytes at doses of 0.5–1.5 mg a.i. l^{-1} (Sills 1964; Hambric 1968; Skender, Filicić, and Steuk 1974; le Cosquino de Bussy 1971; van der Weij, Hoogers, and Blok 1971; Yana 1964; Johnson and Julin 1974). It may be applied as a surface spray or in granular form and may give a long period of weed control.

Buryi *et al.* (1973) reported that diuron or monuron, at 50 kg a.i. ha^{-1} applied to 2000 km of Russian canals, gave 95 to 100 per cent weed clearance which was maintained for 2 to 3 years in large canals, and 5 years in smaller waterways. The lower dose rates, which are more commonly employed, have regularly been noted to give weed clearance for one full growing season or more (Chancellor 1960*a*; Robinson and Leeming 1969; van der Weij, Hoogers, and Blok 1971; Johnson and Julin 1974).

An important reason for this high efficacy is the lengthy persistence of diuron, in phytotoxic form, in the aquatic environment (van der Weij, Hoogers, and Blok 1971; Niehuss and Borner 1971; Bowmer and Adeney 1978). In aquaria studies (van der Weij 1966), the half-life of diuron applied at 0.8 mg a.i. l^{-1} was 70 days. Organic matter content and temperature are important environmental factors influencing the adsorption and desorption of diuron by hydrosoils: more diuron is adsorbed onto cold, organically-rich sediments (Peck, Corwin, and Farmer 1980). In field trials in static drainage ditch waters in southern England, Robinson, and Leeming (1969) noted that 0.09 mg a.i. l^{-1} was still present 42 d.a.t. with diuron granules at 0.4 mg a.i. l^{-1}. The main degradation mechanism for diuron in aquatic environments seems to be microbial action and is clearly not a rapid process (Johnson and Julin 1974; Ellis and Camper 1982).

6. Terbutryne and simazine

Terbutryne is one of the methylthio-triazines (Gast, Grob, and Fankhauser 1965). Simazine, a chloro-triazine, was one of the first triazine herbicides to be discovered (Gast, Knüsli, and Gysin 1956). Several other triazine herbicides have been used for weed control in fresh water (Dowidar and Robson 1971; Payne 1974; Krsnik-Rasol 1975; Funderburk and Lawrence 1963; Sutanto, Widyanto, and Soerjani 1976; Tubea, Hawxby, and Mehta 1981), as well as the related triazinones, such as hexazinone (DPX 3674) (Fowler 1977; Anderson 1981*a*; Langeland, Haller, and Thayer 1983).

The use of the methylthio-triazines as aquatic herbicides has been reviewed by Murphy (1982), while simazine usage in fisheries waters was reviewed by Mauck (1974). The primary mode of action of triazines, at the concentrations used in water, is photosynthetic inhibition with little translocation or activity away from the foliage (Bishop 1962; Sutton and Bingham 1968; Sutton *et al.* 1969; Jones and Winchell 1984). The principal target weeds are submerged and free-floating species (Marks 1974; Blackburn and Taylor 1976; Elamson 1977; Robson, Fowler, and Hanley 1978; Murphy, Hanbury, and Eaton 1981; Dabydeen and Leavitt 1981; Ashton, Scott, and Steyn 1981; Johannes, Heri, and Reynaert 1975; Bowmer, Shaw, and Adeney 1985). The triazines are also potent algicides (Robson, Fowler, and Barrett 1976; Patnaik and Ramachandran 1976; Norton and Ellis 1976; Hawxby *et al.* 1977; O'Neal and Lembi 1983). It is usually necessary to attain an availance of 0.05–1.0 mg a.i. l^{-1} for 7–16 days to control aquatic weeds with terbutryne; for simazine 0.5–2.0 mg a.i. l^{-1} is required for a similar period.

Triazines are quite persistent in water and hydrosoils (van der Weij, Hoogers, and Blok 1971; Maier-Bode 1972; Mackenzie, Frank, and Sirons 1983). The half-life of terbutryne in water is 21–30 days, with 0.5–1.4 μg g^{-1} (dry weight) of terbutryne residues persisting in pond hydrosoils (organic carbon content 1.7 to 4.1 per cent), some 12–30 weeks after standard treatments with the herbicide (Muir *et al.* 1981). Following simazine application at 1.0 mg a.i. l^{-1}, Mauck, Mayer, and Holz (1976) recorded simazine at up to 0.48 μg g^{-1} (dry weight) of pond hydrosoil, 123 d.a.t. At the same time the concentration in the treated pond water was 0.65 mg l^{-1}. The herbicide was still detectable 456 d.a.t. Such extreme persistence is unusual. Tucker and Boyd (1981) found that, 32 d.a.t., a maximum of 20 per cent of simazine, applied at 3 mg a.i. l^{-1}, was lost from pond water alone (in experimental vessels). However, up to 75 per cent of residues was lost from the water, over the same period, if sediments were also present. They suggested that adsorption to clays, uptake by phytoplankton, and chemical/microbial decomposition processes were all important in reducing the persistence of simazine in aquatic systems. Bowmer (1982*a*) has shown the importance of seston as a sink for terbutryne in fresh waters. Burkhard and Guth (1976) suggested that photodegradation mechanisms may also play a role in the breakdown of triazines in water. In the field it is probable that degradation, adsorption, and the presence of numerous sinks for triazine residues tend to reduce their persistence considerably below the periods suggested by laboratory studies.

7. The bipyridinium herbicides

Diquat and paraquat are the most important examples of the bipyridinium herbicides, which are quaternary ammonium compounds. The phytotoxic properties of diquat were reported by Brian *et al.* (1958). Paraquat was first described in 1882 by Weidel and Russo but the phytotoxic properties were not recognized until after the discovery of diquat. Other bipyridinium salts have phytotoxic properties but few have been developed or used in water.

The bipyridinium herbicides are water-soluble salts in which the cation is the active ingredient. They are non-selective, contact herbicides which rapidly desiccate any green plant tissue which they contact. They act by interfering with electron flow in the photosynthetic process in plants and are reduced to free radicals within the cell. In the presence of oxygen, the free radical is oxidized back to the quaternary salts and hydrogen peroxide is generated within the cell and destroys it (Calderbank 1968). Because of this relationship with photosynthesis, these herbicides act more quickly in the light than in the dark. They are only slightly translocated within the plant, although movement is increased if they are applied in the dark.

The principal commercial formulations of diquat and paraquat contain 20 to 25 per cent active ingredient in an aqueous solution. They may also contain wetting agents and corrosion inhibitors. In some countries, formulations containing wetting agents are banned on grounds of the added risk of side-effects for aquatic fauna. Diquat and paraquat are not identical in the range of weed species controlled but have a very similar performance when used for aquatic weed control. However, paraquat has a higher mammalian toxicity than diquat and is often formulated with wetting agents that are toxic to fish. Although paraquat has been used as an aquatic herbicide (Way *et al.* 1971; Booker and Edwards 1973*a*) diquat is preferred and the remainder of this section refers principally to diquat.

Diquat is used for the control of some floating weeds such as *Lemna*, *Salvinia*, *Pistia*, and *Eichhornia* spp. and is applied directly on to the floating vegetation. The normal application rate is 1.0 kg a.i. ha^{-1}, though lower doses have been effective against certain weed species. It is also used for controlling submerged weeds when it is added to the water to produce a concentration of about 0.5–1.0 mg a.i. l^{-1}. The aqueous formulation of diquat dibromide disperses rapidly and may be applied by surface spray, subsurface injection, or simply sprinkled onto the water from a suitable container.

Diquat is most effective in the early part of the growing season when the plants are actively photosynthesizing and the tissues are soft and easily decomposed. It is particularly active against non-rooted species and those which do not have underground rhizomes or storage organs (Hiltibran 1965; Hughes and Meeklah 1977). Species such as *Elodea canadensis* and *Myriophyllum spicatum* can be eradicated by a single application under good conditions. Other plants, such as *Hydrilla verticillata* and *Potamogeton pectinatus* will have their vegetative growth destroyed but the tubers (or subterranean turions), buried in the mud, survive to recolonize the system (Blackburn *et al.* 1976). Repeated applications may help to reduce the numbers of viable tubers but complete eradication is seldom possible. Diquat controls a number of filamentous algae, including *Cladophora* spp., *Rhizoclonium* spp., and *Spirogyra* spp., but some (*Chara* spp. and *Vaucheria dichotoma*) are resistant, at least under field conditions (Robson, Fowler, and Barrett 1976).

Both diquat and paraquat are rapidly absorbed by submerged aquatic plants (Davies and Seaman 1964; Best and van der Wittenboer 1978). Despite this, they are not generally effective in water flowing at more than about 100 m h^{-1}, probably because the availance is insufficient. This problem has been overcome by adding sodium alginate to the formulation, converting it into a sticky, viscous solution (Barrett 1978*a*, *b*, 1981*a*, *b*). The formulation is applied, from a special applicator, as a pencil jet onto the water surface where it breaks up to form discrete strings and droplets in the water. These are dense and sink rapidly into the weedbeds where they stick to the plants and release the active ingredient. The formulation limits the dispersion of the chemical in slow-flowing or static waters (see Fig. 8.3) allowing localized control to be achieved (Barrett and Logan

FIG. 8.3. Selective treatment of 100 m^2 plots, using diquat-alginate, in a salmonid fishery lake (Laird's Loch, Scotland, UK: Murphy and Pearce 1987). Effective clearance of *Ranunculus trichophyllus* and *Potamogeton xnitens* was limited to the treated areas, leaving plant growth elsewhere on the lake unaffected. (Photo: K. J. Murphy.)

1982; Murphy, Fox, and Hanbury 1987). Diquat-alginate effectively controls weeds in fast flowing rivers (Barrett and Murphy 1982; Fox, Murphy, and Westlake 1986).

The bipyridinium herbicides are photodegraded (Slade and Smith 1967) and are strongly adsorbed and inactivated by clays and other organic particles (Funderburk 1969; Narine and Guy 1982). Concentrations of 1.0 mg a.i. l^{-1} typically fall to less than 0.1 mg a.i. l^{-1} within 4–8 days (Yeo 1967; Calderbank 1968;

Brooker and Edwards 1973*a*). Adsorption on to suspended material is very rapid (Coats *et al.* 1964; Fanst and Zarins 1969), so the performance of diquat is greatly reduced in turbid waters, or where the plants are covered by deposits of silt, which act as a barrier to uptake (Way *et al.* 1971; Bowmer 1982*b*). High water hardness also reduces the uptake of bipyridinium herbicides (Parker 1966). This may be caused by calcium ions which are thought to be antagonistic to diquat (P. R. F. Barrett unpublished data; Fox 1987).

In soils, diquat and paraquat residues are strongly adsorbed and so are very persistent though not phytotoxic (Frank and Comes 1967; Funderburk 1969; Simsiman 1974). The long-term breakdown of these residues is due largely to microbial decomposition processes (Baldwin, Bray, and Geoghegan 1966; Hiltibran, Underwood, and Fickle 1972). It is likely that similar behaviour occurs in hydrosoils.

8. Endothall

Endothall is a heterocyclic compound, first described by Tischler, Bates, and Gorgonio (1951), which was introduced to Britain in 1954. Although no longer approved for use in the UK it remains widely used in the USA and elsewhere (Temby 1973; Keckemet 1980; Corbus 1982; Bowmer and Smith 1984).

The mode of action of endothall is still imperfectly understood, but it acts as a contact herbicide which is also translocated quite easily, disrupting a range of physiological pathways including photosynthesis, respiration, and mRNA synthesis (Thomas and Seaman 1968; Simsiman, Daniel, and Chesters 1976; Roberts 1982).

Endothall is usually formulated as a sodium, potassium, or mono (*N*, *N*-dimethyl) alkylamine salt in liquid or granular form (Mixon 1974; Westerdahl 1983*a*, *b*). It is sometimes used in combination with copper (Sutton, Blackburn, and Barlowe 1971).

Breakdown of endothall in aquatic systems is rapid and complete, the rate being dependent upon water temperature and microbial activity. There are no known toxic intermediate breakdown products (Simsiman, Daniel, and Chesters 1976; Yeo 1970; Horowitz 1966; Hiltibran 1962). In water dissipation periods in the range 2.5–30 d are reported, depending on precise environmental conditions (Simsiman and Chesters 1975). There is evidence that some of the applied dose is removed by sorption to hydrosoils, where residues are more persistent (Sikka and Rice 1973; Holmberg and Lee 1976; Reinert and Rodgers 1984).

The main target weeds are submerged species and as the herbicide tends to show a higher efficacy above 18 °C, it is particularly effective against weeds of warmer waters, such as *Hydrilla verticillata* (Westerdahl 1983*a*, *b*; Leonard 1982; Walker 1963; Haller and Sutton 1973*a*). Normal application rates are 0.5–2.5 mg acid equivalent (a.e.) l^{-1}. Controlled dose applications (e.g. using invert emulsion, granular pellets, or polymer carriers) may permit doses as low as 0.2 mg a.e. l^{-1} to be effective (Westerdahl 1983*b*; Keckemet 1980).

9. Fluridone

Fluridone is a heterocyclic compound of the pyridinone family, which was introduced in 1976 (Waldrep and Taylor 1976). It is used for aquatic weed control in the USA and a few other countries outside Europe (Parka *et al.* 1978; McCowen *et al.* 1979). The principal mode of action is inhibition of carotenoid biosynthesis, preventing chlorophyll synthesis in the plant (Anderson 1978, 1981*b*; Bartels and Watson 1978; Devlin, Saras, and Kisiel 1978).

Fluridone is used principally against submerged and floating species (e.g. *Hydrilla verticillata*, *Elodea canadensis*, *Cabomba caroliniana*), but is also effective on some emergent weeds (e.g. *Typha* spp., *Sagittaria* spp.). It is used at application rates in the range 0.1–1.0 mg a.i. l^{-1} or, on a surface area basis, at *c.* 0.56 kg a.i. ha^{-1} against emergent weeds (Parka *et al.* 1978; Marquis, Comes, and Yang 1981; Dechoretz 1980). The weeds die slowly as food reserves in the plant are depleted and not replaced, symptoms appearing some 2 to 4 weeks after treatment. This has the advantage of minimizing ecosystem side-effects, such as deoxygenation of the water or a sudden change of habitat, but is a disadvantage if rapid weed removal is needed.

Fluridone is lost quite rapidly from the water, mainly by photodegradation; the half-life for the parent molecule is 20–30 d (West, Day, and Burger 1979; West and Parka 1981; West *et al.* 1983; Saunders and Mosier 1983). It is more persistent in hydrosoils, with half-lives, under field conditions of 3 to 12 months (Muir and Grift 1982; West and Parka 1981; Muir *et al.* 1980). Formulations include liquid (aqueous suspension), 2.5 per cent a.i. granular products, and a recently-developed controlled-release fibre formulation, (Parka *et al.* 1978; Sanders and Theriot 1979; Van and Steward 1985). Hall, Westerdahl, and Stewart (1984) demonstrated that the minimum sustained (threshold) concentrations of fluridone to give $\geqslant 50$ per cent control of *Myriophyllum spicatum* and *Hydrilla verticillata* were 10–20 μg a.i. l^{-1} and 20 μg a.i. l^{-1} respectively.

10. Acrolein

Acrolein is a simple organic molecule which is very toxic to both plants and animals. It is hazardous to work with, being a lachrymatory agent, and causes skin and lung damage. It is also explosive. Acrolein acts as a contact herbicide, destroying enzyme systems and cell membranes, and kills weeds within minutes or hours of treatment (Ware 1983; Fritz-Sheridan 1982; van Overbeek, Hughes, and Blondeau 1959).

It is less widely used now, being banned in many countries, but is still used in irrigation systems in Australia, Egypt, Argentina, and the USA because of its ability to give useful weed control, with only very short contact times, in fast-flowing waters (Svachka *et al.* 1982; Bowmer and Sainty 1977, 1978; Crossland and Adams 1963; Gaddis and Kissel 1982). One technique, used in the Murrumbidgee irrigation area in Australia, is to inject a high concentration of acrolein in the upstream end of the system. A pulse of herbicide, with a peak concentration of about 15 mg a.i. 1^{-1}, moves downstream killing all aquatic plant growth (and everything else) as it passes (Bowmer and Smith 1984; Bowmer and Sainty 1977, 1978). Concentrations in irrigation water of less than 15 mg a.i. 1^{-1} are generally safe for crops (Bowmer and Sainty 1978; Anon. 1971; USDA 1963). Acrolein is rapidly lost from water, both by degradation to non-toxic residues and by volatilization (Bowmer *et al.* 1974; Bowmer and Higgins 1976).

11. Glyphosate

Glyphosate was first described as a herbicide by Baird *et al.* (1971). It is a broad-spectrum, post-emergence herbicide active against a very wide range of annual and perennial species. The first reports of its use in aquatic situations appeared in 1974 when several workers described control of floating weeds such as *Eichhornia crassipes* and *Salvinia* spp. (Pieterse and Van Rijn 1974), *Nuphar advena* (Blackburn 1974), and *Nuphar lutea* (Barrett 1974*b*). Subsequently, it has been shown to control many bankside, emergent and floating-leaved species (Barrett 1976; Evans 1978; Hanley 1981; Welker and Riemer 1982; Sandberg and Burkhalter 1983; Riemer 1976; Baird *et al.* 1983; Comes, Marquis, and Kelley 1981).

Glyphosate is normally applied as the isopropylamine salt which is more water soluble than the acid. Other ingredients in the commercially-available formulations include a wetting agent to aid uptake by the leaves; details of this and other additives have not been released. Glyphosate is sprayed directly onto exposed foliage, normally at rates of 1.8–2.1 kg a.i. ha^{-1}. Application techniques include conventional hydraulic nozzles, controlled drop application (using low volumes), and rope wicks. All these methods have been used with some success in aquatic situations (Seddon 1981).

Glyphosate inhibits the biosynthesis of aromatic amino acids (Rubin, Gaines, and Jensen 1982) and cells die because they cannot synthesize essential proteins and phenolic compounds. One effect of this chemical is that it inhibits the germination of buds on rhizomes (Balyan *et al.* 1981), which makes it particularly valuable for the control of many perennial aquatic weeds. The symptoms may start to appear within a few days as a yellowing of the treated leaves which die and gradually decay. Unsprayed leaves on the same plant may remain green to the end of the growing season, when natural dieback occurs (Eaton, Murphy, and Hyde 1981). The principal effects become apparent the following season when little or no regrowth occurs in the treated area.

Many floating and emergent water plants have a waxy cuticle which sheds water rapidly. As glyphosate is absorbed into the leaves fairly slowly it is particularly susceptible to being washed off within the first few hours (Caseley and Coupland 1985). Thus rain, or a rise in water level, soon after application can reduce efficacy. Spraying with too high a dilution of the active ingredient can have the same effect, as the spray droplets coalesce on the waxy surface and run off. The passage of a boat through a weed bed during spraying operations, temporarily submerging the leaves, can leave a track of surviving plants which may require a second application to control them.

Many authors have reported poor control if treatments are applied early in the season, when there is insufficient leaf area for herbicide absorption and late-developing shoots may be screened by their more advanced neighbours (Caseley and Coupland 1985). Late treatments also produce poor results if senescence has started before the glyphosate has been fully translocated into the rhizome system. Dense weed growth can be difficult to control, again due to the screening effect, and repeated applications may be necessary to overcome this problem (Hiltibran 1981; Barrett 1985).

A list of aquatic weed species susceptible to glyphosate and the required dose rates for their control was given by Barrett (1985). In general, emergent and floating-leaved species are susceptible to glyphosate while submerged weeds are not. Marquis, Comes, and Yang (1981) showed that both absorption and translocation of glyphosate, in the submerged species which they tested, were poor. Glyphosate is rapidly absorbed onto particulate matter in water and in the hydrosoil, where it is degraded by micro-organisms.

Its rate of degradation depends on the level of microbial activity in the aquatic system, but is generally rapid (roughly the same rate as sucrose) and may drop to the limit of detection within days rather than weeks (Comes, Bruns, and Kelley 1976; Rueppel *et al.* 1977; Bowmer 1982*c*; Brønstad and Friestad 1985). Glyphosate has recently been the subject of a major review (Grossbard and Atkinson 1985).

ECOLOGICAL IMPACTS OF HERBICIDE USAGE IN FRESHWATERS

The impacts of herbicides upon aquatic ecosystems fall into two principal groups of effects (Brooker and Edwards 1975; Newbold 1976; Robson and Barrett 1977):

(1) direct toxicity to both target and non-target organisms;

(2) indirect effects associated with the destruction of target macrophytes.

Some of the interrelated effects of herbicide treatment on the aquatic ecosystem are shown in Fig. 8.4.

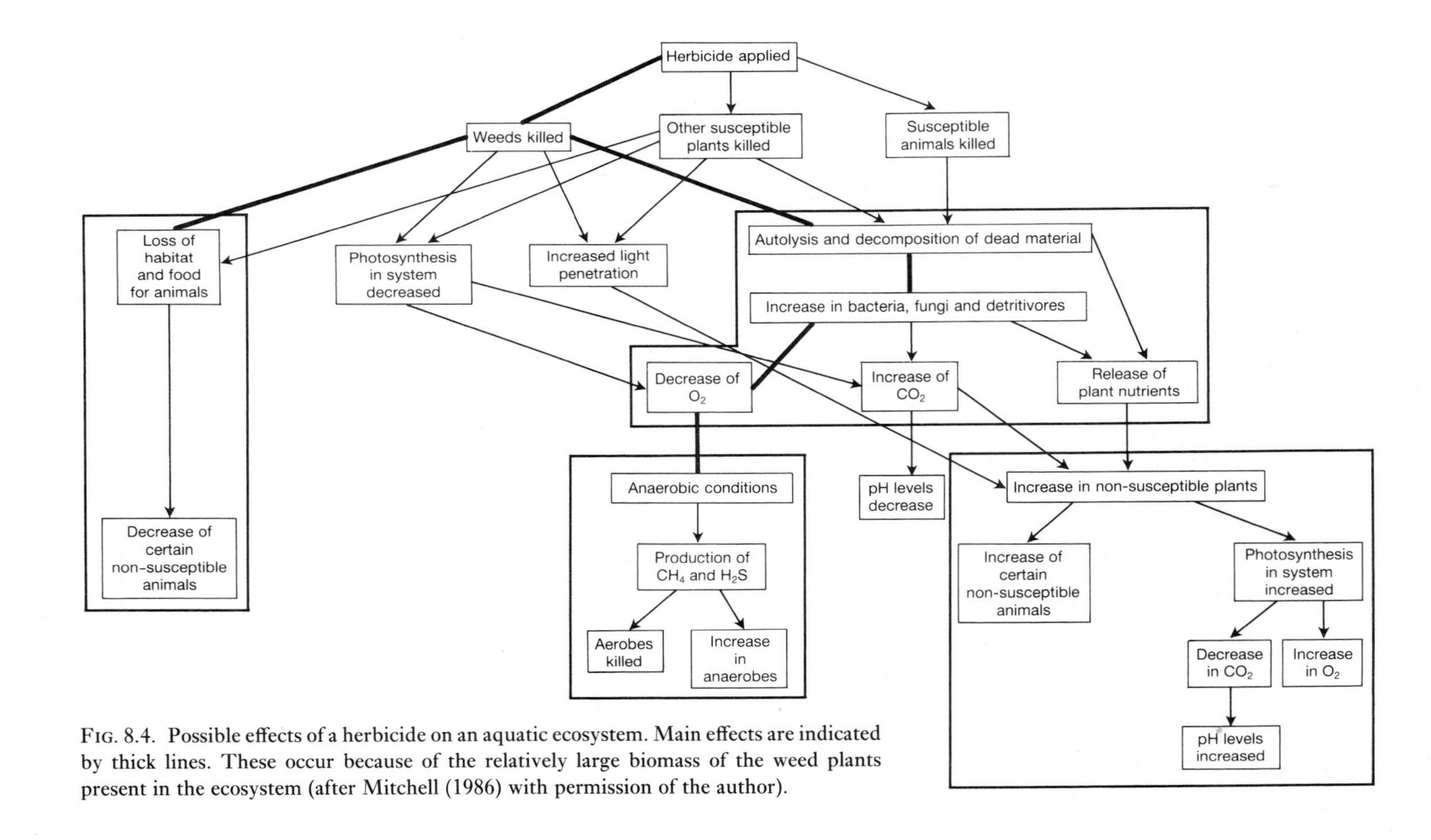

FIG. 8.4. Possible effects of a herbicide on an aquatic ecosystem. Main effects are indicated by thick lines. These occur because of the relatively large biomass of the weed plants present in the ecosystem (after Mitchell (1986) with permission of the author).

Direct toxicity

Most herbicides have more than one potential mode of action, producing differing effects on target macrophytes at differing availances (Hartley and Graham-Bryce 1980) of the herbicide in the freshwater system.

There is always a risk of lethal or sub-lethal injury to non-target organisms in applying a toxic chemical to an aquatic ecosystem (Walker 1960; Holden 1972; Rao *et al.* 1983; Dial and Bauer 1984). The risk may be minimized by enforcing a programme of chemical hazard assessment prior to the commercial release of new aquatic herbicides (AIBS 1978; Goulding 1981; Voyer and Heltshe 1984). Risks can be further reduced by the subsequent enforcement of regulations for aquatic herbicide usage, designed to minimize ecosystem impact of the toxin (Cummings 1977; Ministry of Agriculture, Fisheries and Food 1979, 1986).

The application of one such risk-reduction scheme, used in the UK (Thomas 1981), to aquatic herbicides, was described by Tooby (1978, 1981). The hazard evaluation process involved five test stages of increasing complexity:

(1) preliminary screening;

(2) more detailed acute toxicity testing;

(3) chronic toxicity testing;

(4) bioaccumulation studies;

(5) environmental evaluation.

Chemical, biological, and pattern-of-use data were all required within this overall framework. Tooby (1981) stated that the objectives of all such schemes should be (i) to determine the maximum no-adverse-effect concentration ($NAEC_{max}$), over a defined period of exposure to a given toxin, for a range of representative test organisms and (ii) to compare derived $NAEC_{max}$ values with the predicted likely environmental concentration of toxin. He further contended that $NAEC_{max}$ (for a 'prolonged' exposure period) could be taken as equivalent to 1 per cent of the LC_{50} (the concentration of the toxin which kills 50 per cent of test organisms during a 3 d standard acute toxicity test period).

Since the maximum likely environmental concentration (MLEC) of a herbicide in an aquatic system is usually predictable, then an approximate margin of safety of the herbicide to test organisms could be calculated as the ratio $NAEC_{max}$: MLEC, or, from acute toxicity data, as [0.01 (LC_{50} 96 h)] : MLEC.

Values, for this ratio, of less than 1 would imply a need for more detailed assessment of the potential direct toxicity of the candidate aquatic herbicide to non-target organisms.

An example of the application of this approach is given below. Dichlobenil (MLEC 1.0 mg l^{-1}) has a low margin of safety on this basis (given by Tooby (1981) as 0.06–0.08 for fish, and 0.08 for Crustacea). Fish stopped feeding and lost condition in low concentrations of the herbicide. The chronic toxicity (threshold median lethal concentration) was 1.6 mg a.i. l^{-1} for rainbow trout (*Salmo gairdneri*), very close to the MLEC for dichlobenil. However, the herbicide was rapidly eliminated from fish tissues once ambient concentrations of dichlobenil declined.

Gill-damage to *S. gairdneri* exposed to dichlobenil at 0.31–1.28 mg a.i. l^{-1} for 28 d (giving a tissue concentration of 6.6 mg a.i. kg^{-1}), was reported by Wiersma—Roem *et al.* (1978). Hypertrophy and hyperplasia of gill epithelium cells had potentially serious implications for fish-survival in water of reduced oxygen-content. Tooby and Spencer-Jones (1978) observed, in a lake trial, that roach (*Rutilus rutilus*) bioaccumulated dichlobenil residues at 0.8–6.4 mg a.i. l^{-1}, depending on tissue type. These concentrations were 16 to 33 per cent of those known to kill the fish under test conditions. The pattern of accumulation and elimination closely followed the residue concentrations recorded in the lake water. A major fish mortality occurred some 45 days after treatment in the lake, when daytime dissolved oxygen levels fell to <5 per cent saturation. This followed the crash-decline of a large phytoplankton bloom 3 days previously (open-water chlorophyll concentration *c.* 450 $\mu g\ l^{-1}$) during warm weather, with the water temperature at 20 °C. The authors doubted if the concentrations of dichlobenil found in dead fish tissues would have killed the fish. However, dichlobenil-induced gill damage may have contributed to the poor survival of fish in the lake under low-oxygen conditions.

The direct toxicity effects of aquatic herbicides potentially may have an impact on any non-target biotic component of a treated aquatic ecosystem. Regrettably, much direct toxicity information for aquatic herbicides remains unpublished, being held in the confidential databases of chemical firms and government agencies. This does not increase public confidence in the safety of aquatic herbicides.

Despite the inherent limitations of short-term toxicity assay procedures (Bowmer 1986; Lloyd 1977), studies of herbicide toxicity are usually based on controlled laboratory testing, using the simple LC_{50} test. It is difficult to assess the direct toxicity of aquatic herbicides to non-target organisms under field conditions, mainly because of the difficulties of separating direct effects from indirect ecosystem effects related to the kill of target macrophytes by the herbicide (Brooker and Edwards 1973*a*, *b*, *c*; van Dord, Hoogers, and van Zon 1974; Couch and Nelson 1982).

One attempt to overcome such assessment problems uses controlled-environment micro-ecosystems for herbicide toxicity testing (Wingfield and Johnson 1981; Paterson and Wright 1983; Kersting 1984; Cragg and Fry 1984). These usually incorporate macrophytes and/or algae, micro-organisms, and invertebrates. Large aquaria, or other artificial experimental systems (e.g. ponds, flumes) may be utilized for comparable studies involving fish, alone or with other biota (Yeo and Dechoretz 1976, 1977; Rhodes 1980; Bohl and Wagner 1981).

Direct toxic effects on non-target plants

Herbicide treatments may damage non-target aquatic plant populations in two ways. Firstly, residues may affect plants outwith the target area, by spray drift or residue movement in water currents or sediments (Savage, Truelove, and Wiese 1978; Cooke and Newbold 1986). There has been recent concern over the long-term implications for non-target plant communities of the accumulation of inactivated residues of the bipyridinium herbicides, diquat and paraquat, in the hydrosoils of freshwater systems regularly treated with these herbicides (Birmingham and Colman 1983; Corwin and Farmer 1984). The potential phytotoxicity produced by any re-release in active form of such residues accumulated in sediments is unknown. Secondly, non-target plant species present in the target area may be affected by herbicide treatment. The toxic effects of certain aquatic herbicides on non-target phytoplankton, or other algal populations, are fairly well documented (Robson, Fowler, and Barrett 1976; Setiadarma *et al.* 1976; Hawxby *et al.* 1977). A 45 per cent decrease in phytoplankton productivity, following treatment of a freshwater system with diquat at 1.0 mg a.i. l^{-1}, was recorded by Butler (1963). The algicidal properties of the triazines (Heuss 1972; Veber *et al.* 1981) and ureas (le Cosquino de Bussy 1971) are well known. Other herbicides may affect non-target phytoplankton and epiphytic algal communities to varying degrees (e.g. 2,4-D: Singh 1974; Couch and Nelson 1982; acrolein: Fritz-Sheridan 1982; chlorpropham: Paterson and Wright 1983). Glyphosate, however, had virtually no effect on the diatom populations of forest streams and pools (Sullivan, Sullivan, and Bisalputra 1981).

Susceptibility to herbicides varies widely between algal species and so treatment often results in selection of tolerant species. Algae have a rapid population growth potential. Thus a common result of aquatic herbicide treatment is a rapid growth of algae naturally resistant to the herbicide, to fill the niches vacated by the susceptible target plant species (van der Weij, Hoogers, and Blok 1971; Cassie 1976; Hodgson and Linda 1984). Even where potent algicides such as terbutryne are used, algal blooms may develop rapidly once ambient herbicide concentrations have fallen to sub-phytotoxic levels. This is because algae typically recover more quickly, due to their rapid population growth potential, than do the target macrophyte species (Robson, Fowler, and Hanley 1978).

There are recent reports of the development of herbicide-resistance to diuron within populations of algae (e.g. *Chlamydomonas reinhardii*), normally susceptible to the herbicide, in response to repeated experimental exposure to diuron (McBride and Togasake 1977).

Direct toxic effects on micro-organisms

Paralleling the differential effects of herbicides on microalgae, Yamanaka (1983) showed that certain nitrifying bacteria found in freshwater/hydrosoil systems (e.g.

Nitrobacter agilis) were inhibited by paraquat, whilst others (e.g. *Nitrosomonas europea*) were not. Ramsay and Fry (1976) found after paraquat treatment of *Elodea canadensis* and *Chara vulgaris* in a drainage channel, that populations of paraquat-resistant epiphytic bacteria reached a maximum when herbicide concentrations were highest. Herbicides may also directly affect microbial populations (Cragg and Fry 1984; Chan and Leung 1986). The main importance of the microbial components of aquatic ecosystems which have been treated by herbicides is, however, their role in macrophyte decomposition and associated ecological effects. This is discussed later.

Direct toxic effects on invertebrates

The bulk of the published data on herbicide toxicity to aquatic biota relates to effects on invertebrates and fish. Acute toxicity data are available for a wide range of invertebrate species (Table 8.1). These data have to be considered in the context of MLEC values for the herbicides involved. Terbutryne, for example, with an MLEC of *c.* 0.1 mg a.i. 1^{-1}, has a relatively large margin of safety for invertebrate species (Tooby 1981), despite its high acute toxicity. On the other hand, copper sulphate, with MLEC values around 1.0 mg a.i. 1^{-1}, has an unacceptable margin of safety on most criteria. Hübschman (1967), for example, described the acute toxicity of copper to the freshwater crayfish, *Oronectes rusticus*. The bipyridinium herbicides, diquat and paraquat, are quite toxic to Cladocera, but less so to Copepoda and other macro-invertebrate species (Hilsenhoff 1966; Naqvi, Leung, and Naqvi 1980; Williams, Mather, and Carter 1984). Field conditions commonly reduce the direct toxicity of herbicides to invertebrate fauna (and other non-target biota), for example, by reducing the availance of residues through uptake by target plants, losses to sediment and other sinks, or downstream dilution (Bowmer and Adeney 1978; Weber 1972; Hartman and Martin 1984).

Chronic toxicity effects may occur in non-target aquatic organisms as a result of prolonged exposure to low concentrations of herbicides (Macek *et al.* 1976). Thus, Tatum and Blackburn (1962) considered that, in subtropical ponds, chironomid larvae showed chronic toxicity symptoms as a result of exposure to diquat at 0.5 mg a.i. 1^{-1}. Similarly, Streit and Peter (1978) reported that exposure of mollusc and annelid species (*Ancylus fluviatilis*, *Glossiphonia complanata*, *Helobdella stagnalis*) to atrazine at only 2 per cent of the LC_{50} (96 h) values, for one month, produced a range of significant chronic effects. Survival, food-ingestion, growth, egg-production, gross productivity, and production efficiency (production : ingestion ratio) were all affected, more so in the leeches than in the snail. Kersting (1975) noted that the production efficiency of *Daphnia magna* was reduced by some 66 per cent after 21 days exposure to diuron at 4 mg a.i. 1^{-1}, a rather high concentration, which probably produced direct, as well as chronic, toxicity effects.

Table 8.1. Acute toxicity of aquatic herbicides to freshwater invertebrates

Herbicide (active ingredient)	Organism	LC_{50} (mg l^{-1})	Exposure period (h)	Reference
Copper sulphate	*Daphnia magna*	0.096	(96)	15
Diquat	*Hyalella azeteca*	0.12	(48)	2
Paraquat	*Daphnia pulex*	0.45	(48)	8
2,4,5-T	*Pteronarcys californicus* (nymph)	0.76	(48)	11
Diuron	*Daphnia pulex*	1.4	(48)	11
Terbutryne	*Daphnia magna*	1.4	(48)	1
Terbutryne	*Chironomus thummi*	1.4	(48)	4
Dichlobenil	*Gammarus lacustris*	1.5	(48)	6
Endothall (acid)	*Gammarus lacustris*	2.0	(24)	6
Dichlobenil	*Asellus aquaticus*	2.9	(96)	7
Fluridone	*Daphnia* sp.	6.3	(n.d.)	12
Sodium arsenite	*Daphnia magna*	6.5	(24)	14
Fenac (Na)	*Simocephalus serrulatus*	6.6	(48)	11
Dichlobenil	*Daphnia magna*	7.8	(48)	1
Terbutryne	*Tubifex tubifex*	10.0	(48)	4
Cyanatryn	*Daphnia longispina*	15.4	(96)	3
Dalapon	*Simocephalus serrulatus*	16.0	(48)	8
Diquat	*Callibaetes* sp. (nymph)	16.4	(96)	2
Fenac	*Daphnia magna*	28.5	(96)[a]	16
Cyanatryn	*Asellus aquaticus*	>30.0	(48)	3
Diquat	*Limnephilus* sp. (larva)	33.0	(96)	2
Glyphosate	*Daphnia pulex*	36.0	(48)	1
Paraquat	*Procambarus clarkii*	39.0	(48)	9
Endothall (acid)	*Daphnia magna*	46.0	(24)	13
Glyphosate	*Gammarus pseudolimnaeus*	62.0	(48)	17
Fenac (Na)	*Pteronarcys californicus*	80.0	(48)	11
Diquat	*Libellula* (nymph)	>100.0	(96)	2
2,4-D (Na)	*Daphnia magna*	932.1	(96)	10
Asulam	*Gammarus* sp.	1500.0	(96)	1

[a] Recorded as EC_{50}.
n.d. = No data.
References: 1 = Tooby 1981; 2 = Wilson and Bond 1969; 3 = British Agrochemicals Association 1976; 4 = Tyson 1974; 5 = Haddow, Stovell, and Payne 1974; 6 = Sanders 1969; 7 = Tooby 1976; 8 = Sanders and Cope 1966; 9 = Leung, Naqvi, and Naqvi 1980; 10 = Presing 1981; 11 = Walker 1971; 12 = Parka *et al.* 1978; 13 = Crosby and Tucker 1966; 14 = Sanders and Cope 1968; 15 = McKee and Wolf 1963; 16 = Sikka *et al.* 1982; 17 = Folmar, Sanders, and Julin 1979.

Sublethal effects of aquatic herbicides on invertebrates also include behavioural changes (Folmar 1978). An example is the avoidance response and reduced swimming activity seen in *Daphnia pulex* exposed to 0.2 mg a.i. 1^{-1} cyanatryn for varying periods (Scorgie and Cooke 1979). The snail *Ancylus fluviatilis* tended to lie upside down in 16 mg a.i. 1^{-1} atrazine and was unable to attach to vertical glass aquarium walls (Streit and Peter 1978). They suggested that such sublethal effects would markedly increase the mortality rate of snail populations subject to atrazine in their natural environment, by enhancing their susceptibility to predation. A rather different effect was noted by Tooby and Macey (1977), who found that dichlobenil residues accumulated rapidly in nymphs of the corixid bugs *Sigara dorsalis* and *Corixa punctata*, in dichlobenil-treated ponds. Pigmentation was inhibited at ecdysis, producing colourless nymphs and adults. These individuals were probably more conspicuous, and more heavily predated, than the normal pigmented bugs.

Different stages of the invertebrate life-cycle may be differentially susceptible to a herbicide. Juvenile crayfish (*Procambarus clarkii*), for example, had an LC_{50} (48 h) of only 5.2 mg a.i. l^{-1} paraquat compared with 39.0 mg a.i. l^{-1} for adults, according to Leung, Naqvi, and Naqvi (1980).

Presing (1981) showed that prolonged exposure to 2,4-D Na influenced the reproduction of *Daphnia magna*, at ambient concentrations of <10 per cent of LC_{50} values. Over four generations, kept in water containing 25 and 50 mg a.i. l^{-1} 2,4-D Na, significant reductions in the number of viable eggs per female occurred. These concentrations are, however, much greater than the MLEC for 2,4-D, which is set (in the UK) at 5 mg l^{-1} (Ministry of Agriculture, Fisheries and Food 1979). Under field conditions, effects on life cycles may be further lessened below those seen in the laboratory. Brooker (1979) observed no significant effects of treatment with a dalapon + 2,4-D mixture (which gave maximum water concentrations of 0.029 + 0.117 mg a.i. l^{-1} respectively), on the life-cycle or production of drainage ditch populations of *Sialis lutaria* over two years. Hildebrand, Sullivan, and Sullivan (1980), likewise, noted no significant reduction of *Daphnia magna* populations, even at 100 times the normal field dose of glyphosate, in an experimental forest pond.

Because invertebrates are important intermediate-level food organisms, herbicide uptake and concentration by them is of great importance in relation to herbicide residue bioaccumulation in food chains (Tooby 1981). Streit (1979) found that, in clear water, *Ancylus fluviatilis* eliminates from its tissues 50 per cent of paraquat taken up every 14–24 h. The snail did not accumulate 2,4-D (acid), concentrations in the animal as a whole being lower than the external concentration. In general, herbicides showing a potential for significant aquatic food-chain bioconcentration would be unlikely to get through normal hazard evaluation procedures (Tooby 1981).

Direct toxicity effects on fish

The importance of fish as a human food and recreational resource places particular emphasis on fish toxicity investigations in assessing the ecological impacts of aquatic herbicides. Despite their inherent limitations (Tooby 1976; Lloyd 1977; Brown 1968), short-term acute toxicity tests are the main source of fish toxicity data (Table 8.2). Herbicides such as acrolein, a general cell toxicant which is highly toxic to most living organisms, may have severe effects on fish populations (Burdich, Dean, and Harris 1964; Mullison 1970; Bartley and Hattrup 1975).

Table 8.2. Acute toxicity of aquatic herbicides to freshwater fish

Herbicide (active ingredient)	Organism	LC_{50} (mg l^{-1})	Exposure period (h)	Reference
Acrolein	*Lepomis macrochirus*	0.08	(24)	14
Copper sulphate	*Rasbora heteromorpha*	<1.0	(48)	1
2,4 -D BEE	*Lepomis macrochirus*	1.3	(24)	10,11
Glyphosate	*Pimpehales promelas*	1.3	(96)	15
Terbutryne	*Lepomis macrochirus*	4.5	(96)	5
Fenac (Na)	*Rutilus rutilus*	6.9	(48)	4
Fenac (Na)	*Salmo gairdneri*	7.5	(24)	11
Dichlobenil	*Salmo gairdneri*	8.1	(48)	4
Glyphosate	*Salmo gairdneri*	8.3	(96)	15
Dichlobenil	*Anguilla japonica*	10.0	(48)	3
Fluridone	*Salmo gairdneri*	11.7	(n.d.)	13
Dichlobenil	*Rasbora heteromorpha*	12.0	(48)	1
Silvex (acid)	*Lepomis macrochirus*	14.5	(24)	10,11
Diquat	*Esox lucius*	16.0	(96)	6
Cyanatryn	*Scardinius erythrophthalmus*	>25.0	(n.d.)	2
Paraquat	*Salmo gairdneri*	32.0	(96)	2
Diquat	*Lepomis macrochirus*	35.0	(96)	6
Paraquat	*Rasbora heteromorpha*	35.0	(48)	1
Sodium arsenite	*Lepomis macrochirus*	44.0	(24)	10,11
Glyphosate	*Salmo gairdneri*	48.0	(96)	3
Diquat	*Poecilia reticulata*	50.0	(48)	7
Diquat	*Salmo gairdneri*	70.0	(48)	8
Copper (ethanolamine)	*Gambusia affinis*	170.0	(96)	12
Diuron	*Rasbora heteromorpha*	190.0	(48)	9
Endothall (acid)	*Lepomis macrochirus*	428.0	(24)	10,11
2,4-D amine	*Rasbora heteromorpha*	1425.0	(48)	1
Dalapon	*Rasbora heteromorpha*	3800.0	(48)	1
Asulam	*Rasbora heteromorpha*	>5000.0	(48)	1

n.d. = No data.

References: 1 = Tooby 1976; 2 = British Agrochemicals Association 1976; 3 = Ministry of Agriculture Fisheries and Food, U.K. 1979; 4 = Tooby 1978; 5 = Tyson 1974; 6 = Gilderhus 1967; 7 = van Dord *et al.* 1974; 8 = Tooby 1971; 9 = Tooby, Hursey, and Alabaster 1975; 10 = Hughes and Davis 1962; 11 = Pimentel 1971; 12 = Leung, Naqvi, and Leblanc 1983; 13 = Parka *et al.* 1978; 14 = Mullison 1970; 15 = Folmar, Sanders, and Julin 1979.

Other known fish toxins, widely used as aquatic herbicides, are copper sulphate and dichlobenil (Bolier, van der Maas, and Bootsma 1973; Tooby, Durbin, and Rycroft 1974; Verloop, Nimmo, and de Wilde 1974; Morrice 1977; Bohl, Wagner, and Hoffman 1982).

Water physico-chemistry and herbicide formulation are both recognized as modifying influences on the relative toxicity of herbicides to fish (Hughes and Davies 1982; Brown 1978; Folmar, Sanders, and Julin 1979). Yeo (1964) discussed, in general terms, the implications of water quality interactions with paraquat and endothall toxicity to fish. Hughes and Davis (1962) showed that in bluegill (*Lepomis macrochirus*), the LC_{50} (24 h) values for 2,4-D are highly dependent upon formulation. The dimethylamine salt (2,4-D DMA) with an LC_{50}

(24 h) of 390 mg l^{-1} was much less toxic than the butoxyethanol ester (2,4-D BEE), the LC_{50} (24 h) of which was 2.1 mg l^{-1}.

Combinations of herbicides (Yeo and Dechoretz 1977; Berry 1984), or the addition of adjuvants such as wetting agents, emulsifiers, or anti-foaming agents (Watkins, Thayer, and Haller 1985) may also influence herbicide toxicity to fish or other non-target organisms, but very little published information is available on this aspect.

Long-term exposure to 2,4-D may have chronic effects on fish, as shown by Cope, Wood, and Wallen (1970), who observed lesions on brain, liver, and vascular tissues of *Lepomis macrochirus* exposed to 5 mg a.i. l^{-1} 2,4-D PGBE. In juvenile salmon (*Onchorhyncus nerka*), relatively low concentrations of 2,4-D BEE caused a range of different stress responses (McBride, Dye, and Donaldson 1981). Diuron caused proliferative endocarditis in the heart of carp (Schultz 1972). Numerous other chronic and sublethal toxicity effects of aquatic herbicides on fish are known (Gilderhus 1966; McCraren, Cope, and Eller 1969; Mullison 1970; van Dord, Hoogers, and van Zon 1974; Macek *et al.* 1976; Schulz 1972, 1981; USDI Fish and Wildlife Service 1982).

The effects of aquatic herbicides on fish reproduction tend to be proportional to the overall toxicity of the compound (McCorkle, Chambers, and Yarborough 1977). Over two years following treatment of a pond with 5.0 mg a.i. l^{-1} dipotassium-endothall, Serns (1977) observed no effects on fish survival, or on the number of young-of-the-year of *Lepomis macrochirus* produced by adults present during treatment, or by their first generation offspring. In contrast, Mount and Stephan (1967) found that the maximum prolonged exposure (over 10 months) of 2,4-D BEE which had no effect on fish reproduction was only 0.3 mg a.i. l^{-1}, that is about 5 per cent of the LC_{50} (96 h). Hiltibran (1967) noted that silvex was much more toxic to hatched fry than to fish embryos, which are protected from the herbicide by the egg membrane.

The potential interaction of herbicide-induced sublethal damage to fish gills, and low dissolved oxygen availability, in killing fish has already been mentioned for dichlobenil. McCraren, Cope, and Eller (1969) reported a similar effect of diuron on *Lepomis macrochirus*. Endothall has been shown, by Takle, Beitinger, and Dickson (1983) to influence the maximum thermal tolerance of red shiner, *Notripos lutrensis*. A 7 to 7.5-fold increase in the toxicity of 2,4-D to carp (*Cyprinus carpio*) occurred as water temperatures increased from 17 to 39 °C, winter to summer, in Indian fish ponds (Vardia and Durve 1981).

Dichlobenil at sublethal concentrations can cause behavioural effects in fish, including lethargy and cessation of feeding (Niehuss and Borner 1971; Tooby, Durbin, and Rycroft 1974; van Dord, Hoogers, and van Zon 1974; Bolier, van der Maas, and Bootsma 1973). Beitinger and Freeman (1983) summarized the behavioural avoidance reactions of fish to aquatic herbicides. As with lethal and sublethal toxicity effects, they concluded that behavioural response thresholds in fish are probably strongly influenced by water quality variables such as pH, temperature, and hardness. Folmar (1976) reported that some herbicides, such as copper sulphate are consistently avoided by fish, at concentrations as low as 0.0001 mg l^{-1}. Acrolein produced a violent avoidance reaction in rainbow trout, *Salmo gairdneri* at 0.1 mg a.i. l^{-1}, a concentration likely to cause direct toxic effects (cf. Table 8.2). On the other hand, rainbow trout did not avoid glyphosate or diquat at concentrations below 10.0 mg a.i. l^{-1}.

Although some herbicides, such as diuron (Koeman, Horsmans, and van der Maas 1969; van der Weij, Hoogers, and Blok 1971; Holden 1972) may accumulate in fish tissues, the uptake, accumulation, and elimination of residues usually closely parallel ambient herbicide concentrations in the surrounding water (Rogers 1970; Schultz and Harman 1974; Sikka, Ford, and Lynch 1975; Tooby and Spencer-Jones 1978; Stalling and Huckins 1978; Rhodes 1980; Ministry of Environment, British Columbia 1980; Hoeppel and Westerdahl 1983). The example of dichlobenil has already been given. Bohl, Wagner, and Hoffman (1982) found that long-term periodic applications of copper sulphate at 1 mg l^{-1} to a fish pond, giving ambient Cu^{2+} concentrations of 0.12–0.16 mg l^{-1}, resulted in a relatively low final concentration in trout muscle (0.82 mg kg^{-1} fresh weight). Terbutryne residues accumulated in fish at up to five times external concentrations, but were eliminated as ambient concentrations declined (Muir 1980; Murphy 1982). In carp muscle terbutryne residues of only 0.01 mg kg^{-1} remained 70 days after treatment of fish pond waters at the normal rate of 0.05–0.1 mg a.i. l^{-1} (Maier-Bode 1972). Less than 1 per cent of endothall applied to water at 2 mg a.i. l^{-1} was absorbed by bluegills; reaching a plateau value of 0.1–0.2 mg kg^{-1} 12 to 96 h after treatment (Sikka, Ford, and Lynch 1975).

Direct toxic effects on higher animals

Relatively few of the higher animals, in freshwaters which are likely to be treated with aquatic herbicides, are obligate aquatic organisms. In consequence, higher animals tend to be less susceptible than fish or aquatic invertebrates to toxicity problems arising from aquatic herbicide use. Fewer data are, therefore, available than for fish and invertebrates (Robinson and Morley 1980; South and Robinson 1982; Vardia, Rao, and Durve 1984; Dial and Bauer 1984). Sanders (1970) found herbicide toxicities, as LC_{50} (96 h), to amphibians ranging from 1.2 mg a.i. l^{-1} (endothall: *Bufo woodhousi fowleri*, Fowler's toad) to 100 mg a.i. l^{-1} (2,4-D amine: *Pseudacris triseriata*, western chorus frog). Malformations of tadpoles of *Rana temporaria* (common frog) exposed to high concentrations (500 mg a.i. l^{-1}) of 2,4-D acid were noted by Lhoste and Roth (1946). More recently, Scorgie and Cooke (1979) found an LC_{50} (96 h) value of 30 mg a.i. l^{-1} for *R. temporaria* tadpoles exposed to cyanatryn. Prolonged exposure to cyanatryn at 0.02 mg a.i. l^{-1} (close to the MLEC value) caused behavioural changes in the tadpoles, reducing feeding activity, and hence reduced the rate of weight gain by about 50 per cent compared to untreated controls.

Keckemet (1974) found no effect on mallard ducks given food containing 5000 mg kg^{-1} (a.e.) endothall for 114 d. The acute oral LD_{50} of glyphosate to mallard ducks is 4640 mg a.i. kg^{-1} and to rats 4900 mg a.i. kg^{-1} (British Agrochemicals Association 1976). Aquatic herbicides known to be highly toxic to, or having undesirable effects on, the health or reproduction of birds and mammals, include sodium arsenite (oral LD_{50} 10–50 mg kg^{-1} in rat: Ware 1983), and acrolein (oral LD_{50} 46 mg kg^{-1} in rat: USDA 1963).

Public concern has grown in recent years over the potential teratogenic and chronic toxicity effects of the hormone herbicides, including 2,4-D, 2,4,5-T, and silvex (Gangstad 1983; Turner 1977), particularly in connection with the risks of their associated impurity TCDD (2,3,7,8-tetrachlorodibenzo-*p*-dioxin). This impurity is highly toxic, a potent teratogen, extremely persistent, and liable to accumulate in food chains (Mullison 1970; Brown 1978; Milnes 1971). TCDD adversely influenced reproductive success in *Physa* snails and *Paranais* oligochaetes at 0.0002 mg l^{-1}. Salmon exposed to TCDD at 0.0001 mg l^{-1} for 48 h accumulated enough to kill them within 10–80 d of exposure (Miller, Norris, and Hawkes 1973). Gangstad (1981, 1982*a*, 1983, 1984) discussed in detail the benefits, costs, and risks involved in the proposed banning of silvex in the USA. Dearden (1982) examined the comparative public risk perception of a *Myriophyllum spicatum* weed problem in British Columbia, and of a 2,4-D regime used for its control. He concluded that the perceived hazard of the latter was much greater than that of the former, in the minds of the Canadian public. This led, ultimately, to rejection of the 2,4-D weed control programme.

Indirect effects on aquatic biota

The short-term, indirect, ecological effects of aquatic weed control by herbicides in freshwaters are related to the death and *in situ* decay of the target plants. The important effects include:

(1) alterations in dissolved gas and pH regimes from decaying vegetation;

(2) increases in detritus;

(3) toxin and nutrient release from decaying vegetation;

(4) habitat alteration for non-target organisms; and

(5) the development of replacement growths of opportunist plant species (algae or rapid-growing macrophytes) to fill the niches vacated by the loss of the target macrophytes (Newbold 1975; Brooker and Edwards 1975).

An important effect of aquatic herbicide treatment, of submerged macrophytes in particular, is to cause a reduction in community photosynthesis, with concomitant effects on the balance of oxygen and dissolved CO_2/HCO_3^- present in the system (Edwards and Owens 1962; Edwards, Owens, and Gibbs 1961; Brooker and Edwards 1975; Marshall 1981). Photosynthesis-inhibiting herbicides, such as terbutryne, simazine, and diuron, prevent normal diel photosynthetic oxygen release and carbon fixation by the target macrophytes, but initially have little effect on their respiratory oxygen demand. This can result in rapid deoxygenation of the water (van der Weij, Hoogers, and Blok 1971; Robson, Fowler, and Hanley 1978; Murphy, Hanbury, and Eaton 1981; Wingfield and Bebb 1982; Tucker, Busch, and Lloyd 1983). In extreme cases this can cause the death, by anoxia, of aquatic fauna (Way *et al.* 1971; Price 1976). The use of other herbicides, such as paraquat (Newman 1967; Brooker and Edwards 1973*b*), diquat + endothall (Strange 1976; Strange and Schreck 1976), dipotassium endothall (Serns 1975), 2,4-D (Ahmed, Ito, and Kunikazu 1980), dichlobenil (Walsh, Miller, and Heitmuller 1971) and copper sulphate (Price 1976) may also result in net community oxygen consumption exceeding oxygen supply in the treated system.

The oxygen demand of aerobic decay processes may exacerbate the deoxygenation problem, especially in warm water, if a large mass of nitrogen-rich dead plant material is available to optimize microbial decomposer growth and oxygen consumption (Jewell 1971; Fry, Brooker, and Thomas 1973; Fry and Humphrey 1978; Almazon and Boyd 1978; Cragg 1980). Brooker (1974), and Carpenter and Greenlee (1981) considered that the main factors involved in deoxygenation, following aquatic herbicide treatments, are water temperature, turnover rate of the water column, relative depth of littoral and pelagic zones (in lakes), macrophyte biomass at death, macrophyte shoot nitrogen content, and rates of external oxygen inputs to the treated system.

The net increase in community respiratory CO_2 production, which typically follows herbicide treatments in fresh waters, tends to shift the inorganic carbon equilibrium of the system (Newbold 1975). In poorly-buffered waters this may result in a drop in daytime pH (Tucker, Busch, and Lloyd 1983). Brooker and Edwards (1973*a*) noted a daytime drop from pH 9.3 to pH 8.1, following paraquat kill of a dense *Elodea canadensis* growth in a small lake. Newbold (1974, 1975) observed, in dichlobenil-treated ponds, an increase in free dissolved CO_2 from 5.8 to 21.7 mg l^{-1}, with only a slight decline in pH (7.8–7.6), attributing this to the buffering effects of suspended clay particles.

Short-term recovery from deoxygenation following herbicide treatment is usually closely associated with the development of phytoplankton blooms, or other replacement growth of plants (Strange and Schreck 1976; Wingfield and Johnson 1981; Wingfield and Bebb 1982). Walsh, Miller, and Heitmuller (1971) reported that photosynthetic oxygen release by phytoplankton accounted for 94.5 per cent of total oxygen production (9.1 g O_2 m^{-2} d^{-1}) in the water of a dichlobenil-treated pond. If the replacement opportunist plant growth, following a herbicide-kill of submerged plants, is of surface-floating macrophytes such as *Lemna* spp., daytime dissolved oxygen levels may not recover to pre-treatment levels for prolonged periods (Fig. 8.5; Murphy, Hanbury, and Eaton 1981). The likely reason is that the surface mat of floating vegetation reduces atmospheric oxygen input to the water column (Duffield 1981), as well as shading out submerged macrophyte or phytoplankton growth, thereby minimizing photosynthetic oxygen production in the water beneath the mat (McLay 1976). Carter and Hestand (1977*a*) observed that unusually severe treatments with diquat, cutrine, or endothall, which killed the fish, plants, and invertebrates in small ponds, resulted in explosive growths of sulphur bacteria and phytoplankton, alternating in dominance, and associated with virtually complete deoxygenation of the pond water.

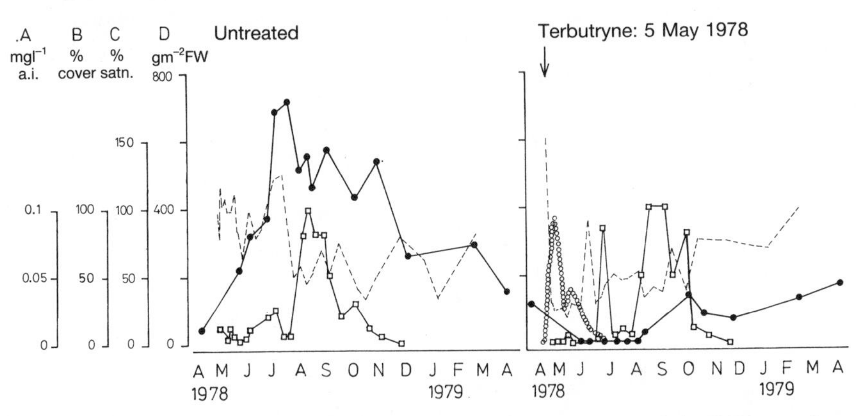

FIG. 8.5. Influences of terbutryne on the freshwater ecosystem of the Monmouth and Brecon Canal, Wales, UK. (A) Terbutryne concentration in water (application rate 0.1 mg a.i. l^{-1}(OOOOOOO); (B) % surface cover of floating plants *Lemna minor* agg. (□——□); (C) midday dissolved oxygen status as % saturation of water (– – – – –); (D) submerged fresh-weight standing crop of vegetation (●——●). Untreated site was located upstream of treated site (after Murphy, Hanbury, and Eaton (1981) with authors' permission).

Release of plant nutrients from decaying macrophytes to the water column tends to favour the development of plants such as phytoplankton or free-floating *Lemna* spp. (Simsiman, Chesters, and Daniel 1972; Daniel 1972; Nichols and Keeney 1973; Newbold 1974; James 1974; Strange 1976; Michaud *et al.* 1979; Hartman, Faina, and Machova 1984; Carter and Hestand 1977*a*; Ahmed, Ito, and Ueki 1981, 1982). Peverly and Johnson (1979) observed that soluble phosphorus concentrations rose dramatically (to *c.* 100 μg l^{-1}) within 24 h of diquat treatment of *Elodea* stands in small ponds, falling back to control levels (10–20 μg l^{-1}) within 3 weeks. The peak concentrations were followed by rapid formation of a phytoplankton bloom (maximum chlorophyll *a* concentration *c.* 200 μg l^{-1}). Ammonium-nitrogen and nitrite-nitrogen, released by mineralization of protein from decaying aquatic vegetation in low-oxygen conditions, may be toxic to fish (Pokorny *et al.* 1971; Hartman, Faina, and Machova 1984; Thurston *et al.* 1984).

Increases in total ammonium-nitrogen of up to 6.7-fold, compared to untreated controls, were noted in simazine-treated catfish (*Ictalurus punctatus*) ponds by Tucker, Busch, and Lloyd (1983). However, unionized ammonia concentrations remained at moderate levels, probably because decreasing pH compensated for the rise in total ammonium-nitrogen. Nitrite-nitrogen levels in the simazine-treated ponds rose to 19 times the values in control ponds within 10 d.a.t. This increase induced potentially dangerous blood methaemoglobin levels in the catfish, likely to lead to respiration problems if the dissolved oxygen status of the water was reduced.

Herbicide treatments of emergent macrophytes tend to produce fewer adverse effects on water quality than do herbicide treatments of submerged plants (Schumacker and Bontwell 1975; Sacher 1978; Murphy and Eaton 1981*a*), although nutrient regeneration from decaying plant tissues may still occur (Boyd 1976). Brooker (1976*a*) observed that control of a *Phragmites australis*-dominated ditch community with dalapon + 2,4-D stimulated rapid replacement growths of phytoplankton, epipelic algae, and submerged macrophytes. There was a concomitant increase in daytime oxygen status of the ditch water, but in this case no discernible alterations of nitrate-nitrogen, ammonium-nitrogen, orthophosphate, silicates, or pH.

In situ herbicide-induced kill and decay of macrophytes usually causes a temporary increase in particulate organic and other detritus in the aquatic system (Fish 1966; Newbold 1975; Strange 1976). Carter and Hestand (1977*a*), for example, noted a 22-fold turbidity increase in endothall-treated ponds. Such habitat alteration may alter the balance of filter-feeding and detritivore organisms in the ecosystem, at least in the short-term, following herbicide-treatment.

Habitat modification, resulting from herbicide-induced loss or reduction of the target aquatic macrophytes, may have a range of ecosystem effects (Katz 1966; Newman 1967; Marshall and Rutschky 1974; Brooker 1976*b*; Denoyelles and Kettle 1980; Scorgie 1980; Hanbury, Murphy, and Eaton 1981; Murphy and Pearce 1987). Loss or reduction of epiphytic algae and plant-associated invertebrates may occur, but there may be compensatory changes in the abundance and diversity of benthic and/or open-water species (Harp and Campbell 1964; van der Weij, Hoogers, and Blok 1971; Walsh, Miller, and Heitmuller 1971; Brooker and Edwards 1974; Hanbury, Murphy, and Eaton 1981; Gordon *et al.* 1982; Clare and Edwards 1983). If, however, there has been little water quality degradation associated with the loss of macrophytes, there may be little or no discernible change in zooplankton and benthic invertebrate community structure, diversity or abundance (Crossland and Elgar 1974; Serns 1975; Mycock 1979; Murphy and Eaton 1981*a*). Grazer zooplankton may benefit from an increased food-supply in phytoplankton bloom conditions, for example, following diquat-alginate treatments in Czechoslovakian fishponds (Hartman, Faina, and Machova 1984).

Changes in invertebrate community structure or abundance may have consequences for organisms higher up the food chain (Brooker and Edwards 1973*b*; Scott, Schultz, and Eschmeyer 1978; USDI Fish and Wildlife Service 1981). Brooker and Edwards (1973*c*) observed that eels in a paraquat-treated lake altered their diet to accommodate changes in the macro-invertebrate community structure, resulting from the destruction of the *Elodea*-habitat.

Hodgson and Linda (1984) reported a large initial increase in the epiphytic algal population on *Hydrilla verticillata*, shortly after treatment with diquat + chelated copper. This increase died away as the host plants deteriorated. The stimulation of epiphyton was attributed to a significant increase in available nitrogen, released from the decaying vegetation; there was no significant increase in total phosphorus following treatment.

Newbold (1976) succinctly summarized the short-term consequences of herbicide treatment for the aquatic plant community, once the constraints of phytotoxicity are removed by residue degradation or dissipation. Essentially, herbicide treatment causes a hydroseral regression to an earlier stage of the freshwater plant succession. Nudation, the initiation of the succession by herbicide disturbance, is followed by immigration of opportunist (disturbance-tolerant) plant species to fill the newly-vacated niche(s). Seral replacement of the opportunists by slower-growing, but more competititive, plant species subsequently occurs.

A classic series of successional plant community events occurred in experimental pools following severe diquat + chelated copper treatments (Carter and Hestand 1977*a*, *b*; Hestand and Carter 1977). Death of the original mixed community of *Myriophyllum spicatum*, *Ceratophyllum demersum*, *Chara* sp., *Vallisneria americana* and *Najas guadalupensis* was complete by 9 d.a.t. An initial bloom of sulphur bacteria (peaking at 1861×10^6 cells l^{-1}, 24 d.a.t.) was followed by a smaller, fluctuating phytoplankton bloom (236×10^6 cells l^{-1}), dominated by green algae (*Selenastrum*, *Gloeocystis*, *Cosmarium*) forming 75 per cent of the population, together with Cyanobacteria (*Anacystis*, *Oscillatoria*). Flagellates (mainly *Euglena*), comprised 10 per cent of the population; the diatoms *Opephora* and *Navicula* were also present. Macrophyte growth recommenced 38 d.a.t. with *Hydrilla verticillata*, followed by *Vallisneria* and *Chara* sp. (68 d.a.t.). *Vallisneria* rapidly became the predominant species, reaching 40 per cent coverage by 280 d.a.t. Phytoplankton populations had crashed by 84 d.a.t., when sulphur bacteria blooms were also declining, reaching $< 1 \times 10^6$ cells l^{-1} by 245 d.a.t.

Crawford (1981) showed that simazine treatment, in May 1979, which killed all the higher vegetation of a lake, was followed by a small regrowth of *Potamogeton foliosus* and *Chara vulgaris* within one month (Fig. 8.6). The initial pioneer regrowth was quickly replaced by a larger growth of *P. foliosus* and *Najas flexilis*.

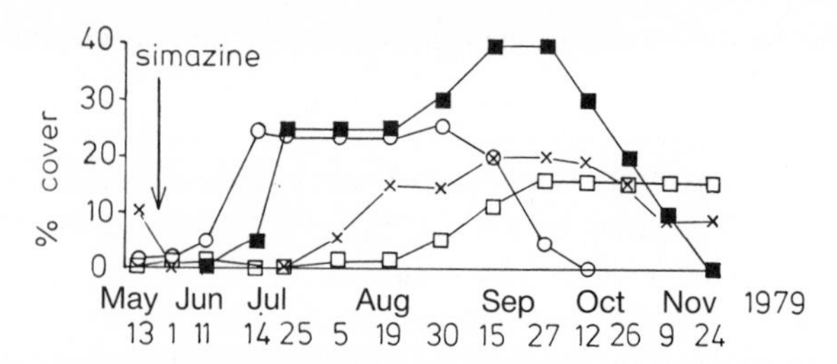

FIG. 8.6. Successional changes in dominant submerged macrophytes and algae (measured as % cover of upper littoral region) following simazine treatment of a farm pond in the USA in 1979. (■) *Najas flexilis*; (X) Zygnematales; (□) *Chara vulgaris*; (O) *Potamogeton foliosus* (adapted from Crawford (1981), with the author's permission).

Later in the season a resurgence of algal growth occurred. Charophyta are often noted as pioneer regrowth species in aquatic habitats disturbed by herbicide treatments (Brooker and Edwards 1973*a*; Getsinger, Davis, and Brinson 1982), as in other regularly-disturbed freshwater environments (Grillas and Duncan 1986).

In the longer term, it has been demonstrated by Wade (1981, 1982) that single herbicide treatments of aquatic systems have very little effect on macrophyte community structure. Over the years following treatment, hydroseral processes eventually lead to re-establishment of the original plant community. Wade suggested that propagule banks, laid down in the hydrosoil prior to treatment, play an important role in this community recovery process. Repeated treatments over several years may, however, maintain the ecosystem at a hydroserally-early stage, dominated by filamentous algae, or other opportunist plant species, season after season (Johannes 1974; Robson, Fowler, and Hanley 1978; Harbott and Rey 1981). Nevertheless, careful use of non-persistent herbicides, such as diquat, within a co-ordinated aquatic weed management programme over a period of years, can maintain a species-diverse, balanced plant community in the target system (Kuthe 1974; De Lange and van Zon 1978; Hiltibran 1984; Clayton 1986).

CONCLUSIONS

Water weed problems arise for a number of reasons but, on a world-wide basis, there are two primary causes, both of which can be blamed principally on human activity. The first is by the introduction of exotic weeds into areas where there are no natural control agents. Examples of this include *Eichhornia crassipes*, *Hydrilla verticillata*, *Salvinia* spp., *Myriophyllum spicatum*, and *Elodea canadensis*. The second is the stimulation of indigenous weed growth, for example, by nutrient enrichment, water impoundment, or channel alteration. Ideally, introduced weeds could be eradicated leaving the indigenous species *in situ*. Where indigenous weeds are causing the problem, total removal is seldom desirable as it would destroy the habitat for aquatic fauna and, subsequently, allow colonization by other, less desirable, species. The herbicides described in this chapter can, in most instances, be used to achieve both types of management. It should, however, be remembered that weeds can be re-introduced to a treated system, by natural or anthropogenic agencies and, if the conditions for growth are suitable, they will return in sufficient quantities to cause further problems. Thus, a single herbicide application seldom provides a permanent solution to a problem and it is likely that a regular management regime will be the usual practice.

It is clear that there are major differences in the properties and modes of action of the different herbicides used in water. They do not control the same weeds, they do not work in the same way, and they are affected differently by environmental conditions. Some are more toxic to aquatic fauna, or more persistent, than others. It is the responsibility of the user to select the herbicide most appropriate for the situation. Even when a weed species is known to be susceptible to a particular herbicide, it is not always appropriate to use that chemical in preference to an alternative approach, such as physical control. Local factors such as high water velocity or turbidity may reduce efficacy in some situations; demand for herbicide-free water for irrigation or potable supplies may prevent use in others. In Europe, the EEC has recommended an upper limit of 1 $\mu g\ l^{-1}$ for *any* pesticide in potable water supplies (EEC Directive 80/778). However, pesticides vary dramatically in their toxicity to man and domestic animals. For example, insecticides, as a class, tend to be far more toxic to mammals than herbicides. There is also increasing evidence to suggest that, at least in low-lying agricultural areas of Europe, background herbicide levels in groundwater commonly exceed the EEC upper limit as a result of leaching from arable and other land areas (e.g. railway embankments) sprayed with herbicides (Croll 1986; Hormann, Tournayre, and Egli 1979). The approach taken in the USA is more realistic. There the Environmental Protection Agency defines an upper permissible limit in potable water for each individual pesticide (e.g. 2,4-D: 87.5 $\mu g\ l^{-1}$).

In comparison with the quantities of herbicides applied on land most aquatic herbicide treatments are made on a very small scale in Europe. Even given the fact that the toxin is being applied directly to the water, it is probable that no detectable increase over background residue levels in potable water supplies could be attributed to aquatic herbicide usage in the typical lowland catchment (in highland catchments aquatic herbicides very rarely need to be used). However, there is a need for further work to resolve the issue of aquatic herbicide usage in the waters of catchments used for potable supply.

It is not surprising that concern has been expressed over the long-term effects of herbicides on the aquatic ecosystem. The majority of data on side-effects suggest that they are transient and sometimes occur only at herbicide concentrations well above those needed to control weeds. So far, there is little evidence of any build-up of herbicide residues, or of chronic toxicity, in natural aquatic systems despite the decades over which some of these chemicals have been used. In particular, fish populations do not appear to have been adversely affected, in the long-term, by aquatic herbicide treatments, except in areas where very toxic chemicals (e.g. acrolein) are still used.

Misuse of herbicides does have undesirable side-effects. Examples of these were given by Price (1976) who reported case histories of overdosing, incorrect timing, or use of formulations not officially cleared for use in water, which resulted in fish mortalities and other undesirable consequences. These consequences were the result of lack of knowledge and training on the part of the water manager or user. They highlight the need to ensure that all those who apply herbicides, or give instructions for their application, should be adequately trained and have a knowledge of the properties of the chemicals, their limitations, and the safety precautions which apply to their use.

When chemicals are developed or adapted for use in water, the regulatory authorities usually require data to show that there is an adequate safety margin between the concentration at which they kill weeds and the concentration likely to harm non-target organisms. There are some chemicals with an unacceptably low, or even no such safety margin (e.g. acrolein, copper sulphate). Nevertheless, there are situations in which these chemicals are used: for example, in freshwater systems which have been specifically constructed for such purposes as crop irrigation. Here, herbicide toxicity to fish and other aquatic life is considered unimportant; safety of the crops is of paramount importance (Anon. 1971).

The major current areas of research interest in aquatic herbicides may be summarized as follows:

(1) development of new uses for existing compounds, for example, by formulation improvements;

(2) improvement of application methodology (e.g. advances in controlled-release technology) to improve the precision of treatments, with the aims of maximizing impacts on target weeds, and minimizing effects on non-target organisms;

(3) development of more rigorous assessment procedures to determine the environmental safety of aquatic herbicides, and the application of improved safety standards to herbicide usage in freshwaters;

(4) finding and developing new selective compounds to control major nuisance species or groups: an outstanding need is for an effective, environmentally-safe selective algicide to control filamentous algae such as *Vaucheria* and *Cladophora*; and

(5) determining the compatibility of herbicides with other aquatic weed control methods, for use in integrated weed management programmes.

In general, the costs of development and registration are now so high that it is unlikely that any new herbicide (excluding rice herbicides), will be developed solely or primarily for use in freshwaters, in the foreseeable future. It is more likely that aquatic uses will continue to be found for compounds developed as herbicides for agricultural and horticultural use.

Aquatic herbicides offer a useful and proven control option within the broad spectrum of aquatic weed management techniques. They are not a panacea solution to all problems, nor are they to be rejected out of hand because of their potential ecological side-effects. As with any weed control measure, aquatic herbicides have their limitations and dangers but the record of their use to date, in terms of both efficacy and environmental safety, is generally good.

ACKNOWLEDGEMENT

We thank Mike Greaves and John Abbott for critically reviewing this chapter.

9

Biological control of aquatic weeds

(a) Introduction to biological control of aquatic weeds

A. H. Pieterse

According to the report on terminology in weed science (Coördinatiecommissie Onkruidonderzoek 1984) a biological control agent is 'a control agent based upon a living organism or virus'. As a consequence biological control of aquatic weeds could be defined as 'activities aimed at decreasing the population of an aquatic weed to acceptable levels by means of a living organism or virus'.

In general, there are three approaches for the biological control of aquatic weeds:

(1) the use of selective organisms, i.e. organisms which attack one or only a few species;

(2) the use of non-selective organisms, i.e. organisms which attack all (or nearly all) weed species present; and

(3) the use of competitive plant species, i.e. plants which compete with a weed species for one or more critical growth factors.

When selective organisms are used two main methods can be distinguished (Wapshere 1982):

(1) classical biological control—this implies the introduction of a biological control agent for the control of an unwanted exotic plant. In general, this is an organism which attacks the weed in its native range but does not occur in the region where the plant has been introduced; and

(2) inundative biological control—this implies an increase in the numbers of a biological control agent which occurs already in the area. It can be carried out throughout the season, or be timed in order to increase effectiveness during certain periods.

Some of the major successes of biological control of weeds have been obtained by means of the classical method. On the other hand, an advantage of the inundative biological control method is that it does not require overseas searches for suitable agents, shipping of the organisms, and quarantine before implementation in the area where the weed problem exists.

The selective organisms which are at present of practical importance for the control of aquatic weeds are exclusively arthropods and fungi. These are discussed in detail in sub-chapters 9b and 9c. Outstanding results have been obtained with:

(1) the Chrysomelid beetle *Agasicles hygrophila* against alligator weed, *Alternanthera philoxeroides*;

(2) the Curculionid weevils *Neochetina eichhorniae* and *N. bruchi*, and the stem-boring Pyralid moth, *Sameodes albiguttalis*, against water hyacinth, *Eichhornia crassipes*; and

(3) the Curculionid weevil *Cyrtobagous salviniae* against *Salvinia molesta*. All are examples of the classical control method.

The most important non-selective organism for the control of aquatic weeds is the grass carp (*Ctenopharyngodon idella*), which, together with other phytophagous fish, is discussed in sub-Chapter 9d. Other non-selective organisms which have been studied in connection with the control of aquatic weeds include snails, such as *Marisa cornuarietis* and *Pomacea australis* (Seaman and Porterfield 1964; Blackburn and Andres 1968; Blackburn, Taylor, and Sutton 1971), and the manatee (*Trichechus manatus*: Allsopp 1960, 1961, 1969; Bertram and Bertram 1966; Blackburn and Andres 1968; Brownell, Ralls, and Reeves 1981; Dill 1961; Pieterse 1981; Vietmeyer 1974).

A disadvantage of the snails is that they cannot be confined to a certain habitat. As a consequence they may enter water bodies which are used for the cultivation of crops, such as rice. The manatee, a mammal which has received wide publicity in connection with aquatic weed control, is an endangered species with a very low production rate (one calf every two to three years). Therefore, it is questionable whether it could ever be widely used as a control agent of aquatic weeds. Although it was described in the earlier publications (Allsopp 1960, 1961) as a voracious feeder on many kinds of aquatic plants, this has recently been disputed (Etheridge *et al.* 1985). This conclusion was based upon the observation that, in a lake in Florida (Kings Bay) as many as 18 manatees would be required to maintain a hydrilla vegetation at a constant biomass. The earlier reports were based on a canal in Guyana where growth of aquatic plants was restricted to the waterway margin.

Although it has been suggested that the phytophagous semi-aquatic South American rodents capybara (*Hydrochoerus hydrochaeris*) and coypu (*Myocaster coypus*) could be used for aquatic weed control (National Academy of Sciences

1976), they should not be recommended for introduction into areas outside their native range as they burrow into the banks of water bodies and may feed on crops. In addition, it seems hardly feasible to use domestic animals, like water buffaloes, goats, ducks, and geese for an effective control of aquatic weeds as management of these animals will be very difficult.

Competition for one or more critical growth factors, such as light, between a weed and other plant species in order to inhibit growth of the weed, has been effected in various ways. An example is planting trees and shrubs alongside streams and ditches (Pieterse and van Zon 1982). In Egypt, growth of aquatic weeds in irrigation and drainage canals is strongly inhibited when the canals are overshadowed by *Casuarina* trees.

In ponds in the USA algal blooms, produced by the addition of organic fertilizers, have been successfully used to outshade submerged weeds (Smith and Swingle 1941; Swingle and Smith 1947). Another means of outshading submerged plants is by encouraging growth of free-floating plants or rooted plants with floating leaves. For example, in rice fields, after the rice plants have reached a certain height, growth of rooted weeds can be inhibited to a large extent when the water surface is covered with a layer of *Salvinia* (Pons 1982). In the Netherlands investigations have been carried out in drainage ditches on preventing the growth of submerged weeds by encouraging growth of the floating leaves of various rooted species (Zonderwijk and van Zon 1976; Pitlo 1978). *Nymphaea alba* and *Potamogeton natans* appear to be the most suitable as the flow resistance of these plants is relatively low (Pitlo 1982) (see also chapters 9d and 10).

Extensive research has been carried out on the use of *Eleocharis* species (spikerush) for crowding out rooted aquatic weeds (Yeo and Fisher 1970; Johannes 1974; Yeo 1980*a*, *b*; Brownell, Ralls, and Reeves 1981; Yeo and Thurston 1984). During the period 1966–1969 Yeo and Fisher (1970) observed that *Eleocharis acicularis* could replace dense growths of *Elodea canadensis* and *Potamogeton crispus*, as well as *P. pusillus* and *P. foliosus*. Yeo and Thurston (1984) have studied the extent to which dwarf spikerush, *Eleocharis coloradoensis*, influences the number of shoots and the dry weight of seven species of submerged aquatic weeds when planted at the same time. The order of susceptibility to *Eleocharis* competition, from high to low, was *Zannichellia palustris*, *Elodea nuttallii*, *E. canadensis*, *Hydrilla verticillata*, *Potamogeton pectinatus* and *Myriophyllum spicatum*. The production of subterranean turions of *H. verticillata* and tubers of *P. pectinatus* was reduced by more than 50 per cent in the presence of *Eleocharis*.

According to Frank and Dechoretz (1980) *Eleocharis coloradoensis*, not only uses mechanical crowding as a means of damaging competitors, but also incorporates allelopathy among its mechanisms of interference. This was based on an observation that leachate from spikerush sod significantly inhibited the production of new shoots of *P. nodosus* and *P. pectinatus*.

Experiments with *E. acicularis* in the Netherlands have not been very successful

(J. C. J. van Zon, pers. comm.). It is conceivable that environmental factors play an important role in this respect. On the other hand, Yeo (1980*b*) reported that *E. coloradoensis* has been found growing in waters with a wide range of qualities and hydrosoil properties. In this context it is worthwhile noting that in field studies in northern California and north-eastern Nevada it was observed that *E. acicularis* growth is more frequently limited by low light and/or low temperature conditions than is *E. coloradoensis* (Ashton *et al.* 1984).

Some of the major successes of biological control of weeds have been obtained against aquatic plants, in particular with alligator weed, water hyacinth and *Salvinia* (Room 1986*b*). In general, however, it may be concluded that, in comparison to other control methods, the importance of biological control of aquatic weeds has been relatively small up to now.

(b) Biological control of aquatic weeds by means of arthropods

K. L. S. HARLEY AND I. W. FORNO

INTRODUCTION

ALTHOUGH insects were first used for biological control of terrestrial weeds over 100 years ago, they were not used for control of aquatic weeds until 1964. Since then programmes using arthropods have been mounted against one emergent weed *Alternanthera philoxeroides*, three free-floating weeds, *Eichhornia crassipes, Salvinia molesta,* and *Pistia stratiotes,* and two submerged weeds *Hydrilla verticillata* and *Myriophyllum spicatum*. Biological control of the floating weeds has been exceptionally successful.

Alternanthera philoxeroides (alligator weed)

Alternanthera philoxeroides is a perennial stoloniferous herb growing primarily as an emergent aquatic plant rooted in soil or in substrate beneath shallow water. Floating mats of interwoven hollow stems develop from rooted segments and extend over the surface of the water. The weed also grows in swampy areas and damp pasture. It is not known to produce viable seeds and reproduction is entirely vegetative (Sculthorpe 1967). When temperatures and nutrient conditions are favourable, growth rates are high. Actual rates have not been reported in the literature.

A. philoxeroides does not reach pest status in its native habitat in South America. It has been spread to Australia, Burma, Indonesia, India, New Zealand, Puerto Rico, Thailand and the USA, (Julien and Broadbent 1980), where it is a problem weed. It is a very serious weed in China where large areas, mainly south of the Chang Jiang (Yangtse) River, are infested (Wang pers. comm.).

Research into biological control of *A. philoxeroides*, instigated by the USA in 1960, resulted in *Agasicles hygrophila* (Coleoptera: Chrysomelidae) being released in that country in 1964, in Australia in 1977, in Thailand and New Zealand in 1982 (Julien 1987) and recently in China (Wang pers. comm.). *Amynothrips andersoni* (Thysanoptera: Palaeothripidae) was released in the USA in 1967 and *Vogtia malloi* (Lepidoptera: Pyralidae: Phycitinae) in 1971, in Australia in 1977, and in New Zealand in 1984 (Maddox *et al.* 1971; Brown and Spencer 1973; Durden, Blackburn, and Gangstad 1975; Spencer and Coulson 1976; Coulson 1977; Julien 1987). Australia also sought agents to control this weed in damp pasture situations and introduced *Disonycha argentinensis* (Coleoptera: Chrysomelidae) in 1980 (Sands, Kassulke, and Harley 1982; Julien 1982). In 1982 *D. argentinensis* was released in New Zealand.

Outstanding success in the control of floating *A. philoxeroides* has been achieved with *A. hygrophila* following release in the USA, Australia, and Thailand (Coulson 1977; Julien 1981; B. Napompeth pers. comm.) and in warmer areas of New Zealand (Roberts, Winks and Sutherland 1984; Julien 1987). *V. malloi* has been of limited value in regions climatically less suited to *A. hygrophila* and in semi-aquatic habitats (Coulson 1977; Julien 1987). Establishment of *D. argentinensis* has not been confirmed in Australia or New Zealand (Julien 1987) and no agent effective in controlling alligator weed plants rooted in soil has been found.

Certain herbicides will kill floating *A. philoxeroides*, but plants rooted in soil are not killed. Regrowth is rapid and repetitive herbicide application is required to keep an infestation in check. Herbicide control is unsatisfactory, expensive and temporary. *A. hygrophila* easily establishes on infestations of floating *A. philoxeroides* except in cooler areas (Coulson 1977; Buckingham, Boucias, and Theriot 1980) and in many regions, costs of establishment would be low and control of the floating weed very satisfactory. Starter colonies of *A. hygrophila* are available from the Commonwealth Scientific and Industrial Research Organization (CSIRO) Australia, National Biological Research Center (NBRC) Thailand or the United States Department of Agriculture (USDA).

Salvinia molesta (salvinia)

The free floating aquatic fern *S. molesta* is native to south-eastern Brazil (Forno and Harley 1979) and occurs from sea level to an altitude of 900 m. The plant is a sterile pentaploid (Loyal and Grewal 1967) and growth is vegetative and rapid with each ramet having the potential to produce new ramets. Ramets bear up to three ranks of side branches and the degree of branching is positively correlated with concentration of nutrients in the ambient water (Room 1983).

Outside its native range *S. molesta* is an aggressive weed which has invaded most tropical and sub-tropical regions of the world. It has been or still is a problem in many African countries including Angola, Botswana, Kenya, Mozambique, South Africa, Namibia (Eastern Caprivi), Zaire, Zambia, Zimbabwe (Reed 1965; Edwards and Thomas 1977; Jacot-Guillarmod 1979; Hattingh 1982) and several west African countries; in Papua New Guinea (Johnstone unpub. report; Mitchell 1979*c*) the Philippines and in Fiji, New Zealand, India, Sri Lanka, Indonesia, Malaysia and Australia.

In the search for suitable organisms for the control of *S. molesta*, studies prior to 1978 were confined to *S. auriculata* (Bennett 1975). In that era three insects were studied and released (Julien 1987): a weevil *Cyrtobagous singularis* (Coleoptera: Curculionidae), released in Botswana between 1972 and 1974 (Edwards and Thomas 1977) and in Fiji in 1979 (Kamath pers. comm.); a moth, *Samea multiplicalis* (Lepidoptera: Pyralidae) released in Zambia in 1970 and Fiji in 1976 (Kamath 1979); and the grasshopper *Paulinia acuminata* (Orthoptera: Acrididae) released at Lake Kariba, Zimbabwe in 1970 (A. C. Whitewell and D. J. W. Rose 1974 pers. comm.), Botswana in 1971–1974 (Bennett 1975; Edwards and Thomas 1977), Kenya in 1970, Sri Lanka in 1973 (Bennett 1975) and Fiji in 1975 (Kamath 1979). Although the insects established in many places (Julien 1987), none successfully controlled weed growth. *C. singularis* is currently only established in the Kwando–Linyanti–Chobe River system in Eastern Caprivi, Namibia and causes no significant reduction in the growth of salvinia (Schlettwein 1988). *S. multiplicalis* is established in Fiji and Australia and although it can cause severe damage to the plant in Australia, it has not reduced weed growth significantly (Room, Forno, and Taylor 1984; Forno 1987). *P. acuminata* is established on Lake Kariba and may have contributed to the decline of salvinia on the lake (Mitchell and Rose 1979).

In 1980, a new species of *Cyrtobagous, C. salviniae* (Fig. 9b.1) (Calder and Sands 1985) collected from the native range of the weed (Forno and Bourne

FIG. 9b.1. The weevil *Cyrtobagous salviniae*, a biological control agent for *Salvinia molesta*.

1984) was released at Lake Moondarra, Australia (Room *et al.* 1981). By August 1981, this weevil had destroyed an estimated 18 000 tonnes of weed on the lake and this event marked the beginning of successful biological control of this weed in the tropics. Since then this weevil has controlled the growth and spread of this weed in other areas of Australia (Room *et al.* 1981; Room, Forno and Taylor 1984; Forno 1985, 1987), Papua New Guinea (Thomas 1985; Room and Thomas 1985), India (Joy *et al.* 1988) and Namibia (Anon. 1985). It is predicted that this agent will successfully control the growth of salvinia in most tropical and sub-tropical areas of the world (Room 1986*b*).

Estimates of losses to man and the environment due to infestations of salvinia are fragmentary, but the following emphasize the economic importance of this weed and the reduction in costs following successful biological control. In Australia during 1976 and 1977, A$116 000 was spent on the control of *S. molesta* in Lake Moondarra, Mt. Isa. This included the cost of building booms to confine the weed and the application of herbicides (Mitchell 1978). Later a further A$23 000 was spent in applying herbicide (Mitchell 1978). These methods could not contain the weed and when *C. salviniae* was introduced to Lake Moondarra it was infested by 100 ha of salvinia. Control was affected twelve months later, (Room *et al.* 1981) and there have been no control costs since 1981. Likewise in Papua New Guinea, in the late 1970s lagoons and ox-bow lakes in the Sepik River system became infested with salvinia and the area covered was estimated at 300 km^2. Physical and chemical control methods proved futile with biomass of the weed at 80 tonnes ha^{-1} and the cost of repetitive aerial spraying at US$50 ha^{-1} (Thomas 1985). *C. salviniae* was introduced to the Sepik system in 1982 and by the end of 1985 there was less than 1 km^2 of salvinia remaining (Room 1985, pers. comm.) with many lakes and lagoons clear except for a few plants on the margin. In India, Sri Lanka and Fiji annual costs have been high in trying to eliminate salvinia using herbicides (Bennett 1975; Varshney and Singh 1976; Senaratna 1943; Sundaresan and Reddy 1979) and in India these costs are being reduced through biological control using *C. salviniae* (Joy *et al.* 1988). In the Eastern Caprivi of Namibia herbicidal application was discontinued following the introduction of *C. salviniae* in 1983. Since then the infestation has decreased (Anon. 1985) and it is predicted that salvinia will no longer be an economic problem in the Eastern Caprivi. *C. salvinae* was released in Sri Lanka in 1986 and control of salvinia has been achieved in many areas (Room *et al.* in press).

Growth of salvinia (Room 1986*a*; Room and Thomas 1986) and the reproductive potential of *C. salviniae* (Sands, Schotz, and Bourne 1983; D. P. A. Sands and M. Schotz: pers. comm.) are positively related to temperature and nitrogen content of the plant. The successful biological control of salvinia on three continents and in Papua New Guinea using *C. salviniae* has demonstrated how an insect species can destroy large areas of salvinia even when conditions for plant growth are optimal. In tropical and sub-tropical areas, *C. salviniae* has contained the growth of salvinia to a few plants on the edge of a water body and maintained the population density of the weed at well below the carrying capacity of the habitat. Starter colonies of *C. salviniae* are available from CSIRO Australia.

Eichhornia crassipes (water hyacinth)

Eichhornia crassipes is a stoloniferous plant with entire leaves, slender or swollen petioles, and conspicuous spikes of largely lilac-mauve flowers. The leaves and flowers rise well above the water surface and plants are usually free-floating. Propagation is mainly by stolons which produce daughter plants at the apices, but is also by seed. Under optimum growth conditions 7.4 to 22.0 g m^{-2} organic matter d^{-1} may be produced. Plant doubling time has been calculated at 5 to 15 d. Growth rates vary according to nutrient availability (Forno and Wright 1981).

E. crassipes is native to Brazil, and perhaps other central and South American countries, but was spread to most tropical and sub-tropical regions of the world where it became a very serious weed (Holm *et al.* 1977, Gopal and Sharma 1981). Research into biological control began in 1961. Control agents were first released about 10 years later (Perkins 1972, 1973) and one or more species have now been introduced into 19 countries (Julien 1982; Limon 1984). Arthropod agents which have been introduced are the weevils *Neochetina bruchi* and *N. eichhorniae* (Coleoptera: Curculionidae), the moths *Acigona infusella* and *Sameodes albiguttalis* (Lepidoptera: Pyralidae), and the mite *Orthogalumna terebrantis* (Acarina: Galumnidae) (Bennett 1970; De Loach and Cordo 1976, 1978; Del Fosse 1977; De Loach *et al.* 1980; Harley and Wright 1984). Attempts to augment populations of the native moth *Arzama densa* (Lepidoptera: Noctuidae) have been made in the USA (Baer and Quimby 1981).

N. eichhorniae, which has been introduced into the USA, Mexico, Panama, Egypt, Sudan, Zambia, Zimbabwe, Republic of South Africa, India, Sri Lanka, Thailand, Burma, Vietnam, Malaysia, Indonesia, Australia, Papua New Guinea, Fiji and Solomon Islands (Julien 1987; Harley and Wright 1984). Adults of *N. eichhorniae* damage the leaves (Figs. 9b.2a,b) and larvae damage the plant by tunnelling towards the base of petioles and into the crown where they cause severe damage (Deloach and Cordo 1976). *N. bruchi* has a similar life cycle to *N. eichhorniae* (Deloach and Cordo 1976). It has established in the USA and Sudan but any impact it may have on *E. crassipes* is overshadowed by the dramatic effects of *N. eichhorniae* (Beshir and Bennett 1988).

The larvae of *S. albiguttalis* feed inside the petioles and buds and attack may be heavy but sporadic as this moth shows a preference for tender, often bulbous, petioles. The other moth species appear to have potential as control agents, but this has not been demonstrated in a field situation outside their native range (De Loach and Cordo 1978).

FIG. 9b.2(a). Water hyacinth, *Eichhornia crassipes*. Note scars on the leaves due to feeding by the biological control agent *Neochetina eichhorniae*.

FIG. 9b.2(b). The weevil *Neochetina eichhorniae*, a biological control agent for water hyacinth, *Eichhornia crassipes*.

The deleterious effects caused by *E. crassipes* on the White Nile system (Beshir and Bennett 1984) typify effects world wide. *E. crassipes* interferes with river transportation, including high operational and maintenance costs of ships; blocks irrigation canals and pumps; restricts access to water by riverine settlements; prevents recreational use of water; increases loss of water by evapo-transpiration; increases the level of damage caused by flooding; causes losses to fishing industries, and the cost of chemical control programmes is high.

In 1963 the Sudan Ministry of Agriculture used a fleet of 42 boats and three aircraft and had three stations on the Nile to try to maintain open water access for commercial steamer traffic and to riverside villages (Beshir and Bennett 1984) by controlling the spread of water hyacinth. In addition, lost time and additional maintenance and repairs to steamers due to *E. crassipes* cost about 500 000 pounds, the cost of control operation about one million pounds and total water loss possibly 7 000 000 000 m^3, or one tenth of the average yield of the Nile annually (Hamdoun and Tigani 1977).

In the Sudan, *N. eichhorniae* was released in 1978, *N. bruchi* in 1979 and *S. albiguttalis* in 1980. *N. eichhorniae* was first recovered in 1980 and was widespread by 1981. The other two insects also became widely established. Whereas in the early 1960s up to 11 350 ha of *E. crassipes* accumulated behind the Jebel Aulia Dam, there have been no accumulations since 1982. Throughout the system virtually every *E. crassipes* plant is scarred by *Neochetina* feeding and the vigour and weight of the plants is lower. As a result, the Ministry of Agriculture has relaxed its chemical control programme (Beshir and Bennett 1988).

This picture of impressive reduction in infestations of *E. crassipes* and in control costs has been repeated in the USA and Australia and is anticipated for other regions, e.g. South-East Asia, India, and southern Africa, where *N. eichhorniae* has been established more recently. The period of five years which elapsed in the Sudan from release of *N. eichhorniae* to noticeable reduction in *E. crassipes* should be regarded as a minimum for control with the period being longer under conditions of lower temperature.

Throughout the introduced range of *E. crassipes* the potential for biological control is excellent. Several insects are available but the most reliable agent for control is *N. eichhorniae*. Starter colonies of this insect are available through the Commonwealth Institute of Biological Control or from national research institutions such as CSIRO, Australia or USDA, USA.

Pistia stratiotes (water lettuce)

Piastra stratiotes is a floating, stoloniferous perennial up to 15 cm tall and 30 cm wide, with fibrous roots. Under tropical conditions where plant nutrient levels are adequate, growth is rapid and many new daughter plants are produced on stolons from the parent plants. Plants may also reproduce by seed (Sainty and Jacobs 1981).

P. stratiotes is cosmopolitan throughout the tropical and sub-tropical regions of the world. Its native range is uncertain, but may be tropical America. It has become a serious weed in some regions and could increase in importance as the more aggressive weeds, *E. crassipes* and *S. molesta* come under control.

A potential control agent, *Neohydronomous pulchellus* (Coleoptera: Curculionidae) was first studied by Deloach, Deloach, and Cordo 1976. Studies were continued by an Australian team and subsequently this insect was released in Australia. Infestations have been dramatically reduced in area, plant size, and weight but an annual resurgence of the weed tends to occur at some sites during late summer (Harley *et al.* 1984; K. L. S. Harley, unpublished data). *N. pulchellus* has recently been introduced into Papua New Guinea, South Africa, and the USA (Julien 1987).

A noctuid moth, *Epipsammia pectinicornis* (Lepidoptera: Noctuidae) is widespread in South-East Asia, and in Thailand it is managed as a control agent for *P. stratiotes* (Suasa-ard and Napompeth 1982; Napompeth 1985). This insect appears to be host specific to *P. stratiotes* (Suasa-ard 1976) and has replaced herbicides for control in reservoirs in Thailand (B. Napompeth, pers. comm.).

A pyralid moth, *Nymphula tenebralis* (Lepidoptera: Pyralidae) has been reported to cause seasonal destruction of *P. stratiotes* in the Northern Territory of Australia (J. D. Gillett, pers. comm.). This insect has not been evaluated as a control agent. Other Lepidoptera have been reported as potential control agents elsewhere but the effectiveness of these species is uncertain (Mangoenddihardfo and Syed 1974; Chaudhuri and Ram 1975). In general, Lepidoptera attract a suite of parasitoids and are less efficient as control agents than insects from other families.

N. pulchellus is known to be host specific and its ability to reduce substantially the severity of problems due to *P. stratiotes* has been demonstrated in Australia. Considerable potential exists for control by this agent in other parts of the world. Starter colonies are available from CSIRO, Australia.

Hydrilla verticillata (hydrilla)

Hydrilla verticillata is a much-branched perennial, submerged, rooted, vascular plant. The stems are long, usually branched, and grow from the hydrosoil to form dense intertwined mats at the surface of the water. It is found growing in waters up to 7 m deep (Aston 1973). Detached portions of hydrilla plants remain viable and are a common mode for infestation of new areas. Growth is vegetative or (rarely) by sexual reproduction (Haller 1976).

Hydrilla is cosmopolitan in its distribution, with Antarctica and South America, the only continents from which it has not been recorded (J. K. Balciunas, unpublished report 1983). It is very common on the Indian subcontinent, many of the Middle-East countries, South East Asia and northern and eastern Australia. Severe problems occur in many of the southern states of the USA (Haller 1982). Infestations can impede water flow and interfere with irrigation, boat movement, and aquatic recreational activities. Economic losses exceeding US$30M have been reported in Texas USA (Guerra 1977). Chemical and mechanical control methods are expensive with in excess of US$8M spent annually trying to control the weed in Florida (Mahler 1979).

The Commonwealth Institute for Biological Control carried out extensive surveys in Pakistan and the US Department of Agriculture and the US Army Corps of Engineers in Kenya, Indonesia, S.E. Asia, and Australia for biological control agents for *Hydrilla* (Balciunas: unpublished report 1983). The moth *Parapoynx diminutalis* (Lepidoptera: Pyralidae) appears native to S.E. Asia but was discovered in Florida in the mid-1970s (Del Fosse, Perkins, and Steward 1976) and is now widespread. It has had some effect in controlling the growth of hydrilla in the USA (Balciunas and Habeck 1981), but further agents are required to reduce the infestations of *Hydrilla*. Currently, the subterranean turion-feeding *Bagous* weevils and the leaf mining fly, *Hydrellia* spp. are being studied as potential agents for the control of hydrilla in the USA and further surveys are being conducted in Australia (Balciunas: pers. comm.).

Myriophyllum spicatum (Eurasian water milfoil)

Myriophyllum spicatum is a submerged macrophyte with long branching stems that grow to the surface and frequently form dense mats. The plant grows in fresh and brackish water in depths to 10 m. The plant spreads vegetatively and any plant fragment containing a dormant lateral bud can develop into a new plant. It produces viable seed but spread by seed is of minor importance (Balciunas 1982).

Eurasian water milfoil is found throughout Europe, most of Asia and in parts of Africa (Balciunas 1982; Jacot-Guillarmod 1977). It has been a problem in the United States since the 1930s where it has infested areas from 20 to 40 thousand hectares (Steenis 1968; Crowell and Steenis 1968). Exploratory studies for biological control agents have been carried out in Pakistan, Bangladesh, and Yugoslavia (Balciunas 1982) with several insects being evaluated (Habib-ur-Rahman, Baloch, and Ghani 1969; Batra 1977; Buckingham and Bennett 1981; Buckingham and Ross 1981; Habeck 1983) including a number of pyralid moths (e.g. *Acentropus niveus*) and several stem-boring weevils (e.g. *Litodactylus leucogaster*). However, many of these insects were found to be non-specific to the target weed or to offer little potential as effective biological control agents. There have been detailed observations of the insects attacking *M. spicatum* in British Columbia

(Kangasniemi 1983) and in the United States (Balciunas 1982). Further exploratory studies are required for agents for the biological control of this weed.

(c) Biological control of aquatic weeds by means of fungi

R. CHARUDATTAN

INTRODUCTION

SINCE 1970, there has been increasing worldwide interest in exploiting the potential of plant pathogens as biological control agents for aquatic weeds (Zettler and Freeman 1972; Freeman 1977; Charudattan 1984). The principles and practices of biological weed control with plant pathogens, which are generally similar to those of other biological control systems, have been previously expounded and will not be repeated here (Charudattan and Walker 1982). Rather, this chapter will deal with the use of fungi as biological control agents for aquatic weeds. The choice of fungi is appropriate; in terms of pathogenic efficacy and technological feasibility of production, formulation, and delivery, fungi are more suitable as biological weed control agents than other types of pathogens (Charudattan 1985). It is not surprising, therefore, that fungi have been, and continue to be, predominant candidates in biological weed control research (Templeton 1982).

Thus far, researchers have concentrated their studies on water hyacinth (*Eichhornia crassipes*), hydrilla (*Hydrilla verticillata*), Eurasian water milfoil (*Myriophyllum spicatum*), water lettuce (*Pistia stratiotes*), salvinia (*Salvinia molesta*), and alligator weed (*Alternanthera philoxeroides*). These constitute the most important threats to efficient utilization of water resources in many countries and will be the focus of this chapter (Table 9c.1).

SOME CONCEPTS

Biological control of weeds with plant pathogens can be accomplished by one of two main strategies: the classical and the microbial herbicide strategies (Templeton 1982; Charudattan 1985). In the classical strategy, which is most suitable for exotic weeds, a pathogen is introduced from the geographic origin of the weed into the weed's adventive range where the control is desired. Usually, the pathogen is released over a small part of the target weed population using non-formulated inoculum. Provided the weed is susceptible to the introduced pathogen, it occurs in high densities facilitating a continuous spatial spread of the pathogen, and the pathogen is aggressive, an epidemic will generally follow the pathogen's release into the new region. The epidemic may be aided by a temporary loss of host resistance resulting from the spatial and temporal separation of the host and the pathogen in the adventive range. But more likely, the high densities of the weed will create a large and continuous susceptible target for the pathogen, enabling the disease to spread throughout the weed population. The host stress from the disease will result in lower weed density provided the epidemic is of sufficient magnitude and remains undiminished by host or environment. Since this strategy relies on a relatively small-scale inoculation and a large-scale epidemic, it is also termed the inoculative technique.

Pathogens utilized for classical weed control have been mostly rust fungi to date, but certain other fungi also capable of self-dissemination through air-borne spores or capable of a sustained residual occupation of soil are under consideration for the classical approach. Host density, availability of susceptible host genotypes, virulent pathogen strains, favourable environmental conditions, and lack of hyperparasites are factors impinging on the success of a classical biological control agent.

Pathogens used as microbial herbicides can be native or exotic, but the former have been used more commonly. Unlike the classical agent, a pathogen to be used as a microbial herbicide is cultured *in vitro* on a large scale and applied in fairly high concentration to the weed (Templeton 1982). The intent is to kill the weed. The need for culturing makes facultative saprophytes and facultative parasites the agents of choice for this strategy. If necessary, microbial herbicides can be applied repeatedly during the growing season or annually using conventional pesticide application techniques. In the United States, microbial herbicides are required to be registered by the Environmental Protection Agency (Charudattan 1982); thus the agent needs to be standardized and formulated to conform to a list of specifications set out by the manufacturer at registration.

Commonly, the classical strategy is regarded as being more suitable for controlling aquatic weeds than the other strategy due to one or more of the following considerations.

1. Many of the important aquatic weeds are exotics in areas where they cause problems and it is a well-recognized concept in biological control that exotics are good targets for classical biological control.
2. Usually aquatic weeds infest vast and inaccessible areas. In such situations, the classical biological control agents, with their capacity for spreading to remote areas of weed distribution are, generally, more practical than microbial herbicides, which must be applied in the target areas.
3. The typical magnitude of aquatic weed problems also imposes a cost consideration that would favour the classical biological control over the microbial herbicide.

Table 9c.1. Fungal pathogens of aquatic weeds of worldwide importance

Name	Type of disease or damage	Important references
1. ***Eichhornia crassipes***		
Acremonium zonatum	Zonate leaf spot, often damaging to the entire lamina	Rintz 1973; Martyn and Freeman 1978; Galbraith and Hayward 1984
Alternaria alternata	Leaf spot	Badur-ur-Din 1978; Galbraith and Hayward 1984
A. eichhorniae	Leaf spot and severe leaf blight	Nag Rag and Ponnappa 1970*a, b*; Maity and Samaddar 1977; Badur-ud-Din 1978; Mangoendihardjo *et al.* 1977; Rakvidyasastra, lemwimangsa, and Petcharat 1981; Stevens, Badur-ur-Din, and Ahmed 1979; Charudattan and Rao 1982; Charudattan 1984
Bipolaris oryzae	Severe foliar blight	Charudattan, Conway, and Freeman 1975; Charudattan 1984
Cercospora piaropi	Leaf spot, leaf necrosis, and general debilitation	Tharp 1917; Thirumalachar and Govindu 1954; Nag Raj 1965; Freeman and Charudattan 1974; Hettiarachchi, Gunasekera and Balasooriya 1983; Balasooriya *et al.* 1984; Galbraith and Hayward 1984; Charudattan 1984; Martyn 1985
C. rodmanii	Leaf spot, leaf necrosis, and general debilitation	Conway 1976*a, b*; Conway and Freeman 1977; Conway, Freeman, and Charudattan 1978; Freeman and Charudattan 1984; Charudattan 1984, 1986; Charudattan *et al.* 1985
Marasmiellus inoderma	Foliar blight	Nag Raj 1965
Myrothecium roridum	Necrotic leaf spot	Nag Raj Ponnappa 1967; Ponnappa 1970, 1971; Charudattan 1973, 1984; Mangoendihardjo *et al.* 1977; Rakvidyasastra, lemwimangsa, and Petcharat 1981; Kasno 1982; Kasno and Soerjani 1981; Hettiarachchi, Gunasekera, and Balasooriya 1983; Gunasekera, Balasooriya, and Gunasekera 1983; Balasooriya *et al.* 1984
Rhizoctonia solani (including *Corticium solani* and *Aquathanatephorus pendulus*)	Foliar blight	Nag Raj and Ponnappa 1967; Joyner and Freeman 1973; Tu and Kimbrough 1978; Freeman *et al.* 1982
Uredo eichhorniae	Chlorotic and necrotic spot and pustule	Charudattan and Conway 1975; Charudattan *et al.* 1976; Charudattan, McKinney, and Hepting 1981

2. *Hydrilla verticillata*		
Fusarium roseum 'Culmorum'	Chlorosis and lysis	Charudattan and McKinney 1978; Charudattan *et al.* 1980, 1984
Sclerotium rolfsii	Discolouration and lysis	Charudattan 1973
3. *Myriophyllum spicatum* and *M. brasiliense*		
Acremonium curvulum	Systematic stem necrosis	Andrews, Hecht, and Bashirian 1982
Articulospora tetracladia	Discoloration; growth retardation	Lekic 1971
Dactylella microaquatica	Discoloration; growth retardation	Lekic 1971
Flagellospora stricta	Discoloration; growth retardation	Lekic 1971
Fusarium acuminatum; *F. oxysporum*; *F. poae*; *F. roseum* 'Avenaceum'; *F. roseum* 'Culmorum'; *F. solani*	Associated with stressed plants, causing die-back	Lekic 1971
F. sporotrichioides	Localized stem lesions	Lekic 1971
Pythium carolinianum	Chlorosis of stem, root loss, and wilting of shoots	Bernhardt and Duniway 1984
Sclerotium hydrophyllum	Associated with stressed plants, causing die-back	Lekic 1971
4. *Pistia stratiotes*		
Cercospora sp.	Leaf spot	Nag Raj and Ponnappa 1967
Sclerotium rolfsii	Blight	Nag Raj and Ponnappa 1967
5. *Salvinia molesta*		
Myrothecium roridum	Leaf spot	Rao 1970
6. *Alternanthera philoxeroides*		
Alternaria alternantherae	Leaf spot	Holcomb and Antonopoulos 1976; Holcomb 1978
Uredo nitidula	Rust	Arthur 1920

Nevertheless, our experience in the south-eastern United States has indicated that a microbial herbicide strategy, rather than the classical, is more suitable for weed control in waterways that are heavily used and intensively managed. In these situations, rapid control is needed, requiring a predictable and effective agent. The cost of control is usually not a concern. These criteria are better met by the microbial herbicides than the classical agents. Moreover, aquatic weeds have the propensity for rapid growth rates and sudden growth spurts triggered by changes in water chemistry and weather, causing them to outgrow disease pressures. Although neither a classical nor a microbial herbicide agent may maintain its effectiveness when the host population increases suddenly, a microbial herbicide, rather than a classical agent, can be more easily augmented through re-application of the inoculum to produce a rapid epidemic.

It is, however, necessary to point out that we need considerably more empirical knowledge about both classical agents and mycoherbicides than we presently have in order to utilize fully the potential of pathogens as weed control agents. For example, we need to know the ways to maximize the capabilities of classical agents through manipulative techniques. Thus far, pathogens have been used singly as classical agents; multiple agents need to be tested as part of the ecosystem. Also, the potential to deploy certain pathogens as classical agents in some areas and as mycoherbicides in other locations and to combine classical and mycoherbicide agents are approaches deserving more study.

TARGET WEEDS AND PATHOSYSTEMS

From a pathological perspective, freshwater macrophytes need to be distinguished as emergent or submerged species since the types of diseases and their epidemiologies differ considerably between aerial and underwater pathosystems. The epidemiology of fungal diseases on aerial parts of emergent vegetation is similar to that of terrestrial plants, and pathologists have considerable experience with these. On the other hand, only limited information is available on the pathology and epidemiology of diseases affecting underwater plants (Lekic 1971; Ridings and Zettler 1973; Charudattan 1983; Charudattan and McKinney 1978; Charudattan *et al.* 1984; Andrews and Hecht 1981; Andrews, Hecht, and Bashirian 1982; Bernhardt and Duniway 1984). Underwater systems impose severe technological and ecological constraints, rendering attempts at biological control of undesirable plants impractical. Future research may elucidate a wide range of possible microbial control mechanisms for submerged species, but at present, fungal control is feasible only for emergent species, as will be discussed here.

THE WATER HYACINTH—*CERCOSPORA RODMANII* SYSTEM

Discovery and taxonomy

In 1973, a disease of water hyacinth characterized by root rot, leaf spots, and leaf necrosis was discovered in the Rodman Reservoir, Florida. It was determined that the leaf spots and the severe leaf necrosis, caused by a *Cercospora* sp., were the primary symptoms, whereas the root rot was a secondary effect. Although a *Cercospora* disease, caused by *Cercospora piaropi,* had been previously observed in the United States and India (Tharp 1917; Thirumalachar and Govindu 1954; Freeman and Charudattan 1974), the Rodman fungus was treated as a new species, *C. rodmanii* (Conway 1976*a*), due to certain taxonomic and pathologic differences between the two fungi.

Pathology

Cercospora rodmanii causes small brown to black spots on the leaves and petioles. As a result of these spots, the leaf dies back from the tip to the base of the petiole until the entire leaf is killed. Plants with severe infections become chlorotic and stressed. At this stage, all leaves on a plant develop dark brown necrotic spots. Eventually, the leaves die and become water-soaked, causing the plant to sink. In advanced stages of disease, root deterioration sets in (Freeman and Charudattan 1984). With the spread of the fungus, the plant population begins to decline and open water appears where previously there had been dense stands of water hyacinth. As the disease intensifies, the mat of vegetation breaks up, and small clusters of heavily diseased plants float away from the mat. Finally, the entire cluster gradually sinks to the bottom. This type of disease progression, which has been observed under controlled conditions, may take several weeks to months and will occur only under severe and sustained disease pressure.

Safety

The host range of *C. rodmanii* was determined in the greenhouse and the field using plants that were taxonomically related to water hyacinth and those economically important to Florida. Eighty-five plants in 22 families were screened but only water hyacinth was highly susceptible (Conway and Freeman 1977). Additionally, based on a fish-toxicity assay, it was determined that *C. rodmanii* will pose no threat to fish (Conway and Cullen 1978).

Field tests

Between 1974 and 1984 *C. rodmanii* was field tested in several locations in Florida and the south-eastern United States using industrial and laboratory-grown fungal inocula. The pathogen was established in test plots usually with one or two applications of inoculum, and the disease typically appeared and established within three weeks, confirming the feasibility of initiating an epidemic in nature. The fungus was capable of secondary spread by wind-borne spores to non-sprayed water hyacinths in the field (Freeman and Charudattan 1984). From these tests, it was determined that *C. rodmanii* can severely affect water hyacinth growth, especially under conditions that favoured a reduced growth rate of the plant. Although the greatest effect of *C. rodmanii* was in reducing plant height and biomass, plant death and plant elimination can occur under severe disease pressure (Conway 1976*b*).

Attempts at registration and commercialization

The use of *C. rodmanii* as a biological control agent for water hyacinth was patented by the University of Florida, and Abbott Laboratories, USA was licensed to develop the fungus as a microbial herbicide for commercial use. Abbott developed wettable powder formulations of *C. rodmanii* and obtained a US Environmental Protection Agency Experimental Use Permit to evaluate their herbicidal use. The possibility of registration and commercialization of *C. rodmanii* appeared good at first. Nonetheless, in 1984 Abbott decided not to register on the grounds of technical difficulties in assuring efficacy of the microbial herbicide and economic uncertainties of the marketplace. Currently, the fungus is not under industrial development, although research on it is continuing.

Efficacy

Water hyacinth is one of the most productive plants and the biological control efficacy of *C. rodmanii* is related to the growth rate of its host. Conway, Freeman, and Charudattan (1978) stated that under conditions favourable for growth, water hyacinth was able to produce one new leaf every 5 to 6 d and, thus, was capable of outgrowing the cercospora disease. Therefore, they hypothesized that when conditions are present that favour disease development and limit leaf production to less than one leaf for every 3 weeks, the pathogen could kill leaves faster than the plant could produce new leaves. The plant would then become debilitated and die unless conditions changed to stimulate its re-growth or conditions became less favourable for the disease.

We tested this hypothesis by determining the relationship between the disease and host growth rate at different nutrient levels (Charudattan *et al.* 1985). It was accomplished by measuring plant growth, disease incidence, and disease severity, and calculating the amount of disease stress and rate of disease progress required to kill water hyacinth.

Although the amount of disease and the speed of disease development obtained in this study were insufficient to kill water hyacinth, there were 20 to 90 per cent reductions in host growth rates, determined by weekly increments of green leaves, as a result of the disease (Charudattan *et al.* 1985). Higher reductions in growth rates occurred at the lower nutrient levels, but when the nutrient concentration was most favourable for growth, water hyacinth grew at a rate faster than the rate at which the disease progressed to newer leaves.

These results indicate that for practical management of water hyacinth by *C. rodmanii*, the fungus should be used under conditions that support only low to moderate host growth rates. Theoretically, however, efficacy of the fungus can be improved by multiple applications of inoculum, when water hyacinth is in the early phase of seasonal growth, to keep disease progress in pace with plant growth. Additionally, a combination of the fungus with other biotic or abiotic agents, such as insects or sub-lethal rates of chemical herbicides, capable of retarding host growth, can be used.

Effects of the combined actions of C. rodmanii and insect biocontrol agents

Until recently, there had been no data available on the level of stress caused by insect biological control agents in the absence of the pathogen's effects and vice versa. Even in cases where the effects of insects (for example, *Neochetina* spp., water hyacinth weevils) have been well documented, the contribution of the ever present microbial agents is usually not duly recognized (Charudattan, Perkins, and Littell 1978). Therefore, we undertook a field study to distinguish the biological control effects contributed by insects (primarily *Neochetina* spp.) from those of *C. rodmanii*. This was accomplished with the use of insecticide and/or fungicide sprays that minimized the insect attacks and/or eliminated the disease according to the experimental design (Charudattan 1986). At the beginning of the experiment, all plants had *Neochetina* spp., and the fungal inoculum was applied thrice in the spring. Data on shoot height, disease incidence and surface cover were gathered monthly, and the experiment was terminated in late fall.

The combination of *C. rodmanii* and insects was capable of eliminating water hyacinth from the test plots (Charudattan 1986). The fungus alone or the insects alone did not completely kill the plants. The fungus had only a slight effect in reducing plant height, but its greatest effect was on leaf browning, debilitation, and death of insect-damaged plants. The insects alone significantly reduced shoot height (by about 50 per cent), but did not kill the plants. On the other hand, plants with fungus and insects were more adversely affected than those with either agent alone. These plants died six months after the first application of the

FIG. 9c.1. An illustration of the kind of control obtainable with a combination of *Cercospora rodmanii* infection and attack from the water hyacinth weevils (*Neochetina* spp.). (a) Water hyacinth plants that were kept free of damage from weevils and pathogen by repeated applications of insecticide and fungicide sprays. (b) Plants subject to stress from the weevils alone. The pathogen was eliminated from this plot by repeated applications of a fungicide. (c) Plants stressed by both weevils and *C. rodmanii*. These floating plots were part of a season-long field trial to quantify the effects of the weevils, *C. rodmanii*, and their combination.

fungus, and dead plants decayed and sank (Fig. 9c.1). After 7 months, the treatments consisting of *C. rodmanii* and the insects yielded 99 per cent control (99 per cent surface clearance) of water hyacinth. Therefore, it appeared that *C. rodmanii* and the insects were compatible and complete control of water hyacinth could be obtained in the field by integrating the pathogen and *Neochetina* spp (see also Chapter 10) .

Current status of biological control with C. rodmanii

Cercespora rodmanii is widely distributed in Florida and occurs in several locations in the south-eastern United States (Freeman and Charudattan 1984). It is particularly noticeable on insect- and nutrient-stressed plants. Along with *C. piaropi,* it is exerting a significant natural biological control of water hyacinth in the south-eastern United States (Freeman and Charudattan 1974, 1984; Martyn 1985). However, the amount of stress is not always sufficient to afford a satisfactory level of control needed by the public. Therefore, reliance on chemical herbicide is still common and control programmes that aim for maintenance of weed densities below the economic threshold depend primarily on chemical herbicides (Charudattan 1986). However, this reliance on chemicals is not without drawbacks. For example, when large populations of water hyacinth are killed rapidly by chemical herbicides, the insect biocontrol agents that subsist on this plant, such as *Neochetina* spp., also die *en masse.* The insect populations do not regenerate as rapidly as the host population, resulting in less insect and pathogen pressure on the new generation of plants. To remedy this situation, it is recommended to leave refuges of the plant from which the insects and pathogens could spread on to new weed growth (Charudattan 1986).

In summary, *C. rodmanii* is a biological control agent of considerable importance and value. To a limited extent, it occurs naturally on water hyacinth in the south-eastern United States, and may yet become available to the public as a microbial herbicide. In view of the close taxonomic and pathogenic relationship between *C. piaropi* and *C. rodmanii*, it is recommended that the former species be considered for commercial development in countries where the latter does not occur (Charudattan 1984; Galbraith and Hayward 1984; Martyn 1985).

SUBMERGED WEEDS

A Fusarium disease of Hydrilla

In 1974, we isolated a strain of *Fusarium roseum* 'Culmorum' from diseased *Stratiotes aloides* (Hydrocharitaceae) plants collected near Wageningen, the Netherlands (Charudattan and McKinney 1978). In laboratory tests, the fungus was pathogenic to *H. verticillata,* also of Hydrocharitaceae and one of the most

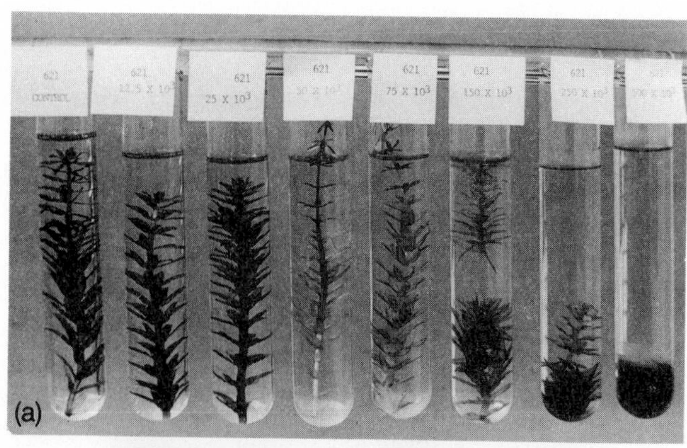

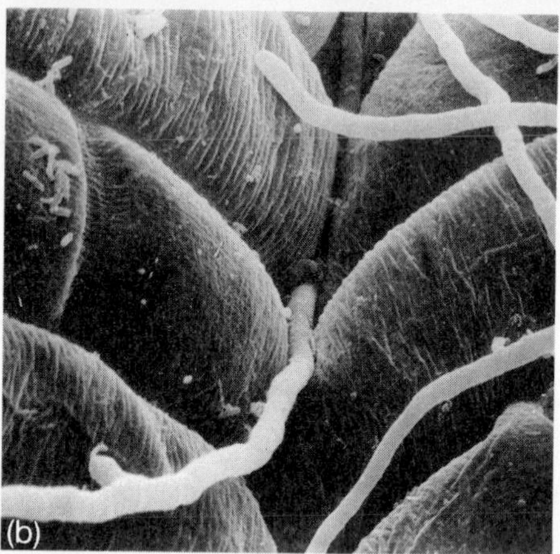

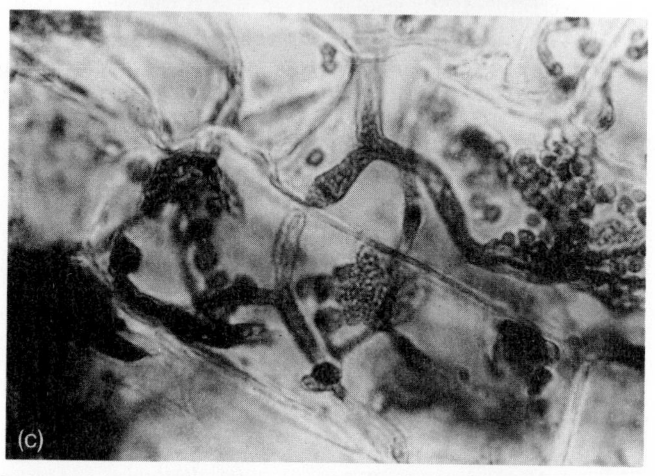

FIG. 9c.2. Pathogenicity of *Fusarium roseum* 'Culmorum' to hydrilla. (a) Effect of increasing concentration of spore inoculum on hydrilla. The tube on the extreme left was a control without the fungus; the other tubes, from left to right, had increasing inoculum levels from 12.5×10^3 to 5×10^5 spores per ml of water. (b) A scanning electron micrograph of growing mycelia of the fungus entering leaf tissue of hydrilla. (c) A transverse section of a hydrilla leaf showing intercellular ramification of mycelium.

important submerged weeds in the United States. A minimum of 25 000 spores per ml of treated water was required to kill hydrilla, but effective kill occurred when hydrilla was grown in different kinds of sterile water in containers ranging from 50 ml tubes to 20 l aquarium tanks (Fig. 9c.2). Following inoculation, shoots turned chlorotic and discoloured in 2 weeks, and disintegrated completely in 3 weeks. Histological evidence of infection (Fig. 9c.2) and the consistent re-isolation of the fungus from dying shoots confirmed pathogenicity. However, the pathogenic effects were seen only under acidic conditions, optimally around pH 5.5 to 6.5 (Charudattan *et al.* 1980). Schoen (1983), working with an isolate of this fungus obtained from Charudattan, found that it did not infect hydrilla at pH 7.0 and above. Spore germination was inhibited *in vitro* at pH 8.0 to 9.0 but not at 5.0 to 7.0 (Charudattan *et al.* 1980), suggesting that the lack of infection at alkaline pH is partly responsible for the lack of efficacy above pH 7.0.

The host range of the Dutch Culmorum strain was tested on 80 plant species including crop cultivars and non-target aquatic plants using seed infestation and/or seedling inoculation methods (Charudattan *et al.* 1980). Results indicated that the fungus infected some non-target hosts, but was not a significant primary pathogen of any terrestrial plant screened and thus a safe agent for field tests.

The strain was then tested for efficacy in several outdoor tests using 3.05 m diameter plastic pools containing hydrilla (Charudattan *et al.* 1984). Although the fungus-treated hydrilla was consistently more damaged than the non-treated plants, a practical level of control was not achieved. Moreover, since the fungus was effective only at pH $\leqslant$5.5 (Charudattan *et al.* 1980; Schoen 1983), its usefulness is further restricted in natural waters of pH $\geqslant$6.0. It was, therefore, concluded that the field efficacy of this strain and the prospects for using it as a microbial herbicide are uncertain. Additional basic studies are necessary to establish the mode of action underlying the lethal effects of this strain seen in the small-scale tests and how it may be rendered effective in large-scale tests.

Pathogens of Myriophyllum species

Eurasian water milfoil, *M. spicatum,* is one of the most troublesome aquatic plants in North America. Hence, attempts have been made to discover pathogens and determine the causes of disease-like conditions affecting this plant.

One such study concerned the disappearance of Eurasian water milfoil populations from the Chesapeake Bay, in the mid-Atlantic United States, between 1965 and 1969. Following its introduction, the plant became abundant in the Bay during the 1950s. But the sudden decline of the weed, by more than 95 per cent, prompted Bayley and others to explore the possible pathogenic agents responsible for the decline (Bayley, Rabin, and Southwick 1968). Bayley (1970) found plants with disease symptoms characteristic of conditions called the North-east disease or the Lake Venice disease and suggested that a virus, virus-like agent, or a toxin was the etiological agent. However, she was unable to infect healthy Eurasian water milfoil plants with any of a number of plant viruses (Bayley 1970).

A subsequent assessment of the problems led Bayley *et al.* (1978) to conclude that fluctuations in milfoil populations were related to several interrelated environmental factors, including tropical storms, turbidity, salinity, and disease. They decided that the evidence did not support the proposed viral etiology of the Northeast disease (Bayley, Rabin, and Southwick 1968; Bayley 1970).

Re-examination of the Lake Venice disease by Bean, Fusco, and Klarman (1973) led to the conclusion that a bacterium, rather than a virus, may have affected plants growing under low light intensities. The bacterial etiology was proposed also by Lammers (1968) who obtained 15 bacterial isolates from plants affected with the Northeast disease. However, none of Lammers' bacteria proved to be pathogenic when tested by Hayslip and Zettler (1973). These authors also tested strains of *Fusarium* spp., *Phytophthora* spp., *Pythium* spp., *Aphanomyces euteiches,* and *Rhizoctonia solani,* but found no or inconclusive evidence for pathogenicity (Hayslip and Zettler 1973).

Lekic (1971) in Yugoslavia, after studying the causes of diseases and the effects of several fungi on Eurasian water milfoil populations, concluded that a large number of micro-organisms were coexistent with this plant, causing its die-back. Species of *Fusarium, Alternaria, Penicillium,* and *Botrytis,* as well as *Flagellospora stricta, Articulospora tetracladia,* and *Dactylella microaquatica* were routinely isolated from different locations, and many of these were capable of causing shoot discoloration on young Eurasian water milfoil plants.

Andrews and Hecht (1981) in Wisconsin discovered a *Fusarium sporotrichioides* that induced localized, elongate necrotic stem lesions and systemic apical chlorosis on wound-inoculated Eurasian water milfoil shoots. Although the fungus was not highly pathogenic and penetrated mainly, but not exclusively, through wounds, it grew extensively in cortical lacunae. Since it lacked destructiveness under natural occurrences, it was not considered a good biocontrol agent.

Another fungus discovered by this group (Andrews, Hecht, and Bashirian 1982), *Acremonium curvulum,* occurred both epiphytically and endophytically in the stems, leaves, roots, and seeds of the plant. Immersion of healthy Eurasian water milfoil shoots in a suspension of spores (10^6 per ml) resulted in systemic basal necrosis that spread to two-thirds of the tissues within one week. However, the fungus was relatively innocuous to vigorously growing, healthy plants.

Another species of *Myriophyllum* that has been the target of pathological studies is *M. brasiliense,* parrot-feather. Bernhardt and Duniway (1984), who surveyed in northern California for endemic diseases of aquatic plants, obtained a pathogenic *Pythium carolinianum* from severe stem rots and root rots on parrot-feather. A test of biological control efficacy of this fungus in a small pond was moderately successful, but additional proof of efficacy appeared warranted.

Bernhardt and Duniway (1982) also found overwintering structures of pondweed, *Potamogeton crispus* and *P. nodosus,* that were frequently rotten when collected from soil in drained irrigation canals in winter. Among fungi isolated from such plant parts were *Fusarium lateritium, Burgoa* sp. (possibly *Papalospora* sp.), *Fusidium* sp., *Pythium* sp., and a *Rhizoctonia* sp. Healthy overwintering structures of *P. crispus* were colonized when incubated at 6°C to 9°C with mycelium of *F. lateritium* and *Burgoa* sp.

AN ASSESSMENT OF THE STATUS OF CONTROL BY MEANS OF FUNGI

Numerous pathologically interesting fungi have been identified, and a recent literature survey by the author revealed that 114 fungal taxa, including many virulent pathogens, have been reported worldwide from water hyacinth alone (Table 9c.1). Although the number and the frequency of reports attest to the worldwide interest in using fungi as control agents, only limited success has been achieved in the sphere of practical control.

Despite the wealth of scientific information that has accrued from surveys for potentially useful fungi from aquatic plants, only *C. rodmanii* has undergone extensive testing and some measure of development. This pathogen has not only served as a model microbial herbicide for aquatic use, it continues to impart limited natural control of the weed in parts of the United States. Thus, from a scientific perspective *C. rodmanii* has been successful, but its use commercially remains an unresolved issue.

There are several reports of pathogenic fungi from submerged aquatic weeds, notably hydrilla and Eurasian water milfoil, but none has been found to be both efficacious and practical in a biological control. Certain interesting diseases, such as the fusarium diseases of hydrilla and Eurasian water milfoil (Charudattan *et al.* 1984; Andrews and Hecht 1981), and the pythium disease of parrot-feather (Bernhardt and Duniway 1984), have added to our understanding of the pathology of underwater plants. These will serve as models for future explorations and research.

PROSPECTS

Whereas we now have expertise to control emergent weeds with fungi, it does not appear feasible to control submerged weeds by the classical or the microbial herbicide strategy in the foreseeable future (Charudattan 1983). The reasons for this are:

(1) the lack of knowledge and experience with diseases of submerged plants, particularly with regard to epidemiology;

(2) limitations (technological and monetary) in production and application of the large amounts of inoculum needed for effective control of underwater weeds; and

(3) unknown environmental consequences resulting from inundative applications of inoculum in large bodies of water.

To assure efficacy in underwater systems, it is expected that the entire body of water (on total volume basis) will need to be treated through inundative application of a fungus. Inoculative application of fungi (as used in classical biocontrol) over limited proportion of the weed population is unlikely to produce the levels of epidemics needed to control large weed infestations. This is likely due to the impedance of inoculum dispersal in water, inefficient deposition of inoculum on the weed, and dilution of inoculum. For these reasons, rates of inoculum up to several thousands or millions of spores per ml of treated water may be necessary (Charudattan and McKinney 1978; Andrews and Hecht 1981), raising questions about the practicability of such use. There are also unknown environmental consequences from use of such high concentrations of inoculum in natural waters. Hence, it is not possible to assure the public that undesirable side-effects will not occur from large-scale applications of fungi over water.

Biotechnological solutions may be possible to solve the environmental and efficacy-related problems through genetic engineering of micro-organisms (Pennington 1986). Although basic research must be undertaken to determine the feasibility of producing bioengineered micro-organism for controlling submerged weeds, I consider the chances of success of this approach to be extremely limited or none at present. First, we do not have suitable pathogens to develop as research models. Second, society does not appear to be ready for utilizing engineered pathogens, due to fears of potential risks that may be associated with this unproven technology.

Manipulation of epiphytic and endophytic microflora associated with submerged weeds to render them stress factors or opportunistic parasites has been suggested (Andrews, Hecht, and Bashirian 1982). Perhaps a more fruitful approach will be to improve the efficacy of pathogens by integration with other control measures. For example, pathogens and insect biocontrol agents can be integrated (Kasno and Soerjani 1981; Kasno 1982; Charudattan 1986), or pathogens may be combined with sub-lethal levels of chemical herbicides or growth retardants capable of reducing the weed biomass (Charudattan 1986).

Much work needs to be done in the case of submerged weeds. Surveys for new pathogens, a greater understanding of the aetiology and epidemiology of diseases of underwater plants, and extensive field trials are all necessary. However, whether such work is undertaken will depend on the economics of weed control and future developments in public ecological concerns. Unfortunately, in the present economic climate, ecological concerns appear to have taken a back seat in the USA, leaving only the chemical alternative.

ACKNOWLEDGEMENT

I thank Dr William Bruckart, Research Plant Pathologist, USDA–ARS, Foreign Disease–Weed Science Research Unit, Fort Detrick, Maryland, USA, for critically reviewing the manuscript and offering helpful suggestions for its improvement.

(d) Biological control of aquatic weeds by means of phytophagous fish

W. VAN DER ZWEERDE

INTRODUCTION

SEVERAL phytophagous fish species have been considered as biological control agents for aquatic weeds. Van Zon (1976) listed 29 species which feed mainly on aquatic macrophytes or phytoplankton. In practice, however, only one species, the grass carp (*Ctenopharyngodon idella*), is used on a large scale for aquatic weed control (Fig. 9d.1). This is because the grass carp is a voracious feeder on a great variety of plants. Most other phytophagous fish species consume less plant material and/or feed on a limited number of plant species. Moreover, the grass carp tolerates a wide range of environmental conditions and, in general, will not breed naturally outside its native habitat (it requires very specific conditions for spawning), which implies that its population can be kept under control.

Species like silver carp (*Hypophthalmichthys molitrix*) and big head carp (*Aristichthys nobilis*) that consume phyto- and zooplankton, and the polyphagous *Tilapia* (e.g. *T. melanopleura*, *T. zillii*) species are sometimes used in combination with grass carp for weed control. Their role in aquaculture, however, often in combination with grass carp, is much more important (Hauser *et al.* 1976; Legner and Fisher 1980; Dimitrov 1984; Smith 1985; Barthelmes 1985; Prinsloo and Schoonbee 1987). The silver carp is also used for the purification of waste water (Oláh, Kintzly, and Varadi 1982; Kovács and Oláh 1982).

In this context it should be noted that the small tropical fish species *Metynnis roosevelti* and *Mylossoma argenteum* (commonly referred to as 'silver dollars'), which occur in South America, feeding on the bases of plant stems, could have a potential for aquatic weed control in tropical regions. However, research is needed on the biology and management of these species before they can be used as a control method.

FIG. 9d.1. Grass carp (*Ctenopharyngodon idella*.) (Photo: Centre for Agrobiological Research.)

The grass carp offers not only a practical and cheap control method, but may also safeguard the 'biological potential' of the aquatic ecosystem. Often a combination of grass carp with other control methods, i.e. in an integrated control programme, is the best way to make optimal usage of the many different functions of water bodies (see also Chapter 10).

An additional advantage is that the grass carp provides a source of protein, which could be of special importance in developing countries. In comparison to physical and chemical weed control methods, however, the use of grass carp is more directly influenced by ecological factors, such as climatological conditions. Moreover, factors such as predatory fish species and water pollution may play a role and, in densely populated areas, it is not always feasible to prevent overfishing and poaching.

For general information on grass carp the reader is referred to Hickling (1965, 1969), Antalfi and Tölg (1971), Jähnichen (1973*a*) and Shireman and Smith (1983).

GENERAL BIOLOGY AND DISTRIBUTION OF GRASS CARP

The grass carp, or 'white amur' (after the River Amur in Siberia, its native habitat), was scientically described in 1844 by Cuvier and Valenciennes and given the name *Ctenopharyngodon idella*. The species belongs to the carp family (Cyprinidae).

The grass carp is a slender fish with a nearly round cross section and a rather broad head. The body is silver-coloured at the belly, shading into golden-brown to dark brown at the back. The scales have dark edges, which marks the pattern clearly. The grass carp differs from common carp (*Cyprinus carpio*) by lacking a high back and barbels.

At maturity, the males have longer pectoral fins than the females: the rays of their pectoral fins are finely serrated and the first ray is thickened. If touched, the fins feel like sandpaper. Mature females have enlarged, soft bellies and reddish marked ovipores. In younger fish sex determination is difficult.

The pharyngeal teeth (which are characteristic of the carp family) are heavy, sawed and in fish longer than 30 cm, have a broadened surface which makes them suitable for grinding hard plant material. The intestine of grass carp is a long coiling structure in the body cavity, of total length about 2.25 times body length. This is rather short for a herbivorous animal and suggests, in fact, an omnivorous diet. Grass carp larvae consume zooplankton initially, but at a length of 2 cm they start to consume aquatic plants. At a length above 3.5 cm the diet is almost completely vegetarian (Hickling 1969; Adamek and Sanh 1981; de Silva and Weerakoon 1981). Experiments in ponds revealed that grass carp above 20 cm consumed only very small amounts (about 0.1 per cent of gut content) of animal food, even when aquatic plants were absent. Grass carp do not search out animal food and prefer even terrestrial plants to animal food (Kilgen and Smitherman 1971; Terrell and Fox 1975; Terrell and Terrell 1975; Colle, Shireman, and Rottman 1978). In field experiments in the Netherlands where aquatic plants were completely removed, only 0.02 per cent of the gut content was of an animal nature (Willemsen *et al.* 1978).

The grass carp does not possess the enzyme cellulase and there are only a few micro-organisms in its intestine that produce this enzyme (Lesel, Fromageot, and Lesel 1986). As a consequence, the grass carp is not able to digest cell wall material, but only the content of plant cells which have been ruptured by the pharyngeal teeth (Hickling 1966; Secer 1976; Stroband 1977).

The conversion rate of plant tissue is variable, but mostly a value of about 30 g plants (fresh weight) for 1 g fish (fresh weight) is found. This rather inefficient use of food makes the grass carp very efficient for weed control (Hickling 1969; Opuszynski 1972; Fischer and Lyakhnovich 1973; Sutton 1974; Kilambi and Robison 1979; Fowler 1982).

The growth rate is very variable because it depends on such factors as type and quantity of food, plant species available, water temperature, oxygen content of the water, salinity, and density of the grass carp population (Shireman and Smith 1983).

The original habitat of the grass carp is in the north-eastern part of China, where it occurs, for example, in the rivers Sung-hau (Sungari), Liao-Ho, Huang Ho, and Chang Chiang, and in the adjacent part of the USSR in rivers like the Amur and Ussuri. In the Amur basin the average temperatures in January are between −20 °C and −25 °C and in July between +20 °C and 25 °C. Precipitation is mainly in summer. In winter dry north-western winds prevail; in summer humid north-eastern winds (Steinberger and Göschel 1979). In the lower reaches of the Amur river the differences in water level are relatively small. For instance the town Chabarowsk, which is about 750 km upstream, is only 60 m above sea level. In this basin the Amur river is a wide weaving river with many bends, swamps and extensive floodplains. In these parts of the river there are good growing conditions for the young fish larvae, because the water is shallow, with high summer temperatures, and there is abundant food (plankton, aquatic plants, and drowned terrestrial vegetation).

Grass carp sexually mature at an age of 3 to 10 years (at a length of 65 cm or more). Normally males mature earlier than females. The spawning season is June–July. Ripe individuals migrate towards spawning areas in fast-flowing tributaries in the upper stretches of the river. To initiate migration a water temperature of 15 to 17 °C is required. The spawning site must have turbulent water (because this provides a high oxygen content and keeps the eggs buoyant) and a clean sandy or rocky bottom.

Spawning itself requires a water temperature of at least 17 °C and is optimal at 26 °C. A sudden rise in water level (1 to 2 m) is a necessary precondition. If this rise does not occur, there will be no spawning, even when the water temperature is suitable.

To keep the fertilized eggs buoyant a flow velocity of about 1 $m\,s^{-1}$ is needed. The development to a free-swimming larval stage (8 mm) takes 4 to 5 d at 21 to 26 °C, which implies that they are transported downstream over a distance of about 400 km, to the inundated floodplains (Gorbach 1961, 1972; Krykhtin and Gorbach 1982).

Man has used grass carp for many centuries. In particular the Chinese have developed very sophisticated systems for fish culture in which grass carp play an important role. This is, in general, in polycultures of fish species with different food preferences (Zweig 1985; Li 1987). In the past young fish were obtained from the natural population in the rivers and transported to new sites locally: fish farming, therefore, remained restricted to the immediate vicinity of those rivers supporting natural populations of the fish. Only recently (during the second half of the twentieth century), have techniques for artificial reproduction been developed. Spawning is induced by hormone injections under narcosis and fertilization of the eggs is carried out *in vitro*. Hatching and the culture of the early larval stages is brought about by means of special equipment (Zonneveld and van Zon 1985). Since facilities for transport of fish by air have become available, the rapid spread of grass carp has become much easier worldwide.

In other parts of Asia the grass carp was introduced mainly for the production of human food (Alikunhi and Sukumaran 1965; Chokder 1967; Singh *et al.* 1967; Chaudhuri *et al.* 1976; Mehta, Sharma, and Tuank 1973; Sinha and Gupta 1975; Tsuchiya 1979*b*). In Eastern Europe food production was the main purpose, but also the potential for aquatic weed control played a role in encouraging the culture of grass carp (Krupauer 1971; Jähnichen 1973*b*; Nikolskiy and Aliyev 1974; Yefimova and Nikanorov 1977; Kokordak 1978; Tölg 1967; Miley, Sutton, and Stanley 1979; Opuszynski 1979; Negonovskaya 1980; Charyev 1984). On the other hand, the introduction of grass carp into the West was mainly for weed control purposes, for example, Western Europe (Bohl 1971; von Lukowicz 1976; Müller 1978; van Zon, van der Zweerde, and Hoogers1978; Jungwirth 1980; Leventer 1981; Stott 1979, 1981; Mugridge *et al.* 1982; Markman 1984; Fowler 1985*a*), New Zealand (Edwards and Hine 1974; Edwards and Moore 1975; Mitchell 1977, 1980, 1981) and North America (Bailey 1975; Ware *et al.* 1975; Custer *et al.* 1978; Guillory and Gasaway 1978; Mitzner 1978; Henderson 1980; Pierce 1983; van Dyke, Leslie, and Nall 1984). In Africa grass carp were introduced both for the production of food and the control of aquatic weeds (El Gharably *et al.* 1978; Pieterse 1979; Dubbers *et al.* 1981; Redding-Coates and Coates 1981; George 1982; El Gharably, Khattab, and Dubbers 1982; Schoonbee, Vermaak, and Swanepoel 1985).

Other benefits of the grass carp are its value for angling and its potential for the control of waterborne diseases (Coates and Redding-Coates 1981; van Schayck 1986).

The use of grass carp in aquaculture is the most important. Farming is practised in special fish ponds, man-made lakes, irrigation canals, and natural waters. The fish ponds are managed in a more or less intensive manner, according to the Chinese experience. In the other waters fish production is to a large extent based on the natural productivity of the system. By introducing grass carp man is able to harvest the plant production of these water bodies in a form which is readily available for human consumption. A combination of aquatic plant control and fish production looks very attractive since the flow of water can be maintained without the use of expensive machinery and, additionally, there is production of valuable animal protein.

CASE STUDIES OF GRASS CARP MANAGEMENT PROGRAMMES

Introduction of grass carp into The Netherlands

The introduction of grass carp into The Netherlands is here taken as an example of a relatively successful introduction under temperate climatic conditions.

It should be noted that in Dutch water courses aquatic weed problems are generally caused by a wide range of species, of varying life forms. It is, therefore, necessary to reduce the biomass of several species simultaneously in most cases.

In 1964 the Organization for the Improvement of Inland Fisheries (OVB) faced a problem with an excessive growth of filamentous algae in newly built fish ponds. In order to cope with this problem grass carp were imported from Hungary in 1966 and from Taiwan in 1968. First stocking took place in experimental ponds in 1968, but the results were disappointing, mainly because the density was too low (25 to 100 kg ha^{-1}). In 1970 a stocking rate of 180 to 360 kg ha^{-1} gave better results. Aquatic plants were eliminated and the fish had to be fed grass to keep them in condition.

Subsequently, in 1972 experiments were carried out in water courses, which also included research into the influence of grass carp on different elements of the aquatic ecosystem. In 1973 a long-term field experiment was started. The results of these experiments were so promising that, subsequently, managers of water bodies obtained official permission to experiment with grass carp under certain defined and restricted conditions (van der Eijk 1978). The research programme was carried out by a working group formed by the various organizations involved in aquatic weed control in The Netherlands. These were the national organization of water boards (responsible for management of the quantity and the quality of the water), the governmental nature conservation organization, the governmental fisheries management body, the national organization of angling societies, the OVB as fish producer, and governmental research institutes working on fisheries, nature management, and management of water courses. The working group operated within the framework of the National Board for Agricultural Research. The results of the working group's efforts are presented in a report entitled 'Graskarper in Nederland' (The grass carp in The Netherlands) (1984). The main conclusions were:

(1) the grass carp was an efficient tool for aquatic weed control under field conditions;

(2) in the Netherlands there is no natural reproduction, which decreases the risk of unwanted spread of the fish; and

(3) grass carp should be released under certain defined conditions.

These conditions were:

(i) grass carp were only to be stocked in water bodies where aquatic plants were causing weed problems;

(ii) the water body must be suitable for fish in general and for grass carp in particular;

(iii) the interests of nature conservation had to be taken into consideration; and

(iv) the maximum stocking rate was 250 kg ha^{-1}.

Permission was to be granted by the Ministry of Agriculture and Fisheries, on the basis of advice from the regional civil servants dealing with fisheries and nature conservation. There were three conditions:

(1) grass carp must be delivered from a disease-free stock;

(2) open connections with other water bodies (where grass carp were not to be stocked) had to be closed by a specific type of fence (Fig. 9.d2); and

(3) the permission to stock was for a period of 5 years at a time.

This procedure for obtaining permission has not since changed.

In The Netherlands the maximum stocking rate of 250 kg ha^{-1} was chosen to avoid too severe a reduction of aquatic vegetation. It has been observed that higher stocking rates (up to 360 kg ha^{-1}) eliminated aquatic plants completely, even in watercourses which showed a luxuriant growth before. A complete elimination of the vegetation is undesirable because this may result in a drift of bottom material and bank erosion may occur when the fish start to graze the zone

FIG. 9d.2. A drainage canal in The Netherlands which is closed by a fence. At the back of the fence the canal is stocked with grass carp. (Photo: J. C. J. van Zon.)

immediately above the water level. Moreover, a total elimination of aquatic plants has a negative impact on the aquatic ecosystem as a whole, because many organisms depend on plants for food, shelter or reproduction.

Plant consumption is, to a large extent, regulated by water temperature, which is not manageable in practice. Therefore, the manipulation of grass carp density is the only means of regulating the amount of plant material consumed. Water boards and municipalities generally aim for a high degree of plant control and, as a consequence, they apply an initial stocking rate near the maximum level. Other users (e.g. angling societies) generally prefer a lower stocking rate (100 to 150 kg ha^{-1}). A high initial stocking rate may imply that even in the first year part of the grass carp population must be removed, in order to prevent complete elimination of the aquatic plants.

The actual density of grass carp in a stocked system is, of course, a function of both the initial stocking rate and the balance between growth and death rate (due to natural death, predation, and poaching). The death rate is usually low in The Netherlands, since water bodies are selected which are suitable for fish and measures are taken against predation (e.g. removal of large piscivorous fish, regulation of angling, and stocking of grass carp with a minimum length of 35 cm). In order to regulate the grass carp population the manager of a water course must know the number and size of grass carp present in each year and whether this corresponds with the particular conditions of the water course and the management objectives.

Costs of maintenance of watercourses vary considerably. To get information on the costs of maintenance by grass carp a comparison has to be made with other control methods in similar watercourses in the same area. Provoost (1981) gives such a comparison for thirteen water boards. In those watercourses where control is by means of grass carp bank vegetation was controlled mechanically (Table 9d.1). In seven water boards the costs of maintenance decreased after the introduction of grass carp, on average 35 per cent. In three water boards the costs remained the same and in three water boards the costs increased. In most instances the costs of the fish were about hfl 0.20 m^{-1} (based on a 5 year depreciation period). The large differences were caused by the costs of fences (10 year depreciation period) and their maintenance.

Table 9d.1. Average costs (hfl m^{-1}) of maintenance before and after introduction of grass carp into watercourses on different soil types in The Netherlands (range in brackets)

Soil type	Before grass carp	After grass carp
Clay	2.06 (0.66–3.36)	1.62 (0.58–2.91)
Peat	1.73 (0.77–2.85)	1.23 (0.13–2.02)
Sand	2.34 (1.30–3.10)	2.40 (1.61–5.09)

Source: Provoost 1981.

Recently, de Vries (1987) carried out a similar survey in sixteen water boards. In nine of these water boards aquatic plant management with grass carp was cheaper (by up to 50 per cent) than previous maintenance (mainly mechanical). In five water boards there was no difference in costs, whereas in only two water boards had the costs increased. The costs of maintenance with grass carp depend on various factors, such as the period of time that the fish are present in that water body, the surface area, and the differences in water level in the watercourses. The costs of fences and their maintenance are a main cause of the variations in total costs. Grass carp offer a cheap method of control in large, relatively flat areas. However, costs rise in relatively small areas (more fences needed per unit area), or when there are marked differences in water level of the different parts of the watercourses (need for fences at the different weirs, stronger fences due to the more variable discharge rate, and more frequent inspection at critical periods). The relative costs are given in Fig. 9d.3. There was a great variability, which is mainly due to the costs of fences. When a water board uses grass carp for the first time the general trend is to start on a small scale, which brings about high

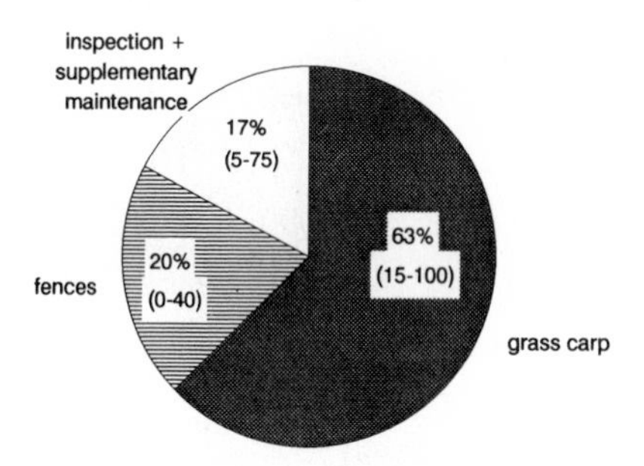

FIG. 9d.3. Average relative share (%) of the elements which constitute the costs of the use of grass carp in The Netherlands (range in brackets). (Source: de Vries 1987.)

initial costs. In one particular water board (Willems 1981) a simple fence has been developed that is self cleaning. However, this type of fence is not absolutely 'fish-proof' and, as a consequence, is not applicable when fences are a legal requirement (if the water body stocked with grass carp is connected with surface waters where stocking with grass carp is not permitted).

In the Netherlands the area stocked with grass carp has steadily increased since 1977. Each year about 25 000 kg are stocked in about 200 ha. In 1987 there were 1092 sites with grass carp and 820 users (a total of 270 000 kg in 2300 ha) (van der Spiegel, OVB, personal communication). As restocking is included in this figure, it may be assumed that the surface area where the fish are currently present is smaller than this (the best estimate is *c*. 1500 ha). Water boards are the

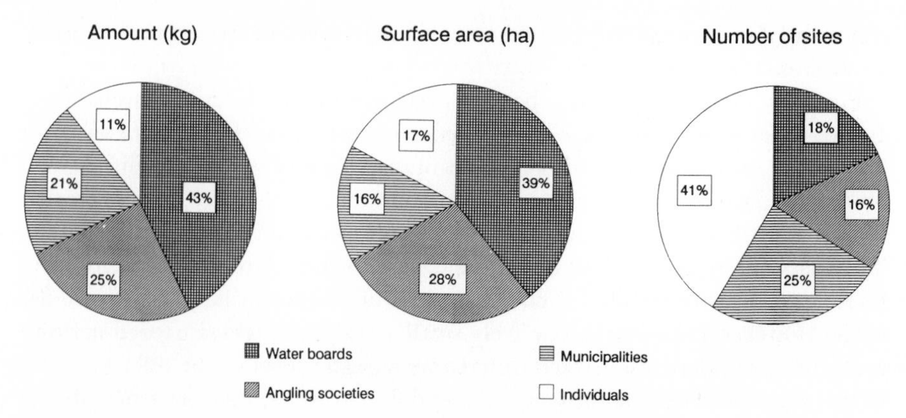

FIG. 9d.4. Relative amount, surface area, and number of sites (%) stocked with grass carp by different categories of users in The Netherlands, after 10 years of stockings (1977–1987). (Source: van der Spiegel, OVB, personal communication.)

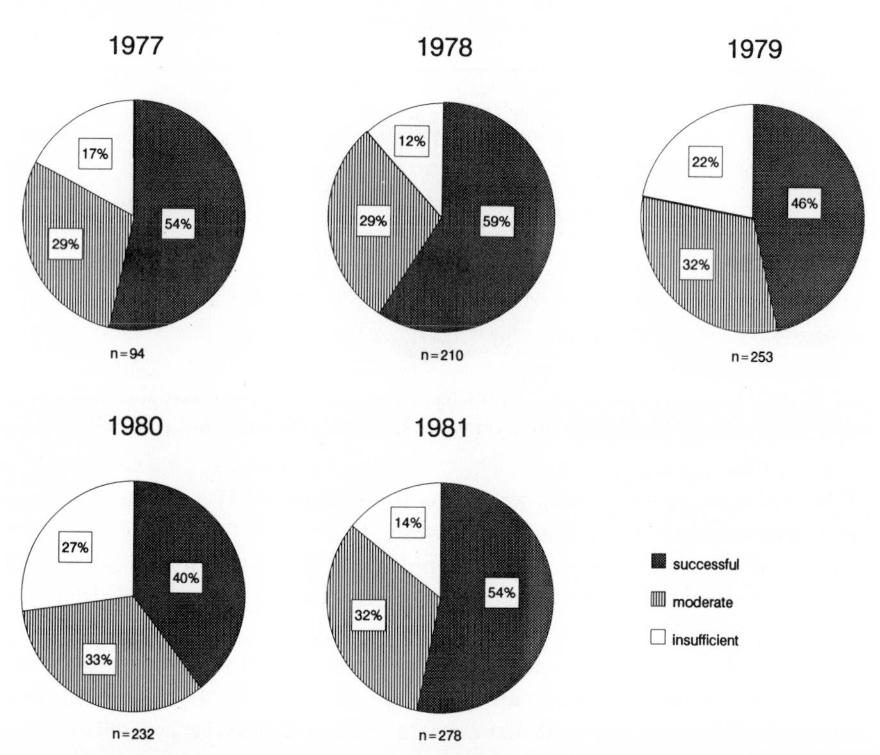

FIG. 9d.5. Relative evaluation (% of water bodies) by users of aquatic plant control using grass carp in The Netherlands. n: number of water bodies. (Source: Graskarper in Nederland 1984.)

major users of grass carp, with other users having a larger number of relatively small sites (Fig. 9d.4).

Evaluation of aquatic weed control by grass carp remains highly subjective as the manager of each area has his or her own objectives. These objectives are partly connected with the conditions in that particular area (width, depth, seasonal discharge pattern) and partly by personal management preferences. In The Netherlands the evaluation of the use of grass carp by managers of watercourses is investigated by means of regular inquiries. Overall results are presented in Fig. 9d.5. In normal years the results were successful in more than half of the experimental sites. In 15 to 20 per cent of the sites there was insufficient control. Both in 1979 and 1980 the winter was relatively severe which led to a high fish kill, especially in shallow waters with a high content of organic material.

A differentiation into categories of managers of watercourses (Fig. 9d.6) reveals that up to 1980 the water boards seemed the least satisfied with the use of grass carp for weed control (especially in 1980, perhaps due to fish kill in the preceding severe winter). Obviously the differences between the categories of managers are caused by their different objectives. Water boards (like their counterparts in other countries, such as Internal Drainage Boards in the UK) are traditionally inclined towards a rather rigorous control of aquatic plants in order to guarantee an optimal discharge capacity. This policy aimed at near-complete control of aquatic weeds is also expressed in the stocking rates (190 kg ha^{-1} on average), whereas other managers use 75–120 kg ha^{-1} (Riemens 1981).

It is interesting to note the more positive conclusion of the water boards in 1986 (Fig. 9d.7). Most probably this is because water board employees have become better acquainted with the use of grass carp: they have learnt to use and manage the fish. In this context it should also be noted that, initially, a large number of water boards stocked grass carp in relatively small-sized plots, while at present only a small number of water boards use grass carp, but on a much larger scale. Various water boards stopped experimenting with grass carp after early disappointments. Others have learned from their initial misfortunes, have adjusted their management accordingly and are now content with this method. In Fig. 9d.8 a canal in The Netherlands is shown in which the aquatic weeds have efficiently been controlled by grass carp.

Introduction of grass carp in Egypt

Although most experience with grass carp, in connection with aquatic plant management, has been obtained in the temperate zones, some studies have been conducted in tropical or sub-tropical regions. An example is Egypt, where grass carp were introduced in 1976.

A preliminary experiment with grass carp for the control of aquatic weeds in Egypt was carried out in a small drainage canal in the Province of Giza near the

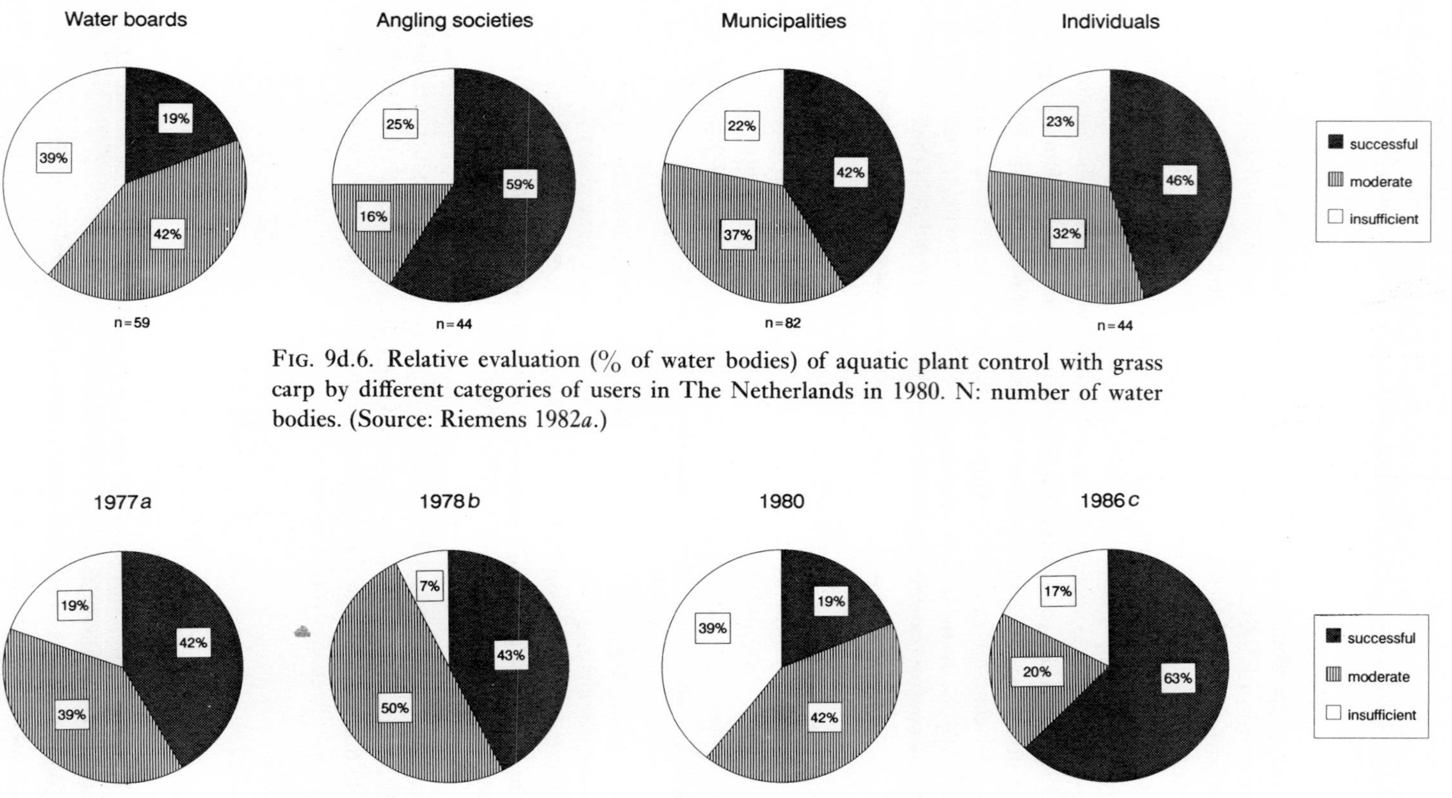

FIG. 9d.6. Relative evaluation (% of water bodies) of aquatic plant control with grass carp by different categories of users in The Netherlands in 1980. N: number of water bodies. (Source: Riemens 1982*a*.)

FIG. 9d.7. Relative evaluation (% of water bodies) of aquatic plant control with grass carp by water boards in The Netherlands in different years. (Sources: 1977: Riemens 1978; 1978: Provoost 1979, 1981; 1980: Riemens 1982*a*,; 1986; 1986: de Vries 1987.)

FIG. 9d.8. A canal in The Netherlands in which aquatic weeds are efficiently controlled with grass carp. (Photo: G. Goodijk.)

FIG. 9d.9. A drainage canal in Egypt: at the left of the fence without grass carp, at the right with grass carp. (Photo: A. H. Pieterse.)

city of Cairo, during the period autumn 1977–summer 1978, in the framework of a Dutch–Egyptian project (El Gharably *et al.* 1978; ILACO 1978; Pieterse 1979). Before stocking, the canal was cleared of weeds with a mowing bucket. It appeared that a stocking rate of about 200 kg ha^{-1} gave an adequate control of submerged weeds. Moreover, the spread of emergent weeds was largely prevented, although at the end of the summer season there were some scattered stands of *Phragmites australis*. On the other hand, in the absence of grass carp (in the control plot) a dense vegetation of aquatic weeds developed which completely filled up the water volume (Fig. 9d.9). In spring, this vegetation mainly consisted of submerged weeds but in summer it was mostly replaced by emergent plants.

These favourable results prompted the Egyptian authorities to continue research on grass carp. In Ismailija a hatchery was built on the campus of the Suez Canal University (Siemelink *et al.* 1982) and a large scale field experiment was conducted in a 10 km stretch of a nearby 25 m wide irrigation canal (Port Said Fresh Water Canal). This canal is an example of an irrigation canal in Egypt which contains water throughout the year. The smaller canals and drains, on the other hand, are generally drained off in the month of January for maintenance purposes. If grass carp are introduced for the control of aquatic weeds in these smaller canals and drains restocking is needed every year.

Breeding of grass carp in the hatchery in Ismailija, which was carried out in co-operation with Dutch specialists, was relatively successful (Siemelink *et al.* 1982). However, growth of fingerlings in cages, which is common practice in The Netherlands, led to a high fish mortality. This was mainly due to diseases. Another problem for culturing grass carp in cages was that, because the cages were placed in large, rapidly flowing irrigation canals, the fish had to be removed when the Ministry of Irrigation applied the herbicide acrolein in upstream parts of the canal.

For the experiment in the Port Said Fresh Water Canal fish were stocked, at a length of about 25 cm, at a rate of about 180 kg ha^{-1} (Dubbers *et al.* 1980). The experiment was started in August 1979 and during the following autumn and winter season weed growth was maintained at a relatively low level. However, in spite of the fact that they had increased in size the fish could not cope adequately with aquatic weed growth in spring. It was concluded that a higher stocking rate was required. Following experience in The Netherlands a stocking rate of 200 to 300 kg ha^{-1} was thought to be effective, but most experiments in The Netherlands had been carried out in relatively shallow water (less than 1 m deep). In the Port Said Fresh Water Canal (*c.* 1.80 m deep), the total water volume is almost completely filled with submerged plants (mainly *Potamogeton pectinatus*). As the stocking rate of grass carp is generally based upon the surface area this implies a larger plant mass per kg fish than in shallow canals. It also appeared that the vegetation was insufficiently controlled at sites which were used by the local people for washing or swimming. Apparently the fish at these sites were disturbed and as a consequence refrained from feeding. Another problem was poaching which took place on a large scale at night.

In 1981 a large hatchery was established in the vicinity of Cairo (the Delta Breeding Station). The survival rate of the fingerlings markedly improved when growth took place in ponds instead of cages. Additional field experiments near Cairo showed that, in contrast to The Netherlands, it was possible to stock relatively small fish, because large predator fish were virtually absent (Pieterse and van Zon 1983; van Weerd 1985; Khattab and El Gharably 1986). Favourable results were obtained with fish between 1.5 and 30 g. However, a disadvantage of the small fish was that they did not feed very well on the more fibrous plant material.

ILACO (1982) made a comparison of the costs of conventional aquatic weed management in Egypt and management by means of grass carp including costs of breeding. In canals which contain water throughout the year the costs of maintenance with grass carp were about half the costs of conventional maintenance. On the other hand, in canals which were drained off in January there was little saving to be had by using grass carp.

It may be concluded that grass carp are an effective and relatively cheap means of control of aquatic weeds in Egypt. Moreover, the fish have the additional advantage of providing an important source of protein for local people. However, field work has shown, especially in the more recent experiments near Cairo (El Gharably, Khattab, and Dubbers 1982), that it will be extremely difficult to prevent overfishing in areas of dense human population. The use of small fish, despite their drawbacks, might be the only way to cope with this problem.

FACTORS WHICH INFLUENCE SUCCESSFUL AQUATIC WEED MANAGEMENT WITH GRASS CARP

Experiences in The Netherlands and Egypt, as well as other regions, have shown that successful aquatic weed management with grass carp depends to a large extent on the following factors.

Water depth

For several reasons grass carp give better results in deep water (> 1 m depth) than in waters < 1 m deep. In deeper water there is less fluctuation in environmental conditions. This is especially important in climatic zones where ice cover occurs in winter. Because submerged aquatic plants tend to grow towards the surface, most biomass is situated in the upper layers of the water column, which implies less plant material in deeper watercourses in relation to the water volume. Consequently, in deeper water even fairly moderate plant consumption by grass carp may suffice to safeguard the required discharge capacity. In this context it should also be noted that in man-made water courses, 'deeper' mostly also means 'wider', which also tends to maximize discharge capacity. Another advantage of deeper watercourses for grass carp usage is that the fish are less vulnerable to predation by birds, such as heron (*Ardea cinerea*).

Regularity of previous control measures

When maintenance of a particular watercourse has been neglected and a dense vegetation of aquatic weeds has developed, stocking with grass carp will not give immediate results unless a very high stocking rate is used. In the latter case this will result in far too high a density of grass carp once the plants have been consumed and a reduction in the number of fish will be needed. It is, therefore, advisable to clean the neglected water body first by other means. The grass carp have then only to control weed regrowth, which reduces the risk of overstocking and improves the chances of successful long-term management.

Water temperature

Grass carp activity is directly influenced by temperature and it is obvious that, in the temperate regions of the world, the seasonal pattern of water temperature determines the success of grass carp usage. A short period of relatively high water temperature in early spring may be decisive for good results. On the other hand, a prolonged cold period in spring may result in insufficient control, because plant growth starts at water temperatures which are too low for grass carp to feed actively.

Grass carp density

The actual density never equals the stocking density, as the fish will grow and some individuals will die. To obtain a favourable result over a long time-span the population density of the fish must be managed. Since the survival of grass carp is generally high, at least in European waters, management mostly involves a reduction of the grass carp density. This can be done by:

(1) removal of fish; and

(2) enlargement of the stocked area.

If the grass carp density becomes too low, grass carp can be restocked, the surface area can be reduced by temporary fences, or an additional clearance can be carried out.

In order to avoid the problem of having to reduce fish density shortly after stocking, it is advisable to start with a somewhat lower than optimal density, in combination with an additional clearance operation. Fish density will then increase gradually, in an approximate equilibrium between plant growth and consumption. In drainage channel networks another possible means of postponing the necessity for removal of fish is to determine the stocking rate only in relation to the surface area of the main channels. Fish growth can then be compensated for by extra consumption of plant biomass from the smaller channels.

Size of the grass carp

The size of the fish is important in relation to predation or poaching. Moreover, fish size is connected with selectivity of feeding. If grass carp are kept in fenced plots, the minimum size is determined by the mesh size of the fences, which should not be too small in order to avoid obstruction of the water flow. A relatively small mesh size is attractive, however, since this permits the use of small grass carp, which have a fast growth rate and consume the largest amount of aquatic plants per kg of body weight (Osborne and Sassic 1981). Moreover, small grass carp are relatively cheap, compared to larger fish.

Predation

Grass carp are rather sluggish, which makes them vulnerable to predation by piscivorous fish. When the vegetation is dense plant cover provides some protection, but the unnaturally high density of the fish in weed control operations makes their presence conspicuous. In areas where a severe predation by piscivorous fish is expected the manager should take measures to minimize the losses. It is possible, for instance, to fish selectively for larger piscivorous fish prior to stocking of grass carp. An alternative is the use of larger grass carp.

Birds are also potential predators. Grass carp are active in the surface layers and often clearly visible. The size of the fish determines the likelihood of predation by birds. Small fish are more vulnerable. They can escape wading birds, such as herons by diving, which is only possible in deeper water.

Poaching can be a serious problem. Shortly after stocking grass carp can easily be caught because in the fish hatchery they are accustomed to being fed by people. Later on, especially after experiences with anglers' hooks it is difficult to catch grass carp. Illegal fishing can be reduced by enforcement of prohibitory rules, or by the use of grass carp of an unattractive size (e.g. very small ones).

Selectivity

Grass carp prefer soft-tissued aquatic plants, filamentous algae, and duckweeds (Lemnaceae). The more fibrous emergent plants are avoided, as well as plants with large floating leaves and plants with toxic or other compounds disliked by the fish in their leaves or stems. All preferences are less marked when water temperature is high and/or the grass carp are of a larger size. When feeding-selectivity poses problems for weed control, which is mostly restricted to water bodies where the environmental conditions are marginal for grass carp (Fowler and Robson 1978), or when small fishes are used, additional control measures will be needed.

Surface area of grass carp stocking in a channel network

Experiments with grass carp have often not produced the desired results in drainage channel networks in The Netherlands, because too small a surface area has been used. Apart from the relatively high costs of fencing, the amount of vegetation in a small area is limited, which makes the stocking rate very critical. In the drainage channel networks in The Netherlands there are, in general, larger main watercourses and smaller secondary ones. Grass carp are mainly active in the main channels, so gross discharge capacity is guaranteed. In the smaller side channels an extra amount of plant biomass is available that can be fed upon in case of food shortage in the main channels. If, in any year, the secondary channels are not sufficiently cleared of plants by grass carp they can be cleaned otherwise.

Integration

Stocking of grass carp is only part of the total management programme of watercourses. The environmental conditions in the watercourse have to be maintained at a level which is suitable for fish and other organisms. Fish density and/or size are the main tools to affect the degree of control of aquatic plants by grass carp. Because in temperate regions weather conditions (especially water temperature) strongly influence the success of grass carp, it is often advisable to combine a low density of fish with another maintenance method. Additional mechanical control is most common, both preventive (for instance by clearing a strip of plants in the middle of the watercourse) and curative (by removing leftover stands of aquatic plants which obstruct the flow of water). An additional biological control method which can be applied is shading of the water body by planting trees or shrubs along the banks, or by stimulating the growth of rooted plants with large floating leaves (such as *Nymphaea* spp.: see Chapter 9a) (Dawson and Kern-Hansen 1978; Hermens 1978; Pitlo 1982; Dawson and Haslam 1983). A combination of grass carp and chemical control (see Chapter 10) is possible if there are no adverse side-effects on the fish.

GRASS CARP AND THE AQUATIC ECOSYSTEM

Macrophytes and phytoplankton, the major primary producers in aquatic ecosystems, provide food for many organisms. Moreover, the macrophytes provide shelter and are a substratum for the deposition of eggs. Between aquatic macrophytes and phytoplankton there exists an equilibrium which is based on competition for light and nutrients, and perhaps also on allelopathic interactions. This equilibrium is disturbed when aquatic macrophytes are removed by an aquatic weed control method. The amount of plants left, as well as the ecological conditions in the particular water body, determine at what level the equilibrium will be re-established. A dominance of phytoplankton will have a negative effect on the aquatic ecosystem as it disturbs the oxygen regime, reduces light intensity and sometimes brings about the production of toxic compounds (see also Chapter 6).

In The Netherlands various studies have been carried out on the ecological side effects of grass carp (van Zon 1974; van Zon, van der Zweerde, and Hoogers 1976; van der Zweerde, Hoogers, and van Zon 1978; van der Zweerde 1980, 1982*a*, 1983; Pot 1986). Marked side-effects were only noted when large amounts of plant material were removed. There are few or no negative effects, provided that 10 to 25 per cent of the surface area of a water body remains covered with plants. In many instances, it appeared that conditions for a large number of organisms improved after the introduction of grass carp. In the western part of the Netherlands, for example, many watercourses are covered during the summer period by a blanket of filamentous algae or a thick layer of duckweeds (Lemnaceae) and/or water ferns (*Azolla* spp.). As these small floating plants do not markedly obstruct the water flow and mechanical removal of these plants is rather difficult, they are generally not controlled. Grass carp, however, preferentially consume these plants and after stocking with grass carp, they usually disappear which improves the oxygen regime in the water.

A large reduction in the aquatic fauna was noted in the studies mentioned above only when the aquatic vegetation was almost completely removed. Under those conditions there was initially a decrease in the biomass of these organisms, subsequently in the number of individuals and finally in the number of species. This implies that there is a high potential for recovery, prior to major community alterations occurring, provided that measures are taken in time to reduce the grass carp density.

In other countries, for instance Sweden (Ahling and Jernelöv 1971), Romania (Cure 1974), Great Britain (Moore and Spillett 1982; Fowler 1985*b*), Germany (Kucklentz 1985), USSR (Aliyev 1976*a*,*b*), New Zealand (Rowe 1984; Mitchell 1986; Mitchell, Fish, and Burnet 1984) and the United States (Beach, Lazor, and Burkhalter 1976; Shireman, Colle, and Martin 1979; Kobylinski *et al.* 1980; Leslie, Nall, and van Dyke 1983; Richard and Small 1984; Cassani and Caton 1985; Richard, Small, and Osborne 1985; Small, Richard, and Osborne 1985) research on side effects of grass carp gave more or less similar results. Ewel and Fontaine (1982) have developed a preliminary simulation model with the ultimate goal of predicting the effects of grass carp in different aquatic ecosystems.

An advantage of the use of grass carp over other methods for aquatic weed control is its gradual effect, which avoids large fluctuations in environmental conditions. Moreover, it is relatively easy for the manager of a water body to determine whether the fish are having too large an impact, and to take the necessary corrective measures before damage to the aquatic ecosystem occurs. It is quite easy to reverse the effects of a grass carp introduction, by removing the fish, which is not the case with most other forms of aquatic weed control.

In discussing the side effects of grass carp, it is essential first to determine what level of weed control is required and secondly whether the remaining plant growth is in accordance with general ecosystem requirements. With other methods of control, especially physical clearance and herbicides, regulation of plant growth over extended periods of time is usually much more difficult as these methods tend to have a more drastic short-term effect, followed by a period of recovery and regrowth. When the same quantities of plants are removed over a similar period of time using different control techniques, the ecosystem side effects of grass carp are comparable to those of other methods, but other techniques may also cause direct ecosystem damage, e.g. toxic effects on non-target organisms of certain herbicides; direct removal of fish by mechanical harvesters (Robson and Barrett 1977; Haller, Shireman, and Durant 1980; see also Chapters 7 and 8).

If there are objections to the use of grass carp because it would negatively affect certain aquatic organisms, owing to the reduction in aquatic plant cover in treated systems, it should be borne in mind that these objections are of course valid for any measure used to clear water bodies (van Leeuwen 1979; van Zon 1985). The only difference is that grass carp, as stated previously, give a more prolonged period of clearance than other measures and will, therefore, tend to maximize long-term impacts on other organisms.

The use of the grass carp outside its native range is sometimes criticized because it is not a native fish species and introductions *per se* are not considered desirable. However, many non-indigenous fish species have been introduced in many areas of the world. During the nineteenth and twentieth centuries, for instance, 27 non-indigenous fish species have been introduced into The Netherlands (Vooren 1972; de Groot 1985) and 25 species into the USA (Lachner, Robins, and Courtenay 1970). Several species were also introduced into Australia (Buckmaster 1980). In order to avoid the risks to the endemic fauna and flora inherent in the introduction of non-native fish species, Vooren (1972) proposed that this should only be carried out when there is a temporary population of non-breeding fish (i.e. the population of the fish can be controlled by man). What then are the prospects for the natural reproduction of grass carp outside its native habitat? The literature on this subject is very contradictory (Tang 1960; Nikolskiy and Aliyev 1974; Vinogradov and Zolotova 1974; Nezdoliy and Mitrofanov 1975; Holčik 1976; Stanley 1976*a*; Pflieger 1978; Stanley, Miley, and Sutton 1978; Tsuchiya 1979*a*; Verigin, Makeyeva, and Zaki Mokhamed 1979; Opuszynski 1982*a, b*; Abdusamadov 1986). A comparison with environmental conditions in the native habitat makes natural reproduction of grass carp in Western Europe seem very improbable. On the other hand, in some other areas natural reproduction cannot be completely ruled out. In the United States this has led to experiments with sterile fish: monosex grass carp (Stanley 1976*b*; Shelton, Boney, and Rosenblatt 1982), hybrids between grass carp and silvercarp or bighead carp (Kilambi and Zdinak 1980; Callahan and Osborne 1983; Harberg and Modde 1985; Osborne 1982, 1985) and triploid grass carp (Allen and Wattendorf 1987) have so far been tried. The hybrids were not very successful because they consumed only small amounts of aquatic plants. On the other hand, promising results have been obtained with the triploid grass carp.

Grass carp could be used for the control of aquatic plants in places where chemical or mechanical weed control are difficult to carry out, are very expensive, or are unwanted for other reasons. An introduction of grass carp, should be considered after balancing the pros and cons of all available methods (van Zon 1979).

The introduction of grass carp is safe when certain general regulations for its introduction are taken into consideration. The fish should be free of diseases and parasites. Precautions must be taken to avoid the simultaneous introduction of any other undesirable fish species. In Germany and Austria a small fish species (*Pseudorasbora parva*) has been introduced which is native to the Amur river (Weber 1984; Stein and Herl 1986). It is assumed that its introduction has occurred together with that of grass carp or silver carp.

It may be concluded that, if sensible precautions are taken, grass carp can be used with no greater risk to the aquatic environment than any other weed control measure and, in many cases, the fish are much to be preferred to alternative measures because of their general lack of ecosystem side-effects, provided that the fish population is managed in the right way.

ACKNOWLEDGEMENTS

Many thanks are due to J. C. J. van Zon (Euroconsult, Arnhem) for his comments. A. van der Spiegel (OVB, Nieuwegein) is thanked for the provision of data on the amount of grass carp stocked into Dutch watercourses.

10

Present status and prospects of integrated control of aquatic weeds

K. J. Murphy and A. H. Pieterse

Integrated control has been broadly defined as a 'management system that utilizes all suitable techniques to reduce pest populations and maintain them at levels below those causing economic injury' (van den Bosch *et al.* 1971). This implies a combination of different control measures instead of an approach based on a single procedure. Although this definition was drawn up in the context of invertebrate pest management, it may equally apply to integrated aquatic weed management (Aurand, Tyndall, and Trudeau 1982). It should be noted, however, that the term integrated control also refers to approaches aimed at minimizing the use of pesticides. The report on Terminology in Weed Science (Coördinatiecommissie Onkruidonderzoek 1984) defines integrated control as a 'control system based upon the population of a harmful organism or virus, taking into account natural resistance factors and based upon a minimal use of techniques and products harmful to the environment'.

In agriculture, integrated control is generally connected with a revaluation and improvement of various kinds of 'traditional' control measures, such as ploughing and mixed cropping, instead of using chemicals or sophisticated machines. For the control of aquatic weeds, however, the common view regarding integrated control is more oriented towards combining management procedures in order to achieve a more cost-effective means of managing aquatic weed problems. The use of herbicides in aquatic habitats is increasingly considered to be less acceptable in environmental terms. In general, chemicals are only applied to fresh waters if other means of aquatic weed control are ineffective or too costly. Consequently, decreasing the use of chemicals is usually not a major objective of integrated control in aquatic systems.

The concept of integrated control in aquatic weed management is long-established (Ranade and Burns 1925; Ahmad 1953; Hofstede 1960) and is generally considered to be a desirable approach. However, up to now the combined application of dissimilar control methodologies, at least when carried out in accordance with a previously planned scheme, has been relatively rare. There are, however, indications that integrated aquatic weed management, aimed at keeping weed growth below a pre-defined threshold nuisance level, is becoming a more widely accepted concept (Clark 1982; Tyndall *et al.* 1982; Henley 1982).

In the USA there has been considerable interest in the integrated approach (e.g. Guerra 1974; Pesacreta, Hodson, and Langeland 1984; Peverley *et al.* 1974). Programmes have been developed which imply the integrated use of a chemical (2,4-D) and a biological agent, the alligator weed flea beetle, *Agasicles hygrophila* against alligator weed, *Alternanthera philoxeroides*, (Foret 1974; Foret, Spencer, and Gangstad, 1974; Gangstad 1975*a*,*b*; Blackburn and Durden 1975; Durden, Blackburn, and Gangstad 1975). In general, these programmes led to improved control. When only chemical means of control were carried out, relatively high herbicide levels were required, while the alligator flea beetle alone did not bring about a satisfactory reduction of floating alligator weed mats. Moreover, in the northern part of the problem area few flea beetles are able to overwinter and the population does not build up to a size sufficient to give effective control before late summer or early winter.

In addition, there have been several studies on the control of water hyacinth, *Eichhornia crassipes*, by combinations of biological and chemical control methods (Center, Steward, and Bruner 1982; Rushing 1976; Perkins 1977; Charudattan, Perkins, and Littell 1978; Haag, Glen, and Jordan 1988; O'Leary and Ouzts 1983) as well as combinations of different biological control methods (Rushing 1976). Although combinations of control measures of a common type (e.g. two biological control agents) are not normally considered to be true integrated control (Dardeau and Hogg 1983), the net effect may parallel the integrated concept if the approaches combine two or more rather different impacts on the target weed in a synergistic manner.

Perkins (1976) obtained promising results when 2,4-D, at a lower than normal dose, was sprayed on water hyacinth plants infested with the mottled water hyacinth weevil (*Neochetina eichhorniae*). More recently Haag (1984, 1985) reported that 2,4-D and diquat do not directly harm water hyacinth weevils at the concentrations commonly used for water hyacinth control. It appeared that adult weevils move from sprayed plants to unsprayed plants in adjacent areas. As a consequence it was suggested that small areas of water hyacinth be left unsprayed at field sites infested with the weevil, if top-up chemical control is required. These areas would provide a food source for the weevils when the sprayed plants have died.

In connection with a potential use of the water hyacinth mite (*Orthogalumna terebrantis*) for integrated control of water hyacinth, Roorda, Schulten, and Pieterse (1978) tested the direct susceptibility of the mite *in vitro* to 2,4-D, glyphosate, paraquat, and diquat. The first three herbicides had little direct effect on the mite whereas diquat appeared to be definitely toxic. Therefore, it seems that 2,4-D, glyphosate and paraquat will have no direct detrimental effect on the mite when used in integrated control programmes. This could be of interest as the effect of the mite alone is, in most cases, insufficient for a complete control of water hyacinth.

Charudattan (1986) reported that the effects of the fungus *Cercospora rodmanii* and the mottled water hyacinth weevil as well as the chevroned water hyacinth weevil (*N. bruchi*) are synergistic. However, treatment with *C. rodmanii* in combination with or followed by a sub-lethal dose of 2,4-D or diquat was not significantly more damaging than treatments which consisted of the pathogen or herbicide sprayed singly. Sorsa, Nordheim, and Andrews (1988) reported that integrated use of the fungal pathogen *Colletotrichum gloeosporioides* could accentuate the effects of endothall activity on *Myriophyllum spicatum*.

Combinations of grass carp (*Ctenopharyngodon idella*), a phytophagous fish, and the mottled water hyacinth weevil reduced the growth of water hyacinth to 20–38 per cent of untreated controls (Del Fosse, Sutton, and Perkins 1976). It should be taken into consideration, however, that water hyacinth is low on the preference list of the grass carp and that it is only eaten when growing in a more or less monoculture situation.

Dubbers *et al.* (1981) proposed an integrated control programme for irrigation and drainage canals in Egypt using a combination of grass carp and mechanical means of control. It appeared from a field experiment that grass carp alone could not cope adequately with the weeds during spring and early summer at a stocking rate of about 180 kg ha^{-1}. However, after the use of a mowing launch during spring, weed control generally remained within reasonable limits. In the absence of grass carp the weed vegetation had to be removed several times during the growing season. Unfortunately, the use of grass carp in Egypt has not been a success due to the fact that overfishing by the local population could not be kept under control.

Tooby, Lucey, and Stott (1980) found that grass carp can tolerate the normal doses of several commonly-used aquatic herbicides. Subsequently, further work in Florida has suggested that the environmental impact of integrated control based on grass carp and herbicides is within acceptable limits (Canfield 1983; Shireman *et al.* 1983; Small, Richard, and Osborne 1985).

A combination of grass carp and a vegetation of aquatic plants with floating leaves (consisting of *Nymphaea alba* and *Nuphar luteum*), in order to suppress the development of submerged weeds, has been studied by Pitlo (1982). It appeared that the submerged weeds were effectively controlled but flow resistance remained fairly high (see also Chapters 5 and 9).

A technique which, under certain conditions, is well suited to use in integrated control programmes is environmental manipulation. It provides a baseline level of weed suppression, which can be topped-up by appropriate use of additional control measures as required. An early approach was the use in fisheries management of plant fertilizer application to fish hatchery ponds, to increase phytoplankton growth. The resulting increase in water turbidity reduced the light available for submerged macrophyte growth in the ponds. Additional herbicide treatments were commonly applied to reduce the growth of submerged weeds to a minimum (Schryer and Ebert 1972; Kirby and Shell 1976). In more recent years it has been suggested that aeration or other water circulation procedures used in fish ponds (and other static waters) may act as a general synergist of herbicide efficacy against *Hydrilla verticillata* and possibly other submerged weed species (Cooley, Dooris, and Martin 1980). Aeration and multiple inversion procedures used in thermally stratified waters, in combination with chelating agents to reduce nutrient availability for aquatic weed growth, was suggested as an integrated control approach by Laing (1979).

In artificial lakes the environmental conditions can be changed by water level manipulation. Manning and Johnson (1975) reported that water level drawdown, in combination with 2,4-D treatment of *Hydrilla verticillata* on exposed substrates, plus diquat application to weed beds in areas still covered by water, formed an effective integrated approach to submerged weed control in a Louisiana (USA) reservoir.

A similar approach to *Myriophyllum spicatum* control in a Tennessee Valley Authority (USA) impoundment, Melton Hill Reservoir, was described by Goldsby, Bates, and Stanley (1978). There, the normal regime of low-amplitude, regular drawdowns, determined by the water requirements of a hydro-electric power station at peak demand periods, was replaced by two different experimental drawdown regimes, during the winter months of 1971–1972 and 1975–1977 respectively (Fig. 10.1*a*, *b*).

The aim of the first regime was to maintain winter water levels between a set upper and lower limit of water levels, but at least 1.1 m below the lower limit applying in summer. This would expose *M. spicatum* beds around the shallow margins of the reservoir to prolonged periods of freezing and/or drying conditions. The second experimental winter drawdown regime involved large amplitude drawdowns, each lasting only 2–3 d, at twice monthly intervals over winter. In both cases the drawdown regimes were supplemented by spring applications of 2,4-D to the weed beds which survived winter drawdown (Table 10.1). Large reductions in the area of *M. spicatum* infestations in Melton Hill Reservoir resulted from the effective aquatic weed management programme (Table 10.2).

In general, it may be concluded that integrated control of aquatic weeds is still in an experimental stage. This is largely due to the fact that under most circumstances aquatic weed control is not carried out according to a preformed programme. It is often an empirical activity which is initiated if the weed vegetation becomes troublesome to optimal use of a water body. However, the

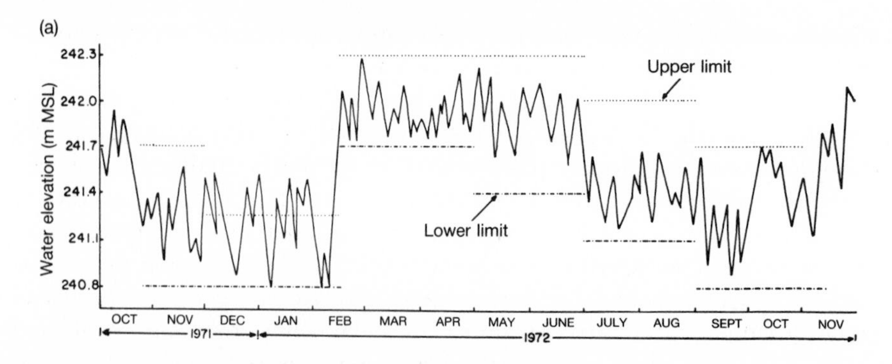

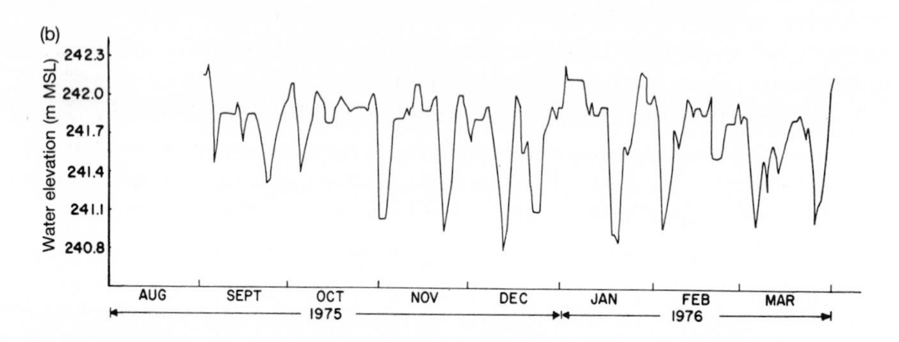

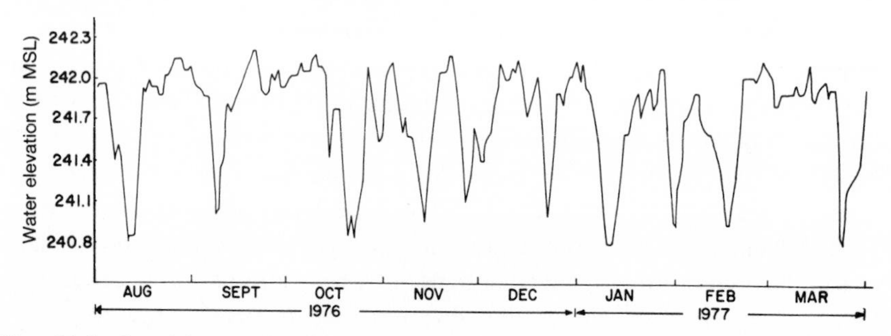

FIG. 10.1. Special water level fluctuation regime for Melton Hill Reservoir, Tennessee, USA (a) October 1971–November 1972 (b) August 1976–March 1977, as part of integrated submerged weed control programme (after Goldsby, Bates, and Stanley 1978). m MSL = metres above Mean Sea Level.

advent of efficient and widely available computer-based data management systems to assist in the management of complex fresh water systems, such as irrigation networks (Foret 1974; Mitchell 1977) has tended to increase the feasibility of operating complex integrated regimes for aquatic weed control. In this context the reader is also referred to Chapter 13.

Table 10.1 Use of 2,4-D in Melton Hill Reservoir 1971–76 (Goldsby, Bates, and Stanley 1978)

Year	Hectares treated	2,4-D applied (kg a.e.)	Cost estimates ha^{-1} (US$)[a]
1971	12	544	84
1972	24	1052	126
1973	567	18366	116
1974	246	11049	101
1975	127	5692	227
1976	142	6350	215

[a] Cost estimates include labour, equipment maintenance, and herbicide expense.

Table 10.2. Estimates of Eurasian water milfoil area (hectares) remaining in autumn after seasonal control efforts in Melton Hill Reservoir 1971–1976 (Goldsby, Bates, and Stanley 1978)

	Eurasian water milfoil	
Year	Total area	Decrease[a]
1971	295	0
1972	179	115
1973	57	238
1974	62	233
1975	134	161
1976	72	219

[a] Based on 1971 area.

11

Survey and monitoring of aquatic weed problems and control operations

K. J. MURPHY

A PROPER understanding of the scale and distribution of an aquatic weed infestation is an obvious prerequisite to effective management of the problem. Information on how long the problem has lasted, or how weed distribution changes with time may also be vital in developing an effective control regime. All too often, however, surveys are inadequately carried out. There is a tendency to attempt weed control on the basis of insufficient survey data. This may, in turn, contribute directly to the lack of success of many control programmes (Mitchell 1979*a*, *b*). On a worldwide basis, data on the scale and distribution of aquatic weed problems are highly variable, largely due to the relatively low priority given to weed survey, in comparison with active weed control measures. To a great extent this situation has arisen because of the difficulties and costs of effective surveying of aquatic weed problems, in general, but particularly so for submerged weeds. Recent advances in the fields of information technology and survey methodology hold out the prospect of major improvements in our overall understanding of aquatic weed distribution.

Even in countries with adequate resources to pay the costs of gathering and collating aquatic weed survey data, existing information may be far from complete. A good example is Great Britain, where quantitative weed survey data have been published for two extensive aquatic systems: the land drainage systems of low-lying farmlands (Robson 1975; Marshall, Wade, and Clare 1978; Marshall 1984), and the national navigable canal network (Murphy, Eaton, and Hyde 1980). Two further reports are on the extent of aquatic weed problems in freshwaters of Scotland (Murphy 1988), and on filamentous algal problems in Great Britain (Fowler 1986). Piecemeal data are scattered through the literature on rivers (Westlake 1968; Westlake and Dawson 1982; Barrett and Murphy 1982) and standing waters (Brooker and Edwards 1973*a*; Dunn 1976; Kelcey 1981; Wade 1982; Mackenzie and Murphy 1986).

Three main factors are responsible for this current state of affairs regarding aquatic weed survey data worldwide.

1. The considerable difficulties involved, and time required to amass quantitative aquatic weed survey data. If submerged weed surveys are needed costs tend to increase above acceptable levels.
2. In consequence, obtaining funding (for extensive surveys in particular) is difficult, unless some central agency is available to conduct the required investigations. Otherwise 'one-off' surveys of isolated aquatic systems tend to be the norm (Coffey 1983). An example of the value of central co-ordination in this aspect is provided by the Aquatic Plant Control Research Progam (APCRP) conducted by the US Army Corps of Engineers. Published reports of the Annual Meetings of the APCRP include data on the status of aquatic weed problems in Districts covering much of the USA (Johnson 1984).
3. The problems of adequately defining 'weed growth' of macrophyte vegetation in a freshwater system. Although there have been many attempts to quantify the degree of infestation or overgrowth which constitutes the threshold level of an aquatic weed problem (Tevyashova and Tevyashova 1973; Wong and Clark 1979; Niemann 1980; Westlake 1981; Kern-Hansen and Holm 1982; Sabol 1984*a*), it is clear that the concept will vary widely between different aquatic systems, to a large part dependent upon species involved and usage made of the system. In biomass terms the differences in tolerable upper limit of macrophyte abundance in different systems are quite striking. For example, in lowland *Ranunculus* streams in Europe, which often support valuable sport fisheries, the maximum dry weight (DW) standing crop of submerged vegetation acceptable for fisheries purposes is usually within the 10^2–10^3 g m^{-2} range (Westlake 1981). If flooding is a potential problem the tolerable limit is lower in such streams, at *c*. 100–200 g m^{-2} DW (Kern-Hansen and Holm 1982). In navigable canals the threshold submerged macrophyte biomass, above which obstruction is caused to boat traffic, is at least an order of magnitude lower than the acceptable range in *Ranunculus* streams (Eaton and Freeman 1982; Murphy, Eaton, and Hyde 1982). In drainage ditches of very low gradient, in flat areas, the increase in channel roughness caused by weed growth may so increase the risk of flooding (by impeding water flow) that even very small standing crops of aquatic vegetation are unacceptable (Powell 1978; Cave 1981; Westlake 1981).

Sabol (1984*a*) provided examples of the use of survey-defined threshold nuisance levels of weed growth in a hypothetical system in relation to the cost effectiveness of control measures (Fig. 11.1). The ratio of treatment cost to

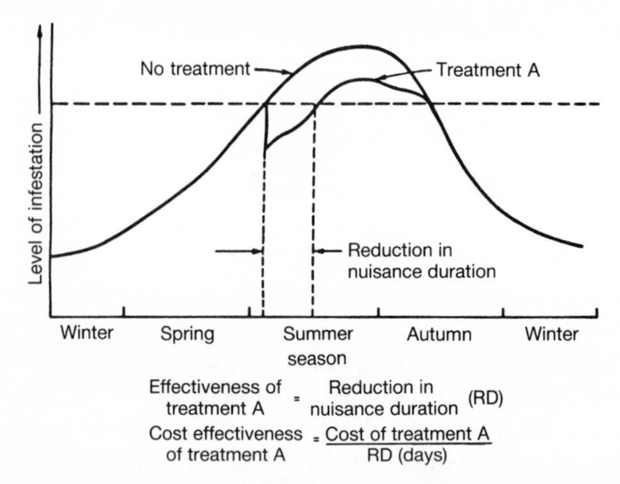

FIG. 11.1. Effects of a hypothetical treatment of an aquatic plant infestation in relation to threshold nuisance level (------), showing reduced duration of nuisance (RD) from that which would occur without treatment, and cost-effectiveness measured as treatment cost/RD (after Sabol 1984*a*, with permission).

reduction in nuisance-duration (measured in days) provides a means of quantifying the cost effectiveness of weed control programmes. Sabol suggests that the methodology for the essential survey procedures involved may vary from simple and cheap visual inspection, or more objective aerial photo-interpretation methods (both providing percentage cover data), to fully quantitative but time-consuming and labour-intensive biomass sampling, depending on the needs of the study in question.

APPROACHES TO AQUATIC WEED SURVEY

Both qualitative and quantitative surveys may be undertaken, on an extensive or intensive basis. One approach to extensive survey is to use procedures based on subjective visual assessment, sometimes backed up by simple on-site biomass estimates. An alternative is the semi-quantitative questionnaire approach, collating data from a wide range of sources, (Robson 1975; Varshney and Singh 1976; Murphy 1980; Hamel and Bhéreur 1982; Murphy 1988; Murphy, Eaton, and Hyde 1980; Janauer 1982, Okafor 1982; Reeders *et al.* 1986; Fowler 1986). Although such techniques have been used with reasonable success hitherto, considerably improved extensive survey techniques are becoming more generally available (Rekas and Bailey 1981; Link and Long 1978; Gustafson and Adams 1973*a*,*b*). These offer the prospect of substantial improvements in both the quality and ease of collection of aquatic weed survey data. The suitability of aerial photography for aquatic vegetation survey has long been recognized (Edwards and Brown 1960). Remote sensing, by aerial photography and satellite (e.g. LANDSAT MSS imaging) survey methods using colour, false-colour, and infra-red photography offers the possibility of rapid, large-scale monitoring of aquatic weed growth virtually anywhere in the world, over whatever time-scale is required (Hilton 1984; Benton 1986; Benton and Newman 1976; Andrews, Webb, and Bates 1984; Long 1979; Lachavanne 1977; Kai and Kai-Yu 1983; Haslam 1979; Dardeau 1983; Breedlove and Dennis 1984; Adams *et al.* 1977).

The effectiveness of remote sensing is dependent on scale, resolution, season of imagery, weather conditions, vegetation type, sensor and spectral sensitivity, processing of data, and precision of data transfer (Best, Wede, and Linder 1981; Jarman, Jarman, and Edwards 1983; Almkvist 1975). In order to maximize the information available from remote survey, it is normally necessary to match the data with results from a set of 'testbed' sites surveyed by ground fieldwork (Raitala and Lampinen 1985). Ground verification may involve the use of a battery of intensive-survey methods (discussed below). These are most difficult to apply in the case of submerged vegetation. However, the newly developed methods of sonar-based plant survey offer a rapid, cheap method of complementing aerial or satellite survey data (Killgore 1984*b*; Killgore and Payne 1984). These techniques permit the accurate determination of submerged macrophyte distribution, stand-height within the water column, and potentially biomass, over substantial areas of water (Maceina and Shireman 1980; Hanley 1982; Shireman and Maceina 1983; Maceina *et al.* 1984; Stent and Hanley 1985; Thomas *et al.* 1985; Brabben 1986). An example of typical sonar-survey data is shown in Fig. 11.2.

Crucial to the success of the new survey approaches is the use of computers for the control of sensors, data processing, and display of results. The spreading availability of low-cost microcomputers permits extensive weed survey data, produced by a combination of advanced and traditional methods, to be rapidly and cheaply collated, analysed, displayed, and disseminated to managers, modellers, or other interested parties, thereby optimizing the maintenance of efficient weed control programmes (Schardt and Nall 1982; Dardeau and Hogg, 1983; Benton, Clark, and Snell 1980; Payne 1982; Ewel, Braat, and Stevens 1975).

Intensive surveys, usually restricted to a small area or a restricted period of time, or both, may be required by water managers or users to provide data on the outcome of a weed control programme; by the agrochemical industry to evaluate novel control measures; and for environmental monitoring purposes to evaluate the ecological consequences of aquatic weed control operations. Their methodology is commonly based upon standard ecological sampling procedures for the estimation of plant population size and production. Commonly used measures are cover, frequency, biomass, or other measure of abundance of macrophyte vegetation, per unit area of water or sediment surface or unit volume

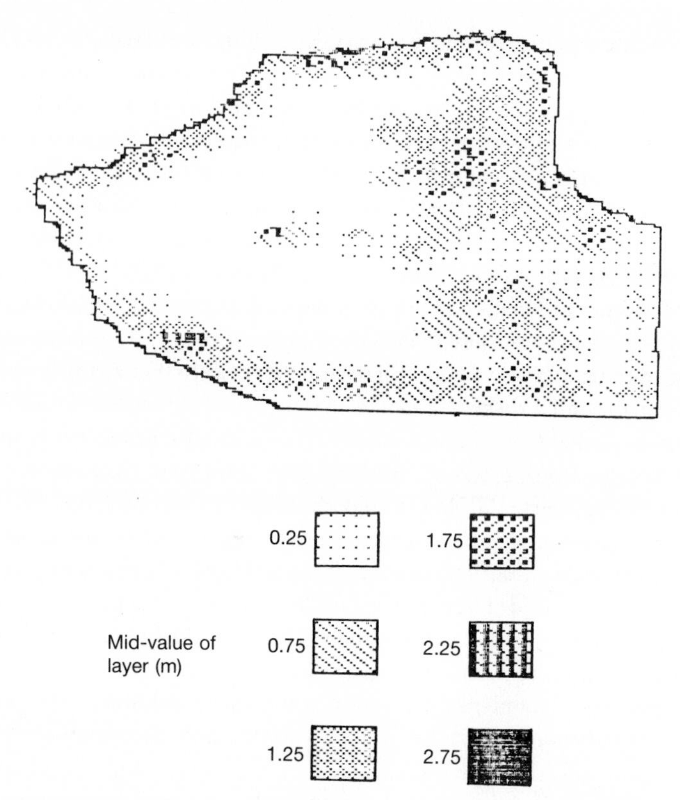

FIG. 11.2. Distribution of *Elodea canadensis* in a reservoir at Great Witcombe, Gloucestershire, England in September 1983, on a 5 × 10 m sampling grid. Mid-values of layers refer to depth strata used by the computer programme to generate maps: e.g. layer mapped as 0.25 m mid-value shows the area of the reservoir where the height of *E. canadensis* above the substratum was between 0 and 0.5 m (after Stent and Hanley 1985, with permission).

of water (Killgore and Payne 1984; Westlake, Spence, and Clarke 1986; Raschke and Rusanowski 1984). The procedures may be destructive or non-destructive of the vegetation. They are slow, and require relatively high labour and other resource inputs (Nichols 1984*a*). Ideally the database should comprise quantitative information on the three-dimensional distribution of plant biomass within the waterbody (Sabol 1984*a*).

Simple quadrat-based sampling procedures, directly derived from terrestrial vegetation-sampling methodology, may suffice for emergent or free-floating macrophyte surveys (Westlake 1966*b*; Haslam 1970; Duffield and Edwards 1981; Dennis 1984). The assessment of submerged vegetation by intensive survey is more of a problem. Destructive sampling from the surface, using mechanical or hand-operated grab, grapnel or cylinder samplers is one approach (Howard-Williams and Longman 1976; Thomas 1978; Wong and Clark 1979; Breen *et al.*

1976; Shapovalova and Vologdin 1979; Murphy, Fox, and Hanbury 1987; Brown 1984; Sabol 1984*b*; Murphy, Hanbury, and Eaton 1981; Hiley, Wright, and Berrie 1981). Combined with this approach may be some form of mapping of plant stands, for example, along transects (e.g. Raschke 1984; Rybicki *et al.* 1985; Hamel and Bhéreur 1982; Wright *et al.* 1981; Dubois *et al.* 1984; Fox, Murphy, and Westlake 1986). In clear-water lakes, stereophotography may be a useful mapping tool (Rørslett, Green, and Kvalvågnaes 1978). An alternative is SCUBA diver-sampling, using suitably modified quadrat-sampling procedures which parallel terrestrial survey methods (Sheldon and Boylen 1978; Wood 1963; Dennis and Isom 1984; Ward and Talbot 1984). Wade and Bowles (1981) compared the costs and efficiency of lake macrophyte surveys carried out by diver-sampling, shoreline survey, and boat-survey. They concluded that in most cases the first two approaches were to be preferred.

SCUBA diver-sampling has the advantage of permitting close contact with the plant stands being surveyed, with all the inherent advantages which this permits, but is particularly slow in operation, and requires fairly substantial backup support in terms of safety personnel and equipment. Where suitably trained personnel are available, however, the SCUBA sampling approach has undoubted merits in terms of the precision of sampling which becomes possible.

CASE STUDIES

Two case studies are outlined below, involving in the first case a questionnaire approach and, in the second, an approach combining both traditional and more advanced survey techniques, of surveys of aquatic weed distribution in freshwater systems in two geographical regions of comparable size (Great Britain and Florida). In both cases data were sought on both the distribution and nature of aquatic weed problems in extensive aquatic systems.

Weed problems in the British navigable canal system (Murphy, Eaton, and Hyde 1980, 1982; Murphy and Eaton 1983)

Of some 2900 km of canals in Great Britain, 2558 km are under the control of the British Waterways Board (BWB), together with 548 km of navigable river and other commercial waterways (Department of the Environment 1975). Though built for commercial usage by barges, the bulk of the BWB canal system is now run as a leisure and amenity system, used by holiday pleasure-craft and for sport angling. The density of boat-usage ranges from zero on derelict ('Remainder') canals to traffic in the range 10 000–15 000 movements ha^{-1} canal surface m^{-1} depth $year^{-1}$ (my), concentrating during the summer months, on the most popular 'Cruising' canals such as the Llangollen Canal in North Wales (Murphy and Eaton 1981*b*, 1983). There is a significant negative relationship between boat traffic density on a canal and the abundance of submerged (and to a lesser extent emergent/floating) vegetation present, owing to the direct plant damage and increased turbidity (due to resuspension of silt particles) associated with heavy usage by powered boat traffic (Table 11.1). The implication is that less-popular canals might be expected to suffer excess growths of aquatic plants, requiring active management to maintain navigation. Survey data for the 1976 season were amassed for all BWB waterways by questionnaire survey of the 49 Sections of BWB, each of which is responsible for a part of the system. Details of the location and length of canal stretches which were considered by Section managers to suffer nuisance weed growths during 1976 were requested, together with information on the type of weeds responsible for the problem, and any control measures used.

Table 11.1. Relationships between macrophyte abundance or standing crop, boat traffic and water turbidity in navigable (traffic $\geqslant$200 my) British canals (after Murphy and Eaton 1983)

Relationship	r	Significance
$\log_e (W+1) = -1.57 \log_e T + 15.67$	−0.69	***
$\log_e (E+1) = -0.53 \log_e T + 7.04$	−0.30	**
$\log_e t = -0.60 \log_e T - 1.68$	0.67	***
$\log_e (W+1) = -1.81 \log_e t + 8.92$	−0.66	***
$\log_e (E+1) = -0.41 \log_e t + 3.96$	−0.23	NS

W, mean summer standing crop of submerged macrophytes (g m^{-2} fresh weight); T, annual traffic (my); E, total abundance of emergent species ($\Sigma \mathrm{SRA_E}$, where SRA is a semi-quantitative measure of abundance of each species present); t, mean water turbidity, as total suspended solids (g m^{-3}); r, correlation coefficient. Significance of association: NS not significant, $P > 0.05$; ** significant, $P \leqslant 0.01$; *** significant, $P \leqslant 0.001$.

Further information was obtained from a parallel survey of a major user-group of canal waters, 73 angling clubs holding sport fishing agreements with BWB for individual canals or canal stretches. Lengths considered weedy by anglers largely coincided with those so-designated by BWB management personnel, although in total 38.7 km were considered to suffer nuisance weed growth by anglers but not by BWB, suggesting that management personnel may not always take fully into account the particular problems of aquatic weed growth in interfering with sport fishing.

Detailed data were reported for 82 canals or canal-stretches, providing 100 per cent coverage of the BWB system, by Murphy, Eaton, and Hyde (1980). Summary results are shown in Table 11.2: submerged macrophytes were the greatest cause of problems. The main reliance for weed control was placed on mechanical weed clearance, using reciprocating-cutter boats and purpose-built weeder-dredger craft (Eaton and Freeman 1982). The survey provided a database for further studies leading to improved management recommendations (and their implemen-

Table 11.2. Aquatic weed problems of British canals controlled by the British Waterways Board in 1976 (after Murphy, Eaton, and Hyde 1980)

Category	Total length (km)	Length with weed problems (km)	Length with weed problems (%)
(1) Waterway type			
Commercial waterways	548.5	40.4	7.4 a
Other canals	2558.0	465.3	18.2 a
(2) Weed group causing problems			
Emergent		120.0	23.9 b
Surface-floating		80.9	16.0 b
Submerged (incl. algae)		329.8	65.2 b
Unspecified		202.3	40.0 b
		Length treated	
(3) Weed control measures used		(km)	(%)
Manual clearance		46.4	9.2 c
Mechanical cutting/removal		146.5	28.9 c
Bottom dredging		17.0	3.4 c
Herbicides		12.8	2.6 c
Untreated		321.9	63.6 c

(a) Percentage of total length per waterway type; (b) percentage of total length with weed problems (in some stretches more than one group of weeds caused problems in 1976); (c) length treated as per cent total length with weed problems (some lengths treated with more than 1 control measure in 1976).

tation) for aquatic plant management in the British canal system (British Waterways Board 1981; Hanbury 1986).

Annual aquatic plant surveys of public-access freshwaters in Florida (Schardt and Nall 1982; Schardt 1983, 1987)

The Bureau of Aquatic Plant Research and Control, Florida Department of Natural Resources has published a series of annual reports on the extent and scale of aquatic weed problems in public-access waters of Florida (USA). The survey approach in recent years has combined advanced techniques (e.g. aerial and sonar surveys, computer-based data-handling and presentation) and traditional survey methodology (e.g. drag-rake sampling, simple visual assessments) in a model example of how to carry out an extensive study of aquatic weed distribution. In 1983, 332 lakes (418 595 ha), 66 rivers (69 253 ha) and 74 canals (37 501 ha) were surveyed, totalling 56 per cent of Florida's total area of freshwater. Altogether 154 plant species were identified, with a total areal coverage of 13 867 ha (adjusted to a constant cover basis of 100 per cent coverage: e.g. 10 ha with 10 per cent plant cover would contribute 1 ha to this total cover figure). Six species accounted for some 50 per cent of the total area covered by aquatic vegetation, three of these (*Hydrilla verticillata, Eichhornia crassipes,* and *Panicum repens*) being alien to Florida.

Figure 11.3 depicts the computer-plotted coverage of *E. crassipes* on a county basis in Florida in 1983. The survey data have permitted more accurate determinations of increases or declines in aquatic weed problems to be made, thereby providing a more quantitative and cost-effective basis for management of weed problems in Florida's public waters.

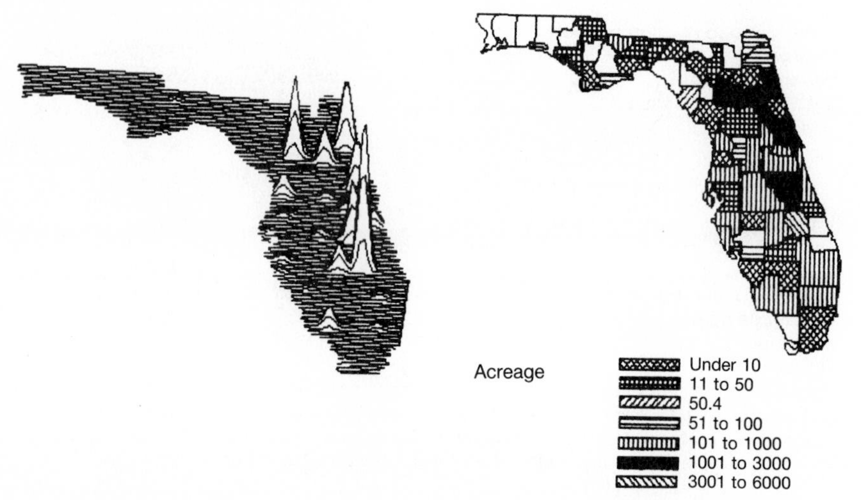

FIG. 11.3. Computer-plotted coverage of *Eichhornia crassipes* in public-access fresh waters in Florida 1983, showing (on left) relative coverage by county; (on right) plant cover as acreage per county: 1 acre = 0.405 ha) (after Schardt 1983, with permission).

CONCLUSIONS

Major improvements in our knowledge of the scale and extent of aquatic weed problems, and of the measures used for their control, have become feasible with the advent of cheap computer-based data capture, storage, analysis, and presentation systems for use in aquatic weed survey programmes. Advances in remote-sensing, sonar-survey and SCUBA sampling technology, which complement and extend more traditional aquatic weed survey approaches, can provide cost-effective means of obtaining quantitive data on aquatic weed problems.

The advantages of improved knowledge of what sort of problems are present, where, and on what scale are fourfold.

1. Government and other funding agencies are more willing to fund both research and management programmes if the 'visibility' of the problem is high, and hard data are available on which to base decisions.

2. Industry likewise bases commercial decisions (for example, on whether to proceed with investment in a promising new herbicide) on data on the potential size and value of the market for the product. At present data for aquatic weeds, in marked contrast to the situation with terrestrial weed problems, are too often either inadequate or simply unavailable.
3. Understanding of the efficacy of a given weed control measure or regime can only be improved by the availability of better survey procedures.
4. Improved databases on the current status of aquatic weed distribution and abundance in freshwater ecosystems would permit much-improved future long-term studies of the environmental impacts of both weed problems and weed control measures used in freshwaters, in turn permitting improvements in our ability to predict and contain the environmental damage associated with aquatic weeds and their control.

12

Relationships between survival strategies of aquatic weeds and control measures

W. VAN VIERSSEN

INTRODUCTION

AQUATIC macrophytes are an increasing nuisance in many areas of the world. The various control methods available to cope with this problem, are usually aimed at killing or removing aquatic plants, without considering the relative vulnerability of the different life-cycle stages of the species involved. Instead, the aquatic weed problems are commonly controlled only when they become troublesome, i.e. when they have formed a dense vegetation.

Since, however, the survival strategy (Grime 1979) of individual aquatic weed species is adapted to particular sets of climatological conditions it is conceivable that control measures could be developed to interfere with the most vulnerable life-cycle stages. In general, this is probably not, in most cases, the period of most rapid vegetative growth.

The timing of growth stages is closely related to the timing of climatological conditions during the course of the year. In temperate regions, there is an initial stage of rapid development of shoots and rhizomes from overwintering organs or plant parts in the spring. The next stage is characterized by maximum vegetative development, followed by the appearance of flowers and fruits. Later, as temperatures gradually fall and daylength shortens, plants or plant parts are confronted with increasingly hostile environmental conditions. The late season period is critical to the survival of many species.

Winter is a critical period in the life-cycle of many species, but unfavourable summer conditions, caused by high temperatures and/or drying up of a water body can be more important for some species. This is commonest in regions with a marked hot and/or dry season. In such environments aquatic plants must develop different survival mechanisms from those necessary for habitats which freeze in winter.

On the basis of these general schemes of plant development and in the context of annual growing season characteristics, two major groups of aquatic macrophytes can be distinguished: annuals and perennials. An annual completes its life-cycle within one growing season and survives unsuitable periods for vegetative growth (cold winters = summer annuals; warm summer periods plus desiccation = winter annuals) by sexual reproduction and seed formation. A perennial, on the other hand, survives periods unsuitable for vegetative growth by means of specialized vegetative perennating organs. In general, less new genetic material (per unit time) is introduced into populations of perennials than is the case for annuals.

For aquatic weeds the distinction between annuals and perennials is essentially whether or not a species reproduces itself sexually within a single growing season. Also of importance is whether or not the major seasonal fluctuation in biomass is derived from sexually produced plants. Some species produce different types of propagules but, in general, only one type is important for the seasonal regeneration of biomass. When this is a sexually produced propagule the species is considered to be an annual, regardless of the characteristics of any other types of propagules which may also be produced.

There are different types of specialized propagules or plant parts which play an important role in the survival of the species during periods unsuitable for vegetative growth. Yeo (1966) distinguished two types of propagules: seeds and specialized asexual disseminules, e.g. tubers and turions. The latter can be divided into subterranean and vegetative axillary propagules. According to Sculthorpe (1967), the most important difference between a turion and a tuber is the fact that the major component of a turion consists of leaves, whereas a tuber lacks leaves and is a typical axial or rhizomatous structure, swollen by the storage of carbohydrates.

Most of the turions are produced above ground (*Hydrilla* is an important exception, producing both axillary and subterranean turions). Most tubers are produced below ground (root tubers), or at plant structures near the sediment–water interface (horizontal runners, stolons; stem tubers). This chapter deals with the influences of environmental factors and aquatic weed control measures impacting the different life-cycle stages of aquatic macrophytes. Environmental conditions in and near the sediments are very different from those in the water. Most aquatic weed control measures differentially affect one of these two habitat compartments. Therefore, a distinction has been drawn between two types of life-cycle, the regenerative phases of which are dominated by above ground propagules versus below ground tubers/turions. The general appearance of turions differs between species. The axillary turions of *Potamogeton crispus* are bud-like, whereas those of *Elodea canadensis* look more like small shoot segments, bearing small buds (Yeo 1966). In addition to seeds and asexual propagules, other plant parts such as rhizomes, stolons or even whole plants may act as survival units (Haag 1979; Brock 1985; Horn 1983; Lytle and Hull 1980).

In Fig. 12.1 a number of schematic situations is presented, from which six basic life-cycle types can be recognized, each represented by a characteristic aquatic macrophyte. As discussed in the next section, a limited set of environmental conditions seems to control the appearance of the different life-cycle stages. Temperature (controlling desiccation and freezing) and light regimes (controlling the photoperiod) during the season are shown in Fig 12.1 in relation to the occurrence of different life-cycle stages.

Most attention will be paid here to the strategies of temperate aquatic weeds

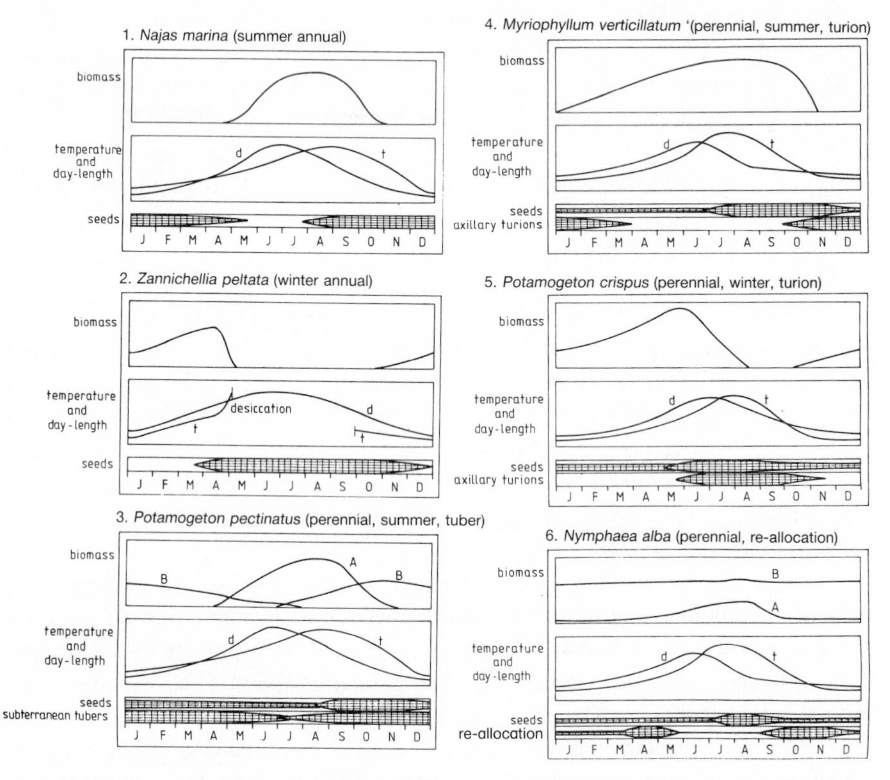

FIG. 12.1. Schematic representation of life-cycles of different aquatic macrophytes in relation to seasonal changes in temperature and daylength. The scales for temperature, daylength and biomass (A = above ground, B = below ground) are arbitrary. The presence of propagules (seeds, tubers, turions) or the process of re-allocation of nutrients during the year is indicated by bars in the lower part of each separate figure (data were taken from, and modified after van Vierssen, 1982b; van Vierssen and Van Wijk, 1982; van Vierssen, unpublished data; Sastroutomo *et al.* 1979; Weber and Noodén, 1974; and Brock 1985; for Fig. 12.1.1–12.1.6 respectively).

because detailed information for the tropics is not available. Life-cycle characteristics are summarized in Table 12.1. It is convenient to consider the strategies of the two basic phases of the macrophyte life cycle (regenerative and established phases) separately in relation to environmental stress and disturbance pressures, and the competition pressures due to the presence of other plants (Grime 1979).

LIFE-CYCLE STAGES IN RELATION TO ENVIRONMENTAL CONDITIONS

Regenerative phase

The regeneration of an established-phase population of aquatic macrophytes from the propagules making up the regenerative phase of the life-cycle, depends on two events. First, environmental conditions must be sufficiently favourable for the established-phase plants in the parent population to produce propagules. Second, environmental conditions must be favourable for the growth and developments of plants from these propagules. In both cases there may be possibilities of influencing the eventual size of the established-phase plant population, by manipulating environmental conditions.

Seeds

Seed production tends to be proportional to the vegetative biomass of the parent population (Grainger 1947), and is influenced by temperature and daylength.

In the literature, two major types of seeds and seed production are distinguished.

1. Some species, annuals in particular, produce high numbers of seeds which easily germinate during the season following the season of their production. Examples include several *Zannichellia* and *Ruppia* species. Under optimum conditions germination may approach 100 per cent (van Vierssen, 1982*b*; van Vierssen and van Wijk 1982; van Vierssen, van der Zee, and van Kessel 1984; Verhoeven 1979). Yeo (1966) mentioned a production of approximately 800 000 *Zannichellia* seeds m^{-2}. Depending on the geographical latitude, both summer and winter annuals can occur. The latter are adapted to survival in habitats which dry out in summer. Examples are *Ruppia drepanensis* and *Zannichellia peltata* (Mediterranean distribution pattern). The seeds of these species must germinate in winter at low optimum germination temperatures (8–12°C) because only during that period of the year does their habitat contain water (van Vierssen and van Wijk 1982; van Vierssen, van der Zee, and van Kessel 1984).

Summer annuals germinate in spring. Their seeds may show innate dormancy during autumn and winter, thereby preventing germination during unfavourable

Table 12.1. Life-cycle characteristics of annual and perennial macrophytes (see also Fig. 12.1).

Life-cycle type	A = Annual P = Perennial	Peak biomass	Main type of propagules produced from vegetative biomass regeneration	Timing of propagule production	Seed germination conditions	Seed longevity	Seedbank size
Summer annuals	A	Summer	Seeds	Summer–autumn	Temperature, light, climate	Short	Large
Winter annuals	A	Winter	Seeds	Winter–spring	Temperature, light, climate	Short	Large
Summer tuber perennials	P	Summer	Tubers	Summer–autumn (shortening photoperiod)	N.D.	Long	Small
Summer turion perennials	P	Summer	Turions	Autumn (short photoperiod)	N.D.	Long	Small
Winter turion perennials	P	Winter	Turions	Spring–summer (long	N.D.	Long	Small
Non-specialized perennials	P	Summer (above ground)	Seeds and re-allocation of carbohydrates to vegetative parts other than specialized propagules (e.g. below ground rhizomes)	Seeds: summer; re-allocation: spring or late-summer–autumn	N.D.	Long	Small

N.D. = no data.

(low temperature, low light availability) environmental conditions for seedling growth. This innate dormancy can be broken by a stratification period, usually of at least 8 weeks. *Zannichellia palustris* seeds exhibit such characteristics (van Vierssen 1982*b*).

2. Certain species generally produce only a relatively low number of seeds and occasionally no seeds at all. Now and then a year of abundant seed production occurs. The number of seeds which germinate is usually very low. Species with such characteristics often produce tubers, turions, or use stem fragments to survive unfavourable conditions. An example is *Potamogeton pectinatus* which, in the Netherlands, sporadically produces large numbers of seeds. Experiments with seeds of this taxon revealed, however, that only a very low percentage could be germinated and only after various pre-treatments (e.g. temperature shocks, salinity shocks, stratification, low redox potentials, desiccation: unpublished data W. van Vierssen; van Wijk 1983). Yeo (1959) found similar results for *P. pectinatus* var. *interruptus*. Such seeds often show higher germination rates after rupture of the seed coat. They have strong innate dormancy. The build-up of a seedbank of seeds with such germination characteristics probably enhances the long-term chances of survival of the population. Haag (1983) mentioned that about one-third of *P. pectinatus* seeds produced in a particular growing season are still present in the seedbank one year later. Similar data are available for other species. Rogers and Breen (1980) found that only 0.001 per cent of the rather abundantly produced seeds ($1450\,m^{-2}$) of *P. crispus* could be germinated. The production of seeds was higher than that of turions ($1450\,m^{-2}$ versus $1130\,m^{-2}$), but, for survival, the species depended mainly on its turions.

Another question is how long seeds can maintain their viability. Seeds of certain species remain viable for long periods. This is especially important when they are added to a long-lived seedbank, as is the case for many species which normally depend on vegetative propagules. In *Eichhornia crassipes* vegetative reproduction is most important (Das 1969), but seeds are, nevertheless, produced which remain viable for 5–7 years, or up to 15 years according to Ueki and Oki (1979).

Important factors which regulate seed germination include temperature, light conditions, desiccation, and oxygen supply (redox potential). However, as briefly indicated before, seeds can remain dormant for prolonged periods, showing one of the three different types of dormancy (after Mayer and Poljakoff-Mayber 1982).

1. Innate dormancy; plant parts/propagules are dormant after being produced. This characteristic should be distinguished from after-ripening.
2. Induced dormancy; plants parts/propagules are capable of germination at production but become dormant after an environmental dormancy-inducing factor has acted on them.

3. Enforced dormancy; plant parts/propagules can germinate immediately after environmental conditions become favourable.

Some examples of these three types can be mentioned. Seeds of *Z. palustris* exhibit innate dormancy after production (van Vierssen 1982*b*), as do those of *Pontederia cordata* (Whigham and Simpson 1982).

There are many examples of enforced dormancy. All macrophyte seeds which do not germinate because temperatures are too low can be considered to be in a state of enforced dormancy. Macrophyte seeds which become dormant because of a shortened day-length (decreasing photoperiod) show induced dormancy. When dormancy has been broken, seed germination success is to a great extent temperature-dependent. Different species often have different optimum germination temperatures. In a number of cases speciation within a genus seems to have occurred on the basis of development of different optimum germination temperatures (e.g. *Zannichellia*, and to a certain extent also *Ruppia*, in western Europe). In general, the winter annuals within these genera show low optimum germination temperatures (8–12°C) and the summer annuals high optimum germination temperatures (20–24°C: van Vierssen, van der Zee, and van Kessel 1984).

Datta (1969), who studied the germination of seeds of *Pistia stratiotes*, found temperature to be of minor importance for germination in comparison with light quality. Blue light inhibited, but orange/red light stimulated germination. Pieterse, de Lange, and Verhagen (1981), however, found a clear influence of temperature on the germination of *Pistia* seeds. Light intensity, quality, and photoperiod, in general, all play an important role in seed germination. There are also species in which temperature and light act together as a combined factor to let germination occur. Non-stratified seeds of *Zannichellia pedunculata*, for example, do not germinate in the dark, in contrast to stratified seeds (van Vierssen 1982*b*). Different species often exhibit quite different light requirements for the seeds to germinate. Seeds of *Najas marina* need dark conditions to germinate (van Vierssen 1982*a*), whereas seeds of *P. crispus* only germinate in the light (Kadono 1982*b*). The light which is needed for germination in *Heteranthera limosa* seeds is probably connected with a red–far red (R–FR) response (Marler 1969). Another important factor is desiccation. Some species produce seeds with a high degree of drought-tolerance; in some species desiccation even stimulates seed germination. This applies, for example, to seeds of *E. crassipes*, which are very drought-resistant (Obeid and Tag el Seed 1976), permitting the survival of water hyacinth during periods of reduced water level (Barrett 1980; Matthews 1967). The timing of germination is related to the dormancy-breaking effects of dry conditions. When vegetative reproduction stops because of drying-up of the habitat, the dormant seedbank maximizes the chances of recolonization of the species upon refilling of the habitat.

When seeds are deposited in the sediment, they are potentially exposed to a whole range of sediment factors. One of the most important such edaphic factors is oxygen status. Highly organic sediments with high pH values may have low to very low oxygen levels. Many seeds are very sensitive to this factor. According to the literature, seed germination is relatively good in sediments rich in organic material as compared to results obtained in tap water. This suggests the involvement of oxygen as an important environmental factor, as shown by Marler (1969), and Obeid and Tag al Seed (1976), for *H. limosa* and *E. crassipes* seeds. Similarly Datta and Biswas (1977) obtained only 1 per cent germination of *Nymphaea nouchali* in the laboratory, but could improve the rate of germination by adding sediments containing organic material. Forsberg (1965*a*) and van Vierssen (1982*a*) showed that the germination of *N. marina* seeds was controlled by the redox potential: a measure of sediment oxygen status.

Vegetative organs

The different types of vegetative perennial organs which play an important role in the survival of species have different morphological and physiological characteristics.

Tubers and turions

Subterranean tubers and turions have been described for a variety of aquatic macrophytes. Two weeds, *Hydrilla verticillata* and *P. pectinatus*, are examples of species which mainly depend on such propagules for their reproduction. According to Van, Haller, and Garrard (1978) and van der Zweerde (1982*b*) the production of such *Hydrilla* turions is controlled by photoperiod. Van Wijk (1986) and Haller, Miller, and Garrard (1976) found an inverse correlation between light availability at increasing water depth and the number of tubers and turions produced for *P. pectinatus* and *H. verticillata*. Bowes, Holaday, and Haller (1979) found up to 300 subterranean turions m^{-2} for *Hydrilla*. For *P. pectinatus* up to 2500 tubers m^{-2} were collected under artificial shading conditions (W. van Vierssen, unpublished data).

The sprouting of tubers and subterranean turions is mainly temperature-controlled (Kadono 1984; Haller, Miller, and Garrard 1976). Light is also considered to favour germination (Miller, Garrard, and Haller 1976). Since most tubers are innately dormant after production, a dormancy-breaking factor is needed to permit germination to take place. A stratification period is a very effective means of breaking this dormancy. Van Wijk (1983) showed this for *P. pectinatus* and Basiouny, Haller, and Garrard (1978) found similar effects for subterranean *Hydrilla* turions.

Two important factors which initiate the production of axillary turions and tubers (above-ground) are temperature and photoperiod (Weber 1974; Aiken

1976; Weber and Noodén 1976*a*). For *Myriophyllum verticillatum* a short photoperiod and a low temperature ($<15°C$) induced turion formation (Weber and Noodén 1974). After production, the turions were dormant. This dormancy could be broken by a stratification period. Quite the opposite results were obtained by Sastroutomo *et al.* (1979) and Chambers, Spence, and Weeks (1980) for *P. crispus*. In this species turion formation occurred in spring and summer (temperatures $>20°C$; daylength: >12 h). A shorter photoperiod inhibited turion formation (the R–FR ratio in the light appeared to be critical). The turions are dormant after production but this dormancy was broken by a decrease of photoperiod. As a result, *M. verticillatum* and *P. crispus* can be considered to be summer and winter annuals respectively. This seasonal pattern obviously depends on the timing of the turion production.

Another factor important under certain conditions for turion formation could be nutrient deficiency. Haller, Miller, and Garrard (1976) and Pieterse, Staphorst, and Verkleij (1984) concluded this from the fact that floating plant fragments of *Hydrilla* produced more turions than rooted ones. Such nutrient deficiency effects are also known from *Spirodela* (Czopek 1963; Augsten and Jungnickel 1983).

The rate of growth from turions and tubers depends on temperature, light availability and photoperiod. According to Weber and Noodén (1976*b*) long-day conditions are necessary to grow turions of *M. verticillatum*. Richards and Blakemore (1975) established the light requirement for germination of turions of *Hydrocharis morsus-ranae*. Kadono (1982*b*) also established a light requirement for germination of turions of *P. crispus*, at relatively high temperatures. At low temperatures, germination can occur in the dark. This fits with the field data which show that turions covered by sediment do germinate in the autumn. Light quality has its influence too. Sastroutomo (1980) found an inhibitory effect on non-dormant subterranean *Hydrilla* turions; R–FR treatment promoted germination.

2. *Rootstocks, tillers, plants, and plant fragments*

Some plant parts do not show a season-dependent external morphological adaptation to environmental stresses. For a number of species, re-allocation of photo-assimilates to rhizomes in autumn can be an effective component of an over-wintering strategy (Lytle and Hull 1980; Brock 1985). During the autumn currently-fixed photosynthate may be preferentially transported from old shoots rather than young tillers and shoots. This permits young plant parts to make a quick start in the next growing season. Even morphologically rather uniform genera, such as *Ceratophyllum*, may exhibit such physiological characteristics (Best and Meulemans 1979).

Established phase

If we want to control only selected plant species, in mixed stands of aquatic vegetation, we might reasonably expect that suppressing the growth of one species could increase resource availability and permit increased production of the remaining species. To be able to predict what will happen under such conditions it is essential to define the level of interaction between macrophytes. Braakhekke (1980) summarized the current terms used to describe different forms of species interaction. From the weed control aspect interference (negative interaction) is of particular relevance because this mechanism might lead to a mixed stand of relatively low biomass compared with the biomass of monospecific stands of similar plant density. Competition and allelopathy both belong in this category.

An appropriate experimental approach might be to use the competition models of de Wit (1960), with replacement series of environmental factors (e.g. nutrients), for partial populations of aquatic macrophyte species. This might show us how the production of macrophytes in the established phase could be influenced by selective control measures affecting the species composition of an aquatic plant community.

Preliminary results with experiments in allelopathy, although often difficult to interpret and sometimes controversial, indicate that this kind of interference might also play an important role in the control of plant populations (Kulshreshtha and Gopal 1983; Szczepanksa 1971).

ENVIRONMENTAL CONDITIONS, LIFE-CYCLE CHARACTERISTICS OF AQUATIC MACROPHYTES, AND WATER QUALITY CRITERIA

Manipulation of single environmental factors

In the previous section the influence of separate environmental conditions on the different life-cycle stages has been discussed. Many of these data were collected in the laboratory under controlled conditions. When we want to influence macrophytes by controlling the production of tubers or turions, indoor experiments tell us that we should manipulate the day-length. For obvious reasons, this is not possible in a direct way. However, indirect methods of manipulation might be sought, or direct action may be taken against the propagule stage. To be able to analyse the problem we must first summarize the field characteristics of the controlling environmental factor and its relationship with possibly manageable physical, chemical or even biological factors.

Temperature, a major environmental factor influencing plant growth is very difficult to manipulate in the field. The only way to do it is to add large supplies of water of different temperature (e.g. thermal effluents).

The quantity and quality of light reaching the macrophytes is manageable to a certain extent. The quantity of incident light can be decreased (for small water bodies) by planting trees on the shores. The introduction of aquatic macrophytes with floating leaves can shade submerged macrophytes. The depth of the water body could be increased by dredging, so reducing the amount of light available for submerged rooted plant growth. Concurrent with the reduction of the amount of light, it would be advantageous if the photoperiod could be shortened. But to increase the photoperiod, an additional light source has to be introduced (photoperiod interruption).

From the observed differences in drought-resistance of propagules of different species we can conclude that drawdown periods might be effective in the selective control of aquatic weeds. The water level of most aquatic plant habitats can be manipulated to a certain extent.

Finally, the redox potential within a habitat, especially if low, could be influenced by regularly disturbing the bottom sediments. Oxygenation of the bottom layers of the water column by pumping in oxygen is also a possibility. Methods to do this have been developed, with the aim of fixing phosphates in the sediments of highly eutrophic waters.

Water quality management

The discussion so far has not taken into account the fact that there are major differences between aquatic ecosystems due to fundamental differences in overall water quality. How might differences in water quality influence the background variability of the plant material present in a given habitat?

For many water bodies, quality criteria are formulated in terms of turbidity figures, light attenuation coefficients, or chlorophyll *a* concentrations. The conceptual model of Phillips, Eminson, and Moss (1978) related such water quality criteria to the growth of aquatic macrophytes. These authors showed that fluctuations in macrophyte biomass influence water quality dramatically, and vice versa, because the biomass of macrophytes is inversely related to the phytoplankton growth. In fact they stated that the quality of the aquatic habitat in its broad sense (including such criteria as the presence of invertebrates, fish production, presence of waterfowl) is reduced considerably when macrophytes disappear.

In their model, the biomass fluctuations of the periphyton and the resulting light reduction for the macrophytes (which can, under eutrophic conditions, be up to 70 per cent) are the main causes of macrophyte biomass fluctuation under field conditions.

An example is the growth and reproduction of *P. pectinatus* (unpublished data, W. van Vierssen) which shows that the level of variation of the plant population is what might be expected, on the basis of difference in water quality (turbidity of water; nutrient-level-related growth of periphyton). This example is an illustration of how water quality management could be incorporated into weed control strategies. Experimental populations of *P. pectinatus* and *Ruppia maritima* were artificially shaded. Light attenuation by the water layer and the periphyton was taken into account, and the relationship between the amount of incident light and biomass production (above ground and below ground) was recorded. *P. pectinatus* and *R. maritima* were artificially shaded at four different levels, including a control (no artificial shading). From the resulting graph, representing the produced biomass plotted against the amount of light which had reached the plants during the growing season (taking into account light attenuation by the artificial shading, the attenuation coefficient of the water layer and the periphyton), it is possible to calculate when light becomes limiting to the growth of *P. pectinatus* and *R. maritima*. It appeared that the light requirements for both species were quite similar. Moreover, it was striking that at different light levels which were not limiting for the growth of *P. pectinatus*, very clear effects on the tuber production were recorded. After the second year of the experiments, these effects were even more conspicuous. In the control, 80 tubers m^{-2} were produced during the first year, whereas at the highest level of shading approximately 1 400 tubers m^{-2} were produced. In general, the number of tubers was inversely correlated with the amount of light received by the plant. The biomass production of the tubers accounted for 5 per cent of the total plant biomass in the control and for 45 per cent of the total plant biomass at the highest level of shading. During the second year of the experiment there was, in general, an increase in total biomass of the plants. In the control, tuber production accounted for 20 per cent of the total plant biomass and for 60 per cent of the total plant biomass at the lowest light levels.

The biomass production of the tubers was also higher over the total range during the second year of the experiment (up to 2 500 tubers m^{-2} at the lowest light levels). It was remarkable, however, that during the second year, the plant biomass production was, in general, positively correlated with the level of shading. Therefore, it may be assumed that the differences in tubers present at the beginning of the growing season to a large extent determine the biomass production at light-saturated levels. From these results it may be concluded that biomass fluctuations of up to 50 per cent can be expected under conditions in which light saturation occurs.

Current research on *P. pectinatus* indicates that under severe light stress, the total biomass decreases as well as the quantity of tubers. These results have clear implications for aquatic plant management, as illustrated by the following example. In Fig 12.2 the results for the light requirements of *P. pectinatus* have been summarized. Three areas (I, II, III) can be recognized indicating situations

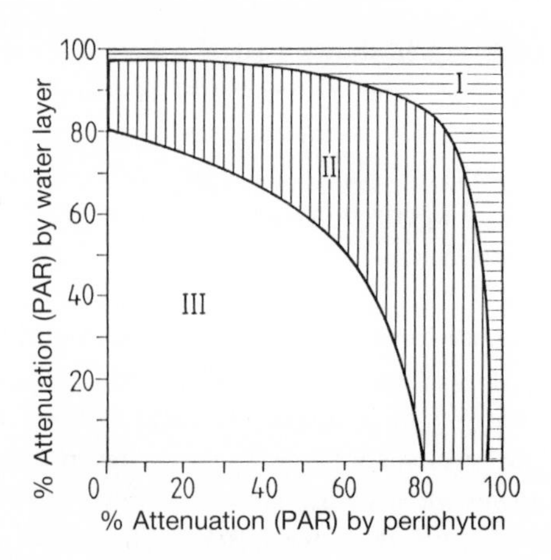

FIG. 12.2. The impacts of light attenuation by the water layer and periphyton on the growth of *Potamogeton pectinatus*. I. No growth is possible because not enough light is available. II. Light controls biomass. III. Light saturation occurs. Total biomass production (below and above ground) is not influenced by changes in amount of light (after van Vierssen, van Dijk and Breukelaar 1985).

in which this macrophyte is unable to grow because light levels are below the compensation point (I), is controlled by the available amount of light (II), or occurs at light-saturated conditions (III). The changes in tuber production discussed above apply to situation III. As indicated before, water quality influences the growth of both phytoplankton and periphyton. From Fig. 12.2 and the results from the above experiment on *P. pectinatus*, it may be concluded that for weed control purposes in a particular water body it could be profitable to know which of the three different situations (I, II, or III) applies. Figure 12.3 shows that different water qualities in the habitat can indicate a requirement for different effective control measures. Here, two different nutrient levels are present, with, as a result, different amounts of periphyton on the plants (Fig. 12.3*a* and Fig. 12.3*b*). These curves were constructed using the data from Fig. 12.2 and the relationship $I_z = I_0.e^{-\varepsilon.z}$ in which I_z is the amount of *PAR* (photosynthetically active radiation) at depth z, I_0 the amount of *PAR* at the water surface, ε is the light attenuation coefficient, and z the depth.

The areas I, II, and III are the same functional areas as those in Fig. 12.2. Three imaginary sites have been chosen (A, B, and C), each with a different water quality in terms of ε and periphyton biomass (Fig. 12.3*a* and Fig. 12.3*b*). What options are available to control *P. pectinatus* in these differing situations?

Site A. Here the growth of *P. pectinatus* is not directly limited by light availability in Fig. 12.3*a* (area III). It is possible that the biomass increases when available light decreases, because tuber production might increase under such

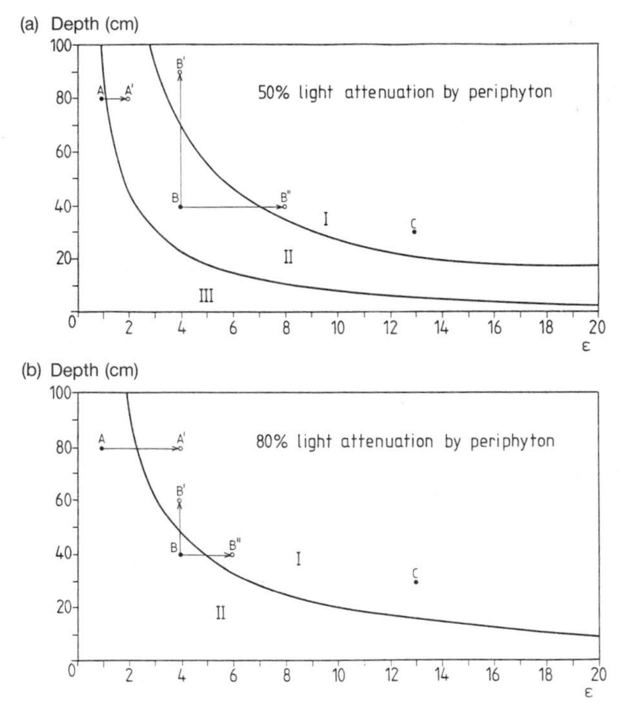

FIG. 12.3. Influence of light conditions produced by waterlayer (attenuation coefficient ε) and two periphyton densities, (a) and (b), on growth of *Potamogeton pectinatus* at different depths. I. No growth is possible because not enough light is available. II. Light controls biomass. III. Light saturation occurs. Total biomass production (below and above ground) is not influenced by changes in amount of light (modified after van Vierssen, van Dijk, and Breukelaar 1985). A, B and C are imaginary stations for which macrophyte control options are being discussed resulting in a changed position within the figures 12.3a and 12.3b (A′ B′, B″). For further explanation see text.

conditions. For shallow waters (up to 1 m) the only option related to Fig. 12.3*a* is to increase the turbidity (influencing ε and bringing station A to A′, area II). The situation represented by Fig. 12.3*b* is less problematic: here biomass is directly limited by light availability. The same strategy applied to the situation in Fig. 12.3*a* will control the vegetation completely (bringing station A to A′, area I). It is important to notice that water level alteration could not solve the problem because of the scale of change required.

Site B. Here the production of *P. pectinatus* is directly controlled by the amount of available light, both in Fig. 12.3*a* and Fig. 12.3*b*. A very effective control in both situations would be to increase the depth, either by local dredging or by increasing the water level (bringing station B to B′). Another option would be to increase the turbidity (bringing station B to B″).

Site C. No problems will arise as long as the water quality does not improve considerably.

This example illustrates that for optimum aquatic weed control measures it is useful to know whether the vegetation is under some degree of biological stress.

CONCLUSIONS

From the preceding sections of this chapter it is clear that, although we basically know how a number of macrophytes survive in their environment, we are still far from being able to control macrophytes by means of this knowledge. It does, however, seem apparent that control strategies can focus either on ecosystem management or on management at the level of the plant population.

In my opinion, because of the worldwide increase in the number of water bodies with multiple functions, aquatic weed control is better discussed at the level of ecosystem management. This means that we urgently need conceptual ecological models which might allow us to formulate the limits of our ability to influence aquatic ecosystems (Mitchell 1986). In this respect it might be useful to consider briefly the concept of 'water quality'. Water quality in the strict sense generally denotes the physical, chemical and biological characteristics of the water, in its broader sense the term has been used to describe the overall biological characteristics (defined rather loosely) of the whole aquatic ecosystem, equating to a 'habitat quality' concept. There can exist no general consensus on the standards which govern 'water quality' (in either sense) because this depends on the uses and management of each individual body of water. Some analysis of the management aims, and uses of an aquatic system should be carried out before weed control measures are considered, particularly if some of the usages conflict with each other.

A selection of non-chemical weed control measures, affecting water quality (in both senses) which would be effective in influencing the survival of aquatic macrophytes at different life-cycle stages is given below (Nichols 1979; Chancellor 1981). This list does not pretend to be complete and only indicates those methods which are most relevant from a practical point of view. There are other, more sophisticated methods; many of them are promising and their development is challenging (Anderson 1986*b*). In many of them herbicides (existing ones or ones to be developed) might play an important role. Non-herbicidal methods which could be further developed are:

1. Influencing seed germination and young seedling growth
   (a) Removal of macrophytes before seed setting. Methods: harvesting.
   (b) Breaking dormancy during unfavourable periods for seedlings. Methods: drawdown, temperature fluctuations by influx of groundwater, effluents or cooling water. Disturbing low redox potential near or in sediments by oxygenation/aeration.

(c) Reducing light availability for young seedlings. Methods: disturbing sediments, temporarily increasing water level.

2. Influencing subterranean tuber/turion formation and germination
   (a) Above ground biomass removal at beginning of tuber production. Methods: harvesting.
   (b) Photoperiod interruption: Methods: additional light input during night (see also Anderson 1986*b*).
   (c) Disruption of dormancy. Methods: increasing temperature by influx cooling water: drawdown period.
   (d) Different strategies in subsequent years in the following order.
      (i) Year 1: Increasing amount of available light, as a consequence low number of tubers. Methods: lowering water level; flushing with non-turbid water; subsequent removal of above ground biomass before tuber production is at its maximum. Methods: harvesting.
      (ii) Year 2: Conditions: relatively small tubers present. During sprouting, diminish light level. Methods: increase water level. Result: from small tubers retarded growth of plants which suffer from light stress (see also Spencer 1986).

3. Influencing axillary turion/tuber production and germination
   (a) Removal of biomass before nutrient deficiency in water occurs. Methods: harvesting.
   (b) Breaking photoperiod. Methods: water level fluctuation to influence light availability, or additional light input during dark period.
   (c) Prevent turion germination. Methods: Increase or decrease temperature by influx of cooling water or ground water.

4. Influencing carbohydrate allocation in vegetative plant parts
   (a) Diminish amount of re-allocated material. Methods: harvesting before reallocation occurs (see also Kimbel and Carpenter 1981).

Many of these listed methods could be more effectively applied if, in addition, herbicides were available which could be specifically targetted against the most vulnerable life-cycle stage(s) of the weed population(s). An improved basic knowledge of the survival strategies adopted by the different phases of the life-cycles of aquatic macrophytes, in response to major fluctuations in environmental conditions, is a clear pre-requisite for further advances in aquatic plant management. Increased research in this field is likely to pay major dividends in improving both the efficiency and ecological safety of aquatic weed control measures.

13

Models on metabolism of aquatic weeds and their application potential

E. P. H. Best

INTRODUCTION

Plant growth models

Models of plant growth are mathematical representations of the physiological processes associated with plant metabolism. It is important that the parameters of the model and the relationships which the model expresses (a) correspond with the biological system which is investigated; (b) agree with the generally accepted scientific theories, the equation coefficients depending on plant group and possibly species; (c) avoid as much as possible empirical equations which often lead to curve fitting. However, when a process is not understood, and theories are lacking, the use of empirical equations is the only possibility to fit the process into the model. Regrettably this decreases the predictive power of the model. Most models contain descriptive and predictive elements. Descriptive usually refers to equations describing an existing system, while predictive refers to equations which can be extrapolated outside the existing boundaries.

The degree to which aquatic macrophytes influence the ecosystem is proportional to plant mass and depends on plant species and physico-chemical factors. Therefore, predictions of the environmental impact of management measures concerning aquatic weeds should be based on accurate estimates of: (a) plant species, mass, and its pertinent physiological activities, (b) the plants' contribution to the various food chains (via grazing and the consumption of detritus), and (c) the contribution of the plants' decay to biogeochemical cycling and oxygen regime.

Models as a tool for management

Modelling can help to orientate macrophyte studies into the larger framework of aquatic ecosystem structure and function. Some models are currently used for the management of aquatic ecosystems. Sometimes independent models on macrophyte metabolism suffice, but often integrative models are useful. Independent models are stand-alone descriptions of the net results of macrophyte metabolism. Integrative models are larger models of ecosystem structure and function of which the macrophyte models are cells.

This chapter summarizes all currently published independent models on the metabolism of freshwater macrophytes and their potential use in the management of freshwater ecosystems. Models on photosynthesis and biomass production of freshwater macrophytes are discussed first. Applications of macrophyte models to studies of freshwater ecosystems and aquatic weed management are subsequently discussed. The chapter concludes with recommendations for research.

INDEPENDENT MODELS ON METABOLISM OF AQUATIC WEEDS

There are several independent models on growth aspects of freshwater macrophytes (Table 13.1). Some can be classified as analytical and most as simulation models. Analytical models are usually associated with linear systems and simulation models with non-linear systems (Patten 1976). The latter models characteristically use differential equations to describe (simulate) the non-linear systems, which can only be effected on a computer. The models are based largely on macrophyte photosynthesis and describe the consequences of this process for the

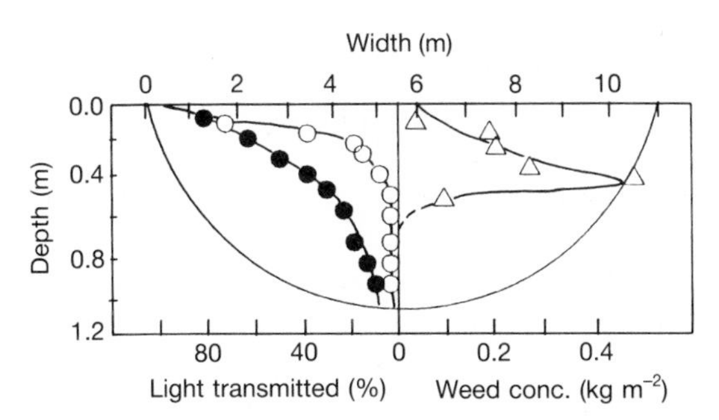

FIG. 13.1. Basic data for a stream considered in July. (●) Light transmitted through the water column when clear (percentage of the subsurface reading). (○) Light transmitted through an average submerged macrophyte bed. (△) Concentration of macrophyte (oven-dry weight) calculated from the two light curves for each 0·1 m depth layer. Plain line, mean cross section of river bed (after Westlake 1966*a*).

Table 13.1. Characteristics of models describing photosynthesis and growth of freshwater macrophytes; A = analytical; S = simulation

A	S	Output	Calibration species	Reference
+		Carbon uptake	Not indicated	Weber *et al.* 1981
+		Carbon uptake	*Egeria densa*	Laing and Browse 1985
+		Oxygen flow	*Potamogeton pectinatus*	Westlake 1966
+		Oxygen flow	*Egeria densa, Lagarosiphon major*	Rutherford 1977
	+	Oxygen flow, biomass	*Elodea canadensis*	Wright and McDonnell 1986*a*
	+	Oxygen and CO_2 exchange	*Elodea canadensis*	Ondok, Pokorny, and Kvet 1984
	+	Carbon flow, biomass	*Myriophyllum spicatum*	Titus *et al.* 1975
	+	Carbon flow, biomass	*Ceratophyllum demersum*	Best 1981
	+	Carbon flow, biomass	*Hydrilla verticillata*	Lips 1985
	+	Mineral (C, N, P) flow	*Hydrilla verticillata, Myriophyllum spicatum*	Collins, Park, and Boylen 1985
	+	Mineral (N, P) uptake, biomass	*Salvinia molesta*	Toerien *et al.* 1983
	+	CO_2 exchange	*Phragmites australis*	Ondok and Gloser 1978*a*

plant or its direct environment (ambient water or air). Most analytical models concern oxygen or carbon flow, while the simulation models are usually tools to calculate or predict plant biomass.

In this paper twelve models on freshwater macrophytes are summarized; ten concern submerged macrophytes, one concerns floating macrophytes and one concerns emergent macrophytes. The models are characterized by their: (a) general features, (b) description of plant processes, (c) calibration, (d) testing, (e) use in connection with other models, and (f) application potential indicated by the author. The plant processes which largely determine the plants' mass balance are taken into consideration.

These are: (a) development, i.e. morphogenetic events, (b) photosynthesis, i.e. the evolution of oxygen, or the uptake and fixation of carbon, (c) respiration, i.e. the consumption of oxygen, or the release of carbon, (d) growth, i.e. the increase in plant mass, and (e) senescence, i.e. the loss of plant mass.

Models of submerged weeds

Weber *et al.* (1981) modified an analytical photosynthesis model for terrestrial plants (Tenhunen, Yocum, and Gates 1976) and applied it to show relationships between the photosynthetic response of submerged aquatic plants and fluctuations in the supply of inorganic carbon (CO_2 and HCO_3^-) and pH. The model centres on the availability and uptake of carbon and the effect of endogenous oxygen concentration or photorespiration.

Net photosynthesis rate (P_n) is described by Fick's law, because the rate of carbon supply to the fixation site is restricted by diffusion:

$$P_n = \frac{C_{ex} - C_{in}}{R_c} + \frac{B_{ex} - B_{in}}{R_b}$$

in which C_{ex} and C_{in} are concentrations of CO_2 in the external medium and at the fixation site, respectively; B_{ex} and B_{in} the concentrations of HCO_3^- in the external medium and at the fixation site, respectively; R_c and R_b the transport resistances for CO_2 and HCO_3^- from the external solution to the fixation site, respectively.

The carbon fixation rate follows Michaelis–Menten kinetics with O_2 acting as competitive inhibitor:

$$P_g = \frac{P_M}{1 + \dfrac{K_c(1 + O_{in}/K_o)}{X_1 . C_{ex} + X_2 . B_{ex} - R_c . P_n}}$$

in which: P_g, P_M, and P_n are gross, maximum, and net photosynthetic rates, respectively; K_c and K_o the concentrations of CO_2 and O_2 at which $P_g = P_M/2$; O_{in} the O_2 concentration at the fixation site; X_1 and X_2 dimensionless terms accounting for the transport resistances of the various carbon species.

The rate of photorespiration depends also on the O_2 and CO_2 concentrations and follow Michaelis–Menten kinetics:

$$W_p = \frac{W_M}{1 + \dfrac{K_o(1 + C_{in}/K_c)}{O_{in}}}$$

in which: W_p and W_m are the rate and maximum rate of photorespiratory production of CO_2, respectively.

Weber *et al.* (1981) showed that CO_2 from photorespiration usually will be recycled rapidly but this recycling depends on endogenous O_2 concentration. Model predictions were usually in agreement with values measured for *Chara corallina*. The model can be used to study further the control of photosynthesis in submerged aquatic plants by CO_2, HCO_3^-, and O_2 concentrations, and the response to changes therein.

Laing and Browse (1985) developed a dynamic analytical model for photosynthesis of *Egeria densa*, which centres on the mechanism of carbon uptake and the role of phosphoenolpyruvate (PEP) carboxylase. The model simulates photosynthesis under non-steady-state conditions, in contrast to that of Weber *et al.* (1981) which pertains to steady-state conditions. It assumed photosynthesis occurs according to the diffusion/enzyme kinetics model of Weber *et al.* (1981) which describes the rates of change of oxygen and various species of inorganic carbon (CO_2, HCO_3^-, and CO_3^{2-}).

Conceptually, the system modelled consists of three compartments: air, water, and plant. The plant exchanges inorganic carbon and oxygen with the environment through a boundary layer, and consumes and produces CO_2 and O_2 by the biochemical processes of photosynthesis and respiration. The equations describing the biochemical processes of photosynthesis are taken largely from the literature and some rate coefficients are derived empirically in the laboratory. Bicarbonate enters the plant in exchange for OH^- via the bicarbonate pumping mechanism which shows a hyperbolic response to HCO_3^- concentration. A set of time-dependent differential equations describes the fluxes and can be solved numerically either using Continuous System Modeling Program (CSMP) or a program by Hindmarsh (1975).

The behaviour of the model was compared with data generated under the same laboratory conditions used to drive the model's rate coefficients. Measured and predicted values were generally similar. The model can be used to examine photosynthesis in the laboratory under steady-state and non-steady-state conditions and may be extended, after some modifications, to photosynthesis in macrophyte beds under natural conditions.

Westlake (1966*a*) developed an analytical model, based on oxygen flow through a plant, to study quantitatively the photosynthesis of submerged aquatic macrophytes. Photosynthesis is calculated on an areal basis integrated over water depth to account for the absorption of irradiation by the water column and self-shading of the macrophyte. Photosynthesis is related to irradiation using an empirical equation originally developed for algae (Talling 1957):

$$ki = P(1 - \mathrm{P}^2)^{-1/2}$$

Photosynthesis (P) depends on irradiation (i) and a macrophyte-specific coefficient, k (e.g. for *Potamogeton pectinatus*, Fig. 13.1). The irradiation (i) reaching the macrophytes' photosynthetic tissues is calculated in hourly averages from daily incident irradiation as the photosynthetically active irradiation (*PAR*, assuming 50 per cent of total irradiation), part of which is extinguished by the overlying water column and average plant biomass (*P. pectinatus*, Fig. 13.1). Respiration is assumed to be constant in time and to occur only in darkness. It increases with temperature.

The model is calibrated on laboratory and field data (Westlake 1961, 1964). The model calculated oxygen in flowing waters which was attributable to *P.*

pectinatus when irradiation, plant standing crop (Fig. 13.1) and water temperature were measured (Table 13.2). Changes in the ratio of respiration rate:light-saturated net photosynthetic rate, over the range normally found in active plants, cause large changes in the daily oxygen balance of the stream. Temperature affects this ratio and is a sensitive variable in the model.

The model can be used to calculate the probable effects of some waterway management measures, e.g. (a) improvement in the clarity of effluents, and (b) controlled cutting of excessive growths of macrophytes, in improving the oxygen balance of a stream. However, the oxygen regimes in flowing waters are usually strongly influenced by the oxygen consumption at the bottom and re-aeration at the water–air interface (Cosby, Hornberger, and Kelly 1984). The present model does not contain descriptions of these processes and, therefore, extrapolation to

Table 13.2. Specimen calculations of net oxygen exchanges (After Westlake 1966*a*)

(A) Depth profiles of oxygen exchange per unit layer		**Depths (m)**		
		0	**0.2**	**>0.65**
At noon (average July day)				
Light transmission (% sub-surface; observed, Fig. 1)	t	100	41.6	< 2.6
Irradiation (t% of i_0; mg cal cm^{-2} min^{-1})	i	293	122	< 7.6
Net photosynthetic rate (g O_2 kg^{-1} h^{-1})	p	10.0	9.10	< −0.7
Relative width (observed; calculated from Fig. 1)	w	1.00	0.972	< 0.8
Relative macr. conc. (calculated from Fig. 1)	c	0.333	1.27	0
Macr. distribution factor	d	0.333	1.23	0
Relative net oxygen exchange per unit layer	n	3.33	11.2	0
At night, i_0 and $i=0$, $p_0=-1.50$ g O_2 kg^{-1} h^{-1}				
Relative respiration rate per unit layer	n_0	−0.5	−1.85	0
(B) Photosynthesis–depth integrals				
Planimetric areas of exchange; n against depth				cm^2
Average noon, 293 mg cal cm^{-2} min^{-1}	a			+210.7−1.3=209.4
Night, 0 mg cal cm^{-2} min^{-1}	a_0			−75.6
Net oxygen exchange per unit area of river				
Night respiration, plant density D.p_0	N_0			$0.123\times1.50\times10^3=184$
Oxygen exchange cm^{-2} of graph, N_0/a_0	f			184/75.6, 1 cm^2=2.44
Integrated oxygen exchange, average noon, a.f	N			$209.4\times2.44=511$

Macr. conc. = macrophyte concentration.

oxygen regimes in other water bodies is difficult. In its present form the model is analytical, but with some adaptations it can be modified into a simulation model for oxygen concentrations in streams.

Rutherford (1977) developed an analytical model for the effects of submerged aquatic plants on dissolved oxygen (DO) levels in rivers. The model is based on oxygen flow through a stream with submerged macrophytes and phytoplankton.

The macrophyte-associated metabolic processes are very briefly described. Effects of irradiation on photosynthesis and those of temperature on plant metabolism are omitted.

The DO concentrations of a river are calculated using one-dimensional biochemical oxygen demand (BOD)–DO equations, and plant (macrophyte and phytoplankton) activity is regarded as contributing to these levels. DO production occurs only in the euphotic zone, but DO consumption takes place in the whole vertical water column.

The rate of change in DO due to macrophyte activity (Q_m) is viewed as being proportional to the surface area of macrophyte beds per unit channel length (A_w) relative to the channel cross sectional area (A), and the average rate of photosynthesis (r_p) minus the average rate of respiration (r_r).

$$Q_m = \frac{A_w}{A}\{r_p . f(t) - r_r\} g(T).$$

DO is produced by photosynthesis in the light and depends on irradiation by an empirical factor $g(T)$ for a period with a length described by an empirical factor $f(t)$. DO is consumed by respiration all day and assumed to be constant in time. The macrophyte vegetation to which the model pertains was composed largely of *Egeria densa* and *Lagarosiphon major*. The maximum macrophyte DO production and consumption rates of macrophytes were taken from the literature, while the coverage of macrophyte beds was recorded *in situ* (Waikato River, New Zealand).

The rate of DO change attributable to phytoplankton in this model is described similarly. However, an additional term is added to account for the phytoplankton concentration in the water column. The model is calibrated by comparing

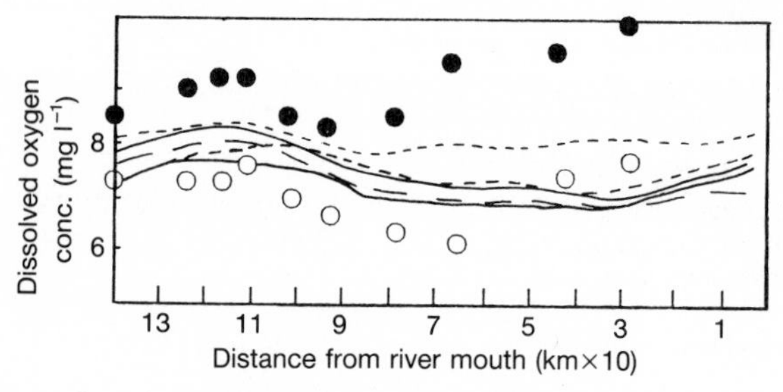

FIG. 13.2. Observed and predicted dissolved oxygen levels in a stream occupied by submerged macrophytes, March 19–21 1974. Observed concentrations; (●) maximum, (○) minimum. Predicted concentrations: (– – – –) Maximum and minimum, assuming high productivity (DO production 5.0, DO consumption 1.5 mg $O_2 m^{-2} h^{-1}$); (——) Maximum and minimum, assuming normal productivity (DO production 1.5, DO consumption 0.75 mg $O_2 m^{-2} h^{-1}$); (— —) Average, neglecting macrophyte productivity (after Rutherford 1977).

observed and predicted DO levels (Rutherford 1975) and was tested on independent data sets. Measured and predicted DO levels at various distances from the river mouth were similar (Fig. 13.2).

The model can be used to calculate the DO concentrations in rivers with or without macrophyte beds, from data on: (a) water temperature, (b) flow and water quality of the tributaries, (c) channel area, (d) average velocity, (e) longitudinal dispersion coefficient, (f) cover area of macrophytes, and (g) concentration of phytoplankton. The model is constructed to predict the effects of farming, urbanization, methods of effluent treatment and thermal stress from power plants. Extrapolation to other water bodies is difficult because oxygen consumption at the bottom is not included in the model.

Wright and McDonnell (1986*a*) developed a simulation model for macrophyte growth in shallow streams based on oxygen flow through a stream containing submerged macrophytes, phytoplankton, and benthic organisms. The DO concentrations of a river are calculated using one-dimensional DO equations, and the activity of the organisms (macrophytes, phytoplankton, and bottom organisms) is regarded as contribution to these levels. In this equation the rate of change in oxygen per area is the sum of oxygen production and oxygen uptake by diffusion both per area, and the oxygen imported by the inflowing water, minus the rate of oxygen consumption per area.

Photosynthesis is described as a function of the maximum rate of photosynthesis which is modified by an empirical temperature coefficient and a seasonal factor for the plants' developmental stage. Limitations by irradiation and nutrient concentration are related to the maximum rate of photosynthesis by Michaelis–Menten kinetics. Irradiation of the macrophytes' photosynthetic tissues is calculated from daily incident irradiation as *PAR* and consideration is given to effects of latitude, reflection at the water surface, absorption by the overlying water column, and self-shading of the stand.

Respiration is described as a function of the maximum rate of dark respiration which is modified by an empirical temperature coefficient and a seasonal factor for the plants' developmental stage. Biomass is seen as a result of photosynthesis, respiration, and senescence with detached mass transported from the system by water streaming. Senescence is described as a function of the maximum death rate (derived from literature) and is modified by a seasonal factor for the plants' developmental stage.

The coverage of the macrophyte beds was recorded *in situ*. The vegetation was mixed, with *Elodea canadensis* and *Potamogeton crispus* dominating.

The model is calibrated on Spring Creek (Pennsylvania, USA: Wright and McDonnell 1986*a*, *b*) by 24 hour recordings of DO within the stream, measurements of photosynthetic and respiratory rates of phytoplankton and sediment with benthic organisms in enclosures. It calculates the macrophyte biomass and its flushing from the system. The model was tested on oxygen data collected at two phosphorus-enriched sites in the same stream. It predicted macrophyte activity successfully (Fig. 13.3) and proved very sensitive to changes in maximum photosynthetic rate. The model can be used to predict changes in macrophyte biomass due to decreased nutrient concentrations of streams.

Ondok, Pokorny, and Kvet (1984) developed a simulation model for diurnal O_2 and CO_2 regimes of submerged macrophyte stands in shallow ponds. The model is empirical, because most functions are derived from curves fitted to laboratory measurements, and it is not based on generally accepted theories. It has four state variables: concentrations of O_2; total CO_2; free CO_2; and HCO_3^-, and three driving variables: irradiation, water temperature, and pH.

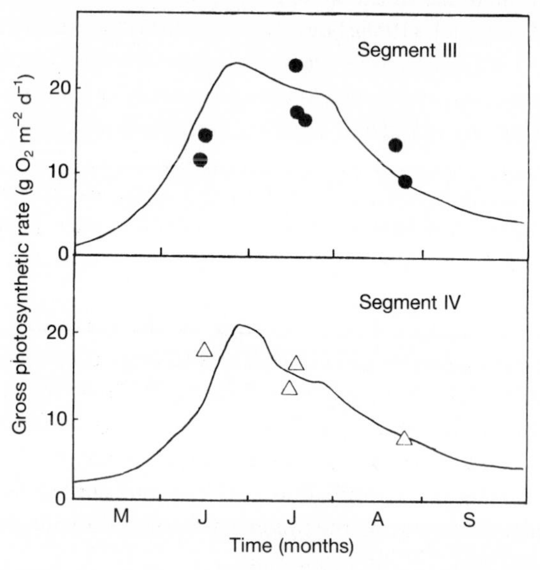

FIG. 13.3. Predicted dissolved oxygen levels in two successive segments of a river occupied by submerged macrophytes. (●) and (△)DO value observed in the segments III and IV, respectively (after Wright and McDonnell 1986*a*).

The temporal changes in O_2 and total CO_2 concentration in the water column containing the macrophyte stand are each described by two equations, one pertaining to the daytime and one to the night-time. During daytime the oxygen concentration is (a) increased by stand photosynthesis as a function of macrophyte standing crop, irradiation, temperature, and pH; (b) increased by re-aeration at the water surface; and (c) decreased by oxygen consumption at the bottom. At night the oxygen concentration is (a) decreased by stand respiration, as a function of macrophyte standing crop and temperature; (b) increased by re-aeration at the water surface; and (c) decreased by oxygen consumption at the bottom. Macrophyte activity is calculated on an areal basis integrated (0.1 m) over the depth of the water column. The evolution and consumption of oxygen are converted into terms of total CO_2 by multiplication with a constant factor (Graneli 1977).

The concentrations of free CO_2 and HCO_3^- in the water column are proportional to the total CO_2 concentration, depending on water pH and temperature.

The model is calibrated on laboratory and field data on an *Elodea canadensis* stand in a Bohemian fish pond (Czechoslovakia). It was tested on an independent data set from the same pond (Pokorny *et al.* 1984). The diurnal changes in O_2, free CO_2, and HCO_3^- were calculated in summer and autumn using macrophyte standing crop and the driving variables. The calculated values were usually similar to the measured ones (Fig. 13.4).

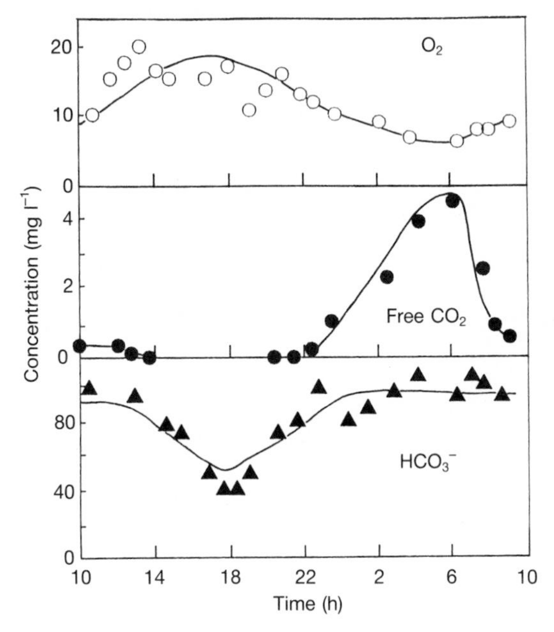

FIG. 13.4. Comparison between measured (symbols) diurnal changes in the concentrations of O_2, CO_2 and HCO_3^- and simulated (lines) diurnal courses in a water column containing submerged macrophyte vegetation (after Ondok, Pokorny and Kvet 1984).

The model can be used to calculate diurnal regimes of O_2, CO_2, and HCO_3^- in stands of *E. canadensis* in shallow, quiet ponds. It cannot be applied to shallow, flowing waters because the effects of water movements are not accounted for in the model.

Titus *et al.* (1975) developed a simulation model for primary production of *Myriophyllum spicatum*, which is based in part on a growth model for terrestrial plants (O'Neill *et al.* 1972: Shugart, Goldstein, and O'Neill (1974)). The model is based on carbon flow through the macrophyte compartment and has four state variables: stems, leaves, roots, and labile carbohydrates and two driving variables:

irradiation and water temperature. All flows are integrated over the depth of the water column (0.1 m depth layer). The conceptual diagram is represented in Fig. 13.5.

Net photosynthesis is described assuming the maximum rate of photosynthesis (P_{max}) per unit of leaf area is not achieved because irradiation and temperatures are limiting. The irradiation reaching the macrophytes' photosynthetic tissues is calculated from daily incident irradiation as *PAR*, which is partly extinguished by the overlying water column and self-shading of the stand. The carbon fixed by photosynthesis is partly lost by excretion and by dark respiration at a rate proportional to macrophyte biomass and temperature.

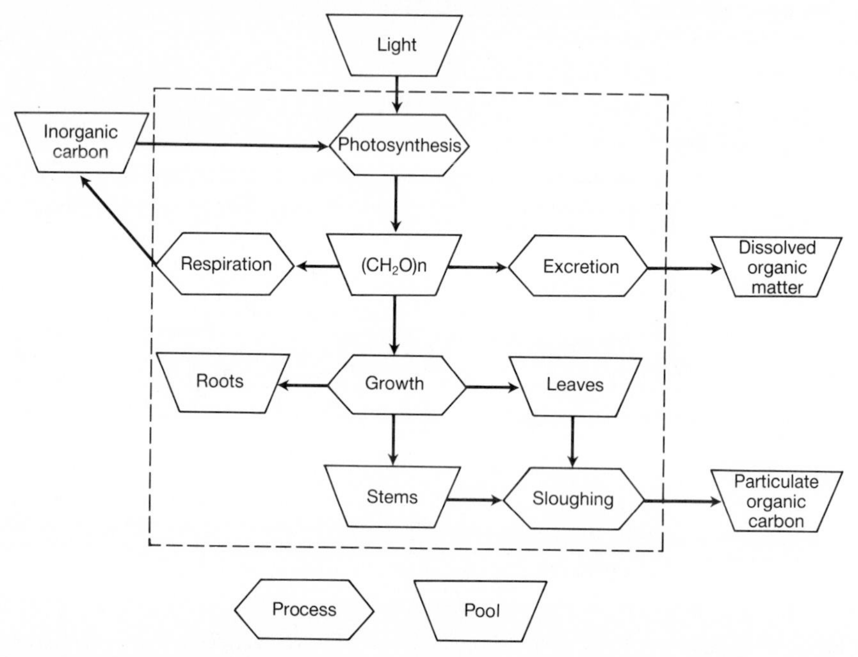

FIG. 13.5. Conceptual diagram of a submerged macrophyte primary production model pertaining to *Myriophyllum spicatum*. The symbol $(CH_2O)_n$ represents the pool of labile carbohydrates (after Titus *et al.* 1975).

The changes in biomass of leaves, stems and roots are assessed individually. Biomass increases by growth and decreases by senescence. The growth rate depends on labile carbohydrate content. The pool of labile carbohydrates increases by net photosynthesis and decreases by respiration and growth.

The model is calibrated on a great number of data collected in the laboratory and in the field (Lake Wingra, Wisconsin, USA: Adams, McCracken, and Schmidt 1971). It is largely written in FORTRAN. It was tested with independent data sets from the same lake (Adams, Titus, and McCracken 1974). The simulated and measured values on biomass and photosynthetic activity were usually comparable. The model can be used (a) as a tool to maximize the information gained from field work; (b) to predict the effects of changes in underwater light climate (e.g. by pollution), and in water temperature (e.g. by warm effluents) on the biomass of *M. spicatum*.

Best (1981) developed a simulation model for growth of *Ceratophyllum demersum*. The model as recently modified (Van der Werf 1986) has three state variables; shoots, labile carbohydrates, and total biomass and two driving variables; irradiation and water temperature (Fig. 13.6). It is based on carbon flow through the macrophytes and all flows are integrated (0.1 m depth layer) over the depth of the water column. The developmental stage is characterized by morphological and physiological characteristics. Gross photosynthesis is defined as the carbon assimilation rate at light saturation (A_{max}) and the initial light use efficiency. The rate of gross photosynthesis is proportional to irradiation and to macrophyte biomass. A_{max} depends on the plant part considered and on developmental stage.

The irradiation reaching the macrophytes' photosynthetic tissues is calculated from daily incident irradiation as *PAR*, less the irradiation which is lost by

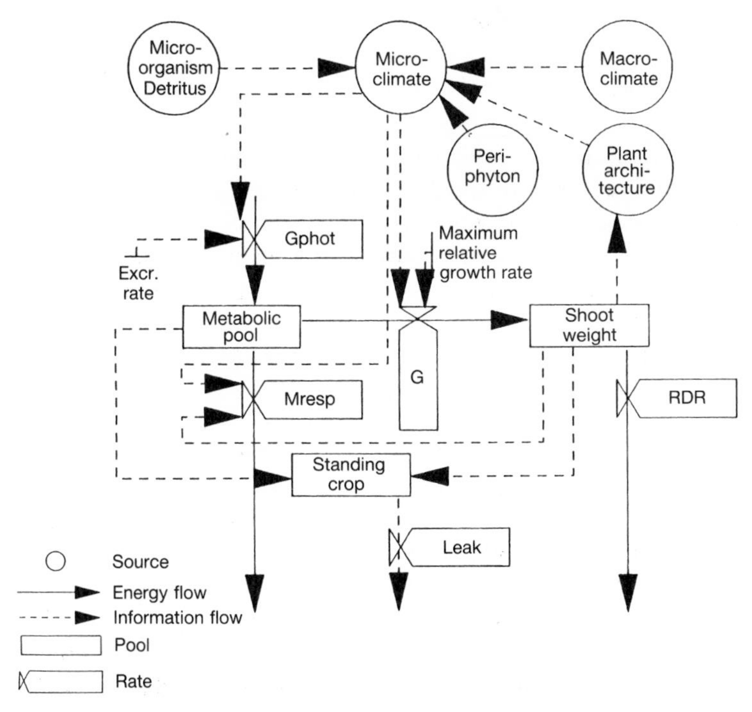

FIG. 13.6. Flow diagram of a submerged macrophyte growth model pertaining to *Ceratophyllum demersum*. G, growth rate; Gphot, rate of gross photosynthesis; Leak, leaching rate; Mresp, rate of maintenance respiration; RDR, relative death rate (after Best 1981).

reflection at the water surface, extinction in the overlying water column, and shading by macrophyte and periphyton. Part of the carbon fixed by photosynthesis is lost by excretion.

Growth and maintenance respiration are calculated individually. Growth respiration is that part of the fixed carbon which is lost in the growth process. Maintenance respiration is proportional to macrophyte biomass and depends on temperature. Macrophyte biomass is the sum of the structural material of the shoot and of the soluble carbohydrates. Growth, i.e. the increase in structural material, is a function of maximum relative growth rate and is proportional to shoot weight. It depends on soluble carbohydrate content and is limited by temperature. The pool of soluble carbohydrates increases by photosynthesis and decreases by growth and maintenance respiration. Senescence causes a continuous loss of soluble material proportional to biomass and a loss of structural material with a species characteristic relative death rate depending on growth stage.

The model is calibrated on a great number of data collected in the laboratory and in the field (Lake Vechten, The Netherlands: Best 1986; Best and Visser 1987). It is written in CSMP III. It was tested with independent data sets from the same lake (Best 1981; Best and Dassen 1987). Biomass, carbon fixation, and respiration were successfully simulated.

The model can be used to relate changes in photosynthetic, respiratory activities, and in biomass of *C. demersum* to changes in underwater light climate and temperature. It pertains to temperate regions.

Lips (1985) developed a simulation model for growth of *Hydrilla verticillata*. It has six state variables; leaves, stems, roots, subterranean and axillary turions, and labile carbohydrates and four driving variables; irradiation, photoperiod, water temperature, and CO_2 concentration. The model is based on carbon flow through the macrophytes and all flows are integrated (0.1 m depth layer) over the depth of the water column (Fig. 13.7).

The developmental stage is characterized by morphological and physiological characteristics. Sprouting and initial growth occur at the expense of the reserves stored in the tubers. Photosynthesis starts when the sprouts have reached the water surface. All assimilates are transported to the subterranean and axillary turions at their initiation, at the expense of all other plant parts.

Gross photosynthesis is defined as the carbon assimilation rate at light saturation (A_{max}) and the initial light use efficiency. The rate of gross photosynthesis is proportional to irradiation and to macrophyte biomass. A light use efficiency characteristic for C_3 plants is used at carbon saturation and one characteristic for C_4 plants at carbon limitation. The irradiation reaching the macrophytes' photosynthetic tissues is calculated from daily incident irradiation as *PAR* less the irradiation lost by reflection at the water surface, extinction in the overlying water column and self-shading.

Growth and maintenance respiration are calculated individually. Growth res-

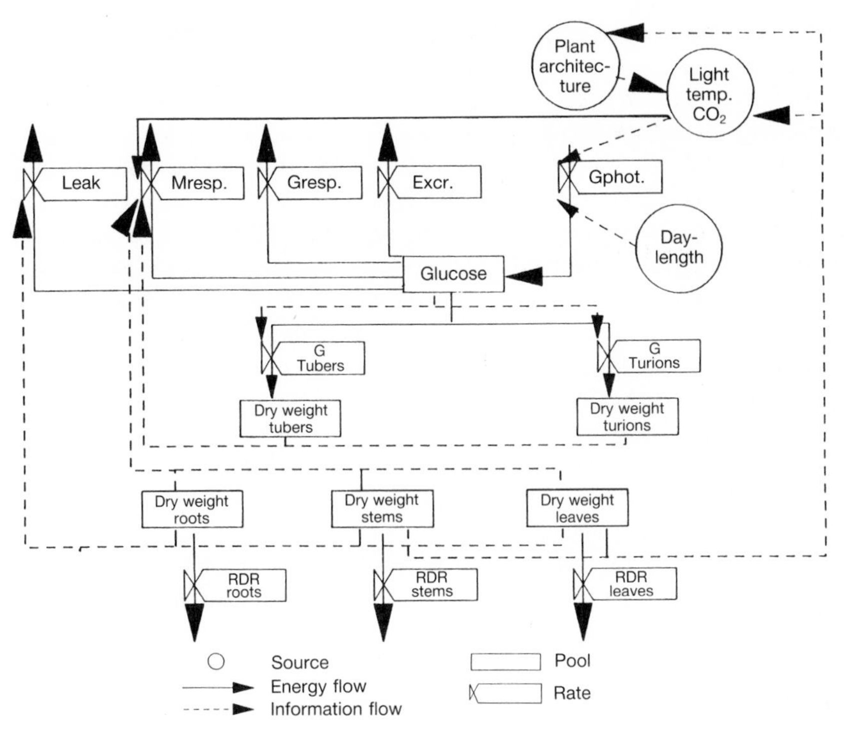

FIG. 13.7. Flow diagram of a submerged macrophyte growth model pertaining to *Hydrilla verticillata*. Excr, excretion rate; G, growth rate; Gphot, rate of gross photosynthesis; Gresp, rate of growth respiration; Leak, leaching rate; Mresp, rate of maintenance respiration; RDR, relative death rate (after Lips 1985).

piration is the carbon lost during the growth process. Maintenance respiration is proportional to macrophyte biomass and depends on temperature.

The changes in biomass of leaves, stems, roots, tubers, and turions are calculated individually. The timing follows a morphogenetically determined pattern and the volume is determined by photosynthesis and respiration. Biomass is increased by (species-characteristic) conversion of net photosynthesis into structural material.

Senescence causes loss of soluble material proportional to biomass continuously and loss of structural material depending on growth stage.

The model is calibrated using literature data on *Hydrilla*, largely collected in Lake Orange, Florida, USA (Van, Haller, and Bowes 1976; Holaday and Bowes 1980; Holaday, Salvucci, and Bowes 1983). It is written in CSMP III. The model was tested on other literature data pertaining to the same lake (Bowes, Holaday, and Haller 1979; Haller, Miller, and Garrard 1976). Biomass and photosynthetic activity were successfully simulated, but the simulated number of turions lagged behind the measured ones, possibly due to incomplete understanding of carbon allocation.

The model can be used to predict changes in biomass of the various plant organs and photosynthetic activity of the *H. verticillata* stock from Florida due to changes in underwater light climate, water temperature, and carbon availability.

Collins, Park, and Boylen (1985) developed a simulation model for growth and decomposition of submerged aquatic macrophytes. The model has seven state variables; macrophyte biomass, dissolved oxygen, particulate organic matter, dissolved organic matter, phosphorus, nitrogen, and organic sediment and two driving variables; irradiation and water temperature. The model is based on mineral (C, N, P) flow through the macrophytes and all flows are integrated (0.1 m depth layer) over the depth of the water column.

Gross photosynthesis is a function of irradiation and temperature. The photosynthetic light response is described by Steele's (1962) equation for photosynthesis of algae. The rate is proportional to macrophyte biomass and depends on water temperature. The irradiation reaching the macrophytes' photosynthetic tissues is calculated from daily incident irradiation as *PAR* less the irradiation extinguished in the overlying water column and by self-shading.

The dark respiration rate depends on the maximum dark respiration rate, is proportional to biomass, and is modified by temperature. Photorespiration depends on the maximum excretion rate and irradiation. Macrophyte biomass

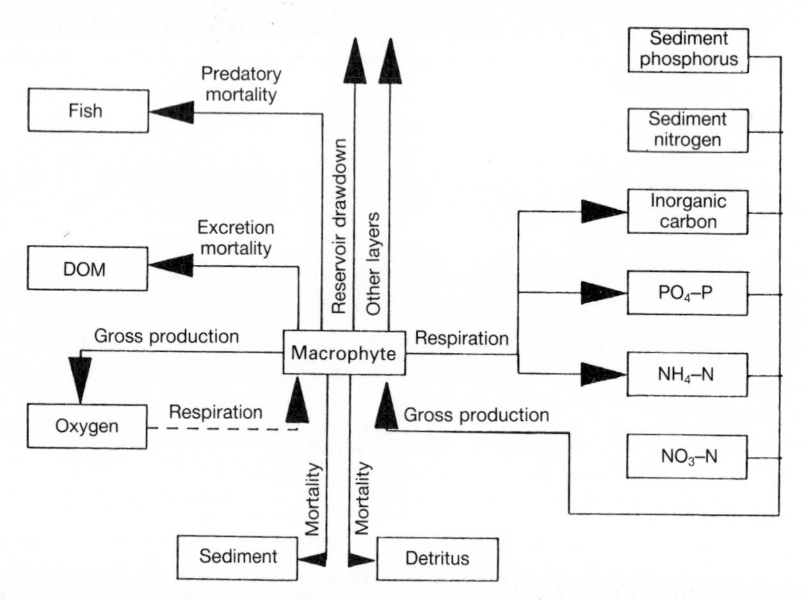

FIG. 13.8. Compartment diagram showing the interactions between a submerged macrophyte model, pertaining to *Hydrilla verticillata* and *Myriophyllum spicatum*, and a one-dimensional water quality model (after Collins, Park and Boylen 1985).

increases by the rate of gross photosynthesis and decreases by the rates of dark respiration, photorespiration, grazing, senescence proportional to biomass and possibly by mechanical harvesting. Senescence causes loss of biomass with a constant death rate and sets in at a temperature-determined moment.

The model is calibrated on a large number of literature data: one version for *Myriophyllum spicatum* and a different version for *Hydrilla verticillata*. It is written in FORTRAN. Several model processes were tested on literature data (Barko *et al.* 1980).

The model can be used to predict the effects of changes in underwater light climate and temperature on macrophyte biomass and detritus formation. It was meant to be run in connection with a one-dimensional water quality model describing the temporal changes in vertical distribution of thermal energy, and biological and chemical material in a reservoir. In the latter case the macrophyte submodel has three additional driving variables, namely the availabilities of C in the water column, and of P and N in water column and sediment (Fig. 13.8). Nutrient limitations to gross photosynthesis follow Michaelis–Menten kinetics. Removal of macrophyte biomass (mechanical and chemical control) can be simulated in a largely simplified manner.

Models of floating weeds

Toerien *et al.* (1983) developed a growth model for the floating *Salvinia molesta*. The model has a very simple structure, is composed of few equations, and deals only with macrophyte biomass.

Growth, defined as the increase in number of leaves, is described as a function of the maximum relative growth rate under optimum environmental conditions and depends on temperature by an Arrhenius relationship. Nutrient (N and P) limitations to relative growth rate follow Michaelis–Menten kinetics.

The model is calibrated on laboratory data describing the effects of N and P on relative growth rate of *S. molesta*. It was tested on field data on relative growth rates of *S. molesta* in sewage treatment ponds (Finlayson 1980) with variable success. Verification of the half-saturation constants for the nutrients is limited by the scarcity of literature data (Mitchell 1970) and should be extended.

The model can be used to calculate growth rates of *S. molesta* in environments varying in nutrient status (sewage treatment ponds) and the nutrient removing potential. It makes the design of appropriate control measures (biological or otherwise) possible because it can be used to estimate biomass of undesired *S. molesta* infestations.

Models of emergent weeds

Ondok and Gloser (1978*a*) developed an empirical simulation model on net photosynthesis and dark respiration in a *Phragmites australis* stand. The model has two state variables; photosynthesis and dark respiration and two driving variables; irradiation and air temperature. It is based on carbon flow through the macrophytes and all flows are integrated (0.2 m layer) over the height of the stand.

The photosynthetic rate of assimilatory organs depends on daily incident irradiation and carbon availability over the vertical air profile. The activities of sunlit and shaded organs are calculated individually. The irradiation reaching the surface of the stand can be calculated as *PAR* in two alternative ways; (a) by a polynom fitted to recorded data, or (b) stochastically.

PAR is extinguished within the stand by the overlying layers of leaf canopy, and per canopy layer a fraction is absorbed which is proportional to the photosynthetic surface area. Daytime is determined by sun elevation. Dark respiration occurs at night and depends on air temperature.

The model is calibrated on data on irradiation (Ondok 1973), temperature (Priban 1973), and gas exchange characteristics of various *P. australis* organs (Gloser 1977). It was tested (Ondok and Gloser 1978*a*, *b*) on field data on *P. australis* stands in a Bohemian fish pond (Czechoslovakia: Gloser 1977). Measured and simulated CO_2 exchange rates over a day were comparable. The model can be used to investigate and calculate the characteristic behaviour of a *P. australis* stand as a photosynthetic and respiratory system under various environmental conditions given the stand biomass and the driving variables.

APPLICATION POTENTIAL OF MODELS OF FRESHWATER MACROPHYTES

Models as tools in aquatic weed control

Man-made water courses, such as channels and ditches, may have two functions: primarily, transporting water of acceptable quality and sometimes also in providing interesting landscape elements. Management is required to maintain these functions. Water courses are generally occupied by aquatic macrophyte communities, the species composition of which depends on site characteristics and complex ecological interactions which may be influenced by management practices. A high biomass of macrophytes can seriously affect water transport capability. Maintenance usually is expensive. A flexible management of the water courses is desirable, allowing a quantity of macrophyte biomass which does not interfere with the primary function of such water courses and takes the more vulnerable function into account. A set of models on (a) macrophyte growth, simulating biomass, (b) hydrology, indicating boundary conditions for open channel flow, and (c) hydraulics, calculating flow resistance as a function of macrophyte biomass can be used to optimize the timing and amount of biomass to be removed. Control measures leading to losses of biomass can be superimposed on the macrophyte growth model (Best, van der Zweerde, and Zwietering 1986).

Aquatic weeds in the tropics and to a lesser extent in the subtropics can produce excessive biomass because growth is not temporarily stopped during a winter season, and, therefore, rigorous control measures are often required. In such cases it is especially important to try and predict the effects of projected control measures on the ecosystem as well as the inherent costs, before their introduction. This is because relatively small modifications in the same control method can have large effects. Models could be useful tools in this case. At present a variety of simulation models exists, which can be used for this purpose independently or in conjunction: on the evaluation of weed control strategies (Ewel, Braat, and Stevens 1975) and on mechanical control (Hutto 1981; Perrier and Gibson 1982), its evaluation (Sabol 1983) and its inherent costs (Mara 1976*b*).

The ecosystem model of Lake Conway (USA, see Whole lake models, below) was used to estimate the impact on the ecosystem of the introduction of white amur as weed controlling agent (Ewel and Fontaine 1982). Simulations indicated effects differing in size and duration on all trophic levels; (a) a decrease in primary production and respiration on lake basis, (b) a two years increase in epipelic algae biomass, (c) an increase in benthic invertebrate biomass, and (d) a nine years or longer increase in biomass and number of secondary predator fish.

Whole lake models

Park *et al.* (1974) developed a general lake ecosystem model. It has sixteen state variables (macrophytes, phytoplankton, zooplankton, bottom dwelling aquatic insects, fish, suspended organic matter, decomposers, sediments, and nutrients) and eight driving variables (irradiation, water temperature, nutrient loadings, wind or changes in barometer pressure, influx of dissolved and particulate material from the surrounding lake basin). The model is divided into modules. The desired modules can be selected and run in conjunction, depending on the questions asked. The macrophyte submodel formed the basis of an independent growth model for *Myriophyllum spicatum* (Titus *et al.* 1975). The model is written in FORTRAN.

It is calibrated on many experimental and field data (nutrient-poor Lake George, New York; nutrient-rich Lake Wingra, Wisconsin, USA). The lake model was tested more or less successfully on field data from Lake George and Lake Wingra.

Mitsch (1976) developed an ecosystem model for water hyacinth management in the highly eutrophic Lake Alice, Florida (USA). The model is based on the concept of energy flow and the effects of control measures are emphasized. It has two sections, one for the marsh and one for the open water. The model contains four state variables (floating macrophytes, benthic plants, phytoplankton, detritus) and six driving variables (irradiation, temperature, and the inflows of water, nutrients, and DO). The benthic plants are not considered in detail.

Gross photosynthesis is stimulated by an autocatalytic function (a characteristic of the energy symbolic language), proportional to macrophyte biomass and depends on water temperature by an optimum curve. Dark respiration is proportional to macrophyte biomass and depends on temperature. Macrophyte biomass increases by the rate of gross photosynthesis and decreases by the rate of dark respiration. Death of the macrophytes eliminates them from the macrophyte compartment and increases the detritus compartment of the lake model. Nutrient concentrations in the water (N and P) limit growth and increase by leaching from the macrophytes. The model is calibrated on experimental data from Lake Alice, USA (Mitsch 1975) and literature data. It is written in FORTRAN.

The ecosystem model was tested successfully on field data from the same lake. Simulation indicated that diversion of secondary sewage from the lake would reduce the water hyacinth population and that exclusion of heated effluent from the lake would allow a fluctuating temperature regime with consequences for DO concentrations in the water, probably causing fish kills.

Ewel and Fontaine (1983) developed an ecosystem model for the highly eutrophic, monomictic Lake Conway (Florida, USA), which is rich in submerged macrophytes. The model is based on C and P flows. It has twenty-one state variables, of which two pertain to submerged macrophytes (above-ground and underground biomass) and three driving variables (irradiation, water temperature, and lake mixing). Three groups of plants are considered: phytoplankton, epipelic algae, and submerged macrophytes. The macrophyte species composition varied annually. Predominant species were: *Hydrilla verticillata*, *Potamogeton illinoensis*, and *Vallisneria americana*.

The rates of change in biomass of the primary producers are described by five functions: gross photosynthesis, respiration, leaching, death, and consumption (herbivory).

The ecosystem model is calibrated mainly on field data from Lake Conway. It was tested on other data sets from the same lake. Simulation suggested that the role of macrophytes in nutrient recycling overshadowed their importance in the grazing food chain, and that slightly more carbon was processed through the grazing food chain than through the detritus food chain in Lake Conway (Fig. 13.9).

CONCLUDING REMARKS

Models on metabolism of freshwater macrophytes have a large application potential. Models should address particular questions and, therefore, vary in many respects. Models, once developed and calibrated, should also be tested on independent data sets.

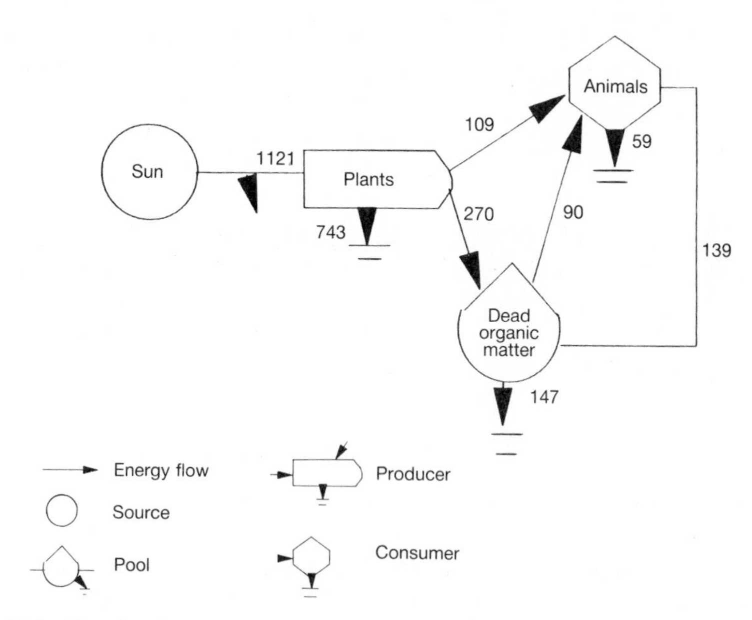

FIG. 13.9. Simulated annual carbon flows ($g\,C\,m^{-2}\,y^{-1}$) in grazing and detritus food chains in the Lake Conway ecosystem model (after Ewel and Fontaine 1983).

Independent monospecies models usually provide a framework for ecophysiological research on aquatic weeds. Compartments and processes usually emerge during development of these models which cannot, or can only marginally, be described because of lack of information. Marginal knowledge of a process forces the modeller into curve-fitting and thus limits potential extensibility of the model. Processes largely unknown in aquatic weeds are: development, nutrient dependence of growth, assimilate allocation (pathways and costs), and senescence. Current research still centres on carbon gain and biomass dynamics, despite the obvious gaps in knowledge on other processes.

Monospecies models can serve as a submodel of an ecosystem model and be used in conjunction with other models to address questions on environmental matters. Initial running of these (series of) models identifies compartments and processes largely outside the weeds which are crucial to the system and cannot be properly described or quantified. Ecosystem research is usually required to accommodate and replenish these models (Wlosinski 1981; Best 1982*b*).

ACKNOWLEDGEMENTS

H. L. Fredrickson (Limnological Institute, Nieuwersluis, The Netherlands) criticized and corrected the English text.

14

Practical uses of aquatic weeds

J. C. JOYCE

INTRODUCTION

IN preparation for this chapter, the Center for Aquatic Plants' computerized aquatic plant information retrieval system was queried in order to obtain a listing of articles on the utilization of aquatic plants on a world-wide basis. This literature search yielded 215 review citations and over 1 764 (1 300 of which were written since 1975) individual citations covering reported uses of aquatic plants, which ranged from human food to wastewater treatment and the production of biogas. Attempts to review this wealth of literature and subjectively determine those uses which are 'practical' is difficult for two fundamental reasons; (a) the amount of material is exhaustive, and (b) the definition of 'practical uses' is subjective and open to potential bias as discussed below.

An overview of the literature indicates that 'practical uses' of aquatic plants is at best a function of the economic base of the country in which the plants are present. For example, aquatic plant species such as water hyacinth (*Eichhornia crassipes*) are the subject of extensive research and operational use for the treatment of sewage effluent to meet stringent surface water quality standards in Western countries. At the same time the water hyacinth is subject to strict control measures in public waters in the south-eastern USA due to its interference with recreational and commercial uses of waterways. In contrast, in less developed countries, where labour costs are less and the manufacturing base is not as dependent upon fossil fuels, much interest has been shown in the use of these plants for fertilizer, animal and human food, biogas, paper, thatching, etc. Thus, for the purposes of this discussion, 'practical uses' will be a function of the economic base and degree of geographical isolation of the areas in which the plants occur. Table 14.1 provides a summary of the uses identified for the various species of plants which are contained within the Center for Aquatic Plants' Literature Retrieval System. Table 14.2 indicates the major identified uses of the plant and the countries in which that use was referenced, along with the number of references of that use for that country. These two tables make no claims as to

Table 14.1. Utilization citations by country in the University of Florida IFAS aquatic plant literature database

	Total utilization	Biogas	Food	Pollution	Miscellaneous uses[a]
Argentina	1	1	0	0	0
Australia	18	0	0	11	7
Bangladesh	24	4	6	2	12
Brazil	14	4	4	3	3
Canada	16	0	8	6	2
Czechoslovakia	9	0	0	1	8
Egypt	3	0	1	1	1
France	14	1	1	7	5
Germany	12	0	0	9	3
Great Britain	15	2	0	6	7
India	75	13	21	8	33
Indonesia	10	1	3	0	6
Japan	10	1	0	5	4
Malaysia	9	0	1	6	2
Mexico	7	0	2	2	3
Netherlands	5	0	1	2	2
Nigeria	4	0	2	2	0
Pakistan	1	0	0	0	1
Philippines	6	1	2	0	3
Poland	9	0	4	2	3
South Africa	2	0	1	1	0
Sri Lanka	5	0	0	1	4
Sudan	8	4	3	0	1
Sweden	7	1	2	0	4
United States	1497	140	278	360	689
Thailand	10	2	3	1	4
Yugoslavia	3	0	0	2	1
Total citations	1764	175	343	438	808

[a] Miscellaneous uses include such uses as thatching, animal feed, paper/pulp/fibreboard, biofertilizer/soil amendment and drug/medicine extraction, etc.

the practicality, feasibility, or economic viability of the listed uses but merely reflect the contents and frequency of references to these uses in the Centre's database of scientific literature. During this discussion reference will be made to comprehensive and useful review papers which provide more detailed discussions of this interesting topic.

HARVESTING

Man's consumptive utilization of aquatic plants generally begins with the harvesting or removal of the plant from the water or wetland area. Developing countries, due to lower labour costs, tend to rely on low energy physical removal systems and the plants are usually processed into end products relatively close to

Table 14.2. Major identified uses or research attempts for various aquatic plant genera

Use	Species
Animal food	*Alternanthera* (crayfish), *Azolla* (cattle), *Brachiaria* (cattle), *Ceratophyllum* (cattle), *Eichhornia* (pigs, fish, rabbits, chickens), *Elodea* (chickens), *Heteranthera* (cattle), *Lemna, Myriophyllum, Panicum* (cattle), *Pistia* (cattle), *Potamogeton* (chickens), *Ruppia* (cattle), *Sagittaria* (crayfish), *Salvinia* (fish, sheep), *Vallisneria* (cattle)
Biofertilizer (compost)	*Azolla. Eichhornia, Lemna, Myriophyllum, Pistia, Salvinia*
Biogas (methane and alcohol)	*Eichhornia, Hydrilla, Myriophyllum, Salvinia, Typha*
Bioindicator	*Azolla, Callitriche, Lemna, Myriophyllum, Ranunculus, Zannichellia*
Biomass	*Cyperus, Eichhornia, Hydrilla, Phragmites, Typha*
Human drugs	*Acorus, Alisma, Elatine, Nelumbo, Nymphaea, Polygonum, Scirpus, Pistia*
Human food (see also Table 14.4)	*Colocasia, Lemna, Ipomoea, Oryza, Scirpus, Trapa, Typha, Vallisneria, Zizania*
Paper, pulp, fibreboard, etc.	*Cyperus, Juncus, Panicum, Pandanus, Phragmites, Salvinia, Scirpus, Typha*
Thatching	*Phragmites, Scirpus, Typha*
Wastewater treatment	*Azolla, Ceratophyllum, Eichhornia, Elodea, Hydrocotyle, Juncus, Lemna, Myriophyllum, Phragmites, Pistia, Potamogeton, Salvinia, Scirpus, Schoenoplectus, Spirodela, Trapa, Typha, Wolffiella, Canna, Pontederia, Sagittaria*

the removal site. Developed countries have produced rather elaborate mechanical systems which attempt to maximize removal rates on the order of several tonnes per hour. These production rates are an attempt to remove the plants at a rate which will provide some level of operational control and alleviate the nuisance nature of the plant. The utilization of the harvested material is for the most part a secondary consideration. In situations where developed countries utilize the plant for commercial products the plants are generally cultured in a specially designed water body and harvesting systems are likewise designed to the configuration of the water body and to maximize harvesting at a rate which supports the production of the end product. Canellos (1981) and Thayer and Ramey (1986) provided excellent reviews of state of the art harvesting technology and provide costs which range from simple in-place mechanical cutting of submerged vegetation, at US \$53 ha^{-1}, to total offsite removal of floating vegetation, at US \$19 000 ha^{-1}.

PROCESSING

Most uses of aquatic weeds have been an attempt to take advantage of the extremely high growth rate of the plants. For example, ten water hyacinth plants have been shown to produce over 655 000 plants and cover over 0.4 hectare in an eight month growing season: a doubling rate of 14 days (Penfound and Earle 1948). Weight gains as great as 4.8 per cent per day and standing crops of 470 tons per hectare have also been reported (Knipling, West, and Haller 1970). One reason for such high growth rates is the fact that the floating and submerged plant species usually have a low fibre content and a high water and air content since they depend upon the water for support rather than plant fibre. The high moisture content of aquatic weeds, usually 85–95 per cent, is a major difference between aquatic and terrestrial vegetation. Substitution of aquatic plants for animal feeds, or to transport them significant distances, usually requires reduction of the water content of the plants. Bagnall (1980*a*) provided a detailed description of various screw presses designed specifically to dewater both floating and submerged aquatic plants. Chopping, pressing, or other mechanical processing of the plants to remove the excess water can be energy or cost prohibitive depending upon the final use of the product.

COMPOSITION OF THE PLANTS

The high moisture content of the plant results in a correspondingly low dry matter content of 5–15 per cent as compared to terrestrial forages of 10–30 per cent. As indicated in the 1976 National Academy of Sciences *Ad Hoc* Committee report, 'The low level of dry matter has been the major deterrent to the commercial use of harvested aquatic weeds. In order to obtain one ton of dry matter, 10 tons of moist aquatic weeds must be harvested and processed; for the water hyacinth, which is usually only 5 per cent dry matter, 20 tons must be harvested and processed to get one ton of dry matter.' Emergent aquatic plants which extend above the water surface tend to be more fibrous and are more amenable to processing into usable products.

The chemical content of aquatic plants is also greatly affected by the chemical characteristics of the water in which the plants are grown. This aspect has been of particular interest in the potential use of the plants as animal feeds and in wastewater treatment polishing ponds to remove nutrients or elemental pollutants. Both of these relationships will be discussed later. Table 14.3 provides a summary of the analysis of the chemical characteristics of a variety of common aquatic plants. However, as noted by Boyd (1978), 'Chemical composition changes as some species mature and variation in chemical composition occurs between different plant structures. No average composition for an individual species or ecological grouping of species based on data from the literature would be reliable. Therefore, actual measurements should be made in studies requiring data on chemical composition. Energy content values are more constant than chemical parameters.'

Table 14.3. Summary of chemical and physical analyses for a variety of wetland plants on a dry weight basis[a]

Constituent	Species (n)	Range of values (all species)	$\bar{x} \pm$ s.e.[b]	C.V. (%)
Ash (%)	40	6.1–40.6	14.03 ± 1.21	53.4
C (%)	28	29.3–48.8	41.06 ± 0.71	9.2
N (%)	27	1.46–3.95	2.26 ± 0.14	32.7
P (%)	35	0.08–0.63	0.25 ± 0.02	58.0
S (%)	25	0.11–1.58	0.41 ± 0.08	98.3
Ca (%)	35	0.20–8.03	1.34 ± 0.24	105.5
Mg (%)	35	0.08–0.95	0.29 ± 0.03	143.6
K (%)	35	0.42–4.56	2.61 ± 0.22	51.0
Na (%)	35	0.07–1.52	0.51 ± 0.08	93.1
Fe (ppm)	33	133–3866	1 420 ± 169	68.0
Mn (ppm)	33	120–5390	1 143 ± 227	113.9
Zn (ppm)	33	20–267	80 ± 10	71.2
Cu (ppm)	33	1–190	40 ± 6	93.8
B (ppm)	31	1.2–112	14.3 ± 4.5	176.2
Crude protein (%)	40	8.5–31.3	15.5 ± 0.83	33.9
Ether extract (%)	39	1.63–8.11	4.2 ± 0.28	42.3
Cellulose (%)	40	10.0–40.9	25.9 ± 0.97	23.7
Energy content (kilojoules g^{-1})	36	10.33–17.99	15.48–2.80	11.3

[a] Data from Boyd (1978).
[b] s.e. = standard error.

BIOFERTILIZERS AND COMPOST

Perhaps the simplest and most practical routine use of aquatic weeds is the use as green manure, compost (Fig. 14.1) and mulch for soil amendment. This is especially true for developing countries where mineral fertilizers are too expensive and inexpensive labour is readily available to harvest and process the plants into compost or mulch.

The most common use of aquatic plants as green manures is the use of the small water fern, *Azolla*, in conjunction with rice culture. A blue-green alga which lives within the frond of *Azolla* has the capability to fix nitrogen from the air. The algae secrete nitrogenous compounds which are incorporated into the tissues of *Azolla*. Moore (1969) reported that the Vietnamese have used *Azolla* as a green manure for rice harvested in May and June. In tropical regions, *Azolla* grows most rapidly in cooler temperatures and is thus planted in the fields in early January. As the rice begins to grow in the warmer temperatures of March and April, *Azolla* begins to die and drop to the bottom releasing its nitrogen to the rice. By the time of harvest in May or June *Azolla* is completely gone. Moore (1969) further noted that in Japan, *Azolla* is considered to be a weed in rice and

FIG. 14.1. Nursery plants grown on 25% water hyacinth compost. (Photo: L. O. Bagnall.)

that since the practice is not used more extensively in S.E. Asia that there may be additional management problems involved.

When utilized as mulch, aquatic plants can either be spread on the soil surface or they can be incorporated into the soil. As a mulch the plants have been shown to suppress weeds, reduce evaporation, increase soil moisture and organic content, and reduce erosion (National Academy of Sciences 1976). The major limitation of use as a mulch is the logistics of removing and transporting the bulk associated with the plants. This can be reduced somewhat by partial solar drying near the waterbody prior to use or transport (National Academy of Sciences 1976).

Composting involves the controlled microbial decomposition of organic plant materials into simpler forms that allows their use as fertilizers. The use of aquatic weeds, water hyacinth in particular, has received much attention with varying results, which appear to depend upon the amount and extent of pre-drying of the plants. The best results appear to occur when the plants are allowed to air dry for several days to reduce the moisture content, otherwise the composted material will form a soggy mass which has little air penetration. Most successful composting operations also involve the addition of soil, ash, fibrous plant material, or dried animal manure.

As pointed out by the National Academy of Sciences (1976) report, the major limitations of utilizing aquatic plants as soil amendments are:

1. Most compost usually contains 1.5–4 per cent nitrogen, 0.5–1.5 per cent phosphorus, and 1–2 per cent potassium which is much less than most mineral fertilizers. Thus large quantities must be used in order to obtain the same benefit as from mineral fertilizers.
2. To receive the most benefit the compost must be made and managed carefully. Some farmers may not be willing to spend the necessary time and effort.
3. Plants havested from waters that contain toxic pollutants may produce compost that is a hazard to humans, animals, or crops.

In spite of the limitations this use of aquatic weeds remains one of the most practical uses of this resource in areas where economic conditions do not allow the investment in mineral fertilizers, and inexpensive labour to process the plants is readily available.

ANIMAL FOODS

As indicated in Table 14.2, the use of aquatic plants for animal forage is the most extensive area of utilization investigation of the major uses listed. The most commonly mentioned animals include cattle (Fig. 14.2), pigs, sheep, and poultry. It appears that the aspects of aquatic plants that determine their acceptability as an animal food on a large scale are moisture, mineral and fibre content, nutrient composition, and digestibility. Two of these factors, moisture and mineral content, have created the greatest difficulties in producing large quantities of readily-usable animal feeds. As previously mentioned, the high moisture content makes harvesting, transport, and storage without spoiling difficult, but not impossible. In drier regions, it is possible partially to sun-dry the plants, but in more humid regions the plants are best prepared by partial dewatering and then ensiling. Successful ensiling of aquatic plants requires chopping the plants and adding a preservative such as propionic, acetic, or formic acids in order to inhibit undesirable bacteria and moulds and to maintain the proper fermentation temperature within the silo. Large scale ensiling operations are most economical and practical if the silos are located near the aquatic weed supply in order to reduce transportation costs of the plants.

The mineral content of the plants is a function of the stage of growth and the quality of the water in which the plants are grown. Most aquatic plants, including water hyacinth and *Hydrilla*, contain adequate levels of the basic elements needed for a quality forage; however, sodium, iron, potassium, and calcium ranged from 3 to 100 times higher than comparable levels in terrestrial forages (Easley and Shirley 1974). Thus, even though various animals will graze aquatic plants, most

FIG. 14.2. Cattle eating silage prepared from water hyacinth. (Photo: L. O. Bagnall.)

appear to refuse to eat large quantities of the plants, because too high a percentage in the diet may cause a mineral imbalance (National Academy of Sciences 1976). Those studies which report the greatest success in using aquatic plants as forage, used the plants as a supplement or mix with other fodder in order to increase the aquatic plants' acceptability.

One method of addressing this concern is to culture intensively organisms which routinely utilize these plants as food. Examples include herbivorous fish species, ducks or geese, and certain ruminants.

Due to attempts to make a resource out of a major pest problem, such as water hyacinth or *Hydrilla*, most attention has been placed on the use of these plants, however, as pointed out by the National Academy of Sciences *ad hoc* Committee in 1976, there may be other aquatic plants which could be grown under aquaculture conditions which would prove to be excellent forage and which could be cheaply grown and processed. As with most major weed species, however, care should be taken before introducing a species into a new region for such a purpose. History is full of examples of such introductions creating adverse ecological and economic impacts.

HUMAN DRUGS

As Sculthorpe (1967) noted, ancient literature and the writings of herbalists refer to a large number of internal and external disorders which are at least partially treatable by extracts or products produced by aquatic macrophytes. In very few cases has any medicinal principle been isolated and identified from the plants, which caused Sculthorpe to question whether or not the 'success of the treatment was achieved through genuine curative properties rather than superstition and psychological reassurance'.

One of the most common species which appears to have been used therapeutically by numerous cultures since at least the time of Hippocrates (*c.* 460–377 BC) is sweet flag, *Acorus calamus*. The plant is native to south-east Asia, but has become naturalized throughout Europe. The rhizome of the plant has been used to make various preparations for the treatment of eye diseases, dyspeptic flatulence, indigestion, toothache, coughs, and colds. Water lettuce, *Pistia stratiotes*, has also apparently been used medicinally for centuries; in Egypt for the treatment of inflammations of the skin and abrasions; and in India, a preparation for the treatment of chronic skin disease is made by boiling *Pistia* leaf juice in coconut oil. Coughs and asthma are treated with leaf extract together with sugar and rosewater. It is claimed that the roots can serve as a diuretic and laxative, while rubbing the ashes of the plant into the scalp is a treatment for ringworm (Sculthorpe 1967). Sculthorpe (1967) lists 25 other plants for which curative properties have been claimed at one time or another. He also noted that the presence of toxic or anaesthetic substances in many of these plants would explain some of the unfortunate results of some treatments utilizing these plants. The example provided was the treatment of hydrophobia, tetanus, and rattlesnake bites by administration of extracts of *Alisma plantago-aquatica*, which was often followed by the complete paralysis of the patient.

HUMAN FOOD

Historically there has been a demand and ready market for certain aquatic food plants. The primary plant tissues used as food are the seeds and swollen vegetative tubers, rhizomes, or stolons. The seeds and fruits which are rich in oils, starch, and proteins, have been eaten raw or ground into flour. Rhizomes and tubers are generally high in carbohydrates and are either eaten raw or cooked. The foliage of certain species is also eaten raw in salads or cooked similar to other vegetables (Sculthorpe 1967).

Oryza sativa, or cultivated rice, is probably the best example of a major aquatic or semi-aquatic food crop. It is grown throughout the tropical and warmer temperate regions of the world and serves as the staple diet for many oriental and African populations. Because it is such a major agronomic crop, the International Rice Research Institute in the Philippines has conducted extensive research into the development of varieties which will tolerate deeper water depths and longer periods of inundation. Various species of wild rice are also edible such as *Zizania aquatica*, which has been harvested by the North American Indians for centuries and is currently being cultivated in Canada and the western USA; and *Oryza perennis* is occasionally used as a grain food in Africa (Sculthorpe 1967; Florida Aquaculture Review Council 1985, unpublished data).

Due, perhaps, to high productivity and its world-wide abundance, particularly in developing countries (which are also least able to cope with the problem) the water hyacinth has received considerable interest as a potential food. Most of the serious efforts at producing a food from the plant have involved the extraction of proteins from the crushed leaves. It appears that use of this source of protein is a function of the availability of other food sources, since terms such as 'are reasonably palatable' and 'little immediate appeal to many people' are used to describe its acceptance (Pirie 1960, 1966).

Table 14.4 provides a listing of various other aquatic macrophytes which the literature indicates are used as a food. Many of these species perhaps form a major part of the diet in areas where food is scarce and the human population is

Table 14.4. Other reported human foods derived from aquatic macrophytes

Plant	Plant part used	Food item
Nymphaea caerulea	Dried or ground seeds	Flour, bread
N. lotus	Dried or ground seeds, rhizomes	Flour, bread
Nelumbo nucifera	Fruits and rhizomes	Vegetable dishes
Cyperus papyrus	Culms (raw and boiled)	Juice
Sagittaria trifolia	Corms (boiled)	Cooked as yams
Eleocharis dulcis	Corms (raw and boiled)	Vegetable dishes
Trapa spp.	Spiny fruits (raw and cooked)	As a vegetable
T. natans		
T. bispinosa		
T. incisa		
Butomus umbellatus	Dried, powdered rhizomes	Flour for bread
Calla palustris	Dried, powdered rhizomes	Flour for bread
Menyanthes trifoliata	Dried, powdered rhizomes	Flour for bread
Colocasia esculentum	Tuberous roots	Potato-like food
Cyrtosperma chamissonis	Boiled or roasted corms	Flour, vegetable
Wolffia arrhiza	Whole plant	Vegetable
Ipomoea aquatica	Young leaves and stems	Leafy vegetable
Nasturtium officinale	Leaves (raw or boiled)	Herb, vegetable
Euryale ferox	Seeds	Flour
Typha spp.	Rhizomes (raw or boiled)	Vegetable
	Seeds, pollen	Flour
Enhalus acoroides	Seeds	Vegetable
Victoria amazonica	Seeds	Vegetable
Ottelia alismoides	Leaves	Vegetable
Spirulina spp.	Microscopic algae	Dried cakes

Information taken from Little (1979), National Academy of Sciences (1976) and Sculthorpe (1967).

isolated either economically or geographically. However, as Boyd (1968) pointed out 'utilization of water weeds as food could probably alleviate protein shortages in local populace of many developing nations, but it is doubtful that these plants could contribute greatly to the total food supply of any nation'.

Two cautions concerning the culture and use of aquatic plants as food crops are (1) the common concern of introducing a potential weed into a new location, and (2) the need to wash and prepare the plants adequately in areas where they may be potentially exposed to micro-organisms, bacteria, and other disease causing organisms.

PAPER AND FIBRE PRODUCTION

This is perhaps one of the oldest uses of certain aquatic or wetland plant species dating back to the Egyptians' invention of paper production utilizing papyrus, hence the word 'paper'. Sculthorpe (1967) indicated that the production of paper from the piths of papyrus (*Cyperus papyrus*) culms dates back to at least 2400 BC. The process involved the peeling off of the superficial fibrous material and slicing the pith longitudinally into strips. The strips were arranged side by side on a flat surface, moistened, and matted together under heavy pressure. After drying, the papyrus sheets were cut and processed as needed. Sculthorpe also reported that the Egyptians used papyrus to construct ropes, baskets, and mats and canvas for use as floor coverings and boat sails. Today, papyrus is considered a pest in some African waters where large floating islands of papyrus interfere with navigation and water movement. The National Academy of Sciences review (1976) indicates that no major commercial uses of the plant are considered today, due to the difficulties encountered in harvesting and processing. However, there is a flourishing, if small-scale, tourist trade in papyrus, used for paintings and drawings, in Egypt.

In stark contrast to papyrus utilization is the Romanian use of *Phragmites australis*, once considered a serious pest in the 434 000 ha Danube Delta of which *Phragmites* occupies 60 per cent. In 1956 the Romanian government, based on botanical, engineering, and hydrobiological investigations at the experimental station at Maliuc, developed specialized systems to culture, harvest, transport, store, and process the plant (Sculthorpe 1967). A high quality pulp is produced from which printing paper, cellophane, cardboard, and various other synthetic fibres are made. Sculthorpe (1967) also reported that the plants and the pulp-mill waste are used for the production of compressed fibre boards, furfural, alcohol, insulation, and fertilizer. The Romanian government-supported factory reportedly pays the relatively high amount of US $85.00 per tonne for the raw reeds, reflecting a partial subsidy of the project (National Academy of Sciences

1976). The raw reeds are also used as thatching, fences, windbreaks, basket making, firewood, fishing rods, and as mouthpieces for musical instruments.

Two other notorious aquatic weed species which have received considerable attention for fibre and paper production are cattails (*Typha* spp.) and water hyacinth, perhaps because of their abundance. Because of its fibrous nature *Typha* does produce a fairly strong paper but it is difficult to bleach. This increases the cost of writing paper production. This would not appear to be a problem in the production of wrapping paper, however. Cattails are also used in localized situations where other fibres are more expensive for the production of mats, baskets, and chair seats. Because they swell when wet, the dried leaves are reportedly used for caulking in houses, barrels, and boats (National Academy of Sciences 1976). The production of paper from water hyacinth is also a labour intensive process which has the most widespread use in localized areas and usually requires the addition of 8–10 per cent cotton or jute to produce a quality paper (Little 1979). An extensive analysis of large scale paper production from these plants was conducted by Nolan and Kirmse (1974) at the University of Florida. These studies indicated only 20 per cent of the plant was available for paper production and that reasonable paper strengths were only available from 65–75 per cent of this remaining material. The most serious disadvantage to water hyacinth pulp was the extremely low freeness or moisture release rates of the pulp compared to more conventional pulps.

Common uses of other aquatic plants include *Scirpus* spp. and *Juncus* spp. which are used for making floor mats in Asian countries such as Taiwan, China, Korea, and Japan. The leaves of the textile screwpine, *Pandanus tectorius*, which forms dense thickets along the shores of lagoons and rivers in south-east Asia and the Pacific islands such as Malaysia are used for weaving mats, handbags, hats and baskets which are extremely durable, lasting several generations. The more elaborate and coloured mats are highly prized and bring excellent prices in tourist markets.

ORNAMENTAL USES

Ornamental uses of aquatic plants are common to both developed and undeveloped countries. As discussed in Sculthorpe (1967), the basic ornamental use of aquatic plants is the growing of the plants for use in garden ponds and aquariums with the beginnings traced back to the rise of Buddhism, when formal lotus pools were an integral part of gardens of Buddhist temples. Elaborate ornamental water gardens are known in China, India, England, Italy, France, the USA, and many other countries. The most common species grown in water gardens include the emergent *Butomus, Cyperus, Menyanthes, Nelumbo, Pontederia, Sagittaria, Typha*; the aroids; *Aponogeton, Hydrocleis, Nuphar,*

Nymphaea, Nymphoides; and the free floating *Azolla, Ceratopteris, Eichhornia, Pistia, Salvinia, Stratiotes,* and *Trapa* (Sculthorpe 1967). Perhaps one of the most famous and notable examples of such species is the gigantic *Victoria amazonica*. Perry (1961) provides a classical work on the propagation of aquatic plants. A somewhat related ornamental use of aquatic macrophytes is the use of the dried seed head of *Nelumbo lutea*, or American lotus, in floral arrangements. A single seed head can bring US $0.50 in the wholesale market in Florida.

Since the end of World War I there has been a great increase in the culture of aquatic plants in aquaria either separately or as a backdrop for ornamental fish (Fig. 14.3). This interest has been most prevalent in the USA, south-east Asia, Europe, Russia, Australia, and New Zealand (Sculthorpe 1967). The Florida Aquaculture Review Council (1985, unpublished data) indicated that the cultivation and harvesting of aquatic plants in the aquarium industry generated between US $3.5–5.0 million annually just in the State of Florida. Due to Florida's tropical climate and excellent air freight system, plants are commercially grown and shipped to all parts of the US, Canada, as well as to many locations in Europe. The predominant aquaria species are the submerged plants; *Sagittaria, Vallisneria, Elodea, Lagarosiphon, Cabomba, Ceratophyllum, Myriophyllum,* and *Limnophila*. There are also many small emergent and amphibious species that will tolerate complete immersion, such as *Acorus, Bacopa, Cardamine, Cryptocoryne, Echinodorus, Heteranthera, Hygrophila, Lagenandra, Micranthemum, Nomaphila, Rorippa, Samolus,* and *Synnema* (Sculthorpe 1967).

FIG. 14.3. Aquaculture of aquatic plants for use in the aquarium trade. (Photo: L. Nall.)

The culture of aquatic plants in man-made ponds, water gardens, and aquaria is a fascinating and enjoyable hobby. However, the history of the water hyacinth, *Hygrophila*, and *Hydrilla*, which were released into the wild in the USA as ornamentals, indicates that once present in natural water bodies, these plants can create severe environmental and economic damage. As a consequence, the import of aquatic plants into the USA is subject to strict laws (Fig. 14.4). Aquaria containing aquatic plants should not be emptied into natural waterways.

FIG. 14.4. Florida Department of Natural Resources inspector checking for compliance with Florida's Aquatic Plant Importation laws. (Photo: L. Nall.)

WASTEWATER TREATMENT

During the past 20 years, extensive research has been initiated to investigate the ability of aquatic plants to remove various inorganic and organic compounds from water. As noted by Little (1979), this fact has long been known by farmers in south-east Asia who utilize a polyculture farming system which recycles nutrients and energy through an aquatic plants–pigs–humans–fish–humans–aquatic plants system.

The scientific basis for the use of aquatic macrophytes in wastewater treatment is the symbiotic relationship between the plant and associated micro-organisms on the plant roots and stems. A major part of the treatment process for the degradation of organic molecules in the wastewater is attributed to these micro-organisms. As degradation products are produced, the plants absorb these products along with N, P, and other minerals. The micro-organisms also utilize the metabolites and oxygen released by the plant roots. Suspended solids are removed by electrostatic charges associated with the plant root hairs where they are also digested and assimilated (Wolverton 1987).

Aquatic plants are also able to absorb and concentrate toxic heavy metals and certain radioactive elements from the water. The plants have also shown the ability to absorb, translocate, and eventually metabolize certain organic waste molecules (Wolverton 1987). If aquatic plants are used for the removal of toxic materials, the plants must be disposed of safely (National Academy of Sciences 1976).

The two types of aquatic plants which have received the most attention in this area are floating and emergent species. The water hyacinth has been used operationally in several locations for over 11 years, but the technology appears restricted to warmer climates. For this reason and the fact that the harvesting process is easier, the smaller and more cold-tolerant duckweeds (*Lemna, Spirodela,* and *Wolffia*) are receiving increased attention. The emergent species which have been investigated and used operationally are *Phragmites australis, Typha* spp., *Scirpus* spp., *Pontederia cordata, Canna flaccida,* and *Sagittaria latifolia.* Recently the integration of emergent plant species with microbial filters has shown promise. This system consists of a series of lateral flow channels containing rooted aquatic plants with microbial communities on rocks serving as a trickling filter (Wolverton 1987).

Anyone interested in an up-to-date assessment of this complex subject is referred to the excellent compilation of over forty research efforts discussed in *Aquatic Plants for Water Treatment and Resource Recovery* (Reddy and Smith 1987).

BIOGAS PRODUCTION

The relatively high production rates of the major aquatic weeds has created interest in the use of this organic material for the production of biogas. The process, which has been used for about 80 years to transform sludge in sewage treatment plants, involves the controlled decomposition of the plants by methane-producing anaerobic bacteria. The chopped plants are placed in a sealed tank with an inoculum of the bacteria which after a period of time produces a mixture of gas that is approximately 70 per cent methane and 30 per cent carbon dioxide (Wolverton and McDonald 1976).

This is one consumptive use of aquatic plants in which the moisture content of the plants is an advantage, because the moisture is needed for the fermentation process. The only processing involved (depending upon the species utilized) is harvesting and chopping. The greatest emphasis for this type of work occurred during the early 1970s due to high prices and unpredictable availability of fossil fuels. As with other attempted uses of aquatic weeds, the hope was that the development of a commercially valuable product would assist in the management of the problematic nature and over-abundance of the plants. The US National Aeronautics and Space Administration (NASA) has undertaken a major research effort to utilize water hyacinth and other species in the production of biogas (Wolverton, McDonald, and Gordon 1975). The overall system attempts to utilize the water hyacinth wastewater treatment potential, as discussed in the previous section, in conjunction with biogas production (Wolverton and McDonald 1976; Wolverton 1987) (Fig. 14.5 and Fig. 14.6). Wolverton and McDonald (1976) estimated that one hectare of water hyacinth would produce 70 000 m^3 of biogas with each kilogram of dried plant producing about 370 l of biogas. The relative fuel value of this gas is 22.16×10^3 kJ m^{-3} as compared to 33.34×10^3 kJ m^{-3} for pure methane. This gas can readily be used for cooking and heating; however, before the gas can be used in combustion engines, the carbon dioxide must be removed, which to date has rendered the process impractical for use as a fuel for engines (National Academy of Sciences 1976).

There is much interest in utilizing this technology in rural areas of developing countries. Research has shown that it takes skill and continual supervision to produce methane successfully. Thus the National Academy of Sciences (1976) specified that requirements for such a system are:

(1) make maximum use of indigenous and homogenous materials;
(2) require only simple maintenance;
(3) be cheap and easily constructed; and
(4) be safe and foolproof to operate.

This conclusion has been partially verified by researchers at the Asian Institute of Technology in Thailand, who concluded that 'there is little potential for manually operated biogas plants if their substrates are to be heterogeneous and consist mostly of plant matter (Polprasert *et al.* 1986).

The US Gas Research Institute and the Institute of Food and Agricultural Sciences' Center for Biomass and Energy Systems are conducting extensive research directed towards refining these systems and significant progress has been made in reactor design, plant species selection, and processing procedures. Chynoweth (1987) discussed biomass conversion options and listed the major research needs required to make further significant advances in this potentially promising aspect of aquatic plant utilization.

CONCLUSIONS

This discussion of the practical uses of aquatic plants is by no means meant to be a complete listing of all recorded uses of this potential resource. It should be obvious, however, that based on the original premise, the economic base of the

FIG. 14.5. Water hyacinth in channels containing secondary sewage effluent: harvested for biogas production, at Walt Disney World Resort, Orlando, Florida. (Photo: V. Ramey.)

FIG. 14.6. Methane generator utilizing water hyacinth at Walt Disney World Resort, Orlando, Florida. (Photo: V. Ramey.)

area in which the aquatic plant occurs determines to a large extent the 'practicality' of the various uses developed for the plant. As man continues to search for methods to find a use for weeds while also expending vast amounts of resources on the management of nuisance species such as *Hydrilla* and water hyacinth in order to minimize their adverse impacts, the quote from Lucius Anneaus Seneca with which Sculthorpe ends his classic work seems appropriate:

'Nature does not reveal all her secrets at once...... Those secrets of hers are not opened to all indiscriminately. We imagine we are initiated in her mysteries: we are, as yet, but hanging around her outer courts..... Of one of them this age will catch a glimpse, of another, the age that will come after.'

ACKNOWLEDGEMENTS

The assistance of Mr Victor A. Ramey, IFAS Aquatic Plants Information Retrieval Service in the compilation and retrieval of information and statistics for this chapter is most appreciated. Critical reviews by Dr William T. Haller and Dr Kenneth A. Langeland are also acknowledged and appreciated.

PART III

AQUATIC WEED PROBLEMS AND MANAGEMENT IN VARIOUS PARTS OF THE WORLD

15

Aquatic weed problems and management in Europe

K. J. Murphy, T. O. Robson, M. Arsenović, and W. van der Zweerde

INTRODUCTION

Europe comprises the western part of the Eurasian landmass and associated islands, lying between longitude 30° E and the shores of the Atlantic Ocean (approximately 10° W). It stretches from the northerly parts of Scandinavia (*c.* 70° N) within the Arctic Circle, to the Mediterranean Sea (*c.* 35° N) in the south, and shows a wide range of climatic conditions, geomorphology, land use, industrialization, agricultural systems, and population density. Such factors all influence the scale and nature of aquatic weed problems in Europe, and approaches to their management. Description of the attitudes to aquatic weed management in Europe is made more complex by the fact that there are, in total, 26 nations (including the European part of the USSR) within the area covered by this chapter. This account provides an overview of the main aquatic weed problems currently occurring in Europe and illustrates, with selected examples, some of the different approaches to the management of aquatic plants in European waters.

The range of fresh waters (including inland brackish waters found, for example, in the Netherlands) in which aquatic weed problems occur in Europe may be conveniently divided into four groups:

(1) alpine, highland, and arctic waters;

(2) lowland rivers and canals;

(3) lowland lakes, ponds, and reservoirs; and

(4) irrigation and land drainage channels.

There is a large and burgeoning literature on the aquatic vegetation of each of these classes of European water bodies. A comprehensive overview of flowing-water plant communities in western Europe was given by Haslam (1987). Table 15.1 cites selected references on the vegetation of each class of waters in the countries and states of Europe. This literature is a useful source of information on actual or potential aquatic weed problems in Europe.

Precise information on the scale, nature, and distribution of nuisance growths of aquatic macrophytes (i.e. aquatic weed problems *per se*) in Europe is less easy to obtain. Robson (1986) provided a brief summary of European approaches to aquatic weed management. Table 15.2 gives some examples of typical problems. Květ and Hejny (1986) discussed some of the principal weed species in relation to their survival strategies (Grime 1979), and suggested management approaches be tailored to the vulnerable points in strategies of individual species. There have been few attempts to summarize the status of aquatic weed problems on a national scale in Europe, let alone on a continent-wide basis. The data which are available tend to be piecemeal and often published in limited-circulation reports or minor journals. For certain areas of Europe (e.g. Albania, Bulgaria, Byelorussia, Estonia, Latvia, Greece, Spain) we have located no or very little information on aquatic weeds and their management. This chapter of necessity gives an incomplete picture of the situation with regard to aquatic weed problems in Europe.

OVERVIEW OF AQUATIC WEED PROBLEMS IN EUROPEAN WATERS

Alpine, highland, and arctic waters

With relatively few exceptions the waters included under this heading tend to be oligotrophic. In consequence, they tend to support fairly low levels of macrophyte production. The waters included in this group are those found at high altitudes (>500 m) throughout Europe, together with lower-altitude water bodies in Scandinavia and the Highlands of Scotland (Fig. 15.1 and Fig. 15.2). In most cases there is a reasonably low level of direct human disturbance, or usage of these waters. Consequently, whatever plant growth exists usually causes few problems of interference with human activities: aquatic weed problems are, therefore, rare (e.g. Light 1975; Spence 1964; Schlott and Malicky 1984; Solander 1982; Blake, Dubois, and Gerbeaux 1986).

Exceptions are found mainly where there are problems of eutrophication (notably in Finland, Sweden, Austria, Switzerland, and Scotland) or where plant growth interferes with hydroelectricity schemes (Lachavanne 1985; Holzner 1974; Lachavanne and Wattenhofer 1975; Rautava 1972; Solander 1982; Juge *et al.* 1985). For example, in recent years *Ranunculus fluitans* has caused the blockage of intake screens of water pipes feeding Swiss hydroelectric power stations.

FIG. 15.1. Haapajärvi, a lake in the municipality of Kirkkonummi, about 30 km from the city of Helsinki, in southern Finland. The vegetation in this oligotrophic lake consists mainly of *Nuphar lutea, Phragmites australis* and *Equisetum fluviatile*. (Photo: A. H. Pieterse.)

FIG. 15.2. Loch n'Achlaise, Rannoch Moor, Scotland. The vegetation of this oligotrophic lake consists mainly of *Carex* spp., *Menyanthes trifoliata, Juncus bulbosus* and isoetids. (Photo: K. J. Murphy.)

Table 15.1. Aquatic vegetation of four classes of European water bodies: published references for individual countries

Country	Alpine, highland and arctic waters	Lowland rivers and canals	Lowland standing waters	Irrigation and land drainage channels
Austria	Schlott and Malicky 1984; Malicky 1984	Janauer 1982	Schiemer and Prosser 1976; Imhof and Burian 1972	Blümel 1974; Bulcke 1978
Belgium	n.d.	Dethioux 1982; Ska and vander Borght 1986; Haslam 1987; Leclerc 1977; vander Borght *et al.* 1982	n.d.	Haslam 1987
Czechoslovakia	n.d.	n.d.	Dvorak 1970; Husak and Otahelova 1982; Květ, Svoboda, and Fiala 1969; Straskraba 1963	n.d.
Denmark	n.d.	Larsen 1978; Thyssen 1982; Kern-Hansen and Dawson 1978	Sand-Jensen and Sondergaard 1979; Fjerdingstad, Fjerdingstad, and Kemp 1979	Haslam 1987
Eire	n.d.	Caffrey 1982, 1986; Praeger 1934	Caffrey 1982	Haslam 1987
Finland	Rintanen 1976	Sirjola 1969	Uotila 1971; Toivonen and Lappalainen 1980; Toivonen 1983	n.d.
France	Blake, Dubois, and Gerbeaux 1986; Dawson 1973	Dupont and Visset 1968; Ortscheit *et al.* 1982; Haslam 1987	Dutartre 1980, 1986; Guerlesquin and Podlejski 1980; Grillas and Duncan 1986	Portier 1971; Haslam 1987
Germany (BRD)	Schröder and Schröder 1982; Krause 1977; Lottausch, Buchloch, and Köhler 1980	Kohler, Wonnenberger, and Zeltner 1973; Weber-Oldecop 1970, 1971; Wiegleb 1984; Haslam 1987; Fricke and Steubing 1984	Weber-Oldecop 1970, 1971; Heckman 1982; Melzer, Haben, and Kohler 1977	Reschke 1978; Reschke and Blaszyk 1974
Germany (DDR)	n.d.	Sukopp 1971	Sukopp 1971	Kramer, Schmaland, and Nanzke 1974

Great Britain	Spence 1964; Light 1975; Holmes 1983; Stokoe 1983	Mycock 1983; Holmes 1983; Murphy and Eaton 1983; Murphy, Eaton, and Hyde 1980; Wright *et al.* 1982; John and Moore 1985; Harding 1979; Brooker, Morris, and Wilson 1978; Haslam 1987	West 1905; Lambert 1946; Ho 1979; Spence 1964; Spence and Allen 1979; Britton 1974; Jupp, Spence, and Britton 1974; Stokoe 1983; Mackenzie and Murphy 1986	Miles 1976; Courtney 1974; Robinson 1986; Haslam 1987; Bowker and Duffield 1981; Wade and Edwards 1980; Wolseley 1986
Greece	n.d.	n.d.	n.d.	Schiele 1986
Hungary	n.d.	n.d.	Dinka 1986; Schiemer and Prosser 1976; Rejmankova 1975	Sarkany 1982
Italy	Ravera, Garavaglia, and Stella 1984	Haslam 1987; Baudo and Varini 1976	n.d.	Schiele 1986; Piccoli and Gerdol 1981
Luxembourg	n.d.	Haslam 1987	n.d.	n.d.
The Netherlands	n.d.	Beltman 1979; Haslam 1987; Zonderwijk and van Zon 1978	Best 1982*a*; Best, de Vries, and Reins 1984; van der Velde and Peelen-Bexkens 1983; Clason 1953; van der Velde 1978; van Dam and Kooyman van Blokland 1978	Pitlo 1986; van den Linden 1980; Beltman 1984; Zonderwijk and van Zon 1978
Norway	Økland 1974; Rørslett 1984	Rørslett 1969, 1977, 1984; Haslam 1987	n.d.	n.d.
Poland	n.d.	n.d.	Bownik 1970; Ozimek and Kowalczewski 1984; Szmeja 1979; Tomaszewicz 1969; Ochyra 1979	n.d.
Portugal	n.d.	n.d.	n.d.	Figuereido *et al.* 1984; Moreira *et al.* 1983
Romania	n.d.	Kavetskiy, Karnaukhov, and Paliyenko 1984; Pallis 1916	n.d.	Sarpe *et al.* 1974
Spain	n.d.	Penuelas 1983	n.d.	n.d.
Sweden	Johnson 1978; Solander 1982	Johnson 1978; Erixon 1979	Lohammar 1965; Grahn 1977; Jensen 1979; Wallsten 1981, 1982; Hertzman 1986; Carlsson 1980	n.d.

Table 15.1. (*contd*)

Country	Alpine, highland and arctic waters	Lowland rivers and canals	Lowland standing waters	Irrigation and land drainage channels
Switzerland	Juge *et al.* 1985; Schröder and Schröder 1982; Lachavanne 1979a, b, c	n.d.	n.d.	n.d.
Yugoslavia	n.d.	Vrhovsek, Martincic, and Kralj 1981	n.d.	Skender *et al.* 1982; Cancek 1969; Stanković *et al.* 1982; Arsenović *et al.* 1982
USSR: Byelorussia	n.d.	n.d.	Prischepov 1974	n.d.
Lithuania	n.d.	n.d.	Stepanaviciene 1985; Antanyniene and Trainauskaite 1985	n.d.
Moldavia	n.d.	Kavetskiy, Karnaukov, and Paliyenko 1984	n.d.	n.d.
Ukraine	n.d.	Dubyna and Chornaya 1984; Belokon 1971	Shelyag-Sosonko and Semenikhina 1984; Pligin 1983; Korelyakova 1970; Zerov 1976; Yaschenko 1984	n.d.

n.d. = No data (no published papers located; or this class of waters not found in country specified).

Table 15.2. Examples of aquatic weed problems, and approaches to management in European water

Location (reference)	Weed group or species	Length or area affected	Lengths treated with different control measures	Control measures used
UK				
Internal Drainage Board areas – main channels (1,2)	E/FF/S	32 180 km (1974–75) (*c.* 10 000–20 000 ha)	9 650 km	M: Manual
			13 840 km	M: Mechanical
			1 930 km	M: Dredging
			4 830 km	H: Dalapon, 2,4-D, paraquat, terbutryne, copper, diquat, cyan-

British Waterways Board navigable canals (3)	E/RF/FF/S	506 km (1975–76) (*c.* 600 km)	207 km 55 km	M: Manual, mechanical, dredging H: Dalapon, glyphosate, diquat, cyanatryn, terbutryne, dichlobenil
Netherlands				
Water Board watercourses (4,5)	RF/FF/S/E	50 000 km (1982)	Whole system cleared annually	M: Mechanical (mainly), manual; B: Grass carp; H: paraquat, diquat (up to 1987 only), dalapon (limited use)
Germany (DDR)				
Waterways (6)	E	34 000 km (1973)	Total length treated with herbicides	H: Dalapon, simazine, aminotriazole, 2,4-D
Germany (BRD)				
Weser–Ems drainage ditches (7,8)	n.d.	25 000 km (1978)	5 297 km (1975 n.s. (1975)	H: Paraquat, dalapon, others H: Mechanical (increasing usage)
USSR				
Ukraine canals (9)	S/E	*c.* 40 ha (1969) (approx 15 per cent of shallow water areas of 3 major canals infected)	Grass carp introduced, successful control in early 1970s	B: Grass carp
Tuzda reservoir (10)	E/S	540 ha (1978)	—	—
France				
Aquitaine lakes (11)	Lm	95 km^2 affected	Experimental trials only	P: Fragment barriers; H: diquat

References: (1) Marshall, Wade, and Clare 1978; (2) Robson 1975; (3) Murphy, Eaton, and Hyde 1980; (4) Arts and van Wijk 1982; (5) Zonderwijk and van Zon 1978; (6) Kramer, Schmaland, and Nanzke 1974; (7) Blaszyk 1977; (8) Reschke 1978; (9) Belokon 1971; (10) Korsak and Myakushko 1980; (11) Dutartre 1980.

Weed groups and species: E = emergent; S = submerged (including filammentous algae); RF/FF = rooted floating-leaved and/or free floating; n.d. = no data available; Lm = *Lagarosiphon major*.

Control measures: H = herbicide; M = mechanical/manual clearance (incl. harvesting); P = other physical control; B = biological control.

In the Lake of Zürich (Switzerland) a dense population of filamentous algae (mainly *Cladophora* spp. and *Rhizoclonium* spp.) developed between 1960–1965, causing foul-smelling deposits of decaying plant material along the lake shores. The cause was nutrient enrichment from sewage effluent derived from the city of Zürich. Improved sewage purification schemes implemented since 1965 resulted in a decline of the filamentous algae, but *Potamotegon* spp. have replaced the algae as dominants in the submerged vegetation. Both here, and in Lac Leman (with similar problems) cutting and harvesting are now used to manage vegetation (Lachavanne 1977, 1982, 1985).

Of considerable importance in upland lakes (and some lowland oligotrophic waters, for example in moorland areas of the Netherlands) is the question of acidification and its effects on aquatic plant communities. There is evidence for major and deleterious community changes in the aquatic vegetation of acidified waters. Effective management systems are needed, if the problem is not tackled at source, to manage the aquatic macrophyte communities of acidified waters, in order to minimize environmental damage (Grahn 1977, 1985; Roelofs 1983; Raven 1985).

Fast-flowing rivers draining highland catchments may fall into this first group of European waters, even though their altitudes may be less than 500 m, because their high water velocities, tendency to spate, and rocky or coarse gravel substrates severely limit the growth of macrophytes. Again, however, eutrophication can lead to problems, at least in the slower-flowing reaches of such rivers. In some Scottish salmon rivers (e.g. R. Spey) it is now necessary to control the growth of *Ranunculus* spp., by cutting or use of the herbicide diquat-alginate, as a result of the enhanced growth caused by sewage effluent inputs from newly-developed mountain holiday resort towns (Murphy 1988). In the upper reaches of the R. Rhine in Switzerland *Ranunculus fluitans* has developed into a major cause of weed problems since 1970, following increases in the phosphate content of the river water. In Norway mass growths of *Juncus bulbosus* are a serious problem in some regulated rivers (B. Rørslett, pers. comm.).

Weed-cutting in shallow fast-flowing streams and rivers is not always solely necessitated by eutrophication. In Britain, for example, weed-cutting in such rivers used for trout and salmon fishing is often carried out by hand (using scythes) as part of a traditional fishery management regime, despite the high cost of such labour-intensive work (Birch 1976; Wright 1973; Murphy and Pearce 1987).

Lowland rivers and canals

Lowland rivers in Europe are normally of a gentle gradient and low velocity, with a substrate of fine clay, silt, and organic detritus in varying proportions. They can be quite turbid and high loadings of suspended solids are common, but this is rarely sufficient to prevent the growth of submerged macrophytes, at least in the shallower marginal zones (Haslam 1987). The largest rivers of Europe (e.g. Rhine, Danube) tend to be too deep, too polluted, and often carry too high a density of heavy boat traffic to permit much in the way of aquatic plant growth. Weed problems, therefore, tend to be restricted to smaller rivers. The genera most often reported to cause problems are *Cladophora, Potamogeton, Ranunculus, Callitriche, Sparganium, Typha, Phragmites, Nuphar, Glyceria, Berula, Apium, Rorippa,* and *Sagittaria* (Bolas and Lund 1974; Caffrey 1982; Cernohous 1980; Dawson 1978; Dawson and Robinson 1984; Harding 1979; Haslam 1982, 1986; Janauer 1982; Jorga and Weise 1977; Kern-Hansen and Dawson 1978; Murphy 1988; Larsen 1978; Kelsall 1981; Mycock 1983; Ortscheit *et al.* 1982; Ska and vander Borght 1986; Thommen and Westlake 1981; vander Borght *et al.* 1982; Wharfe, Taylor, and Montgomery 1984).

Lowland canals included in this second class of European waters are those man-made waterways originally constructed to carry commercial boat traffic. These fall into two categories:

(1) large navigable canals carrying heavy barges or ships (e.g. the Kiel Canal, Germany; Caledonian Canal, Scotland), with very little or no macrophyte growth, and no aquatic weed problems (Haslam 1987; Murphy, Eaton, and Hyde 1980); and

(2) canals of smaller cross-sectional area originally used for freight or passenger transport, but now generally either disused, or with a new role in providing recreational and other amenity functions (Hyde 1977; Taekama 1980; Gerritsen 1979; Pearce and Eaton 1983; Murphy and Eaton 1981*b*).

Canals (of the latter type) occur in Britain, Ireland, France, Belgium, and the Netherlands (Fig. 15.3). Their aquatic flora tends to resemble that of small rivers of similar water chemistry, but genera typical of static waters (e.g. *Lemna* spp., *Ceratophyllum demersum*) are also found (e.g. Ross, Doughty, and Murphy 1986). Because they are very slow-flowing or even virtually static in some cases (notably in the Netherlands), usually mesotrophic–eutrophic, and relatively little polluted such canals can suffer very serious aquatic weed problems (Murphy, Eaton, and Hyde 1980, 1982). However, powered boat traffic at moderate–high traffic densities is itself an effective means of controlling plant growth. The disturbance of the water by propeller action, and by boat wash, damages plant tissues and stirs up sediments which reduce the available light for submerged plant photosynthesis. This disturbance is the main mechanism of boat damage to plants, although some direct damage, caused by contact with moving hulls or propeller blades also occurs (Murphy and Eaton 1981*b*, 1983).

The main problem which aquatic macrophytes cause in European rivers is obstruction of the water flow, which leads to an increased risk of flooding (Dawson

FIG. 15.3. Weed growth (*Cladophora* spp., *Elodea canadensis* and *Phalaris arundinacea*) in the Bridgwater and Taunton Canal in England. The canal is no longer a navigable waterway. (Photo: K. J. Murphy.)

1978*a*; Dawson and Robinson 1984; Brooker, Morris, and Wilson 1978). Attempts have been made to define, in biomass terms, the maximum tolerable limits of plant growth in European lowland waterways. In the British canal system the maximum submerged biomass which seems to cause no problems for boat passage, in a canal of average cross-sectional area, is only 300 $g\,m^{-2}$ (fresh weight). If submerged biomass exceeds 1 000 $g\,m^{-2}$ (fresh weight), weed growth causes serious obstruction of boat movement (Eaton, Murphy, and Hyde 1981; Murphy, Eaton, and Hyde 1982). In Danish streams, Kern-Hansen and Holm (1982) suggested that the tolerable 'flood-safe' biomass of aquatic vegetation is 100–200 $g\,m^{-2}$ (dry weight).

In Europe it is commonplace for lowland rivers to have their discharge capacity increased, as part of flood-prevention or land-drainage schemes. This involves excavation of the river bed and alteration of the course of the river. Such canalization of rivers requires a plant management input at the planning stage if the conservation, fisheries, and amenity value of the river is not to be reduced or destroyed (Newbold 1982; Dutartre and Gross 1982; Raven 1986; Brookes 1984; DVWK 1984, 1987).

Weed control in lowland rivers and canals in Europe continues to depend heavily on manual and mechanical techniques, supplemented by herbicides in some countries, and limited usage of grass carp (Moore and Spillett 1982;

Mugridge *et al.* 1982; Swales 1982; Soulsby 1974; Hermens 1978; Fox, Murphy, and Westlake 1986; Ham, Wright, and Berrie 1982; Robson 1976*a*; Eaton and Freeman 1982; Westlake 1968; Westlake and Dawson 1982, 1986). Herbicide treatments are normally limited to spraying of emergent weeds, using dalapon, 2,4-D and glyphosate. Diquat-alginate, a viscous formulation suitable for use in water flowing at up to $0.5\,m\,s^{-1}$, is a herbicide gaining acceptance for submerged weed control, for example, in Britain, Ireland, and Czechoslovakia (Barrett 1981*a*, *b*; Barrett and Murphy 1982; Hanbury 1986; Fox, Murphy, and Westlake 1986). Other herbicides (e.g. terbutryne, dichlobenil) have been used for submerged weed control in British canals, but are only effective if the flow can be halted for several days after treatment (Murphy, Hanbury, and Eaton 1981; Murphy, Eaton, and Hyde 1982). Shading, by natural or purpose-planted bankside vegetation, or by artificial screens, has been put forward as a stream weed management technique in Europe (Dawson and Kern-Hansen 1979; Dawson and Haslam 1983; Dawson and Hallows 1983), and is actively encouraged in some countries (e.g. the Netherlands and Denmark). The role of bankside vegetation for bank protection against erosion is also increasingly recognized (Bonham 1980).

Lowland lakes, ponds, and reservoirs

European lowland standing waters include natural lakes, ponds, and lagoons (though often these are considerably altered by human activities, such as drainage schemes), and artificial water bodies including reservoirs, man-made fishponds, and flooded excavations produced by gravel extraction or peat-cutting (e.g. the Norfolk Broads in England). There is a wide diversity of physical shape, depth, surface area, water chemistry, substrate characteristics, catchment use, and management regime. The range is illustrated by extremes such as the shallow, ephemeral, saline-influenced lagoons of the Camargue, on the south coast of France, which support dense aquatic plant populations (Guerlesquin and Podlejski 1980; Grillas and Duncan 1986); deep, steep-sided, oligotrophic lakes such as Loch Ness in Scotland, virtually devoid of macrophyte growth (Bailey-Watts and Duncan 1981); large, fairly shallow, nutrient-rich lakes, supporting highly productive and extensive beds of *Phragmites australis* and other macrophytes, such as Rescobie Loch, Scotland; L. Mikolajskie, Poland; and Neusiedlersee, Austria (Imhof and Burian 1972; Ozimek and Kolaczewske 1984; Scheimer and Prosser 1976; Bownik 1970; Burian and Sieghardt 1979; Murphy 1988; Gunatilaka, Broebhart, and Aranjo de Oliveira 1983). Because of the difficulties of generalizing about lowland standing waters, only the major issues concerning their macrophyte communities, weed problems, and approaches to plant management are highlighted.

Eutrophication is a principal cause for concern in many lowland water bodies in Europe, and is often pinpointed as the cause of increased weed problems, at least during the early stages of nutrient-enrichment (Holden 1976; Forsberg and Ryding 1980; Stewart, Tuckwell, and May 1975; Vollenweider 1968; Murphy 1988; Giulizzoni *et al.* 1984). Severe eutrophication is, however, commonly associated with macrophyte diebacks, rather than increases: periphytic and planktonic algae replace the macrophyte vegetation (e.g. Lachavanne 1985; Jupp and Spence 1977; Best, de Vries, and Reins 1984; Phillips, Eminson, and Moss 1978; Boorman, Sheail, and Fuller 1978; Boorman and Fuller 1981). The reduction of nutrient levels is frequently advocated as a sensible long-term strategy for the management of nutrient-enriched lowland water bodies. In practice it is difficult, though not impossible, to achieve, as shown by Weidenbacher and Willenbring (1984) and Moss *et al.* (1986).

Other, more direct, means of limiting the growth of aquatic plants tend to be used in place of (or alongside) nutrient-reduction schemes in lowland water bodies suffering aquatic weed problems. As in river management, weed-cutting, and other physical removal techniques are the standard approach throughout Europe (Hertzman 1986; Husak, Květ, and Plasencia Fraga 1986). Grass carp have been introduced to a limited number of standing waters in Western Europe (Ahling and Jernelov 1971; von Menzel 1974; Neururer 1978; Markman 1982; Riemens 1982; van Zon 1977). In Eastern Europe grass carp gained earlier and more widespread acceptance, not only for weed control purposes, but also as a food fish, often in mixed culture with other species (Opuszynski 1982; Krzywosz, Krzywosz, and Radziej 1980; Krupauer 1971; Charyev 1980; Sobolev 1970; Vovk 1976; Cure, Snaider, and Chiosila 1970; Mestrov and Tavcar 1973; Pinter 1980).

The main herbicides used in static waters (in those European countries permitting aquatic herbicide use) are diquat, dichlobenil, paraquat, terbutryne, dalapon, glyphosate, and 2,4-D (Evans 1982; Wade 1982; Kelcey 1981; Neururer 1978; Robson, Fowler, and Hanley 1978; Tooby and Spencer-Jones 1978; Barrett and Logan 1982; Dunn 1976; Hartman, Faina, and Machova 1984; Brooker and Edwards 1973*a, b, c*). Copper sulphate is also widely used as an algicide in trout ponds in West Germany (Bohl, Wagner, and Hoffman 1982).

Some common aquatic weeds in European lowland standing waters are the submerged genera *Elodea, Potamogeton* (especially *Potamogeton pectinatus, P. perfoliatus* and *P. lucens*), and *Myriophyllum.* In acidified lowland waters (e.g. in the Netherlands), *Juncus bulbosus* can cause severe problems. Nymphaeids such as *Nuphar lutea,* and other floating-leaved rooted species, for example, *Potamogeton natans,* may also cause nuisance if their floating foliage covers too much of the surface area of a lake, interfering with recreation and other uses of the water. Filamentous algae are a further cause of problems, notably *Cladophora, Rhizoclonium*, and *Enteromorpha* spp. In the shallow littoral zones of lakes and other water bodies emergent species such as *Phragmites australis, Typha latifolia,* and *Carex* spp. commonly form large stands. In normal circumstances these cause no problems, and indeed are beneficial to the functioning of the lake ecosystem, but excessive growth encroaching on the open water may necessitate active management. Dredging (Hertzman 1986) or spraying with dalapon or glyphosate (Barrett and Robson 1971, 1974; Barrett 1985) are common management techniques.

In static waters, as in other aquatic systems, physical obstruction is the normal problem caused by excessive plant growth. However, there may be other problems, such as interference with sport fisheries (Kelsall 1981). Tölg (1971) reported the effects of decaying plant material in reservoirs, in adversely affecting the taste of drinking water in Hungary. In Scotland aquatic weeds may cause taint problems in reservoirs supplying whisky distilleries, where the purity and quality of water source is of paramount importance. Species such as *Equisetum fluviatile*, *Menyanthes trifoliata*, and some *Potamogeton* species release organic substances into the water which may be sufficient to taint the distilling process, leading to the rejection by quality control of whole batches of whisky. Weed control measures can be remarkably difficult in these circumstances, as herbicides cannot be used, and physical measures rarely remove enough of the plant growth to obviate the problem. Some distilleries are now investigating the potential of grass carp for weed control in supply reservoirs (Murphy 1988).

Irrigation and land drainage channels

In low-lying areas of Europe, notably coastal alluvial plains (e.g. in the Netherlands and Belgium), and the flood plains of large rivers such as the R. Po in Italy, there are extensive networks of land drainage channels. In southern Europe irrigation channel systems are a common feature of agricultural areas. Throughout Europe, both types of artificial waterway provide a habitat well suited to highly-productive plant growth, especially since nutrient levels are often enhanced by the leaching of fertilizers from surrounding fields (Marshall, Wade, and Clare 1978; Haslam 1987). In consequence, aquatic weed problems are common in such waters (Fig. 15.4 and Fig. 15.5). In drainage channels aquatic weeds create a high risk of flooding and waterlogging because they increase channel roughness and block water flow (Agusti, Duarte, and Guimaraes 1984; Powell 1978; Robson 1986; Robinson 1986; Moreira *et al.* 1983; Wolseley 1986; Murphy 1988; Miles 1976; Cave 1981). In irrigation channels the reduction in discharge by aquatic weeds can increase the risk of drought damage to irrigated crops (Oksiyuk 1976; Portier 1971; Figuereido *et al.* 1984).

Annual programmes of ditch and channel maintenance are the norm in both drainage and irrigation systems in Europe, using manual and mechanical techniques (Best, van der Zweerde, and Zwietering 1986; Wade and Edwards 1980; Robson 1975; George 1976; Blümel 1974; Kemmerling 1978; Miles 1976), or herbicides, particularly terbutryne, dichlobenil, dalapon, and glyphosate, although others such as 2,4-D, MCPA, MCPP, atrazine, dicamba, aminotriazole,

FIG. 15.4. Weed growth in a drainage ditch in the Province of Drente in the north-eastern part of The Netherlands. The vegetation consists mainly of *Alisma plantago-aquatica*, *Sparganium erectum*, *Glyceria maxima*, *G. fluitans* and Lemnaceae. (Photo: H. G. van der Weij.)

FIG. 15.5. Weed growth in a drainage ditch in the Province of Utrecht in the western part of The Netherlands. The vegetation consists mainly of *Phragmites australis* and filamentous algae. (Photo: H. G. van der Weij.)

and imidazoline may also be used on embankments and slopes, and in ditches which dry out seasonally, for example, in Yugoslavia (Arsenović *et al.* 1982; Arsenović, Dimitrijević, and Konstantinović 1986; Stanković *et al.* 1982; Sarkany 1982; Spagnoli 1973; Kramer, Schmaland, and Nanzke 1974; Fernandes *et al.* 1978; Robson 1975; Kovacs 1983; Murphy 1988; Blaszyk 1977; Morrison and Courtney 1981; Courtney 1974; Heuss 1971; Robinson 1971). Biological control using grass carp is an important approach to drainage ditch weed control in polders in the Netherlands (Bouquet 1977; van der Zweerde 1983) and in irrigation and drainage channels elsewhere, e.g. Britain, Yugoslavia, USSR, East Germany (Aliev 1976; Vovk 1976; Jähnichen 1973) (Fig. 15.6).

FIG. 15.6. Mass growth of *Salvinia natans* and *Nymphoides peltata* in a drainage canal of the Dunaw–Tisa–Dunaw system in Yugoslavia containing grass carp. Control of submerged weed growth beneath the floating foliage was sufficient to permit acceptable discharge rates for water flow in the canal. (Photo: K. J. Murphy.)

Typical weed taxa causing problems in irrigation and drainage channels are the filamentous algae (notably *Cladophora*, *Enteromorpha*, and *Vaucheria* spp.) and submerged macrophytes characteristic of slow-flowing eutrophic waters: *Ceratophyllum demersum*, *Elodea nuttallii*, *Elodea canadensis*, *Potamogeton pectinatus*, *Zannichellia palustris*, and *Lemna minor* agg. Other free-floating species may infest some drainage and irrigation networks in southern Europe, e.g. *Eichhornia crassipes* in Portugal (Moreira *et al.* 1983).

Haslam (1987) noted that filamentous algae are less of a problem in the extensive ditch networks of the Netherlands than in comparable drainage ditches of eastern and southern England. There is some evidence (Harbott and Rey 1981) that the vegetation of ditches subject to long-term treatment with herbicides may be maintained by the management regime at an early-successional stage, dominated by opportunist plant species, of which filamentous algae are a good example. In the UK there is a considerable reliance on herbicides for ditch maintenance (Courtney 1974; Cave 1981; Robinson 1969, 1986; Murphy 1988; Robson and Fillenham 1976). Although speculative, this may be a partial explanation for the observed differences in filamentous algal abundance in Britain and the Netherlands. However, it should be noted that *Vaucheria* and *Enteromorpha* spp. were a serious problem in the ditches of the Fens and other low-lying areas of England before the advent of herbicides.

In Yugoslavia Skender *et al.* (1982) stated that *c.* 15 per cent of the artificial 'hydroameliorative' channel network in low-lying agricultural areas was affected by aquatic weed problems, mainly submerged and emergent species. Much research is devoted to developing cost-effective methods for the control of these problems (Arsenović *et al.* 1982; Arsenović, Dimitrijević, and Konstantinović 1986; Stanković *et al.* 1982). Glyphosate and dalapon are mainly used against emergent weeds in drainage canals (with atrazine, 2,4-D, MCPA/MCPP and dicamba on dry embankments and slopes).

There are serious weed problems in the 38 000 km of drainage channels of the coastal region of northern Germany (BRD) according to Reschke and Blaszyk (1974). Weed problems and their control in irrigation channels of the Stoinesti–Visina system in the Olt region of Rumania were described by Sarpe *et al.* (1974). Characteristic species included *Typha* spp., *Carex* spp., *Cyperus* spp., *Sparganium erectum, Glyceria maxima, Myriophyllum verticillatum, Ceratophyllum* spp., *Elodea canadensis, Potamogeton* spp., *Lemna gibba,* and *Ranunculus aquatilis.*

In the irrigation channels associated with rice fields in Southern Europe a specialized weed flora develops, often characterized by *Najas* spp. (Piccoli and Gerdol 1981; Triest 1986). Both native species and the adventive *Najas gracillima* (Triest 1986) are a cause of problems in rice-growing areas of Spain, France, Italy, Portugal, and Hungary.

IMPORTANT AQUATIC WEED SPECIES IN EUROPE

Submerged weeds

The major problems caused by the adventive submerged macrophytes *Myriophyllum spicatum* and *Hydrilla verticillata* elsewhere in the world do not occur in Europe. *M. spicatum* is in its home range in Europe (Faegri 1982). Although it can and does cause weed problems (Dutartre and Dubois 1986) these are not as severe as those found in other continents such as North America, where the plant is a rapidly-spreading invader. In southern Europe other *Myriophyllum* species (e.g. *Myriophyllum verticillatum, M. aquaticum*) can cause serious nuisance (Teles and Pinto da Silva 1975; Sarpe *et al.* 1974). *Hydrilla verticillata* has two areas of occurrence in Europe, i.e. Poland and the north western part of the Soviet Union, and one lake in Ireland, but causes no problems (Wolff 1980; Verkleij and Pieterse 1986; Pieterse, Ebbers, and Verkleij 1983). This is probably due to the fact that the European strains of *Hydrilla* are genetically very different from strains occurring in other continents.

There are, however, several examples of introduced aquatic macrophytes which have spread widely through Europe, and continue to cause major problems. Most are submerged species. The best known and documented is *Elodea canadensis*, introduced in the late nineteenth century from North America (Siddall 1885; Walker 1912) and still a serious problem (Simpson 1984; Bolman 1977; Pokorny *et al.* 1984; Cook and Urmi-König 1985). Norway is one of the most recent countries to have experienced invasion by *E. canadensis* (Rørslett 1969, 1977; Rørslett, Berge, and Johansen 1986). Recently, *Elodea nuttallii* has shown signs of similar expansion of its range through the eutrophic waterways of northern Europe (Chandler 1975; van der Ploeg 1966; Weber-Oldecop 1977; Lindner 1978).

Other warm-water submerged species from Africa, Australia, and South America, including *Lagarosiphon major* (Dutartre 1980, 1986; Dutartre and Capdevielle 1982; Kent 1955; Triest 1982; Ravera, Garavaglia, and Stella 1984), *Egeria densa* (Dupont and Visset 1968; Kent 1955; Cook and Urmi-König 1984), and *Crassula helmsii* (Pain 1987) have established successfully in European waters, and show signs of acclimatization permitting them to occur as far north as Scotland.

Of native European macrophytes, the most important cause of weed problems in faster-flowing waters is *Ranunculus* subg. *Batrachium* (e.g. *Ranunculus fluitans, Ranunculus penicillatus* var. *calcareus*: Westlake and Dawson 1986; Mycock 1983; Dawson 1976; van der Borght *et al.* 1982). In slower-flowing and static waters *Potamogeton* spp. are common weeds throughout Europe. *Ceratophyllum demersum* is also often reported to cause flow-obstruction in slow-flowing waterways such as Yugoslavian drainage channels (Skender *et al.* 1982), Rumanian and Italian irrigation networks (Sarpe *et al.* 1974; Piccolo and Gerdol 1981), and British canals (Murphy, Hanbury, and Eaton 1981).

As already noted filamentous algae are a major problem in drainage channels and other eutrophic waters throughout Europe, the main problem taxa being *Cladophora, Spirogyra, Vaucheria, Rhizoclonium,* and *Enteromorpha* (van Himme, Stryckers, and Bulcke 1977; Bulcke 1978; Whitton 1971, Bolas and Lund 1974; Kovrizhnykh 1978; Dowidar and Robson 1971, 1972; Fowler 1986).

Free-floating weeds

Large free-floating species such as *Eichhornia crassipes* are reported to cause problems in areas such as the Ribatejo irrigation system in South Portugal (Guerreiro 1976; Fernandes *et al.* 1978) but are not a widespread problem even in southern Europe. In the Netherlands *Pistia stratiotes* established a foothold in one region, grew to nuisance proportions during the course of one summer, but was wiped out by winter frosts (Pieterse 1977*a*, *b*; Pieterse, de Lange, and Verhagen 1981). *Salvinia natans* can cause weed problems in rice irrigation channels, for example, in Italy (Piccoli and Gerdol 1981), and also occurs in drainage channels such as the Dunav–Tisa–Dunav system in Vojvodina region of Yugoslavia. *Stratiotes aloides,* free-floating for at least part of its life-cycle, is a rare, protected species in much of Europe, but in drainage and irrigation channels in East Slovakia (Czechoslovakia) has become a serious pest species in recent years (Husak and Otahelova 1982).

More of a problem are the small free-floating plants, such as the Lemnaceae, widely distributed in Europe (Landolt 1982), and *Azolla* spp., which are more localized but can produce dense surface mats in areas such as the western part of the Netherlands (Pieterse, de Lange, and van Vliet 1977), and southern England (Murphy, Hanbury, and Eaton 1981). Growths of these plants sufficient to cause nuisance are usually confined to static or very slow-flowing waters. These include drainage ditches, ponds, and reservoirs, and navigable canals carrying low densities of powered boat traffic, such as the Monmouth and Brecon Canal in South Wales (UK) (Murphy, Hanbury, and Eaton 1981). *Lemna* problems can also occur in rivers, particularly in times of drought and reduced flow—an example being the R. Kelvin in Glasgow (Scotland) which has developed a mass growth of *Lemna minor* agg. during hotter summers of recent years. *Lemna minor, Lemna gibba, Spirodela polyrrhiza,* and *Wolffia* spp. are all problem-causing taxa in various parts of Europe (Rejmankova 1975; Bowker and Duffield 1981; Wolek 1974; Duffield and Edwards 1981; Murphy, Hanbury, and Eaton 1981). *Lemna minuscula* is an adventive species with the potential to cause weed problems, recently reported in Britain (Moore 1983; Leslie and Walters 1983). Ross, Doughty, and Murphy (1986) described a typical *Lemna* problem, where growth of *Lemna minor* agg. in a Scottish canal was stimulated by a point-source input of phosphate-enriched water, from a polluted canal feeder stream. Wind action caused the plants to accumulate in mats up to 1 m thick at obstructions in the channel (e.g. bridge-narrows). The decay of the *Lemna* mats resulted in deoxygenation of the canal water, with adverse environmental effects, and vociferous complaints from the local population about the foul smell caused by the release of hydrogen sulphide from the decaying vegetation and anoxic water. Management of the problem was effected in this case by reducing the input of polluted feeder water during the summer months.

Rooted floating-leaved weeds

Compared with the other major groups of aquatic weeds, rooted floating-leaved species are less of a problem in European waters. Indeed, in the Netherlands there has been interest in promoting the growth of species such as *Nymphaea alba* in slow-flowing drainage channels, because this is an effective means of shading out submerged weeds, but does not in itself greatly increase the roughness coefficient of the channel (Pitlo 1982, 1986) (see also Chapters 9a and 10).

However, rooted floating-leaved species can cause weed problems. *Potamogeton natans* is a good example, causing problems in static waters, canals, and slow-flowing rivers (Barrett and Murphy 1982; Murphy, Fox, and Hanbury 1987; Murphy 1988). Other species reported to cause problems in various parts of Europe include *Sparganium emersum, Glyceria fluitans, Hydrocharis morsus-ranae,* and nymphaeids (including *Nymphaea alba, Nuphar lutea,* and *Nymphoides peltata*). The status of these species varies considerably between different countries: *Nymphaea alba* is a case in point. This is a plant with highly attractive flowers, commonly introduced to gardens, parks, and other amenity locations as an ornamental species, throughout Europe. In Austria wild populations of *N. alba* are fully protected by law, as befits one of that country's rarest plants. In contrast, in the Netherlands *N. alba* and other nymphaeids are common, occurring in some 23 per cent of Dutch canals, where they can interfere with boat traffic, and are managed as necessary by cutting (van der Velde and Peelen-Bexkens 1983; van der Velde, Custer, and de Lyon 1986; Smits and Wetzels 1986). In the Highlands of Scotland *N. alba* is, again, quite common in freshwater lochs, but even in abundance rarely interferes with human usages of these waters (mainly fishing, hydroelectricity supply, and cage fish-farming of salmonids).

Emergent weeds

Perhaps the commonest cause of nuisance amongst emergent macrophytes in Europe is *Phragmites australis,* which can also spread from the aquatic habitat into terrestrial fields, where the water table is close to the surface (e.g. in low-lying grazing in Dutch polders, areas of Bayern in southern Germany, and drained marshlands used for grazing and arable crops in southern England, such as Romney Marsh). Several other emergent species (*Schoenoplectus* spp., *Glyceria maxima, Phalaris arundinacea, Typha* spp., *Sparganium erectum*) are also responsible for aquatic weed problems in European waters (Barrett and Robson 1974; Neururer 1974*a*; Hertzman 1986; Husak, Květ, and Plasencia Fraga 1986; Rodewald-Rudescu 1974; Eaton, Murphy, and Hyde 1981; Arsenović, Dimitrijević, and Konstantinović 1986). In general, dicot species are less of a problem than monocots, although genera such as *Berula, Alisma,* and *Sagittaria* are all known to cause occasional problems. Species poisonous to livestock (e.g. *Cicuta virosa,*

Oenanthe spp.) are sometimes also singled out as weeds needing active control, especially in ditches bordering on pastures. *Nasturtium amphibium* was reported by Skender *et al.* (1982) to cause severe channel blockage in drainage networks in Istria, Yugoslavia. *Mentha aquatica* was one of the three most important aquatic weeds in the Bačka region of Vojvodina, Yugoslavia, the other two being submerged species (Arsenović *et al.* 1982).

Schiele (1986) reported that exotic aquatic weeds of the Pontederiaceae (*Heteranthera* spp. from North America and *Monochoria korsakowii* from Asia) have recently successfully invaded the freshwater systems associated with rice fields in Italy and Greece, and are spreading.

Excessive growth of emergent plants is not the only management problem with this type of vegetation. There is much concern about the dieback of reeds (*Phragmites australis*) and other emergent plants, which form an essential ecosystem component of the great wetland areas of Europe, such as the Danube delta, major lakes (e.g. Bodensee and Neusiedlersee in Austria), the Norfolk Broads in England, and the Havel river and lakes of Berlin. In the Netherlands marginal beds of *Phragmites* play an important role in minimizing bank erosion in the larger waterways. The reed beds require active management to keep them in a healthy condition. The issue of 'reed-death', and effective management aimed at reversing losses of emergent vegetation is an important one in Europe at present (Haslam 1973*a*; Ho 1979; van den Linden 1980; Britton 1974; van der Toorn 1978; Danell 1979; van der Toorn and Mook 1982; Sukopp 1971; Klötzli 1971; Klötzi and Züst 1973; Jackson 1978; Nikolayevski 1971; Boorman, Sheail, and Fuller 1978; Boorman and Fuller 1981).

APPROACHES TO AQUATIC WEED MANAGEMENT IN EUROPE

A wide range of attitudes and approaches to aquatic weed management currently exists in Europe. Certain countries permit no usage of aquatic herbicides (e.g. Denmark: Kern-Hansen and Holm 1982), or severely restrict herbicide use in freshwater systems (e.g. the Netherlands: Zonderwijk and van Zon 1974, 1978), even though they suffer serious problems of aquatic weed growth (Larsen 1978; Best, van der Zweerde, and Zwietering 1986). These countries rely instead on mechanical control, supplemented by the use of biological agents, almost exclusively grass carp (Markmann 1982; Zonderwijk and van Zon 1978; Riemens 1982*a*). Such countries usually have influential nature conservation lobbies, generally tend to restrict the use of herbicides in non-agricultural systems, to minimize environmental contamination, and are prepared to pay for plant management regimes which exclude the use of herbicides.

In contrast, the countries of southern and south-eastern Europe such as Italy,

Portugal, Hungary, Rumania, and Yugoslavia take a more pragmatic approach to the control of aquatic weeds in areas of vital economic importance, such as irrigated cropland, employing a full range of chemical, mechanical/manual and biological control measures (Spagnoli 1973; Kryzywosz, Kryzywosz, and Radziej 1980; Teles and Pinto da Silva 1975; Guerreiro 1976; Fernandes *et al.* 1978; Sarkany 1982; Kovacs 1983; Moreira *et al.* 1983; Figuereido *et al.* 1984; Sarpe *et al.* 1974). In Yugoslavia, for example, herbicides are used for *c.* 40 per cent of aquatic weed control, mechanical clearance 40 per cent, manual clearance 18 per cent, and grass carp 2 per cent. In drainage canals glyphosate and dalapon are the main herbicides used. No herbicides may at present be used in irrigation canals. Active research is underway in Yugoslavia (at the Faculties of Agriculture in Novi Sad and Osijek, financed by the Water Management Organizations) to improve aquatic weed control. Major current themes include description of the major aquatic weed communities of Yugoslav waters; studies on potential herbicides of lower environmental toxicity and persistence than currently-used compounds for use in freshwaters; the potential impacts of herbicide-residues in irrigation water on agricultural crops; and the potential of biological control by micro-organisms, insects, and fish in Yugoslav waters (a joint programme of research with the US Department of Agriculture).

Most European countries adopt an attitude and approach to aquatic weed control intermediate between the extremes described above. For example, Czechoslovakia, West Germany, East Germany, France, Switzerland, Austria, Great Britain, and the European states of the USSR all suffer serious aquatic weed problems, as described earlier. These states permit at least some herbicide usage in aquatic systems, but tend to enforce strict regulations on chemical and other forms of aquatic weed control (Beitz *et al.* 1982; Bryan and Hellawell 1980; Engelhardt 1974; Neururer 1974*b*; Thomas 1981), whilst devoting a considerable research effort to monitoring the environmental effects of aquatic weed management (Pearson and Jones 1978; Robson and Fearon 1976; Brooker and Edwards 1975; Robson 1978; Hartman, Faina, and Machova 1984; Wade 1978, 1982; Holz 1963; Swales 1982; Moore and Spillett 1982; Harbott and Rey 1981; Bohl and Wagner 1981; Hanbury, Murphy, and Eaton 1981; Murphy and Eaton 1981*a*; de Lange and van Zon 1978; Statzner and Stechmann 1977; Svobodova *et al.* 1984).

Table 15.3 compares herbicide usage in six European countries. In Britain, it proved difficult to gain acceptance of both herbicides and grass carp for aquatic weed control purposes, in part due to the caution of the Nature Conservancy Council (NCC), the government agency with statutory responsibility for environmental conservation in Britain (Newbold 1975). Recently, however, the NCC has changed its opinion of aquatic herbicides to the extent of recommending two compounds, diquat-alginate and dalapon for usage in water in designated nature reserves suffering aquatic weed problems (Cooke and Newbold 1986). There are

Table 15.3. Permitted aquatic herbicide usage in seven European countries

Herbicide (active ingredient)	Denmark[a]	Norway[a]	Sweden	Netherlands	UK	Yugoslavia	Belgium
2,4-D					+	+[e]	+
Dalapon			+	+[b]	+	+	+
Diquat				+[c]	+		+
Glyphosate			+		+	+	+
MCPA/MCPP						+[e]	
Paraquat				+[c]			+
2,4,5-T							+
Terbutryne			+[d]		+		
Atrazine						+[e]	
Dicamba						+[e]	
Dichlobenil					+		

[a] No herbicide use permitted in water.
[b] Not before 15 July; no surfactant additives permitted.
[c] Not before 1 June; no surfactant additives permitted (use of diquat and paraquat likely to be phased out from 1989).
[d] In fireponds only.
[e] In ditchbanks, and in dry ditches only.

still, however, severe limitations on the use of grass carp in British waters, because a separate licence is required for each introduction, necessitating a detailed investigation of each site. The NCC is consulted by the licensing authorities prior to introduction of grass carp. Grass carp are currently used in about 1 per cent of the watercourses of the Netherlands, with satisfactory results in the majority of cases. Grass carp have also been introduced in Belgium, West Germany, Denmark, and Sweden in northern Europe. Norway has not permitted grass carp introduction.

Published data on the costs of aquatic weed control in Europe are few. Jähnichen (1973*c*) reported that the introduction of grass carp to 152 ha of waterways in the Magdeburg region of East Germany resulted in an 80 per cent reduction in weed control costs, and successfully cleared two-thirds of the system. Newbold (1975) estimated the 1972 costs of drainage channel maintenance in Great Britain at UK £8.6M, of which UK £6.0M went on mechanical clearance and the rest on herbicides. More recently, Cave (1981) provided information on the costs of weed control measures used in British drainage channels of varying dimensions (Table 15.4). Using these figures, and estimates of drainage channel lengths treated annually with different weed control techniques (Marshall, Wade, and Clare 1978), a conservative estimate of expenditure on aquatic weed control (at 1981 prices) in the main channels of the major drained agricultural areas of Britain is UK £16–20M. Such costs are roughly in line with expenditure on aquatic weed control in other comparable areas of Europe. In the 25 000 km of ditches in the Weser–Ems district of West Germany, for example, Reschke (1978)

Table 15.4. Costs of three different aquatic weed control measures used in drainage channels in Great Britain (data from Cave 1981)

Weed problem	Cost (UK £) per 100 m of channel of different weed control methods (ditch cross-sectional area m^2)					
	Manual		Mechanical[a]		Herbicide[a]	
	(2)	(7)	(2)	(7)	(2)	(7)
Filamentous algae	100	200	13	20	8	22
Submerged	16	100	11	13	8	24
Emergent	50	50	11	13	4	4

[a] Averages of all appropriate techniques used for mechanical control, and for all herbicides in use in British drainage channels up to 1981.

estimated annual aquatic weed control costs at DM 25–30M. In the Netherlands average costs are Hfl 150 per 100 m of watercourse treated with the standard twice-yearly mechanical clearance.

ACKNOWLEDGEMENTS

Thanks are due to Maud Wallsten (Sweden), J. Stryckers, R. Bulcke, and P. van der Borght (Belgium), B. Rørslett (Norway), G. Wiegleb (Germany BRD), J. B. Lachavanne and C. D. K. Cook (Switzerland), P. O. Jennings (Ireland), I. Moreira (Portugal), and all others who contributed data for this chapter.

16

Aquatic weed problems and management in Asia

B. GOPAL

INTRODUCTION

THE Asian landmass, the largest continent on earth, is unique in many respects. It is the most populated part of the world with more than half of humanity living there. It extends from the Equator to the Arctic: all climatic zones are represented. The continent exhibits a great diversity of landscapes: high mountains, vast alluvial plains, dense tropical and temperate forests, hot and cold deserts; innumerable ephemeral streams and seasonal ponds, mighty rivers, lakes, and reservoirs are all represented (Fig. 16.1).

The great diversity of natural and man-made aquatic habitats is matched by the diversity of aquatic vegetation, not always desirable from the viewpoint of human interests. Aquatic plants often interfere with the utilization of water resources. From an economic point of view the most important aquatic habitats in Asia are those associated with the production of food for man, either directly (fish ponds and paddy fields) or indirectly (irrigation reservoirs and irrigation channels). Naturally, the aquatic weeds in these habitats have attracted greater attention than in those used for navigation or recreation. During the past few decades, many new reservoirs have been constructed for irrigation, hydroelectric power generation, and flood control in all Asian countries. Many have become infested with native and/or exotic weed species.

In this context, aquatic weeds have been the subject of scientific study for about a century. There is now a vast literature on the kinds of weeds, their distribution, ecological attributes, diversity of the problems they cause, and the efforts made to control aquatic weed problems in Asia. Some areas of Asia are less well studied, e.g. arid hot deserts (Afghanistan to Saudi Arabia), the mountains of northern India, Nepal, Tibet, and China, and the cold desert of Siberia. In these areas there are few problems caused by aquatic weeds.

Studies on aquatic weeds in south and south-east Asia are often reported at

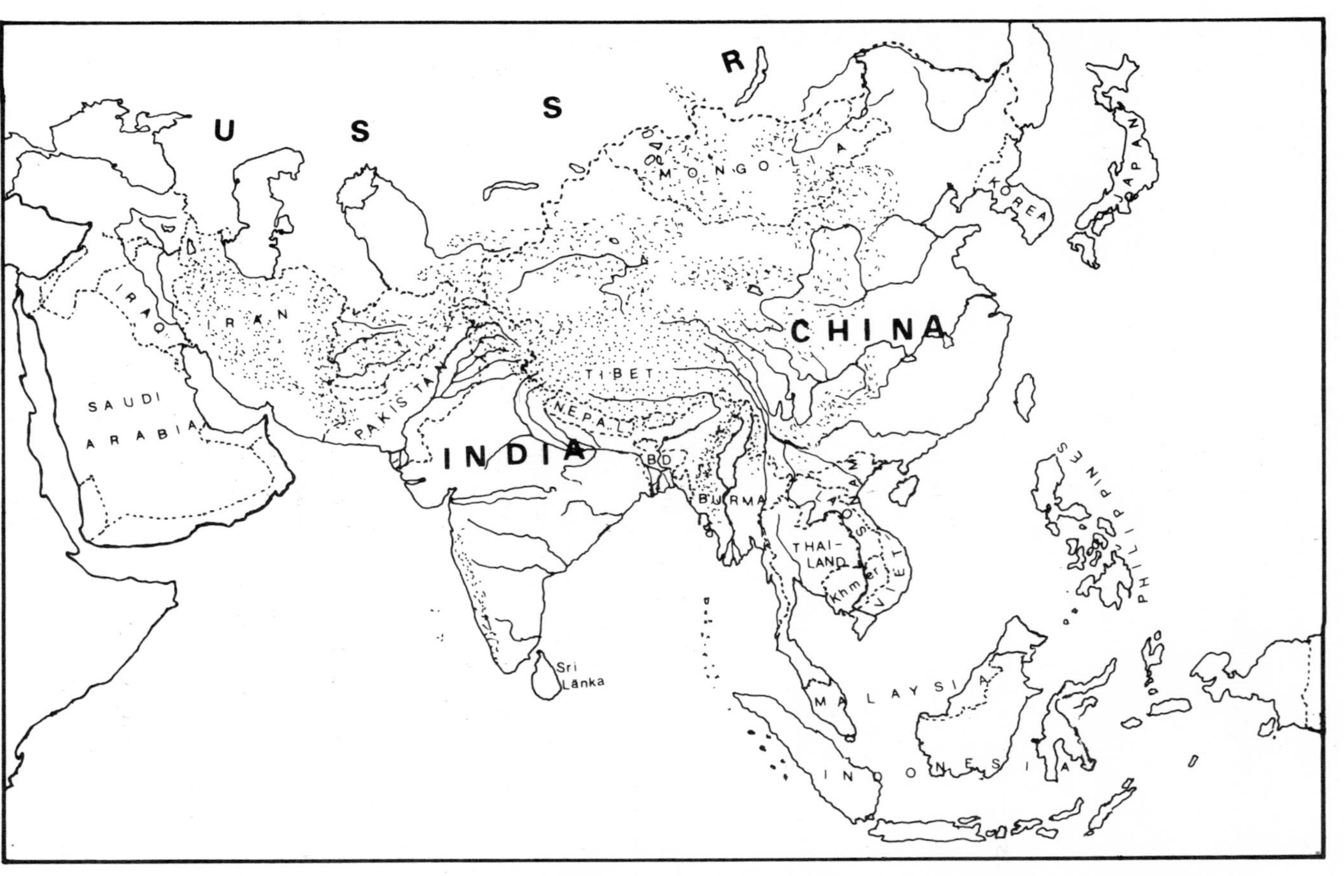

FIG. 16.1. The Asian continent (excluding USSR) showing major rivers and mountainous regions (stippled).

the meetings of the weed science societies of several countries, e.g. India (Gupta 1979), Indonesia (Soerjani 1971; Soerjani, Comber, and Tjitrosoepomo 1979), Thailand (Thamasara 1982), Bangladesh (Majid 1986), the Philippines (Pancho, Vega, and Plucknett 1969), and Japan, and at the biennial conferences of the Asian Pacific Weed Science Society. However, they received major attention in 1973 at a UNESCO-sponsored Regional Seminar held in New Delhi (Varshney and Rzoska 1976). This was followed by another Regional Seminar in Indonesia at Malang (Soerjani *et al.* 1976) where reports were presented from almost all south-east Asian countries. The two meetings arrived at rather similar conclusions. Since then, despite continuing interest in aquatic weeds little new insight has been gained into the problem, or breakthrough achieved for its management. Water hyacinth has received special attention, focusing on the possibilities of its utilization. An international research project involving India, Sri Lanka, Bangladesh, and Malaysia from Asia, and an international conference on water hyacinth (Thyagarajan 1984) supported by the Commonwealth Science Council, United Nations Environment Programme and other organizations reflect the magnitude of the problem and importance of aquatic weeds in the region.

IMPORTANT AQUATIC WEEDS

Almost all aquatic and marsh plants known to occur in Asia have been considered weeds in one or the other situation. Some sixty-eight families of flowering plants are represented in the aquatic weed flora of south and south-east Asia (Pancho 1979; Pancho and Soerjani 1978; Gupta 1979; Hong-jun and Xue-ming 1986), plus eight families of ferns. The list includes a large number of grasses (Poaceae), sedges (Cyperaceae), and other weeds which invade paddy fields and interfere with crop growth and yield: Table 16.1 lists the more common species, with their distribution in Asian countries.

Aquatic weeds are broadly grouped into four categories according to their growth habit: free-floating, submerged, rooted floating-leaved, and emergent weeds. Though many of them tend to form monospecific stands by their rapid vegetative growth, often weeds of different growth-form occupy the same water body. It would be useful to examine the weeds of the four groups separately as they also differ in the nature of problems caused by them.

Free-floating weeds

The most widespread and obnoxious of all aquatic weeds in Asia are two free-floating weeds of south American origin. Introduced by man at different times around the end of the nineteenth century (Gopal 1987) in Japan, Indonesia, India, Sri Lanka, China, and other neighbouring countries, the exotic water

Table 16.1. Common aquatic weeds in paddy fields in south and south-east Asia

Alisma canaliculatum (J)	*Limnanthemum indicum* (K, I)
Alternanthera sessilis (V, I, In)	*Limnophila conferta* (K, I)
Aneilema keisak (J)	*Lobelia chinensis* (J)
Blyxa aubertii (V)	*Marsilea quadrifolia* (J, I)
Blyxa japonica (V)	*Mimulus orbicularis* (V)
Callitriche verna (J)	*Monochoria vaginalis* (J, V, I, In)
Ceratopteris thalictroides (V)	*Nitella* spp. (V, I, T, J, In)
Chara spp. (V, I, T, J, In)	*Nymphoides indica* (V)
Cyperus difformis (J, V)	*Oenanthe javanica* (J)
Cyperus procerus (V)	*Oryza rufipogon* (V, In, I)
Cyperus serotinus (J)[a]	*Ottelia alismoides* (V, I)
Diplachne fusca (V)	*Ottelia japonica* (K, J)
Dopatrium junceum (J)	*Paspalum distichum* (J, I, In)[a]
Echinochloa colonum (K, I, In, J)	*Paspalum vaginatum* (V)
Echinochloa crus-galli (J, V, I, In)	*Pentapetes phoenica* (V)
Elatine triandra (J)	*Pistia stratiotes* (V, I, P)
Eleocharis acicularis (J)	*Potamogeton distinctus* (J)
Eleocharis acutangula (V)	*Rotala indica* (J, I, In)
Eleocharis dulcis (V, Ch)	*Rhynchospora corymbosa* (V)
Eleocharis kuroguwai (J)	*Sacciolepis* spp. (V)
Fimbristylis miliacea (K, J, V, I, In)	*Sagittaria pygmaea* (J)[a]
Hydrilla verticillata (V, I, In, T)	*Sagittaria trifolia* (J, I)[a]
Ipomoea aquatica (V, I, In, T)	*Salvinia cucullata* (V, I, B, In)
Isachne globosa (V, I, In)	*Scirpus grossus* (V, T)
Jussiaea suffruticosa (V, I, In)	*Scirpus hotarui* (J)[a]
Lagarosiphon spp. (V)	*Scirpus mucronatus* (V, I)
Leersia hexandra (K, J)	*Sphenoclea zeylanica* (V, I)
Leptochloa chinensis (J, V)	*Utricularia flexuosa* (V, I)

[a] More than 42 per cent of cropped rice area in Japan was infested by these weeds in 1971.
I = India, J = Japan, K = Khmer (Cambodia), T = Thailand, In = Indonesia, P = Philippines, M = Malaysia, L = Laos, S = Sri Lanka, B = Bangladesh, V = Vietnam, Ch = China

hyacinth (*Eichhornia crassipes*) is today unchallenged as the worst aquatic weed throughout the tropics and subtropics. Much more has been published about water hyacinth than any other aquatic weed (Thyagarajan 1984; Gopal and Sharma 1981; Gopal 1987). The second ranking free-floating weed, *Salvinia*, is represented in Asia by three species (Fig. 16.2). *Salvinia molesta* (earlier identified as *Salvinia auriculata*) is an exotic which was introduced possibly in the early 1930s in tropical Asia through botanical gardens in Colombo, Sri Lanka (Williams 1956); Bogor, Indonesia (Nguyen-van-Vuong 1973*b*); and Calcutta, India (Kammathy 1967). It spread widely in Sri Lanka and possibly from there into Kerala, Andhra Pradesh, and Tamil Nadu in India. In the north it spread from Calcutta to the whole of West Bengal, Orissa (Philipose *et al.* 1970), and Bangladesh; from Java to the rest of Indonesia (Fig. 16.3), and neighbouring countries of the Indochinese Peninsula (Nguyen-van-Vuong and Sumartono 1979). It has also started colonizing waters in more northern latitudes (e.g. near Delhi). *Salvinia cucullata*, a native of tropical Asia (perhaps eastern India or Burma) has spread to the whole of south-east Asia due to introduction by man (Nguyen-van-Vuong

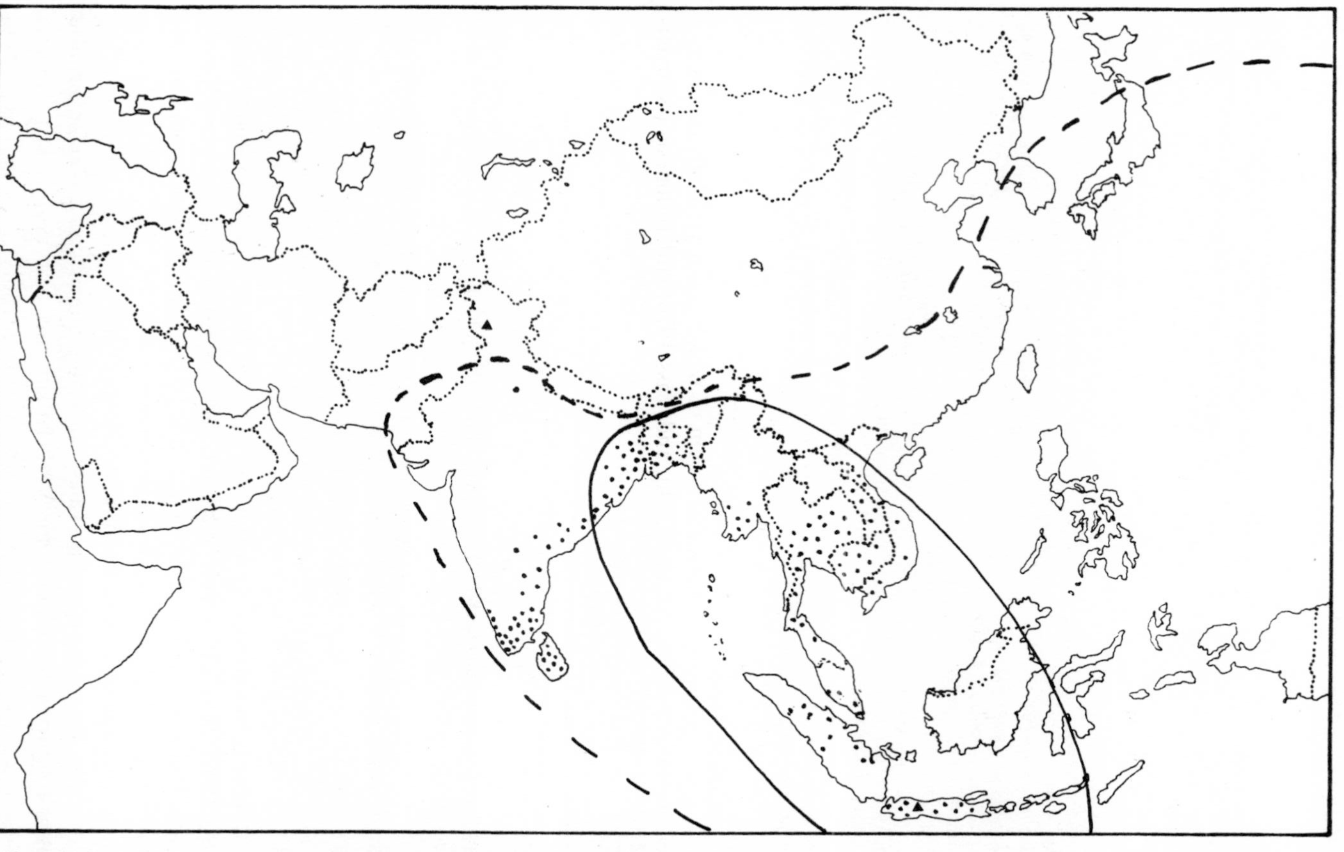

FIG. 16.2. Distribution of water hyacinth (broken line) and *Salvinia* species. Continuous line—*S. cucullata*: dots—*S. molesta*; and solid triangles—*S. natans*.

FIG. 16.3. A dense mat of *Salvinia molesta* in the Curug lake in West Java, Indonesia. (Photo: A. H. Pieterse.)

1973). *Salvinia natans*, the place of origin of which is not yet clear, occurs as a weed in Kashmir lakes (Kaul, Zutshi, and Vass 1976) and Indonesia (Nguyen-van-Vuong 1973*b*).

Among other free-floating weeds, water lettuce (*Pistia stratiotes*) is most important. It ranks second to water hyacinth in Thailand (Chomchalow and Pongapangan 1976) and is a major cause of problems in irrigation channels in rice-growing areas of the Philippines (A. B. Higgins, ICI Agrochemicals, pers. comm.). Several lemnids, particularly *Spirodela polyrrhiza* and species of *Lemna* are considered important in several countries as they may form a thick, complete cover over the surface of water bodies.

Submerged weeds

Among the submerged weeds, *Hydrilla verticillata* is a particular problem (Dhahiyat *et al.* 1982; Probatova and Buch 1981; Pieterse 1981), together with *Vallisneria spiralis, Potamogeton pectinatus, Ottelia alismoides, Ceratophyllum demersum,* and *C. muricatum*. Species of *Utricularia, Lagarosiphon,* and *Zannichellia,* and the fern *Ceratopteris thalictroides* are often found in weedy proportions. *Elodea nuttalli* has been a major weed in Lake Biwa in Japan since 1961 but has been partly replaced by another submerged weed, *Egeria densa,* since 1971 (Ikusima 1983, 1984). Species of *Najas, Myriophyllum, Potamogeton,* and *Ceratophyllum* are also widespread weeds in the Soviet Union and Middle East Asia.

Rooted floating-leaved weeds

Several members of the Nymphaeaceae, such as *Nelumbo nucifera, Nymphaea nouchali,* plus *Nymphoides cristata,* and species of *Marsilea* are considered as weeds since their leaves can cover the water surface completely. Many other species creep over the water surface while rooted in shallow water on the shores of the water body. These include species of *Ipomoea* (*I. aquatica, I. crassicaulis*) and *Jussiaea repens.*

Emergent weeds

The rooted emergent weeds dominate shallow lakes, fish ponds, margins of reservoirs, and irrigation channels (Fig. 16.4). They include a large variety of plants of which the dominants are species of *Typha, Monochoria, Alternanthera, Cyperus, Eleocharis, Scirpus, Juncus,* and a number of grasses. *Typha angustata* (Mehta 1979) is widespread throughout Asia including southern USSR. *Typha elephantina* occurs throughout northern India and extends into the USSR (Saha 1968). *Typha javanica* is a common weed in Sri Lanka (Kotalawala 1976). *Monochoria vaginalis* occurs widely in northern India. In eastern India and south-east Asia, China, Japan, and Korea, *Monochoria hastata* is a common weed. Other dominant taxa include *Eleocharis plantaginea, Scirpus grossus, Hygrophila auriculata,* wild rice (*Oryza rufipogon* and *Hygroryza* spp.), *Paspalum distichum,*

FIG. 16.4. *Typha angustata* and *Phragmites karka* (flowering on the left) occupying shallow water margins of a water body. (Photo: B. Gopal.)

Panicum repens (Siregar and Soemarwoto 1976), *Isachne globosa*, and *Limnocharis flava* (Kotalawala 1976). *Alternanthera sessilis* is a native of Asia commonly occurring as a weed. However, in recent years, *Alternanthera philoxeroides* has been introduced and is spreading fast. In the countries of the Middle East *Phragmites australis* is the most important aquatic weed species (Robson 1976*b*).

Weeds of rice paddy fields

Paddy fields are very important habitats because aquatic weeds interfere directly with the crop and cause serious problems. There are numerous reports from each country listing rice paddy field weeds and the problems caused by them (Suvatabandhu 1950; Hoq 1955; Chakravarty 1957; Datta and Maiti 1963; Numata and Shinozaki 1967; Shankar 1966; Noda 1969, 1979; Heckman 1980; Hanafiah, Sisomba, and Sathal 1979; Sathal 1979; Nguyen-van-Vuong 1979; Sisounthone and Sisombat 1979; Moody 1981, 1985; Majid 1986). Among the most notorious weeds are the free-floating *Salvinia* species widespread in Kerala state (south India), Bangladesh, and south-east Asia. Many aquatic and marsh species, e.g. *Fimbristylis, Echinochloa colonum, Echinochloa crus-galli, Sphaeranthus indicus, Eclipta alba, Alternanthera sessilis, Paspalum distichum, Panicum repens, Leersia hexandra,* and a number of Cyperaceae are reported (Table 16.1).

Relative importance of different weed species

Depending upon the extent of distribution and nature of the problem caused by a weed in a water body, it may be ranked differently in different regions. The relative importance of the major weeds was evaluated through a questionnaire survey conducted in India in 1973 and the weeds were ranked in decreasing order of their importance on the basis of the number of districts in which they were present and causing serious problems. A similar survey was made for Indonesia and the south-east Asian region. Table 16.2 shows that water hyacinth unequivocally ranks first among them but considerably different importance is attached to other weeds in these surveys.

PROBLEMS CAUSED BY AQUATIC WEEDS

The problems caused by aquatic weeds in Asia are not much different from those in other parts of the world. Two points, however, need to be made. Firstly, rice is a major crop in the region and aquatic weeds are responsible for large losses in crop yield. Secondly, most of the water bodies are put to multiple use in the rural environment. Very often the same water body is used for drinking water, irrigation, recreation, and other domestic purposes. Excessive growth of aquatic plants can thus create a wide range of problems.

Table 16.2. Noxious aquatic weeds of south and south-east Asia and their ranking in different countries

Species	India (a)	Indonesia (b)	S.E. Asia (c)
Eichhornia crassipes	1	1	1
Salvinia spp.[a]	10	2[a]	2[a]
Salvinia cucullata			10
Pistia stratiotes	9		3
Lemnaceae	6		
Hydrilla verticillata	4	3	4
Najas indica		5	
Ceratophyllum demersum		6	
Vallisneria spp.	7		
Potamogeton spp.	8		
Potamogeton malaianus		9	
Nymphaea stellata	2		
Nelumbo nucifera	3	7	5
Typha spp.[b]	5		8[b]
Scirpus grossus		4	6
Panicum repens		8	7
Monochoria vaginalis			9
Mimosa pigra		10	

[a] *Salvinia molesta*.
[b] *Typha angustata* (misidentified as *T. angustifolia*).
(a) Varshney and Singh 1976, (b) Soerjani 1980, (c) Soerjani *et al.* 1976.

Aquatic macrophytes in reasonable quantities have several important functions in aquatic ecosystems. The submerged plants oxygenate the water, serve as food for many animals, and provide suitable sites for shelter and spawning of aquatic fauna. They may also reduce nutrient loads and, as a consequence, minimize development of algal blooms. Excessive growth rates of aquatic macrophytes can, however, lead to accumulation of organic matter and subsequent depletion of the oxygen content of the water, causing distress or death of fish and other aquatic animals. Anaerobic conditions render the water undrinkable and sometimes lead to the production of hydrogen sulphide, which may seriously affect generating equipment in hydroelectric power installations. Among the many organisms supported by aquatic vegetation are the vectors of several waterborne diseases. Excessive weed growth can also result in silting of rivers, canals, lakes, and reservoirs, reduction in water flow rate, and, consequently, flooding and erosion of the shores of the water body. Luxuriant growth of emergents on the shores and of submerged or floating vegetation in the open water may also hinder recreation. Masses of dead plants reduce the aesthetic value of the water body. All these and many more problems associated with the utilization of freshwater resources are caused by aquatic weeds in Asian countries. Some of them are examined in detail below.

Interference with irrigation

As mentioned earlier, agriculture is the mainstay of the economy in the countries of south and south-east Asia. It depends on irrigation, for which numerous reservoirs have been constructed and a huge network of irrigation channels has been laid out. In many countries these reservoirs are heavily infested with weeds, sometimes choking the entire water surface (e.g. Nguyen-van-Vuong 1973*a*; Pieterse, Siregar, and Soemarwoto 1975; Junk 1977; Soerjani 1976, 1978). Some examples are given in Table 16.3.

Weeds constitute the biggest problem in the maintenance of irrigation channels for efficient utilization of irrigation water. Submerged or emergent weeds not only impede water flow in irrigation channels (Mehta and Sharma 1976), but more importantly reduce the capacity of reservoirs and tanks to store water, and of channels to carry it, due to increased siltation and organic matter deposition (Haque and Rahman 1976). Besides the loss of a scarce water resource at the point of desired use, overflowing water from blocked water courses causes bank erosion and flooding of adjoining fields or villages. The magnitude of the problem can be judged from the following widespread examples: the Bhakra canal system in northern India (Malhotra 1976); Damodar valley region in eastern India (Kachroo 1959); the Chambal irrigation system in central India (Brezny and Mehta 1970, Reeders *et al.* 1986); creek systems in Japan (Shibayama 1981); the Mekong river basin in Laos and Thailand (Nelson, Gangstad, and Seaman 1970); irrigation and drainage systems in Turkey (Altinayar and Onursal 1982). Huge sums of money are spent on the regular removal of weeds and maintenance of channels.

Emergent weeds, through their extensive food-storing rhizome systems, can create further serious problems. The death and decay of the rhizomes leaves small tunnels through which water seepage causes breaches in the canals. Such tunnels are also created by the rodents which feed on the rhizomes, and the crabs feeding on organisms living among the weeds (Brezny and Mehta 1970; Mehta and Sharma 1976).

Development of sudds/floating islands in lakes

The addition of large amounts of organic matter by rapidly growing aquatic weeds often results in the development of sudds (floating islands). As the dead plants decay slowly, remaining buoyant due to air trapped in their tissues, they also trap large amounts of wind blown dust. The resulting floating islands may gradually become colonized by a variety of emergent and marsh plants (Fig. 16.5). The sudds are quite frequent in most of the more heavily infested major water bodies: e.g. Rawa Pening in Java (Vaas 1951; Polak 1951; Little 1969*b*; Hardjosoewarno *et al.* 1972; Adam 1979) and Lake Loktak in India. The develop-

Table 16.3. Examples of major infestations in lakes and reservoirs in Asia

Country	Lake/Reservoirs	Weed	Reference
India	Kakki reservoir (Kerala)	*Salvinia molesta*	Cook and Gut 1971
	Lake Loktak (Manipur)	Water hyacinth	Yadav and Varshney 1981
	Lake Kitham (U.P.)	Water hyacinth	Personal Observation
	Dal lake (Kashmir)	*Salvinia, Typha*	Kaul, Zutsi and Vass 1976
	Kolleru lake (A.P.)	*Nymphaea, Hydrilla*	Seshavatharam and Venu 1981
Indonesia	Jatiluhur	Water hyacinth	Soerjani 1976
	Rawa Pening	Water hyacinth	Vaas 1951; Achmad 1971
	Curug reservoir	Water hyacinth	Pieterse, Siregar and Soemarwoto 1975
		Water hyacinth, *Salvinia*	Soerjani 1976
Sri Lanka	Maduru Oya reservoir	*Salvinia molesta*	Anon. 1986
Thailand	Lake Songkhla	Water hyacinth, *Salvinia, Pistia,* and others	Anon. 1982
	Bung Borapet and other reservoirs	Water hyacinth	Junk 1977
Japan	Lake Biwa	*Elodea nuttallii* *Elodea canadensis*	Ikusima 1983, 1984

FIG. 16.5. A large lake infested by water hyacinth forming large sudds overgrown with *Phragmites karka* and other grasses. (Photo: B. Gopal.)

ment of natural sudds has been described by Polak (1951) and Trivedy *et al.* (1978). In Kashmir lakes the floating islands are encouraged and aided by man for the cultivation of vegetables (Sahni 1927; Kaul and Zutshi 1966). These sudds and the increased evapo-transpiration from weed mats often result in the rapid filling in of the water body, gradually turning it into a marsh, unfit even for fish culture.

Interference with fisheries

Whilst macrophytes may provide food for fish in some measure, the deoxygenation problem associated with their excessive growth can cause serious interference with fisheries (Gupta 1979; Jhingaran 1982). Fisheries are also indirectly affected by other changes in environmental physico-chemical characteristics, and the density and composition of phyto- and zooplankton in weed infested water bodies (Gopal, Trivedy, and Goel 1984). There are numerous reports of aquatic weed problems in fishery reservoirs and fish ponds (Chokder 1968; Chokder and Begum 1965, Philipose *et al.* 1970; Jhingaran 1982).

Reduction in crop yields of rice

The weeds in rice paddy fields adversely affect seed germination, interfere with the establishment of seedlings, and later compete with the crop for nutrients, thereby reducing crop yields (Kasahara 1961; Noda and Obayashi 1971; Soemartono 1979; Majid and Akhtar Jahan 1984; Assemat, Morishima, and Oka 1981). Some of the weeds (e.g. *Leersia hexandra*) also have an allelopathic influence on the crop (Chou, Lee and Oka 1984).

Interference with navigation

Aquatic weeds in waterways cause serious problems for the movement of boats and steamers in several areas. Soon after water hyacinth had been introduced into India, it started choking streams and making them impenetrable to boats. In many parts of Bangladesh, West Bengal, and north-east India—in the lower reaches of rivers Ganges and Brahmaputra and their tributaries—water hyacinth interferes seriously with navigation. Both water hyacinth and *Salvinia molesta* create similar problems on a large scale in Kerala (S. India) where waterways are the main means of transport for the rural people. Significant transport problems are reported also in many parts of Thailand and Indonesia.

Similar interference with the movement of boats, either for transport of people and materials or even for removal of these weeds, is experienced in numerous lakes and reservoirs in all south Asian countries. The development of sudds only aggravates the problem further.

Problems for public health

The health problems caused by the presence of aquatic weeds in water bodies are well known (see Chapter 6). Water hyacinth, *Salvinia, Pistia,* and several other aquatic plants support populations of mosquitoes such as *Mansonia* and *Anopheles* species (Burton 1960; Chow, Thevajagayam, and Wambeck 1955; Gass *et al.* 1983; Neogy, Kachroo, and Biswas 1957). In the Damodar Valley area in eastern India, *Anopheles* breeding and malaria have been serious problems (Kachroo 1959). The incidence of several water-borne diseases like filariasis and schistosomiasis is greatly increased by the presence of noxious weeds as they provide ideal conditions for the growth of vector mosquitoes and gastropod molluscs (Krishnamoorthi 1976). Recently, Spira *et al.* (1981) have found evidence that cholera epidemics in Bangladesh may be positively related to water hyacinth infestation because *Vibrio cholerae* tends to concentrate around the roots.

General summary

The magnitude of the problems caused by aquatic weeds varies not only with the magnitude of their growth but also according to the importance of the water body and its use. Since agriculture is the most important activity in Asian countries, the impacts of aquatic weeds on crops like rice, and on its associated irrigation systems, are most important. A survey by Soerjani and Pancho (1974) confirms this. They listed the problems caused by aquatic weeds in order of importance as:

(1) retard growth of crops;
(2) interfere with irrigation and drainage systems;
(3) interfere with hydroelectric schemes;
(4) cover the surface of impounded water;
(5) prevent fishing and recreation;
(6) interfere with navigation;
(7) cause water loss through evapo-transpiration; and
(8) impart an unpleasant taste and odour to drinking water.

However, Varshney and Singh (1976) showed that, in India, the greatest concern is caused by weed cover on impounded water surfaces, followed by the weed problems in fisheries. Interference with irrigation schemes, reduction in crop yields, and drinking water pollution rated next in order of decreasing importance. Evapo-transpiration losses of water and disease problems were rated higher than the problems for navigation and recreation.

MANAGEMENT OF AQUATIC WEEDS IN ASIA

Ever since aquatic weeds were recognized as harmful plants, in water bodies used by man, they have been manually removed and destroyed (Fig. 16.6). With the increasing magnitude of the weed problem, manual removal alone did not suffice because only small tools were used. Machines are relatively rarely used in Asian countries. Mechanical removal presents technological and operational problems. There has been little effort to develop indigenous equipment (Velu 1976); most of the imported machines often do not suit the field conditions. Chemical and biological control methods (discussed below) have their own limitations, and have rarely yielded the optimum desired results. Experience in several areas shows that in the Asian context there is no better alternative than a sustained effort of manual removal and strict vigilance. Water hyacinth, for example, has been kept under good control in the Chambal irrigation system near Kota by employing a work force to removing all floating plants along the canals. In the Bharatpur Bird Sanctuary (now a National Park), the menace of water hyacinth was almost completely eliminated for several years by manual removal and regular surveillance. Only as the effort slackened has the infestation reappeared.

FIG. 16.6. Manual removal of water hyacinth along the banks of river Yamuna near Delhi. (Photo: B. Gopal.).

Water level management

The growth of aquatic plants, particularly submerged species, is greatly affected by changes in water level. A sudden drop in the water level results in death or at least reduced growth. Flowering and formation of vegetative propagules (turions or tubers) may be induced. If the water body dries quickly, the reproductive process is often not completed and regeneration of growth on refilling is reduced. Management of weeds by artificially regulating water levels exploits these effects. Submerged weeds have been effectively controlled in irrigation channels (e.g. in the Chambal irrigation system, Bhakra canals and several canal systems in Uttar Pradesh) by periodically stopping the water flow and allowing the bed to dry for a few weeks (Garg 1968; Brezny 1970; Malhotra 1976).

The growth and reproduction of other weeds is also influenced, to different extents, by water level changes. Most emergent plants exhibit a decline in growth in deep water, particularly if submergence continues for long periods. Singh, Pahuja, and Moolani (1976) reported that *Typha angustata* can be controlled by cutting shoots at the flowering stage and submerging the stubble under water for four weeks. Sharma (1978), however, could not confirm this.

The practice of water level manipulation is risky because it may result in an undesirable growth of emergents which are, of course, favoured by shallower water. Field observations, confirmed by laboratory experiments, suggest that seed germination and seedling establishment in *Typha angustata* is facilitated under conditions of waterlogging and/or shallow water (Sharma and Gopal 1979*a*). This weed invaded the internationally famous bird sanctuary at Bharatpur after attempts were made to fill in the impoundments gradually by pumping, following a serious drought.

Chemical control

Many herbicides are used on different scales against different weeds in almost all Asian countries. Philipose (1976) has reviewed the Indian literature on the subject. Chemical control efforts against water hyacinth are summarized in Gopal (1987). In the early days arsenites, copper sulphate and similar inorganic chemicals were tried. Ammonia has been recommended against a variety of submerged weeds (Ramachandran and Ramaprabhu 1976). Herbicides such as 2,4-D, methoxone, diquat, paraquat, aminotriazole, and others, either alone or in combinations, have proved effective in varying measures against submerged, free-floating and/or emergent weeds, both in rice fields and other water bodies (Nakagawa and Miyahara 1967; George 1976; Soerjani *et al.* 1976; Gupta 1979; Table 16.4).

Herbicides do offer a good prospect of controlling aquatic weeds rapidly and at low cost in Asia but, as elsewhere, they also pose some problems. The chemicals must be applied at regular intervals to prevent regeneration and regrowth because they do not offer a permanent solution. Further, the multiple use of water bodies is a constraint to the widespread use of large doses of chemicals. Through irrigation and drinking water, chemicals can affect crops and human beings, if irresponsibly applied. In many areas, the spraying of chemicals is not recommended if crops are cultivated in adjoining areas because the aerial drift of spray droplets can damage the crops as well. In most cases where herbicides have been applied and satisfactory control obtained, no follow up action in the form of maintenance control has been taken. Consequently, the weeds have often regenerated with increased vigour. For example, in Rawa Pening water hyacinth has persisted for several decades despite best efforts to control it using herbicides.

Biological control

Starting with the efforts of the Commonwealth Institute for Biological Control in the West Indies, there have been several projects to identify organisms (insects and phytopathogens) which attack different aquatic plants and apply stress to their growth both in their native and introduced ranges. Long lists of such organisms are now available from many countries, particularly India, Pakistan, and Indonesia, besides those from South America (Ghani 1965; Rao 1970; Sankaran and Krishna 1967; Sankaran and Rao 1972; Bennett 1984; Nag Raj

Table 16.4. Chemicals used on controlling different aquatic weeds in Asian countries

Chemicals	Dosage	Weeds
2,4-D amine, 2,4-D Na, and other related compounds alone or in combination with other chemicals	2–4 kg a.e. ha^{-1} (in one or more applications)	Water hyacinth, *Monochoria* spp., *Ludwigia, Scirpus grossus,* sedges
Diquat and paraquat	0.6–0.8 kg ha^{-1}	Water hyacinth, *Salvinia* spp., *Pistia stratiotes, Hydrilla verticillata, Typha* spp.
Dalapon	5–15 kg ha^{-1}	*Panicum repens, Typha* spp., *Echinochloa* spp., other grasses
Ametryne, cyanatryn, terbutryne	28 g a.i. ha^{-1}	Water hyacinth
Glyphosate	2–4 kg ha^{-1}	Water hyacinth
Amitrole, amitrole-T	0.5–2 kg ha^{-1}	Water hyacinth
6-aminotoluic acid		*Pistia stratiotes*

and Ponnappa 1970*a*; Ponnappa 1977; Mangoendihardjo and Soerjani 1978; Mangoendihardjo *et al.* 1977; Kasno and Soerjani 1979; Syed 1979; Napompeth 1982; Gopal 1987). However, most of these organisms are either not efficient in controlling the weed growth appreciably or they are not host specific so that their biological control potential is nil. Many of the insects attack not only noxious aquatic plants but also crops such as rice, taro, etc. Most studies have been carried on in connection with water hyacinth, *Salvinia* and *Pistia*. The more important biocontrol agents of these weeds are listed in Table 16.5.

There have been several glasshouse and field trials involving *Neochetina* species in India, Thailand, and other countries, and recommendations have been made for field releases. However, no significant successes have yet been achieved. It should be recognized that insects are just one of the components in a control strategy to stress the plants to the level where pathogens may also infect the damaged plants and kill the weed. The limitations imposed by climatic and other natural factors on the survival and naturalization of the insects are not yet fully understood.

Many studies have focused on grass carp (*Ctenopharyngodon idella*) for controlling submerged weeds and water hyacinth, and trials have been carried out in almost all Asian countries (Ahmad 1968; Pheang and Muchsin 1975; Chaudhuri *et al.* 1976; Soewardi 1979; van Zon 1981). Only limited success has been achieved in controlling the growth of submerged weeds with grass carp. The most important problem is the selective feeding behaviour of the carp and its preference for tender-leaved plants. It feeds very little on water hyacinth, *Salvinia,* and emergent weeds. Further problems are presented in rearing the fingerlings and regular re-stocking of the water bodies, as grass carp fail to reproduce in field conditions outside their native range (N. China and Siberia). In several cases, the removal of fish, before they grow to any appreciable size, by local people has been reported.

Utilization

Though the problems caused by aquatic vegetation in water bodies and waterways are very old, only relatively recently have they become serious. Native plants were never considered to be a problem; instead, the aquatic vegetation in ponds and lakes was valued in the past from the aesthetic viewpoint and these plants served various human needs. Where they grew out of proportion, man could keep them under control by manual removal. The aquatic plants provided important food for man. Lotus was sacred in India and the flowers were offered in temples: the spongy petioles served as a vegetable, the seeds were eaten raw or roasted (Fig. 16.7). *Cyperus esculentus* was cultivated for tubers, *Trapa bispinosa* for fruits (Malik 1961), wild rice for grains (Steeves 1952). *Ipomoea aquatica* is used as a vegetable in many parts of south-east Asia (Satpathy 1964, Edie and

Table 16.5. Native and introduced biological control agents of important aquatic weeds used in trials in Asian countries

Weed species	Biocontrol agent	Country	Reference
Water hyacinth	*Neochetina eichhorniae*	Thailand (1979)[a]	Napompeth 1984
		Indonesia (1979)	Kasno, Aziz, and Soerjani 1979
		Sri Lanka[b]	Napompeth 1984
		Burma[b]	Napompeth 1984
		India (1983)	Jayanth 1987
	Neochetina bruchi	Indonesia (1979)	Kasno and Soerjani 1979
		India (1983)	Jayanth 1987
	Gesonula punctifrons	India	Sankaran, Srinath, and Krishna 1966
	Orthogalumna terebrantis	India (1986)	Jayanth 1987
	Sameodes albiguttalis	India	Jayanth 1987
	Acigona infusella	India	Jayanth 1987
	Ctenopharyngodon idella	Most countries	
Salvinia molesta	*Nymphula responsalis*	Indonesia	Subagyo 1975
Hydrilla verticillata	*Parapoynx diminutalis*	Indonesia	Balciunas 1983
	Hydrellia pakistanae	Pakistan	Baloch and Sana-Ullah 1973
	Bagous sp. nr *limosus*, *Bagous* sp. nr *lutulasus*	Pakistan	Baloch, Sana-Ullah, and Ghani 1980
Pistia stratiotes	*Proxenus hennia*	Indonesia, Malaysia	Mangoendihardjo and Nasroh 1976
	Episammia pectinicornis	Thailand	Napompeth 1982
Myriophyllum spicatum	*Bagous geniculatus*, *Bagous vicinus*, *Phytobius* spp.	Pakistan	Habib-ur-Rahman *et al.* 1969

[a] The year in parentheses refers to the year of field release.
[b] Introduced from Thailand, field release not confirmed.

FIG. 16.7. Rhizomes and petiole (foreground) and fruits (right) of lotus (*Nelumbo nucifera*) and the fruits of *Trapa bispinosa* (left) removed from the large pond (in the background) being sold near Delhi. (Photo: B. Gopal.)

FIG. 16.8. A heap of *Typha elephantina* leaves for thatching/mat making. (Photo: B. Gopal.)

Ho 1969, Ochse and van den Brink 1931, National Academy of Sciences 1976). The rhizomes of *Eleocharis dulcis* (matai of China) are eaten widely in China (Hodge 1956). The leaves of *Typha* spp. are widely used for mats, ropes, thatching, and other purposes (Saha 1968) (Fig. 16.8). Several aquatic plants are used as animal feed (Dolberg, Saadullah, and Haque 1981) and also yield important medicines. The problems multiplied with the growing eutrophication of water bodies due to an ever increasing population and with the introduction of exotics like water hyacinth and *Salvinia molesta*.

Now the utilization of aquatic weeds is being increasingly advocated again (e.g. Majid 1986) as a control strategy on two assumptions. Firstly, if a suitable use could be found for the weeds, they would be removed from the water bodies purposefully and, therefore, their growth would be reduced. Second, such utilization would also offset, at least partly, the costs of weed removal and therefore, removal would become attractive. The utilization is still more appealing if the weeds could serve as substitutes for other resources in short supply as feed for cattle, swine and poultry (Shahjahan *et al.* 1981) and for paper and energy. Weeds are being promoted as new resources more suitable for utilization in developing countries (National Academy of Sciences 1976).

In most Asian countries, the use of native plants has always been very common and perhaps nowhere else have the aquatic plants been so intensively used for food, feed, fertilizer, and fibres. The exotic weeds like water hyacinth were also put to use in the very first years of facing the problem. Asians were the first to suggest uses like cattle feed, pig feed, mulch, compost, and energy from water hyacinth (see Gopal 1987 for a detailed review), but the problem continued to grow faster than the use. During recent years numerous studies have been made to explore the possibilities of utilization in a more profitable manner and the literature in the past ten years or so is full of such references, at the expense of control studies. Much of this literature is full of rhetoric, and projects the potential for utilization out of all proportion from the results of small laboratory experiments.

I have argued earlier of the dilemma between utilization of weeds (considering them as alternative resources) and the need for their control (Gopal and Sharma 1979; Gopal 1982, 1984). We must distinguish between two types of utilization practices. One arises out of the needs of economically poor people whose interest is in a continual supply of the resource, who are tempted to use any alternative material, and who have little or no interest in controlling the weed growth. Such utilization continues on a small, unorganized scale everywhere in Asia. The other type of practice is aimed at reducing the growth drastically by one-time exploitation or at a rate greater than the natural growth rate, so as to diminish the problems caused by the weed. This requires large-scale effort, employment of machines and suitable technologies to process huge quantities of biomass which contains up to 90 per cent water. This alternative is economically non-viable and certainly not efficient in terms of energy costs required for harvest, transport, and processing. It is essentially a short-term or periodic exercise which fails to generate commercial interest. Irving and Beshir (1982) have, with reference to water hyacinth in Sudan, very rightly concluded that 'utilization even if shown to be commercially and economically viable is unlikely to reduce the plant to below the status of a noxious weed'.

Integrated management

Integrated management refers to employing several suitable methods together or in succession to achieve the desired goal. Several suggestions have been made for combining such control methods as low doses of herbicides, suitable biological control agents, and mechanical harvest with the utilization of the weed biomass (Soerjani 1977). For example, a chemical spray in small amounts, below the safe limits, may help retard the growth, cause the plants to senesce and make it easy to harvest them. The harvested biomass can be used for biogas generation and the slurry recovered from the digesters used as manure. Biocontrol agents may be used in conjunction with small doses of herbicides for reducing the weed populations which can then be more easily harvested and utilized (see also Chapter 10).

I should add a word of caution here. All too often laboratory results have been exaggerated to paint a rosy picture, but successes in the field are rather small. For example, the trials to manufacture paper from water hyacinth on a pilot plant scale in India and Bangladesh made it clear that it cannot be developed into a small-scale rural technology and at least a medium-scale factory would be necessary to make the project viable. To feed the medium scale factory, the water hyacinth would have to be harvested and transported from distant areas adding to the costs, and making it unsuitable for the rural areas. Likewise, despite great interest in the idea of biogas production, water hyacinth alone cannot be used successfully in biogas digesters but has to be mixed with other organic wastes. Grass carp is often projected as a panacea for all weed problems; rarely are solutions that simple under field conditions.

CONCLUSIONS

The problems caused by aquatic weeds have received attention from all quarters over the past few decades but with the growing need for water storage for irrigation and hydroelectric power, the weed problem has also grown. Attention has now been diverted in Asia from control to utilization, thus giving weeds the status symbol of a resource. Though the problems are realized, the interest has shifted from removing weeds to putting them to use, thereby aggravating the problems further. For example, in Java (Indonesia), many people around Rawa Pening grow rice on the floating islands and use weeds as fodder for cattle, but are these benefits really enough to compensate for losses to fisheries, hydroelectric power generation, and water for irrigation (Adam 1979)? Similar situations exist throughout Asia. In small water bodies, weeds are allowed to flourish because the water body is not considered important, and in the larger water bodies control is not fully achieved. The basic problem is that the small populations surviving in small ponds and pools act as reservoirs for the inoculum. In the absence of long-term planning for weed control with concerted effort, and while we try to find appropriate ways for utilization, aquatic weeds continue to stress both natural and man-made freshwater ecosystems and cause severe damage to the economy of the countries affected. We need to assess more objectively the importance of water and land resources *vis-à-vis* the poor economic returns and the socio-economic problems caused by aquatic weeds.

17

Aquatic weed problems and management in Africa

D. S. Mitchell, A. H. Pieterse, and K. J. Murphy

INTRODUCTION

Aquatic weed problems in freshwaters in Africa are caused by two groups of aquatic macrophytes.

1. Plants native to Africa which grow in nuisance proportions because the environment has been disturbed or altered by man by the construction of artificial canals, drains, and man-made lakes, or by anthropogenic enrichment of the water by plant nutrients.
2. Plants alien to Africa which are able to exploit aquatic habitats and build up large populations because of the absence of environmental controlling factors present in their native environments (Mitchell 1985*a*).

In both cases the plants can only be considered as weeds in terms of man's use of the water or water body, though it may be important to recognize potential weeds especially in the case of alien plants. It is also important to recognize that aquatic weeds also have some benefits, though these are usually outweighed by their deleterious effects.

The problems caused by aquatic weeds in Africa are broadly similar to those caused by aquatic weeds elsewhere in the world. In Africa aquatic weeds:

(1) interfere with water flow in rivers, canals and drains, thereby imperilling irrigation schemes and slowing drainage of water from floodlands;

(2) impede the movement of boats for transport, fishing and recreation;

(3) interfere with various methods of catching fish;

(4) compete with rice in paddy systems;

(5) degrade water quality by adding taints and odours to the water and by decreasing dissolved oxygen content;

(6) alter the flora and fauna of aquatic ecosystems by providing new habitats, removing others and by affecting the light climate in the water;

(7) favour the spread of diseases such as malaria and schistosomiasis by providing habitats for the intermediate vectors of the parasites causing these diseases;

(8) threaten engineering structures such as bridges, weirs, and devices to control and measure water flow, especially when large mobile mats of aquatic vegetation are present;

(9) block pump intakes;

(10) impair the access of stock and wildlife to drinking water;

(11) decrease the useful capacity of reservoirs by occupying useful volume and by increasing water loss through evapo-transpiration; and

(12) entrap sediment and thus cause a progressive decrease in the capacity of reservoirs.

The true cost to human society in Africa of aquatic weed problems is near-impossible to estimate, although some minimum estimation is possible from consideration of the amounts of money spent on controlling individual aquatic weed infestations (El Tigani 1979). This is discussed further. As development proceeds human populations increase and the demand for water resources is becoming steadily greater throughout Africa. It is not surprising that the adverse effects of water weeds, which interfere with human uses of water resources, are also perceived as a steadily increasing problem in Africa.

As in other parts of the world, the literature dealing with ecological studies of aquatic macrophytes in freshwater systems is a valuable source of information on actual or potential aquatic weed problems, for example, in Nigeria (Chachu 1979), South Africa (Musil, Grunow, and Bornman 1973; Howard-Williams 1979), Sudanese Nubia (Ahti, Hämet-Ahti, and Pettersson 1973), Malawi (Howard-Williams and Walker 1974; Proctor 1980), and Uganda (Denny 1973). Particularly useful publications are the major reviews of African wetland vegetation edited by Denny (1985) and Pieterse *et al.* (1987). General information (some now outdated but still of interest) on aquatic weeds in regions, or defined freshwater systems, in Africa is available from Wild (1961) and Robson (1976*b*) for the continent as a whole; Ivens (1967) and Ferguson (1971) for East Africa; Piaget and Schliemann (1973) and Jacot-Guillarmod (1979) for Southern Africa; Khattab and El Gharably (1984, 1986), Simpson (1932), Andrews (1945), Beshir and Gadir (1975), Desougi (1979), Freidel and Beshir (1979), Pieterse (1979), Gay (1958, 1960), and Gay and Berry (1959) for Egypt and Sudan; Okafor (1980, 1982) for Lake Chad, Nigeria; and Pierce and Opoku (1971) for Volta Lake, Ghana.

Further information on potential nuisance genera in Africa can be gleaned from Dandy (1937), El Hadidi (1965) and van der Bliek *et al.* (1982) on *Pota-*

mogeton; Rantzien (1952) on *Najas*; El Hadidi (1968) on *Vallisneria*; Fayed (1985) and Jacot-Guillarmod (1977*a*) on *Myriophyllum*; Imevbore, Odu, and Adebona (1968) on *Salvinia*; and Hall and Okali (1974) on *Pistia*. Recent major reviews of the aquatic weed status of *Eichhornia crassipes* (Pieterse 1978; Gopal 1987) contain information on problems in Africa caused by this species. Pieterse (1978) reported that *E. crassipes* was a cause of problems in fourteen mainland African countries, and was also established in the Indian Ocean islands of Zanzibar, Mauritius, and Réunion.

FRESHWATER SYSTEMS AFFECTED BY AQUATIC WEED GROWTH IN AFRICA

Rivers

Most African rivers are liable to annual floods because of the seasonal nature of rainfall over much of the continent. For most of the year, therefore, these systems provide a hostile environment for plant growth, and aquatic vegetation is usually sparse. Sand bank rivers, typical of this type, do not usually have aquatic weed problems. Some macrophytes may occur in pools along such rivers but seldom constitute a weed problem and indeed enhance a limited biological productivity.

The weed problems that do occur are generated in the non-flood periods, but may only be manifested as problems during the flood season. Thus *Eichhornia crassipes* populations which build up rapidly in the quiet backwaters and swampy areas of the Zaire (Congo) River are loosened by rising water levels and flushed out onto the main river by increasing currents. Lebrun (1959) estimated that about 150 tonnes of the plant passed through Kinshasha during the floods of 1958, threatening engineering structures and making navigation hazardous. The situation in the Nile is similar, though here the population of the plant builds up in the immense swamp regions of the Sudd in the Upper Nile (Rzoska 1974, 1976*a, b*; Denny 1984; Freidel 1978).

Eichhornia crassipes has been reported to infest a number of other rivers in Africa, such as the Pangani River, Tanzania, Incomati River, Moçambique, Sabi and Makabusi Rivers, Zimbabwe, Kafue River, Zambia, and Swartkops and Vaal Rivers in South Africa (Wild 1961; Scott, Ashton, and Steyn 1979; Jacot-Guillarmod 1979). While occasionally a nuisance, it does not appear to be a major problem in such rivers unless these flow into a man-made lake.

The infestation of the Chobe–Linyanti system by *Salvinia molesta* is also promoted by flood action, but in a different way. The Chobe is a small river draining a large floodplain adjacent to the Zambezi above the Victoria Falls. When the Zambezi floods, the Chobe River reverses its flow and the weed is carried upstream. Between floods it completely covers the river, making navigation impossible.

By contrast, the infestation of the Bree River near Cape Town, South Africa, by *Myriophyllum aquaticum* (Jacot-Guillarmod 1979) is exacerbated more by pollution of the river than by floods. For the most part this river is slow-flowing through clear sandy beds. *Myriophyllum aquaticum* has infested those reaches of the river where there is nutrient rich inflow. This is a rooted emergent plant that has the capacity to grow out from the bank over the water. Rafts of floating material eventually break free and are carried down the river, where they interfere with fishing and other activities, as well as fouling otherwise clear reaches.

Canals, drains, and irrigation channels

In Africa, construction of canals and drains is usually related to the movement of water to and from irrigation areas (van Aart 1985). In Egypt, for example, there are approximately 44 000 km of irrigation canals (van der Bliek *et al.* 1982; Pieterse and van Zon 1983). However, canals may also be constructed to provide a passage for boats and/or to redistribute water from one region to another (e.g. the Jonglei Canal, which was intended to by-pass the Sudd in the Sudan, for both purposes). Similarly, channels may also be constructed to drain water from agricultural areas liable to flooding (usually these are 'reclaimed' from wetlands by drainage to begin with), or to drain water from a swamp for use in industry, as in Botswana.

The most serious problems occur in the smaller canals of irrigation systems, especially if the water is clear, so that there is sufficient light penetration for the growth of submerged plants, and there are periods of lower flow, so that plant populations may establish themselves. Problems in Egyptian irrigation canals have increased since 1975, when much of the silt load previously carried in the water began to be deposited instead in Lake Nasser (El Gharably, Khattab, and Dubbers 1982; El Gharably *et al.* 1978; Pieterse 1979), and the light climate for submerged plants was consequently improved. Irrigation supply canals are generally kept full of water and thus provide a stable environment for colonization by submerged aquatic plants (the most troublesome in Egypt, for example, being *Potamogeton pectinatus*, *P. crispus*, *P. nodosus*, and *Ceratophyllum demersum*). Filamentous algae, such as *Cladophora*, can become associated with these weed beds and increase the resistance to the flow of water. Emergent plants, such as *Typha domingensis*, *Echinochloa stagnina*, and *Phragmites australis* can colonize the canal banks but will only slowly invade the deeper water in neglected supply canals (Figs 17.1 and 17.2). Free-floating weeds, such as *Eichhornia crassipes*, are less commonly a problem while flows are maintained, unless there is an invasion of a large number of plants into the system from another source. Terry (1981)

FIG. 17.1. An irrigation canal in the Nile Delta in Egypt which is overgrown by *Echinochloa stagnina*. (Photo: A. H. Pieterse.)

FIG. 17.2. A dense vegetation of *Phragmites australis* in a drainage canal in the Nile Delta in Egypt. (Photo: A. H. Pieterse.)

mentioned the problems of *Pistia stratiotes* and *Nymphaea* spp. in irrigation supply channels of ricefields in the Gambia.

In contrast with irrigation supply channels flow in the drainage channels of irrigation systems is generally slower, and the system is likely to be stagnant more frequently. Also water is more shallow, and nutrient and salt levels are higher in drains than in supply channels. Such an environment is more favourable for emergent weeds such as *Phragmites australis, Echinochloa* spp., *Paspalum* spp., *Panicum* spp., *Polygonum* spp., *Typha* spp., and *Cyperus* spp. Submerged weeds occur but may be shaded out, while free-floating plants are generally not swept away, because of the absence of strong flows, and so can accumulate in nuisance proportions.

Natural lakes and ponds

The large Rift Valley lakes of Africa: Lakes Malawi, Tanganyika, Kivu, Turkana, and Edward, have only sparse development of aquatic macrophyte vegetation and no weed problems. However, several African lakes are associated with large swamps, which interface with the lake in a number of ways. In most cases there is an interchange of water between the systems (Carter 1955; Gaudet 1975, 1976*a*, 1977, 1979*a*; Beadle 1981) and, in areas sheltered from wave action where water is clear, there is the development of classical hydroseral zonation from the emergent plants of the swamp to submerged and floating plants in the lake (see Chapter 2). Large lakes associated with swamps in this way include Lakes Victoria, Chad, Chilwa, Bangweulu, Kyoga, and Naivasha. The lake : swamp interface is important for the biological productivity of the lake (Beadle 1981). Although vegetation may interfere with navigation, and areas of papyrus may break free and drift around as floating islands, the plants are not usually considered as weeds.

Under certain circumstances, however, there may be a marked change in the extent of the aquatic vegetation, which interferes with well-established patterns of human exploitation of the lake. For example, the excessively low levels of Lake Chad following the Sahelian drought of 1972/73 provided conditions for the establishment of extensive areas of aquatic vegetation and the extension of the fringing swamps. Okafor (1980) estimated that 2000 km^2 of the lake were then infested with aquatic vegetation, in the period since the drought, during which the lake varied from 6000 to 25000 km^2 in area. Okafor (1982) reported on the problem in the Nigerian section of the lake where weeds drastically interfere with transport and fishing, in one of the potentially most productive inland fisheries in Africa. Similar problems obviously occur throughout the lake (Beadle 1981). Journeys by boat which previously took one hour could take over four hours. Fishing was directly impeded and the delay in transporting fish from where they are caught to the points of export resulted in a high proportion of spoilage.

The main plants involved are the emergent swamp species, *Cyperus papyrus, Echinochloa pyramidalis, Nymphaea guineensis, Phragmites australis, Typha domingensis*, and *Vossia cuspidata*. The floating plant, *Pistia stratiotes*, has also increased in extent and is particularly troublesome on two intake channels for irrigation schemes: the 15 km channel at Baga feeding the 20 000 ha Baga Polder Project; and the 30 km channel at Kirinowa feeding the 67 000 ha South Chad Irrigation Project. *Pistia* completely covered the channels and impeded flow (Okafor 1982). Beadle (1981) also noted that even at times of more 'normal' lake levels during the 1960s submerged vegetation, consisting mainly of *Potamogeton schweinfurthii, Vallisneria spiralis*, and *Ceratophyllum demersum*, blocked the passage between many of the swamp islands that are a feature of the lake. There is also extensive development of aquatic and marsh vegetation on a flat plain along the southern coast east of the Chari River, depending on the extent of inundation which, in turn, depends on the season.

The plants causing these problems are all native to Africa. They are exploiting opportunities presented by naturally occurring environmental changes. They can only be assessed as weeds because man's exploitation of the resource is less adaptable than the plants, in that a decrease in utilization or in fisheries production with the consequent decrease in income and change of lifestyle, is unacceptable to a significant number of people. Maembe (1981) reported that fish production in Lake Chad used to support about 10 000 fishermen whose way of life and livelihood is now threatened.

Natural lakes can also be invaded by alien plants as exemplified by the infestation of *Salvinia molesta* on Lake Naivasha, Kenya (Gaudet 1976*b*). The plant was first reported to be present in 1963 and initially interfered with recreational boating. The plant was controlled with herbicides and was not further reported until 1969, when further control measures had to be undertaken when the weed was found growing inshore and on the lakeward side of the papyrus stands. Further attempts at control have not succeeded, and *S. molesta* is now a permanent constituent of the lake vegetation. Similar problems have occurred in the much smaller and more remote Lake Liambezi, part of the Chobe–Linyanti system (Edwards and Thomas 1977).

Among the characteristic standing water bodies of the savanna and grassland regions of Africa are the shallow depressions (pans or dambos) which are filled with water in the rainy season. They provide ideal growth conditions for aquatic vegetation, which is not usually regarded as a problem. However, the vegetation can become so dense that it may interfere with the access of stock and wildlife to the water.

Man-made lakes and reservoirs

Much of the development of Africa's resources has involved the construction of dams across rivers, for the conservation of water to ensure supply for agricultural, industrial and domestic purposes and for hydroelectricity. The lakes created by these dams include some of the largest in the world, namely Kariba, Kainji, Nasser, Volta, and Cabora Bassa. Others are smaller, though still big by normal standards, e.g. Verwoerd and Strijdom dams in South Africa, Kyle in Zimbabwe, Kafue in Zambia, Barrage d'Inga in Zaire, and Nyumba ya Mungu in Tanzania. There are many other man-made water bodies in Africa, including a very large number of small catchment dams and other small reservoirs in agricultural areas. All these waters can provide good conditions for plant growth, often leading to nuisance aquatic weed problems.

Although farm dams are frequently relatively shallow, which encourages plant growth, they may also experience regular fluctuations in water level, which tends to be deleterious to rooted aquatic plant growth. Only if water levels are reasonably stable, at least for most of the growing period, can submerged and emergent plants establish in sufficient numbers to constitute a weed problem. However, if floating plants are present, especially aliens such as *Eichhornia crassipes* and *Salvinia molesta*, considerable difficulties may occur. The main problems are interference with stock access for drinking, tainting of water, occupation of useful volume, evapo-transpirative loss, and fouling of pump intakes.

The growth of aquatic plants in man-made lakes has been described in general by Little (1969*a*) and Mitchell (1973). Gaudet (1976*b*) reviewed the topic of aquatic weeds in African man-made lakes. Other authors have reported on the vegetation of specific lakes and their catchment areas, such as the Volta (Hall *et al.* 1969; Hall and Okali 1974), Kainji (Cook 1968; Chachu 1979), Cabora Bassa (Bond and Roberts 1978), the Aswan Reservoir (Pieterse 1977*a*, *b*), and Lake Nasser (El Hadidi 1976; Entz 1976, 1980; Ahti, Hämet-Ahti, and Petterson 1973).

Unpublished survey data from Aswan Faculty of Science (I. Springuel, pers. comm.) suggest that aquatic weed problems are increasing rapidly in Lake Nasser, *Najas, Potamogeton*, and *Zannichellia* being the dominant genera. Fluctuating water levels and increasing siltation may be one factor in this increase. In man-made lakes aquatic vegetation has generally established sparsely, though a succession of species may occur as the lake matures (Gaudet 1979*b*). However, the growth of *Salvinia molesta* on the first of the large lakes built in Africa, Lake Kariba, was a spectacular example of the invasion of a man-made lake by an aquatic weed (Schelpe 1961; Boughey 1963; Mitchell 1969, 1970; Mitchell and Rose 1979; Marshall and Junor 1981). Increasing rapidly in the newly-filled lake, from the small areas first reported in 1959 to over 1 000 km^2 in 1962, the plant quickly became a severe cause of problems in Lake Kariba interfering with navigation, recreation, and fishing (though the fish fauna are favoured by its presence: Mitchell 1976). These deleterious effects are exacerbated when floating mats are colonized by emergent aquatic plants such as *Ludwigia* spp. and *Scirpus cubensis* (Boughey 1963). It was also feared that the hydroelectric output of Lake Kariba would be adversely affected and concern was expressed that a similar explosive growth of aquatic weeds might occur in other man-made lakes then being planned for Africa.

Several factors contributed to the early explosive growth of *Salvinia* on Lake Kariba. The most important were the presence of a large proportion of calm water sheltered from strong winds, waves, and currents; the plant's capacity to rise with the rising waters of the filling lake; the continuing supply of plant nutrients made available from the decay of plants and animals drowned by the rising waters of the lake; the absence of a complex aquatic flora and, therefore, of competition from established plants; and the absence of animals which grazed on the weeds themselves. However, as the lake increased in size to reach full capacity in 1962, wave action increased in extent and in frequency, the proportion of sheltered areas suitable for the establishment of stable weed populations decreased, and the first marked lake drawdowns left large areas of *Salvinia* stranded. Over the next ten years there was a progressive decline in suitable habitats with the increasing break-up of partially submerged trees, which had previously anchored weed mats. Increasing competition from a complex of other aquatic plants which colonized the lake during this period and a decrease in the availability of plant nutrients (tending toward increasing oligotrophy) also had an adverse effect on the weed. During this period the areas of weed on the southern bank of the lake fluctuated at around 400 km^2 (areas could not be measured on the northern bank but the total area of weed in the lake was estimated at about 750 km^2). In 1970 the biological control agent, *Paulinia acuminata*, a South American Acridid grasshopper, was introduced to the lake. Over the following two years populations of this insect increased, and in 1973 the area of weed on the southern bank had decreased to 77 km^2. It has remained at about this level, or below, ever since (Mitchell and Rose 1979). It is considered that plants took longer to recover from the destructive action of seasonal floods in river inlets (a favoured habitat) and other agents, such as wave action during a storm, because of the decrease in the availability of plant nutrients and the presence of *Paulinia acuminata* which tends to feed particularly on young plant tissue (Mitchell and Rose 1979; Marshall and Junor 1981).

The invasion of Lake Kariba provides a spectacular and reasonably well documented example of an aquatic weed problem in an African man-made lake, even though the infested area eventually declined. It should be stressed that the rise and decline in the weed was due to particular factors present in Lake Kariba and to changes in these factors as the lake matured. The same events may not occur in other lakes and, particularly if nutrients are being supplied to a lake, populations of floating and submerged aquatic plants may continue to grow. For example, Bon Accord Dam near Pretoria, South Africa, which receives treated sewage effluent has been covered by *Eichhornia crassipes* on several occasions, despite chemical control measures that have reduced the cover to a small percentage of the whole lake each time they are applied. *Eichhornia crassipes* has also required vigorous chemical control measures to bring it under control on Lake McIlwaine, Zimbabwe. Here it was necessary to sustain a control programme for several years to eliminate germinating seedlings which otherwise would have re-infested the lake. Furthermore, as plants may occasionally enter the lake from Makabusi River, a relatively small river flowing into its headwaters, it is necessary to maintain controls to remove any plants that are seen. Build-up of an excessive population of the weed occurred when these measures were suspended for several years (Jarvis, Mitchell, and Thornton 1982).

Hartebeespoort Dam near Pretoria, South Africa is another significant impoundment that has been severely infested by an invasion of *Eichhornia crassipes* (Scott, Ashton, and Steyn 1979). The plant was first reported on the lake in 1959, and for a number of years it was successfully controlled by mechanical means. However, the dam wall was raised in 1971, and when the lake filled to its new capacity for the first time following heavy rains in 1974/75, the weed invaded areas where boat access was difficult. Here it proliferated rapidly (it was recorded to double in 30 days in September, 11.7 days in November, and 17 days in December). This resulted in the development of large populations that eventually covered 60 per cent of the lake, severely interfering with angling and boating. The decision was taken to spray the weed with terbutryne, which was licensed for use in aquatic environments in South Africa. It is noteworthy that the presence of *Eichhornia crassipes* had reduced the numbers of bacteria, algae, and zooplankton in the water. This could be regarded as an improvement in water quality, but any benefit was more than offset by the interference with recreation, and possible accelerated loss of water by evapo-transpiration. Following treatment there was a reduction in oxygen concentration to between 2 and 3 mg l^{-1}, which probably reduced zooplankton numbers, but did not adversely affect the fish fauna.

Even though large quantities of plants were killed by the application of the herbicide over three to four months, extensive follow-up treatments with diquat were necessary to eliminate the rapidly-proliferating survivor plants.

The fears expressed during the early stages of the weed infestation in Lake Kariba have not materialized there, nor on any other major man-made lake in Africa. However, these are no grounds for complacency, as the potential for serious aquatic weed problems to occur continues to exist, particularly from the surface floating plants, *Salvinia molesta* and *Eichhornia crassipes*. For this reason it is important that pre-impoundment weed surveys be undertaken during the planning stages for major new reservoirs and that frequent surveys of lake basins be undertaken in existing lakes. These surveys should have the objective of looking for new growths of aquatic vegetation, particularly of aliens, which, potentially, could pose problems in the future. Due to the rapid vegetative growth of the plants a lot of time and money can be saved if the problems are tackled at an early stage and early warning of the presence of these plants is very important.

Wetlands and floodplains

At the outset of this section it is important to distinguish between the plants in a wetland, which may interfere with the utilization of that system by human society, and the vegetation characterizing the wetland itself, which may be regarded as a potential threat to a human enterprise nearby. Also the wetland, as well as the vegetation in it, may be seen as undesirable in comparison with an alternative use for the area that is seen as more productive, such as hydroelectric power generation or drainage for agricultural land (Chabwela and Siwela 1986; Drijver and Marchand 1985; Scheppe 1986). The latter situation, though of critical importance to the future existence and management of wetlands in Africa, is a separate issue in the context of this chapter. However, the wetland itself can markedly interfere with the utilization of water resources and inhibit the development of whole regions on the continent (Pieterse *et al.* 1987). Perhaps the outstanding example of this is the Sudd of the Upper Nile. There is no question that this vast area of wetland has been an effective barrier to transport, travel, and trade between communities on either side of it for many centuries. This situation underlies proposals to remove part of the system, or to establish a by-pass canal around it such as the Jonglei Canal (El Sammani 1984). In such cases the system rather than the plants is regarded as a nuisance, and the designation of individual plants as weeds is inappropriate.

The situation would be different, however, if islands of vegetation break free and interfere with human use of a water body away from the wetland. In some ways such plants could be regarded as weeds, but as control of the situation would require massive management of the wetland from which they originate with all the consequent environment deterioration that would occur, it is more sensible to adjust to the situation rather than attempt to prevent it.

There are relatively few instances of weeds in natural wetlands in Africa. The presence of *Eichhornia crassipes* in the Sudd, *Mimosa pigra* in the Kafue Flats, Zambia (Pieterse *et al.* 1987) and the growth of submerged vegetation which interferes with the passage of boats between islands of emergent vegetation in Lake Chad, provide appropriate examples.

However, paddy rice constitutes a managed wetland with a well defined purpose. Any plants, other than rice, are clearly weeds. Most of these are emergent aquatic plants of which the most common infesting rice fields in Africa are grasses (e.g. *Echinochloa* spp., *Paspalum* spp., and *Panicum repens*) and sedges (*Cyperus* spp.).

APPROACHES TO AQUATIC WEED CONTROL IN AFRICA

Manual and mechanical control

Weed control using simple manual cutting tools is common in Africa (Druijff 1973, 1979). It is often slow and brings the operators into direct contact with water all too often infested with schistosoma-carrying snails (Vercruysse 1985; see also Chapter 6). On the other hand, in small canals the use of hand tools such as a chain scythe, can be effective and relatively cheap. Most mechanical devices for controlling aquatic vegetation are expensive and require the use of a trained operator. They are, therefore, usually only feasible for relatively large state-aided or managed projects such as irrigation systems. Maintenance of complex equipment in remote situations also provides a problem. Both cutting and dredging equipment for use in irrigation systems have been employed in Africa (Druijff 1979; El Sayed, Tolba, and Druijff 1978; Lubke, Reavell, and Dye 1984). Most of this equipment consists of heavy machinery that is very effective when operated from the bank of a canal or drain. Such equipment includes dragline excavators, hydraulic dredgers, and back-hoes. Lighter machines which can be operated from boats have also been developed. Most of these are fitted with various forms of cutting blades that can be very effective in maintaining part of a lake free of weed to allow access of boats to open water and for other purposes. They can also be used effectively in channel systems. Burning is sometimes used as a support treatment, for example, against water hyacinth in the Sudan (Mohamed and Bebawi 1973*a, b*).

There are a few papers giving the costs of manual and mechanical aquatic weed control programmes in Africa. Bruwer (1979) gave the total cost of controlling *E. crassipes* on Hartebeesport Dam, South Africa, using these techniques, at R600 000–1 000 000 between 1959 and 1977. El Tigani (1979) estimated the annual cost of mechanical weed control in the main (>2 m wide) irrigation channels of Egypt at US $232 per km, with manual clearance costs for smaller channels at US $52 per km. On this basis, given the figures for total extent of the Egyptain irrigation channel network in El Tigani (1979), the annual cost of aquatic weed control in the network would be approximately US $2 million, at 1979 prices.

Chemical control

Some indication of the scale and costs of chemical control of aquatic weed problems in Africa can be gained from the limited literature on use of herbicides in African waters. Koch *et al.* (1978) described attempts to control *E. crassipes* in the Sudan using herbicides, while Khogali and El Moghraby (1979) detailed some of the impacts of 2,4–D on the biota of the White Nile.

Approximately US $1.5 million was spent annually controlling *Eichhornia crassipes* in the Nile in the early 1960s (Holm, Weldon, and Blackburn 1969) whilst the figure for 1974–1979 was about 1 million Sudanese pounds (El Tigani, 1979). About US $1 million (50 million Belgian francs) were spent on attempts to control the same plant in Zaire (Congo) River in 1956/57 (Lebrun 1959). In both these cases, herbicides were used together with limited manual and mechanical control. Although considerable quantities of weed were destroyed successful control was achieved in neither case. In 1977 *Eichhornia crassipes* on Hartebeesport Dam, South Africa, was successfully controlled with herbicides at a cost of R220 000, approximately one-third to one-fifth of the previous cost of physical control measures. The cost in decreased use of recreational facilities, and reduced land values, as a consequence of the weed infestation was estimated to be of the order of R2 000 000 over about 20 years (Bruwer 1979).

Chemical control of *Salvinia* on Lake Kariba was investigated: field trials with paraquat and sodium arsenite confirmed that these herbicides could be used at a cost of US $11–22 per ha and US $2–4.5 per ha respectively (Hattingh 1962, 1963). However, if the whole infestation was to be controlled by even the least expensive method, the cost would have been US $350 000 and this was not considered feasible, especially as the area of the plant then appeared to be declining. *Salvinia molesta* in the Chobe River system, Botswana, was sprayed with paraquat and glyphosate at a cost of R200 000 in a pilot control project in 1974 and 1975 (Edwards and Thomas 1977). Again control was only partially successful.

In Egypt the very poisonous herbicide acrolein is commonly used for the control of submerged weeds in the larger irrigation canals. The costs were in 1980 US $800 per 200 l barrel (Pieterse and van Zon 1983). Approximately two barrels are needed to treat a distance of 6 km, and treatments, by injection of acrolein, are carried out 5 to 10 times per year. This implies that the annual costs are US $1 500–2500 per km.

These figures can only give an indication of the cost of water weeds in Africa, and in any case only represent a small part of the total adverse effect of water weeds to man in the continent.

Biological control and the utilization approach

Insects (e.g. the grasshopper *Paulinia acuminata*, discussed earlier) have been used to control *Salvinia molesta* in Zimbabwe (Bennett 1974; Mitchell and Rose 1979) and Botswana (Edwards and Thomas 1977; see also Chapter 9b). Grass carp (*Ctenopharyngodon idella*) have been introduced into Africa for weed control in Egyptian irrigation systems (Pieterse and van Zon 1983; Pieterse 1979; El Gharably *et al.* 1978; Dubbers *et al.* 1981; El Gharably, Khattab, and Dubbers

1982; Siemelink *et al.* 1982). Although the results were promising it has proved impossible to prevent overfishing of the grass carp population, in the densely-crowded Nile Delta (see also Chapter 9d).

The harvesting and utilization of *E. crassipes* to generate biogas has been advocated for use in Africa (National Academy of Sciences 1976; Philipp, Koch, and Köser 1983; Philipp, El Tayeb, and Hag Yousif 1979). In addition it has been investigated for its nutritive value as foodstuff for ruminants in the Sudan (Chalmers 1968; Philipp, El Tayeb, and Hag Yousif 1979; Osman, El Hag, and Osman 1975; Philipp, Koch, and Köser 1983). In the Sudan, and Malagasy Republic, the fresh plant is grazed by cattle in the dry season (Davies 1959; Ramarokoto 1968).

The use of *E. crassipes* as a mulch and for fertilizer in the Sudan appears more promising (Kamal and Little 1970). Suitable compost can be made from piles of partially-dried plants mixed with cow dung and covered with earth, though more research is required to ascertain the best method. Also Abdalla and Abdel Hafeez (1969) showed that, when the plant is used as a mulch, it suppresses the growth of one of the major crop weeds in the Sudan, *Cyperus rotundus*, as well as conserving soil moisture.

Salvinia molesta is another weed that has been investigated for possible forms of use. It is most suitable as a compost or mulch, and around Lake Kariba was used for the latter purpose when mixed with the dung of game animals (Boyd 1974).

ACKNOWLEDGEMENT

The text of this chapter is partly based upon the chapter 'African aquatic weeds and their management' by David S. Mitchell in *The ecology and management of African wetland vegetation*, edited by Patrick Denny and published in 1985 by Dr W. Junk Publishers, Dordrecht, the Netherlands, a member of the Kluwer Academic Publishers Group. We are very grateful to Kluwer Academic Publishers and Dr Denny for permission to use part of their copyrighted text, which has been updated and modified.

18

Aquatic weed problems and management in Australasia

D. S. MITCHELL AND K. H. BOWMER

INTRODUCTION

AUSTRALIA, New Zealand, and Papua New Guinea, which together comprise Australasia, are remarkable for their degree of phytogeographic isolation from the rest of the world. They are, therefore, particularly at risk from invasions of plants and animals, which reproduce rapidly in the absence of their natural enemies. Thus, the growth of alien aquatic plants is a major cause of aquatic weed problems in the region. Other aquatic weed problems are caused by human impact on natural hydrogeological cycles, leading to permanence of otherwise intermittent water bodies and nutrient enrichment of aquatic ecosystems.

Also many of the physical and chemical properties of the waters of the region are unique, and influence aquatic weed management, so that experience elsewhere may not be applicable.

The purpose of this chapter is to review the broad nature of aquatic weed problems in Australasia in relation to the above general features by reference to specific examples. The development of sensible strategies for managing aquatic vegetation in the region will then be discussed.

PLANTS CAUSING AQUATIC WEED PROBLEMS IN AUSTRALASIA

Lists of aquatic weeds have been compiled for Australia by Mitchell (1978) and for New Zealand by Graham (1976), while Mitchell (1979*c*) has discussed aquatic weeds in Papua New Guinea. Monographs or extensive reports on aquatic plants have been written for Australia by Aston (1973, 1977), Sainty (1973), and Sainty and Jacobs (1981); for New Zealand by Mason (1970) and Hughes (1976); and for Papua New Guinea by Leach and Osborne (1985).

The senior author has compiled, from the above sources, a composite list of aquatic vascular plants which regularly cause problems in at least one Australasian country. Of the 69 species or groups of species listed, 39 are alien to the region or to one of the countries in it. By contrast, six (*Myriophyllum salsugineum, M. propinquum, Potamogeton cheesemanii, P. ochreatus, P. tricarinatus,* and *Damasonium minus*) are endemic to one or other of the countries or the region as a whole. These proportions emphasize the significance of alien invasions as a causative factor in the generation of water weed problems in Australasia. The effect is further enhanced if the weeds which pose either the most serious and/or the most widespread problems are selected. Eight plants stand out: *Ceratophyllum demersum, Eichhornia crassipes, Salvinia molesta, Elodea canadensis, Lagarosiphon major, Potamogeton tricarinatus, Alternanthera philoxeroides,* and *Echinochloa* spp. Only one of these, *P. tricarinatus,* is native to the region in which it is a weed.

The weeds of rice paddies (such as *Echinochloa* spp., *Damasonium minus, Cyperus difformis,* and *Diplachne fusca*) and those of irrigation supply canals and drains (exemplified by *Myriophyllum* spp., *Potamogeton* spp., *Phragmites* spp., *Typha* spp., and aquatic grasses such as *Paspalum* spp.) cause serious economic problems in Australia, but in restricted areas only. Most of these plants are indigenous.

CHARACTERISTICS OF AUSTRALASIAN FRESHWATER SYSTEMS

The nature of freshwater systems in Australasia is affected by a combination of latitude, climate, and topographic factors.

The land masses that constitute Australasia extend from 2° S to 47° S and climatic conditions vary from tropical to temperate oceanic. Annual rainfall can vary from an average of above 4 000 mm in high rainfall areas such as the highlands of Papua New Guinea, New Zealand and south-west Tasmania to a median of less than 100 mm in the centre of Australia. Catchment topography is also variable, ranging from exceptionally flat to rugged mountains. Water bodies are thus diverse in both physico-chemical and hydrological characteristics.

It is, however, possible to identify certain characteristics by which Australasian water bodies differ in degree, if not in kind, from the temperate waters of the northern hemisphere which have provided the basis of most current understanding of freshwater science.

In significant areas of mainland Australia, the aridity and consequent poor cover of vegetation combine with repeated sedimentary stratigraphy and associated flat topography to produce extremely fine textured soils. When these erode under the influence of water, particles of approximately colloidal size become widely dispersed and cause high levels of persistent turbidity. This has a profound effect on the penetration of solar radiation into the water. For instance, the presence of suspended particulate matter in Lake Burley Griffin, Canberra, Australia can reduce the euphotic depth from about 9 m to 2 m (Kirk 1985).

Suspended solids also affect the chemical characteristics of waterbodies. They provide sites for the adsorption of heavy metals (Hart 1982), herbicides (Bowmer 1982*a*) and some plant nutrients (Melack 1985). In addition, suspended solids provide micro-environments for bacteria and algae which contribute to the dynamic physical, chemical, and biological processes that occur at the interfaces between these surfaces and the water (Melack 1985). Fulvic and humic acids also determine the surface charge density of particulates, solubilize non-polar substances such as pesticides, and are strong complexing agents for trace metals. Characterization of fulvic and humic acids in Australian waters is currently being investigated (Beckett 1988).

Australian fresh waters are dominated by sodium and chloride ions, in contrast to North American and European freshwaters, which are usually dominated by calcium and bicarbonate (Bayly and Williams 1973). This is due to the dominance of oceanic aerosols in the supply of salts to inland waters as a result of their proximity to the ocean.

Widespread catchment development also has a marked influence on the nature of surface waters. There are two main reasons for this. First, large water bodies are relatively rare and small ones are readily modified. Second, the area has been profoundly affected by western colonialism with all its emphasis on exploitation. The fragility of the environment was not recognized by early settlers and many of the techniques of land management they employed were derived from Europe, and were quite unsuited to the soils and climate of the new territories. These practices have had a profound impact, which still continues. Indeed, the recognition of an aquatic plant as a nuisance and, therefore, as a weed, has its origin in western exploitative attitudes to nature that are classically displayed in colonial development of newly acquired regions.

These characteristics of Australasian waters have affected the nature and spread of water weed problems and their management in the region. Furthermore, they are so unlike those of northern temperate waters that studies there have limited applicability in Australasia.

THE IMPORTANCE OF PLANT BIOLOGY AND ECOLOGY

Management of aquatic weeds should be based on an understanding of the biology of the plants and of the ecology of the systems they inhabit. This is particularly important in developing national procedures for weed management and in formulating long-term programmes to control intractable weed problems. The capacity to predict likely problems or the environmental consequences of control measures will improve the effectiveness and acceptability of water weed management programmes. Research on a number of Australian water weeds has sought to develop this understanding and capacity. Cary and Weerts (1983*a, b*) investigated the interactive influence of nutrients and water temperature on the growth of *Salvinia molesta* and *Typha orientalis*. Denny *et al.* (1983) developed a technique for measuring the carbon dioxide flux of submerged leaves of aquatic weeds, such as *Potamogeton tricarinatus,* under different pH, temperature, and light conditions. Sale *et al.* (1985) and Sale and Orr (1986) measured gaseous exchange and photosynthesis of communities of *Eichhornia crassipes* and *Salvinia molesta*, and of *Typha orientalis* respectively under enclosed canopies. Relevant existing knowledge on selected important species has also been compiled. These include *Eichhornia crassipes* (Forno and Wright 1981), *Salvinia molesta* (Harley and Mitchell 1981), *Typha* spp. (Finlayson *et al.* 1983), *Phragmites australis* (Hocking, Finlayson, and Chick 1983), and *Hydrilla verticillata* (Swarbrick, Finlayson, and Cauldwell 1981). However, further work is required on all these and other plants.

The particular impact of alien invasions has commanded special attention. Arthington and Mitchell (1986) demonstrated a common framework for alien plant and animal invasions of aquatic ecosystems in Australia. They pointed out that many introductions have been made and that most of these have been intentional because of some desirable attribute of the organism concerned. Mason (1970) considered that as many as 90 aquatic macrophytes had been introduced into New Zealand by 1970. However, many of the plants which have been introduced have not become naturalized.

Two groups of factors determine whether a plant will become a weed in a particular area; the environment, and aspects of the plant's life-history. In general, life-history features such as high reproductive rate, rapid growth, short life-span, multiple reproductive strategies, and effective dispersal mechanisms for propagules are associated with weediness. Other attributes which may be important under certain conditions are physiological adaptations such as an ability to withstand water stress and tolerance of a wide range of temperatures or nutrient levels.

Important abiotic environmental factors are physical and chemical parameters such as temperature and nutrient availability, while the presence of competitors, predators, and diseases are the major biotic environmental factors of importance.

It is the complex interaction of these factors which determines whether or not a plant will be a weed in any habitat and prediction is, therefore, very difficult. However, an improvement in long-term management of aquatic weeds in Australasia makes it imperative that research on understanding the causes of aquatic weed problems be given as much emphasis as research on control techniques.

EXAMPLES OF WATER WEED PROBLEMS

Surface-floating plants

Only two plants, *Eichhornia crassipes* and *Salvinia molesta* have been declared noxious aquatic weeds in each country in the region. In Australia they are two of the four Tier 1 species regarded as undesirable throughout Australia by the National Committee on Management of Aquatic Weeds (1982). In New Zealand they are the only two aquatic species gazetted as Class A noxious plants for which the objective is eradication with costs to be borne by the central government (Coffey 1988). In Papua New Guinea they are also the only two aquatic species to be declared noxious weeds (Mitchell 1979*d*).

Eichhornia crassipes was probably introduced to the region in the 1890s, being first found in Brisbane, Sydney, and Grafton (New South Wales) in 1894 (Aston 1973). Within a few years it was successfully established east of the Great Dividing Range in Queensland and northern New South Wales. It was first reported from New Zealand in 1914 and from Papua New Guinea in 1962.

Within each country, *E. crassipes* has spread to new sites or has re-infested areas from which it had previously been eradicated. Spread between aquatic systems appears to be brought about by humans carrying the plant accidentally or intentionally from place to place, as there is no evidence of dispersal by natural agents such as birds or wind. Recently it was discovered for the first time in the Kakadu National Park in the Northern Territory of Australia some 200 km east of the nearest previously known locality, and in Papua New Guinea it has recently spread to Madang and Wewak when the only previously known current infestation was at Bulolo (Leach and Osborne 1985). *Eichhornia crassipes* has also been introduced into the States of Victoria (Australia) and South Australia on a number of occasions, but a stringent policy in these States of eradicating infestations as soon as they are reported has prevented further spread. Aston (1973) recorded 21 separate infestations that were reported in Victoria between 1939 and 1969, each being apparently successfully eradicated.

The spread of *Salvinia molesta* in Australasia has been similar, though it was introduced much later, being first recorded in the region in 1952 from Ludenham, New South Wales. Authorities quickly became aware of the weedy nature of this plant and its spread in Australia has been reasonably well documented. For the most part the plant was introduced to new areas as an ornamental plant for indoor aquaria or outdoor fishponds. Thus the plant has been reported in cultivation from localities as widely dispersed in Australia as Griffith (New South Wales), Melbourne (Victoria), Waikerie and Adelaide (South Australia), Alice Springs (Northern Territory), Derby (Western Australia), and Mount Isa (Queensland) (Harley and Mitchell 1981; Finlayson and Mitchell 1982/3). In Mount Isa the plant escaped into Lake Moondarra, the water supply for the city and associated mine. It has recently been reported in Kakadu National Park (Northern Territory), indicating that it is still spreading. However, it does not seem to survive south of Nowra, New South Wales (35°S).

Salvinia molesta was first recorded in New Zealand (misidentified as *S. natans*) in 1963 and spread exponentially until 1984 when it was gazetted as a Class A weed. It appears to be unable to survive south of Hamilton due to cooler temperatures there (Howard-Williams *et al.* 1987).

In Papua New Guinea *S. molesta* was first reported in 1973 from Lae and is now present in three localities (Leach and Osborne 1985). At its most extensive, the infestation of the Sepik River, Papua New Guinea, by *S. molesta* was the most serious aquatic weed problem in the region and possibly in the world (Mitchell 1979*c*; Mitchell, Petr, and Viner 1980; Thomas and Room 1986). The plant was first reported from the River in 1977 and Mitchell (1979*b*) deduced that it had probably been present in the system for about 5 years. By the early 1980s it had spread to cover approximately 250 km^2 of the system and was having a severe effect on the lives of the local inhabitants (Gewertz 1983).

Both *Eichhornia crassipes* and *Salvinia molesta* cause problems primarily because of their well-documented capacity for rapid vegetative reproduction (Gopal and Sharma 1981; Mitchell and Tur 1975). Both are obligate acro-pleustophytes and exhibit special adaptations to the surface-floating habit (Mitchell 1985*a*). This gives them a particular advantage as primary colonizers of open waters which are liable to extensive and/or rapid fluctuations in water level. Many such waters are important for human use, such as water storage reservoirs and irrigation dams, where the presence of the plants interferes with water use. Experience has shown that the presence of even a small number of these plants in a water body poses a potential problem and that early action aimed at eradication is the best course to follow. Experience has also shown that determined, sustained action can be successful in eradicating the weeds, especially from small, confined water bodies with accessible shorelines. In these circumstances, manual removal, or initial chemical control followed by manual removal of survivors, are appropriate courses of action. *Salvinia molesta* is the most difficult in this regard as it breaks readily when handled and it is capable of surviving in a reduced form less than 1 cm in length.

Submerged plants

Problems resulting from the excessive growth of submerged plants occur in the clear temperate lake waters in New Zealand (Mason 1975) and the warm, slow flowing, sometimes nutrient rich water of drainage or supply canals in irrigation systems and flood mitigation schemes (Mitchell 1978).

Chapman *et al.* (1974) have documented the problems caused in New Zealand lakes by the introduced species, *Ceratophyllum demersum, Lagarosiphon major,*

Egeria densa, and *Elodea canadensis.* For example, massive growths of *C. demersum,* sometimes associated with other plant species, drifted onto turbine screens of hydro-electric dams in such large quantities that, on some occasions the mechanical clearing devices were unable to keep the screens clear. The turbines would then have to be closed down, resulting in reduced power generation and increased operating costs. The Ministry of Energy (Electricity Division) subsequently improved techniques for mechanical collection and clearance from the turbine screens and these now provide a reliable means of preventing the weeds from interfering with the operation of their hydroelectric power installations (Johnstone 1982).

Sheltered areas on the shoreline of Lake Taupo, New Zealand, form particularly suitable sites for the growth of aquatic plants (Howard-Williams and Vincent 1983). The plants sometimes cause problems, particularly around boat ramps, where dense growth of the plants interferes with their use. The problems have been exacerbated by introduction of the alien weeds *Elodea canadensis* and *Lagarosiphon major.* The latter is now the major weed species dominating the zone between 2 m and 6.5 m deep, where it sometimes reaches lengths of 4 m, *Elodea* grows in the depth zone below the *Lagarosiphon* beds to a depth of 10 m.

However, Howard-Williams and Vincent (1983) also point out the value of dense beds of submerged plants which, in association with adjacent *Typha* swamps, provide food and habitat for waterfowl and other fauna. The decline of submerged plants in some New Zealand lakes, as a result of management of the water body or its catchment is, therefore, regarded with some concern (Gerbeaux and Ward 1986; Coffey 1988).

Mitchell (1978) listed nine submerged species, causing the most frequent problems in three extensive irrigation and flood mitigation systems in Australia, that were considered representative of others, at least in the south-east of the continent. It is significant that seven of the nine species are native to Australia. This confirms that the main ecological factor promoting the development of nuisance populations of indigenous aquatic plants is the creation of permanent water bodies such as irrigation distribution systems. These provide suitable conditions for the growth of aquatic plants, where prolonged periods without water and occasional heavy flooding had previously inhibited the development of stable aquatic plant populations.

Alien plants that are introduced to the system can sometimes cause more severe problems than the native species. The occurrence of *Elodea canadensis* in the irrigation systems of Victoria and south eastern New South Wales in 1958 (Sainty 1973), and its subsequent rapid spread through the system in the early 1960s (Aston 1973; Mitchell 1978), gave rise to considerable concern. Bowmer *et al.* (1979) and Bowmer, Mitchell, and Short (1984) carried out studies of the biology and management of the weed in the Murrumbidgee Irrigation Areas (New South Wales). They demonstrated that the main vegetative growth of the plant occurred in spring and early summer so that the plant reached nuisance proportions requiring some form of control by midsummer at the latest. However, dispersal of the plant occurred mainly in late summer when the plants break into many viable fragments that are carried downstream. Overwintering propagules in the form of swollen stem apices are produced in large numbers (up to about $5000\,m^{-2}$), mostly in the substratum, with the onset of cold weather in autumn. These provided the basis for rapid growth in the following spring. The control strategy for this species should, therefore, include control just before dispersal and before propagule formation, even if the population is small.

Emergent plants

Emergent plants also cause problems in irrigation supply and drainage canals and in flood mitigation drains by obstructing flow. Populations of plants that previously had to contend with long periods without free water are now able to grow until water is drained from the system. Populations can, therefore, be large and troublesome particularly in the summer. Design of canals and drains should take account of their potential for weed infestation. Steep sides and a good gradient to ensure complete drainage will minimize the occurrence of serious weed problems.

Mitchell (1978) listed fourteen species or groups of species typically causing such problems in south eastern Australia. As with submerged weeds of such systems, most of the plants are native to Australia, only four being alien. Recently, however, another alien species, *Alisma lanceolatum* has begun to spread in the rice paddies of south eastern Australia and is causing some concern (McIntyre 1987).

The most frequent emergent species causing problems are *Typha orientalis* and *Phragmites australis* in Australia and New Zealand and *Typha domingensis* in Australia, though several of the aquatic grasses, such as *Brachiaria mutica* and *Paspalum* spp. in Australia and *Zizania* spp. in New Zealand can also be troublesome.

Emergent aquatic plants also provide the major weeds of paddy rice, of which about 110000 hectares are grown every year in southern New South Wales. Chemical weed control predominates and, in 1985, approximately 3.9 million dollars was spent on chemicals for this purpose, indicating the economic significance of this problem. McIntyre (1987) investigated the weed flora of the New South Wales rice paddies. She recorded a total of sixty-eight species of which 79 per cent were emergent, presumably because the light regime of the dense mature rice crop competitively inhibits the growth of submerged and floating leaved plants.

Another feature of the rice weed flora is the high proportion of native species. Six plants were listed as being economically most important and of these, four

(*Cyperus difformis, Damasonium minus, Diplachne fusca,* and *Typha* spp.) are native, while only two (*Echinochloa crus-galli* and *Rumex crispus*) are alien (McIntyre 1987).

Problems from emergent plants, like those mentioned above, have caused many authorities to regard all such aquatic macrophytes as potential, if not actual, weeds, even when they are growing in beneficial wetlands. Fortunately, this tendency is now seen as incorrect and policies are being formulated to conserve native wetlands (New South Wales Water Resources Commission 1986). Regrettably, considerable areas have already been destroyed usually by so-called 'reclamation' processes. In New Zealand it is estimated that only 10 per cent of the country's original wetlands now remain (Howard-Williams *et al.* 1987). The situation in Australia is probably similar but because of the wide extremes of climate, it is not always possible to demarcate wetlands from dry lands. Many wetlands are transitory, though they may develop a sizeable aquatic flora and fauna during wet periods. Furthermore, it is probably true to say that most, if not all swamps that are dominated by emergent aquatic macrophytes, such as the Macquarie Marshes in New South Wales, no longer experience natural conditions of water level fluctuations. The effect of this on some of the rarer aquatic macrophytes is unknown.

The alien emergent plant, *Alternanthera philoxeroides,* is recognized as a Tier 1 species in Australia (National Committee on Management of Aquatic Weeds 1982) and is thus classified as a noxious weed in every State and Territory in the country. In New Zealand, it is a Class B noxious plant, so that its sale and distribution is prohibited and eradication on a national or regional basis may be recommended (Coffey 1988).

In Australia, research on the physiology and control of *Alternanthera philoxeroides* has been given high priority because of the difficulty of controlling it by current chemical and biological methods when it is growing in terrestrial ecosystems, on roadsides and damp pasture. Its current distribution is restricted to the Georges River in Sydney, pastures and ditches in the vicinity of Williamtown, near Newcastle (New South Wales) and a site near Woomargama near Holbrook, west of the Great Dividing Range in New South Wales. Current concern is generated by the fear that it could spread from there to the irrigation systems in Victoria and southern New South Wales. The research to be undertaken will aim to establish why chemicals are less effective against the terrestrial than the aquatic form of the plant, which is well controlled in Australia by the insect *Agasicles hygrophila* (Julien 1981).

STRATEGIES FOR THE MANAGEMENT OF AQUATIC WEEDS IN AUSTRALASIA

There are two main components to the management of aquatic weeds in Australasia: control of nuisance populations of weeds, and prevention of problems caused by weeds. An integrated management strategy requires good co-ordination and proper planning of both these components as well as a good understanding of the ecological nature of water weed problems, and a careful evaluation of the problems caused by water weeds (Mitchell 1978; Mitchell 1979*a*).

Chemical control

The application of herbicides is generally used in the first instance to control water weeds, if the funds are available and the benefit to be obtained outweighs the cost. Consequently, there is considerable expertise in both New Zealand and Australia on the use of herbicides to control aquatic weeds, and official guidelines and approval procedures are published by both countries (Australian Water Resources Council 1985; New Zealand Ministry of Agriculture and Fisheries 1975).

Chemicals are used widely to control submerged and emergent weeds in Australian irrigation systems (Bowmer 1979). It has been clearly shown that this is the most cost-effective means currently available for controlling the multi-species infestations in the flowing water conditions in supply and drainage canals. There are a number of features in which these flowing waters differ from similar water bodies in other countries. Most of the canals are earth-lined and carry warm, highly turbid water during the irrigation season. Also, weed control operations are not restricted by the need to protect aquatic biota and fish since the systems are maintained strictly for irrigation water supply and drainage. Another feature is that most supply canals run close to capacity in the growing season, so that there is very little tolerance for aquatic plants and almost no possibility of taking parts of the system out of commission for more than a few days.

In these circumstances it has been shown that the injection of acrolein into the flowing water of canals is easily the most effective means of control of submerged weeds. Many thousands of kilometres of channel are treated with acrolein each season (Bowmer and Sainty 1977; Bowmer 1979). The disadvantage of acrolein, that it is highly toxic to aquatic life and fish, is compensated for by its high volatility. Also, within 48 h of treatment the water can be used for irrigation without putting crops at risk.

O'Loughlin and Bowmer (1975), Bowmer (1975), and O'Loughlin (1975) investigated the dispersion and dissipation of acrolein downstream of the injection point in order to establish the most economical methods of addition. They derived an equation to calculate the length of canal for which a given addition of the chemical is effective and showed that a number of small injections uses less chemical to achieve control than a single large one.

Because of the high toxicity of acrolein, its unstable nature and the uncertainty of continuing supplies, other chemicals have been investigated. Diquat is rapidly inactivated by the turbid particulate matter and by epiphytes on the leaves of plants like *Elodea canadensis* (Bowmer 1982*a, b*). Endothall-amine used at high concentration for a short time, was also shown to be ineffective against *E. canadensis* (Bowmer and Smith 1984). Terbutryn which is effective in Europe, is another chemical that appears to be incapable of controlling submerged weeds in Australian irrigation systems (Bowmer, Shaw, and Adeney 1985). The most likely explanation for the lack of success of this herbicide in these situations is the short contact time that is possible in irrigation distribution systems. Also the volume of weed material may be so great as to impede flow through the weed bed and, therefore, limit the contact between weed and chemical. Dichlobenil has been tried as a soil residual treatment during winter (Bowmer *et al.* 1976) but it was not successful.

Emergent weeds are mainly controlled by a combination of ditch-bank sprays, and use of soil-residual herbicides applied after drawdown of irrigation water in the autumn.

The foliage-acting herbicides, glyphosate, and dalapon, are widely used for banks and marginal plants, including *Phragmites australis,* species of *Typha* and aquatic grasses. The degree of contamination and fate of these compounds are described, respectively, by Bowmer (1982*b*) and Bowmer (1987*a*).

The herbicides TCA and dalapon are often mixed together and applied to exposed sediments and emergent plants in the autumn; and diuron is used similarly by many farmers on smaller on-farm channels. Both TCA and diuron persist in the sediments and are washed out into the irrigation water at concentrations above the legal tolerance in New South Wales, though the implications for damage to crops and aquatic biota are uncertain. The use and fate of diuron is described by Bowmer and Adeney (1978); and of TCA by Bowmer (1987*a*).

There is a notable lack of information on the sensitivity of aquatic macrophytes to different patterns of exposure for even the most frequently used chemicals (Bowmer 1986). Also, little is known about possible synergistic effects on aquatic biota of exposure to combinations of different pollutants. The decay kinetics of the herbicides in water and sediments are also poorly documented. Many herbicides are rapidly dissipated from the soil, but their persistence in standing water, where microbiological activity is less, and reactive surfaces are sparse, may be much longer. The whole subject of the fate and effect of herbicides in surface water has recently been reviewed by Bowmer (1987*b*).

In spite of these difficulties, herbicides remain the most cost-effective way of maintaining the considerable lengths of channels involved. It is significant that most of the development and evaluation work on the application of herbicides to these systems has had to be carried out by government authorities in Australia as the market is too small to interest commercial companies. However, the expenditure on chemicals of about A$1 million to control weeds in these systems protects the irrigation agriculture industry which is worth A$1 billion annually.

Mechanical/manual control

Mechanical control measures are also widely used but have been shown to have major disadvantages in terms of high cost and potential damage to canal linings. It is important to avoid the latter for two reasons: excavation below bed level can leave non-drainable areas in which permanent weed beds can establish; and standing water in these depressions and damage to the linings can increase seepage to the water table, increasing the regional problems of salinization of the surrounding soil. Harvesting has been considered as a possible control method on Australian lakes infested by aquatic weeds (e.g. Finlayson, Farrell, and Griffiths 1984) but little progress has so far been made with this approach.

Biological control

The Australian Commonwealth Scientific and Industrial Research Organisation's Division of Entomology has achieved outstanding success in the biological control of *S. molesta* and *A. philoxeroides* resulting in their virtual elimination from some localities, while control of *E. crassipes* is effective in some areas (Julien 1981; Room *et al.* 1981; Wright 1981; Roberts, Winks, and Sutherland 1984). Spectacular results were obtained in eliminating *S. molesta* from the Sepik River, Papua New Guinea, (Figs 181a, b) using the weevil *Cyrtobagous salviniae,* which was collected from southern Brazil (Forno and Harley 1979) and subsequently described as a new species (Calder and Sands 1985). Following an initial period during which the weevil population became acclimatized and built up in one site, weevil-infested weed was distributed throughout the system. After about a further six months of weevil population growth and establishment in each release site, damage to *S. molesta* became increasingly obvious and the weed was virtually eliminated from the water body during the next 4 months. Thus within the whole system, 250 km^2 of *S. molesta* in mid-1984 was reduced to less than 2 km^2 by the end of 1985 (Room 1985*b*; Thomas and Room 1986).

The outstanding success of the biological control of *Salvinia molesta* has markedly reduced the plant's significance as a threat to utilization of tropical waterways throughout the world. However, the weevil is apparently less tolerant of cold winters than the weed (Room 1986) and it is possible that *S. molesta* may continue to infest temperate waters. Such conditions are, however, marginal for

FIG. 18.1. Binatang Lagoon (Sepik River, Papua New Guinea) (a) shortly after release of *Cyrtobagous salviniae* in 1983, showing no effect yet discernible on 100 per cent cover of *Salvinia molesta*, and (b) in 1985, showing weed cover reduced to <10 per cent of lagoon surface area as a result of successful biological control. Total area of the lagoon is *c*. 50 ha. (Photos: P. Thomas.).

the weed as growth rate is strongly related to temperature (Cary and Weerts 1981) and chemical or mechanical control is likely to be cost-effective.

Biological control of submerged weeds has been little researched in Australasia. The Chinese grass carp, *Ctenopharyngodon idella* has been introduced to New Zealand on a trial basis but there has been strong resistance, largely from the fish authorities, to its introduction into Australia.

Preventative measures

As a consequence of the cost of water weed control and the limited range of suitable techniques available, strong emphasis is given throughout the region to preventative measures. Every attempt is made to prevent the entry of alien weeds and to control infestation of any that are present. In addition, managers need to be aware of the likely costs, in terms of water weed control, of any modifications to water bodies, including eutrophication of catchments. For example, the role of catchment management in eutrophication of water bodies, including the growth of macrophytes has been studied in Australia. Important case studies include Lake Burley Griffin, Canberra (Cullen and Rosich, 1979), and the Peel Harvey Estuary, Western Australia (Hodgkin *et al.* 1985). The subject has been reviewed by Cullen and Maher (1987), Cullen and O'Loughlin (1982), and McComb and Lukatelich (1986). If such costs can be predicted it may be possible to introduce ameliorative measures to minimize problems or at least ensure that control measures are properly planned and funded (Mitchell 1974). All these are important factors in determining the current philosophies and practice of weed control particularly in Australia and New Zealand.

The present system in Australia is extremely complicated, with responsibilities for various aspects of weed control being divided between Federal Government and State Governments. However, since the establishment of the National Coordinating Committee on Aquatic Weeds (formerly the National Committee on Management of Aquatic Weeds), it has been possible to develop a national strategy for management of aquatic weeds:

(1) the most serious weeds were identified as Tier I or Tier II species with the former being declared noxious throughout the country (National Committee on Management of Aquatic Weeds 1982);

(2) guidelines for the use of herbicides in or near water have been produced (Australian Water Resources Council 1985);

(3) lists of aquatic plants prohibited for import have been revised and procedures developed for evaluating requests for new imports;

(4) priorities for research have been identified and funding facilitated for appropriate research projects;

(5) posters identifying Tier I species have been distributed nationwide;

(6) a national public awareness campaign has been approved and is in the process of being implemented.

New Zealand's strategy for the management of its aquatic flora is a mix of three elements: conservation of native vegetation in a range of unmodified waterways; acceptance of non-native aquatic plants where they are beneficial in providing habitat diversity or in other ways; and control of alien plants where they are a nuisance or have the potential to be so. Emphasis is placed on preventative weed management which requires that managers can recognize potential weeds, that vigilant surveys identify potential problems at an early stage, that contingency plans are available to deal successfully with outbreaks of nuisance plants when they occur, and that weed control methods are environmentally acceptable (Coffey 1988).

In Papua New Guinea eutrophication of waterways is limited and there have been relatively few introductions of alien plants. As a result, the country has very few water weed problems. The Plant Disease and Control Act of 1953 provided for 3 categories under which a plant could be gazetted: noxious plant (prohibited for cultivation), notifiable noxious plant (occupier or owner required to notify a plant inspector of the presence of the plant on a property), and restricted plant (it is an offence to move the plant from one designated area to another). The purpose of this legislation was to deal both with problem infestations and to prevent spread. Both *E. crassipes* and *S. molesta* have been gazetted under this Act which has enabled authorities to attempt rigorous control of troublesome infestations of these plants.

CONCLUSIONS

Australasia's separation from other continents and consequent development of a unique flora and fauna has made it particularly vulnerable to invasion from alien aquatic plants. Furthermore, its relatively low human population density contrasts with a surprisingly extensive modification of natural aquatic regimes (in those areas where there are sufficiently reliable supplies of water to make development worthwhile!).

These factors promote the development of aquatic weed problems. However, they also provide the basis for designing effective measures to prevent and control water weed problems. Stringent attempts to prevent further invasions certainly decrease the entry of weeds in Australia and these are possible because of Australasia's geographic isolation. Also the high degree of control over developed systems such as irrigation areas minimizes some of the environmental hazards that attend weed management measures in natural waterways.

The national management of water weeds in Australia and New Zealand is soundly based in existing formally constituted bodies. It is effective and appears to have sufficient flexibility to deal with unexpected events and to adapt to changing situations as necessity demands. National management of aquatic weeds in Papua New Guinea appears to be more *ad hoc,* though one of the best examples of the successful biological control of an aquatic weed occurred in that country.

ACKNOWLEDGEMENTS

The authors are very grateful for the assistance they received from Fiona Rogers, whose constructive criticism greatly improved the cohesion of the final compilation.

19

Aquatic weed problems and management in North America

(a) Aquatic weed problems and management in the western United States and Canada

L. W. J. ANDERSON

INTRODUCTION

AN understanding of aquatic weed impacts and solutions in the western United States and Canada is best gained from a general view of water catchment, storage and distribution. Although there are many thousands of kilometres of natural riverine systems and lakes, the most frequent and severe problems associated with aquatic macrophyte vegetation occur primarily in the man-made lakes, reservoirs, and canals throughout this region.

Much of the inhabited and agricultural lands in the West (especially in the far West) are arid or semi-arid, receiving only 25 to 50 cm annual precipitation. This fact, coupled with the generally excellent soils and favourable climate for temperate crops, together with the westward expansion of populations during the late nineteenth and early twentieth centuries, prompted the construction of immense, extensive (and expensive) networks of reservoirs, major high-capacity canals, and lateral (subsidiary) canals for agricultural, domestic, and industrial uses. In concert with these uses, the development of hydroelectric power production and recreational facilities has been an integral part of the overall water-delivery systems. Beginning in the late 1800s, but accelerating dramatically in the 1930s–1960s, federal (US Bureau of Reclamation (USBR) and US Army Corps of Engineers), and various state water projects were designed, financed and constructed to carry tremendous volumes of water from sources such as mountainous areas that receive heavy snowpack, and major rivers, to the more central and southern areas where population demands and optimal climates for crops exist. For example, in California, 70 per cent of the state's water resources are from north of latitude 38° (San Francisco) (Gurmukh, Gray, and Seckler

1971). Table 19a.1 summarizes some of the major western water projects and illustrates the magnitude of the water resources. The Bureau of Reclamation alone transports about 37 billion (10^9) m^3 of water each year, of which 80–85 per cent is used for agriculture (Snyder 1983; USBR 1985).

Understandably, the storage and transport of these vast amounts of water provides excellent opportunities for the establishment and proliferation of aquatic vegetation, including algae. This opportunity is further enhanced since 60–70 per cent of the canals and laterals are earthen and thus provide suitable substrates for rooted plants. Even those reaches which are concrete-lined support aquatic plant growth because sufficient sedimentation occurs within a few years after

Table 19a.1. Features of selected government operated water storage and conveyance facilities in the western US

Agency	Storage volume: ($\times 10^6$ m^3)	Area (ha)	Canal length (km)	Water delivered ($\times 10^6$ m^3)
US Bureau of Reclamation[a]	172887	688500[b]	11984	36660
US Army Corps of Engineers[c]	n.d.	1028570	n.d.	n.d.
State of California (California Water Project)[d]	8350	22275	671	2730

[a] From: 1985 Summary Statistics, Volume 1, US Dept. of Interior, 319 pp.
[b] Recreational surface area.
[c] From: Aurand (1983) and Tyndall (1982): includes only reservoirs greater than 2585 ha surface area.
[d] From: California Department of Water Resources Bulletin 132–85, 'Management of California State Water Project'. pp. 269.
n.d. = no data available.

construction. For example, the recently-constructed Central Arizona Project (CAP), which will eventually convey Colorado River water some 480 km into Arizona, has begun to support the growth of aquatic weeds even though the entire system is concrete lined (Winn Winkyaw 1987 pers. comm.). When this new system is operating at full capacity in 1990 it will carry *c.* 85 m^3s^{-1}. At this volume flow, water velocities should reach nearly 1 $m\,s^{-1}$, which may be sufficient to lessen the siltation problem, at least in the upstream half of the system.

Thus, in addition to the natural aquatic habitats, the topographically diverse western landscape is further punctuated by many thousands of hectares of storage reservoirs and many thousands of kilometres of canals with flow volumes ranging from 0.5 to 100 $m^3\,s^{-1}$. As an example, Fig. 19a.1 shows only the US Bureau of Reclamation projects throughout the seventeen western states. The importance of these systems is apparent when one considers that the total storage capacity in California alone is about 95 billion m^3, and that about 2 million ha are irrigated

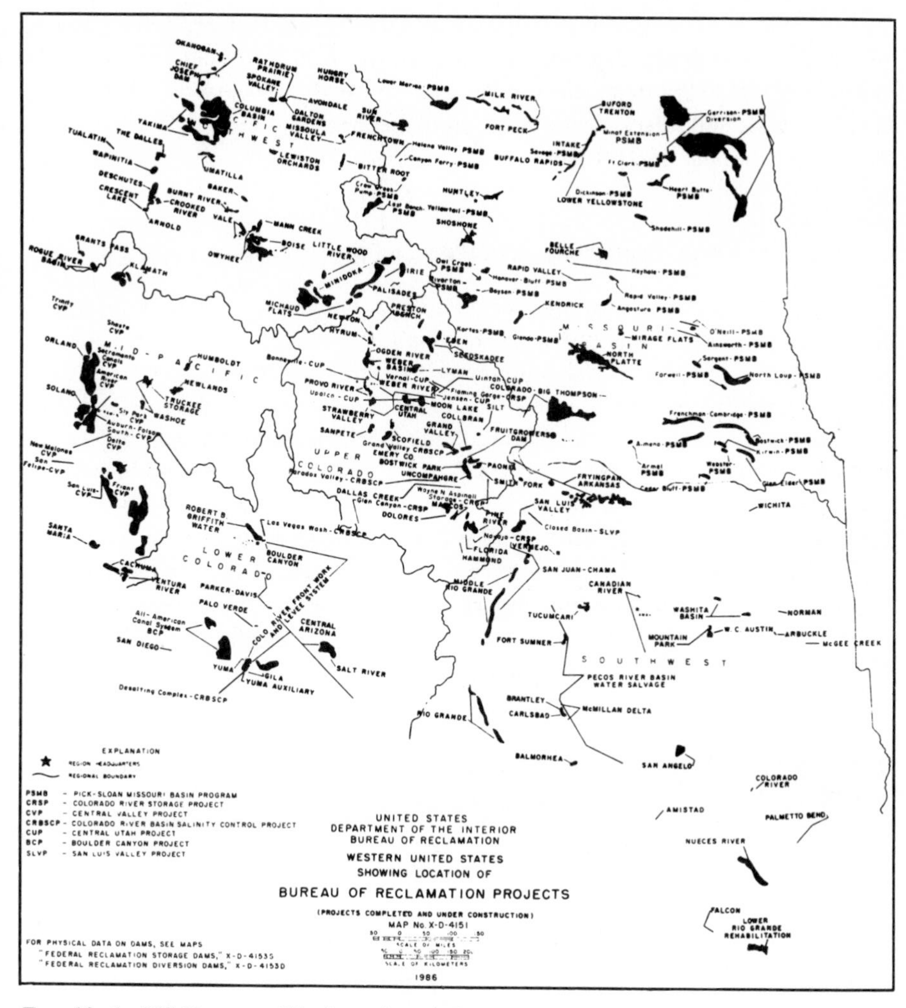

FIG. 19a.1. US Bureau of Reclamation projects (black areas) in the seventeen western states from Texas (far right) to California (far left). State boundaries are excluded to emphasize regional aspects.

from these sources (Snyder 1983). In all the seventeen western states of the USA, the USBR system provided irrigation water for over 48 million ha in 1985 which produced a crop value (US \$ unless otherwise stated) of about \$7.4 billion (USBR 1985). These same aquatic resources also provided about 50 billion 12 hour recreational visitor days with a value of over \$470 million (USBR 1985).

It should be pointed out that the systems listed in Table 19a.1 represent 'bulk' or 'wholesale' water storage and delivery. From each of these major trunks (and other delivery systems not listed), myriad private, quasi-public and municipal districts purchase, treat, sell and distribute water for agriculture, industrial, and domestic needs. When these supply systems are included (where, it should be noted, the most persistent and serious aquatic weed problems occur), then the total of canals and laterals with weed problems probably exceeds 100 000 km in the west (Timmons 1960, 1966).

AQUATIC WEEDS OF PRIMARY IMPORTANCE

The most detailed and extensive survey of aquatic weed problems and impacts in the USBR system was published by Timmons (1960). This survey covered seventeen Western states and represented conditions in the late 1950s to early 1960s. Although there have been some increases in numbers of canals and their total length, particularly with the completion of the Central Arizona Project, the Delta Mendota Canal and California Aqueduct, and though recent problems with *Hydrilla verticillata* and *Eichhornia crassipes* did not exist then, many of Timmons' findings are probably still valid. Table 19a.2 summarizes data from Timmons' report and emphasizes the importance of the moderate to rapidly-flowing conveyance systems. Note that this report did not designate weeds by species but rather categorized them by 'life form' or 'ecological classification' (Sculthorpe 1967; Hutchinson 1975). No floating weed problems were noted in these irrigation systems at that time.

Table 19a.2. Summary of western US aquatic weed infestations in Bureau of Reclamation Systems *c.* 1960[a]

Type	Percent of total types	Canal Length[b] (km)	Infestation by canal type (%)[c]		
			Canal	*Lateral*	*Drain*
Algae (Primarily filamentous)	15.5	32 944	25	11	18
Submerged (*Potamogeton* spp.,[d] *Myriophyllum* spp., *Elodea canadensis, Ceratophyllum demersum*)	33.1	76 282	28	36	31
Emergent (Primarily *Typha* spp., *Phragmites australis, Phalaris arundinacea*[e])	16.5	36 911	6	16	25

[a] From Timmons 1960.
[b] Timmons' extrapolation to western United States based on 47 USBR Districts.
[c] Average capacities ($m^3 s^{-1}$): canal: 13.7; lateral: 1.1; drain: 0.6.
[d] Species indicated in italics are present author's additions based on other reports and personal observations. Averages for species are weighted by length of infested channel type.
[e] From Bruns 1973.

Based on a more recently published series of surveys and interviews (Tyndall 1982; Aurand 1983) a fairly complete listing of problem aquatic weeds is presented in Table 19a.3. Some additional weeds have been included based on other reports and on personal observations of the present author. No ranking of 'economic impacts' or 'frequency' by species based on surveys or published observations is available. However, from contacts with major irrigation districts, field observations and general frequency of inquiries for assistance, the most consistently

Table 19a.3. Aquatic plants causing problems in western US states

North-western United States[a]	**South-western United States**[b]
Brasenia schreberi	*Alternanthera philoxeroides*
Ceratophyllum demersum	*Azolla* spp.
Chara spp.	*Ceratophyllum demersum*
Elodea canadensis	*Chara* spp.[c]
Fontinalis spp.	*Cladophora glomerata*[c]
Lemna spp.	*Egeria densa*[c]
Myriophyllum spicatum	*Eichhornia crassipes*
Myriophyllum spp.	*Elodea canadensis*[c]
Najas spp.	*E. nutallii*[c]
Nitella spp.	*Enteromorpha intestinalis*[c]
Nymphaea odorata	*Hydrilla verticillata*
Phalaris arundinacea[c]	*Hydrocotyle umbellata*[c]
Potamogeton amplifolius	*Lemna* spp.[c]
P. crispus	*Ludwigia repens*[c]
P. natans	*Myriophyllum aquaticum*[c]
P. nodosus	*M. exalbescens*
P. pectinatus	*M. spicatum*
P. richardsonii	*Najas marina*
P. praelongus[c]	*N. guadalupensis*
Ranunculus spp.	*Phragmites australis*[c]
Scirpus spp.[c]	*Polygonum amphibium*[c]
Typha spp.	*Potamogeton* spp.
Utricularia spp.	*P. amplifolius*[c]
Veronica spp.	*P. foliosus*[c]
Zannichellia palustris[c]	*P. gramineus*[c]
	P. nodosus[c]
Filamentous algae (N.W. & S.W.)	*P. pectinatus*[c]
Chara spp[c]	*P. pusillus*[c]
Cladophora glomerata[c]	*P. richardsonii*[c]
Enteromorpha intestinalis[c]	*Ranunculus* spp.[c]
Pithophora spp.[c]	*Ruppia maritima*[c]
Rhizoclonium spp.[c]	*Typha latifolia*[c]
	T. angustifolia[c]
	Zannichellia palustris

[a] States: Alabama, Idaho, Montana, Nebraska, North Dakota, Oregon, South Dakota, Washington, Wyoming (Aurand 1983).

[b] States: Arizona, California, Colorado, Hawaii, Kansas, Nevada, New Mexico, Oklahoma, Texas, Utah (Tyndall 1982).

[c] Species which are current author's additions, based upon other reports and personal observations (Otto 1975; Bruns 1973).

problematic weeds appear to be: *Potamogeton pectinatus, P. nodosus* (Fig. 19a.2), *P. crispus, P. foliosus, Elodea canadensis, Myriophyllum spicatum, M. aquaticum, Ceratophyllum demersum, Typha latifolia, T. angustifolia, Ludwigia repens, Chara* spp. and *Cladophora glomerata*. Although both *Eichhornia crassipes* and *Hydrilla verticillata* are significant problems, they are not widespread in the western United States (except in Texas) and are discussed as special cases below.

FIG. 19a.2. A typical irrigation canal in the western United States, heavily infested with American pondweed (*Potamogeton nodosus*). These canal systems are seasonally drained from about November until April.

These lists do not include all weeds causing problems in rice, a major commodity in California (*c.* 145 000 ha were grown in 1987: US Department of Agriculture 1987). An excellent review of rice weeds was published by Barrett and Seaman (1980). These authors reported that the most frequently encountered weeds species were: *Sagittaria montevidensis* spp. *calycina, Ammania coccinea, Bacopa rotundifolia, Heteranthera limosa*, and *Echinochloa crus-galli*, the last of which probably causes the most serious problems. These weeds (and others in rice) are controlled primarily by variously timed applications of the herbicides molinate, thiobencarb, basagran, and MCPA. Another new herbicide, bensulfuron methyl (Londax^{Tm}), is currently under intense testing for use in rice and possibly for other non-crop aquatic uses.

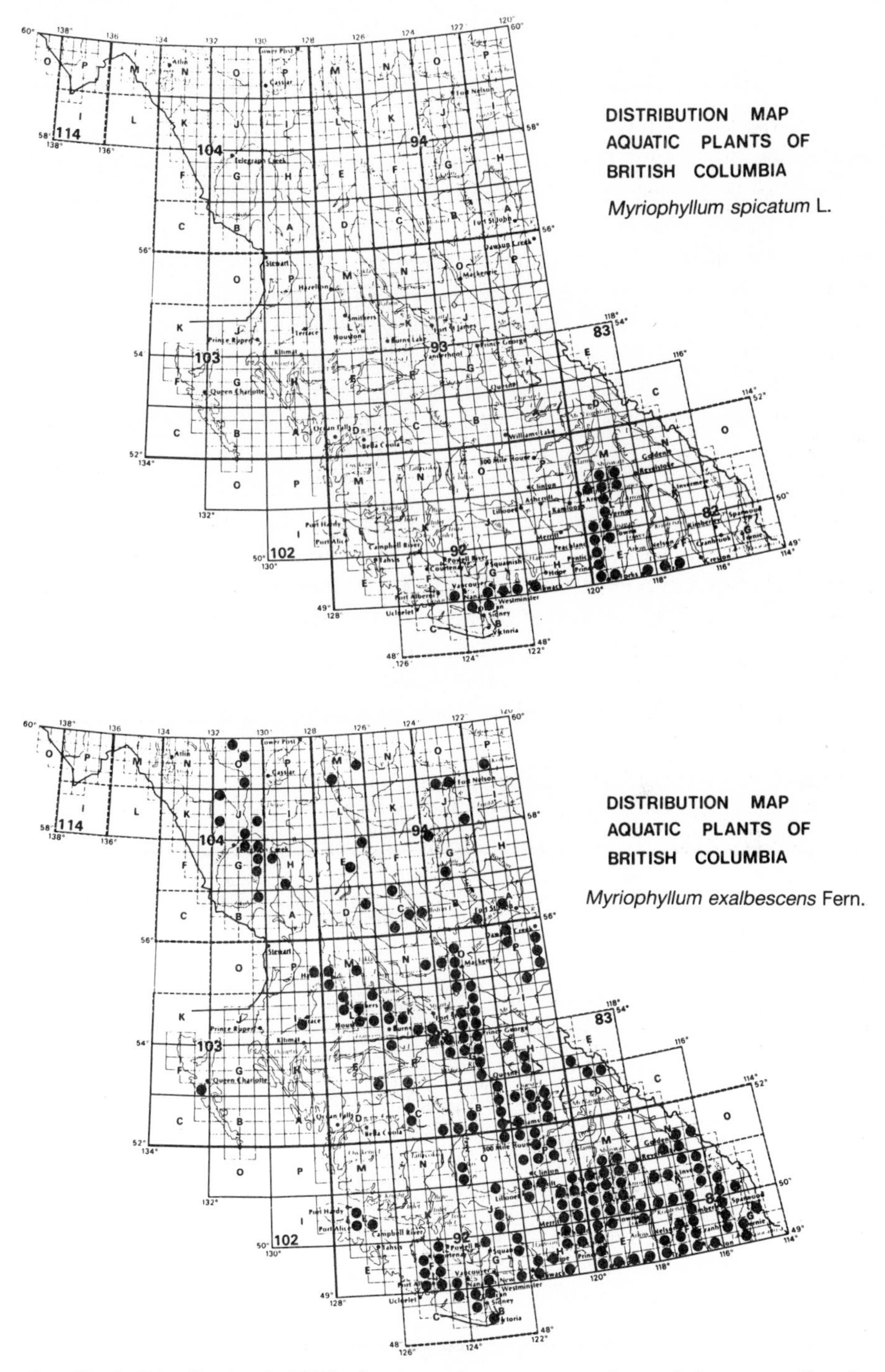

FIG. 19a.3. Distribution (*c.* 1987) of two most important aquatic weeds in western Canada, *Myriophyllum spicatum* and *M. exalbescens* (from P. D. Warrington and P. Newroth, pers. comm. 1985).

CANADIAN AQUATIC WEED PROBLEMS

Canadian aquatic habitats support a wide variety of vascular plants (Warrington 1980), but the most troublesome species are limited primarily to the genus *Myriophyllum*. In fact, *M. spicatum* and *M. exalbescens* account for perhaps more than 90–95 per cent of all economic losses and control costs associated with aquatic vegetation in Canada, and that is primarily focused in and around British Columbia and around Toronto (Anderson 1985). *M. spicatum* was reported in British Columbia in 1971 and by 1980 about 1 000 ha were infested (Dove and Wallis, 1981). As of 1985, estimates were 2 000–3 000 ha for British Columbia and 100 000 ha for the Toronto/Quebec area (Symposium discussion comments by P. Newroth and S. Painter in Anderson, 1985; Newroth 1985). Distribution of these two species in British Columbia as of 1987 is shown in Fig. 19a.3 (P. D. Warrington, pers. comm.)

Several recent papers presented at the First International Symposium on Watermilfoil document the general spread and occurrence of *M. spicatum* in the western-Canada/US region (Rawson 1985; Warrington 1985). It is apparent that from 1977 to 1978 the infestation in the Okanogan River breached the northern United States border and has been progressively invading the state of Washington ever since (Fig. 19a.4). However, other populations of *M. spicatum* were documented in 1973 in Washington state (Falter *et al.* 1974) and even perhaps in the 1960s (Couch and Nelson 1985).

In an archival survey done by the British Columbia Ministry of Environment

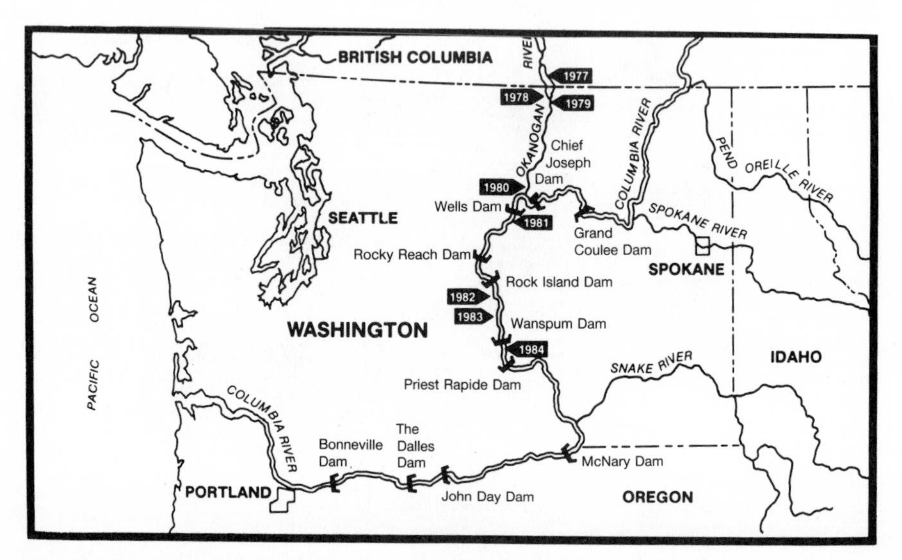

FIG. 19a.4. Chronology of *Myriopyllum spicatum* spread into Washington state from the Okanagan River system (from Rawson 1985).

for 1975–1981, *M. exalbescens* and *Potamogeton* spp. accounted for 60 per cent of logged aquatic weed complaints (*M. spicatum* was not included). Of lesser incidence were problems from *Elodea canadensis* (12 per cent) and *Ranunculus aquatilis* (7 per cent) (P. Newroth and P. Warrington, pers. comm.) Table 19a.4, which shows the frequency of occurrence of aquatic plants in British Columbia and Alberta, further illustrates the dominance of a few genera. It should be noted that unlike British Columbia, southern Alberta has thirteen irrigation districts whose 11 500 km of canals can support abundant aquatic plants much like those in the western United States (Burland and Catling 1986).

Table 19a.4. Relative prevalence of twelve aquatic weeds in British Columbia and Alberta, Canada

Rank	Species		
	British Columbia	Alberta: Irrigation Systems	Alberta: Recreational lakes
1	*Nuphar polysepalum*	*Potamogeton pectinatus*	*Myriophyllum exalbescens*
2	*Utricularia vulgaris*	*Potamogeton richardsonii*	*Potamogeton pectinatus*
3	*Myriophyllum exalbescens*	*Typha latifolia*	*Potamogeton richardsonii*
4	*Potamogeton gramineus*	*Phalaris arundinacea*	*Potamogeton vaginatus*
5	*Ranunculus aquatilis*	*Myriophyllum exalbescens*	*Ceratophyllum demersum*
6	*Potamogeton natans*	*Potamogeton vaginatus*	*Potamogeton praelongus*
7	*Typha latifolia*	*Ceratophyllum demersum*	*Lemna trisulca*
8	*Scirpus lacustris*	*Alisma gramineum*	*Potamogeton pusillus*
9	*Potamogeton pectinatus*	*Ranunculus circinatus*	*Elodea canadensis*
10	*Sparganium emersum*	*Elodea canadensis*	*Ranunculus circinatus*
11	*Polygonum amphibium*	*Potamogeton pusillus*	*Potamogeton zosteriformis*
12	*Potamogeton richardsonii*	*Scirpus spp.*	*Ruppia maritima*

Note: Data are from P. Warrington and P. Newroth (pers. comm.) for British Columbia and from Burland and Catling (1986) for Alberta. Both sources list other less prevalent species not included here. For Alberta, rankings take into account abundance and severity of problems caused by the weeds.

IMPACTS AND MANAGEMENT

It is not surprising that the main problems with aquatic plants in the western United States and Canada are caused by rooted plants (both emergent and submerged), given the extensive reaches of moderate to rapidly-flowing waters, both natural and man-made. Rapid spread of introduced plants occurs, facilitated by downstream proliferation and seasonal nutrient inputs from run-off and sedimentation. These flowing waters, together with the thousands of static-water storage reservoirs and natural lakes, sustain and disseminate plants causing severe economic impacts, which may be described as follows:

(1) direct blockage of flow (loss of canal capacity) and subsequent canal bank damage. These affect irrigation as well as drainage and flood control;

(2) loss of storage capacity in reservoirs, particularly small ones;

(3) increased wear on water pumps due to abrasion by plant parts (for example *Cladophora* filaments);
(4) blockage of drip-irrigation emitters;
(5) reduction of rice yield and quality;
(6) increased mosquito habitat, particularly in slow-flowing drains and flood control channels;
(7) interference with hydroelectric production;
(8) interference with recreation and related businesses (fishing, boating, swimming);
(9) distortion of canal design features: increased sedimentation, formation of berms and hummocks, channelling of flow through weed beds (*Typha* spp., *Phragmites* spp., and *Phalaris arundinacea* are the main problem species.); and
(10) direct and indirect effects on fisheries: dissolved oxygen; predator habitat, spawning habitats, habitats for secondary producers.

The health-related impact (e.g. 6 above) is exemplified by the current increasing problems with dense populations of *Myriophyllum aquaticum* (parrotfeather) and *Ludwigia repens* (water primrose) in central and northern California. In this and other US states, mosquito control is conducted through abatement districts, the personnel of which are responsible for ensuring that the numbers of various mosquito species are kept in check by using chemical adulticides and larvicides, biological control with the mosquito fish (*Gambusia affinis*), and proper management of breeding habitats. A 1985 survey of irrigation, abatement, flood control and reclamation agencies in California revealed that 24 per cent of 337 agencies were trying to control *M. aquaticum*, which infested almost 1 000 km of waterways plus over 200 surface hectares. Most (58 per cent) reported some success. Table 19a.5 summarizes this survey and indicates that both species are on the increase in California.

Although it is extremely difficult to attribute specific costs to many of these impacts, some estimates have been made, based primarily on the cost of management. However, Timmons (1960) used information from 47 USBR districts to calculate that in the late 1950s aquatic weeds (not including ditchbank weeds) in the whole Bureau system, caused approximately \$2.7 million from 'water lost' plus another \$1.3 million losses from 'other damages', totalling \$4 million annually. Ditchbank weeds caused another \$1.7 million in losses. Timmons also reported that with about half the infested canals being treated in some form, this cost was about \$1.6 million for aquatic weeds and another \$6.5 million for ditchbank weeds. A grand total of \$13.8 million was reported for losses plus control costs, from both aquatic and ditchbank weeds. The average cost for

Table 19a.5. Status and impacts of *Myriophyllum aquaticum* and *Ludwigia repens* in California

		M. aquaticum	L. repens
Occurrence:			
% Systems infested		24	23
Length of waterways infested (km)		940	820
Additional surface area infested (ha)		214	208
% Agencies showing increases:		48	29
% Agencies showing decreases:		3	5
Impact/costs:			
% agencies attempting control		35	33
Methods used:	Herbicides	52%	
(both species)	Dragline	21%	
	Dredging	16%	
Costs: (2 year period)		$215544	$170800

Note: Data are results from 1985 California-sponsored questionnaire sent to 600 agencies of which 337 responded.

aquatic weed control was *c.* $26 per km in 1957. In an unpublished survey of nine California irrigation districts in 1984, the overall average cost for mechanical, chemical, and combinations of these methods was *c.* $250 per km (USDA Economic Research Service 1985, unpublished data). Considering inflation from 1960 to 1987, this figure is not unreasonable, and totals for the West probably exceed $50 million annually.

MANAGEMENT APPROACHES

Western United States

The early days of aquatic weed management in the West (e.g. USBR 1949) were characterized by an almost complete reliance on mechanical methods including steam or diesel-powered dredgers, draglines, chaining devices, and cutters. Copper sulphate was (and still is) used extensively. However, the development of organic herbicides in the 1940s to 1960s led to a shift towards these less labour-intensive approaches. In particular, for flowing systems, the aromatic solvent xylene, applied as an emulsion up to 740 mg a.i. l^{-1}, and later acrolein and endothall, applied at *c.* 10 mg a.i. l^{-1} provided control of the submerged species.

Though these developments in the 1960s and early 1970s provided more management tools, there has been a reduction in the numbers and types of herbicides available, due to economic and environmental concerns over the last 8–10 years. Few new aquatic herbicides have been approved by the US Environmental Protection Agency recently, but two notable exceptions are glyphosate (Rodeo™) and fluridone (Sonar™) which are used primarily for ditch-bank (emergent) and submerged weeds, respectively. Unfortunately, neither is effective in moderate to rapidly flowing water conveyance systems where the submerged weeds create problems, and Sonar is not currently approved for use in California even in static water. It is approved for use in several other western states.

It should be apparent that the rapid flows and multi-use nature of western water systems makes EPA registration for aquatic herbicides difficult. The central concerns are the potential impacts on 'downstream' non-target species and use-sites which include crops, urban vegetation, municipal (potable) water, fisheries, recreational sites, and livestock. It is not surprising that few sufficiently active chemicals can fit these constraints and still provide effective aquatic weed control. In Table 19a.6 the herbicides currently used in the Western United States are listed.

These circumstances have meant that new strategies for aquatic weed control are needed, particularly in flowing systems. Since most western canal systems are drained during the winter (from about November to April), this window of opportunity is being explored for several potential methods. Among such methods are: (1) use of better soil-active herbicides (for example, bensulfuron methyl); (2) use of selective re-watering to induce precocious germination; (3) water

Table 19a.6. Herbicides used to control aquatic weeds in the western United States

Aquatic herbicides registered for use in static water

Trade name	Common name
Aquathol-K	Endothall (K salt)
Hydrothol 191	Endothall (mono N,N-dimethyl alkylamine) salt
Norasac	Dichlobenil
Aquazine	Simazine
Sonar	Fluridone
Fenatrol	Fenac
Komeen	Copper EDA (ethylene diamine)
Cutrine	Copper TEA (triethanolamine)
Ortho Diquat	Diquat
Aqua-Kleen	2,4-D
Rodeo	Glyphosate
Copper sulphate	Copper sulphate

Aquatic herbicides registered for use in flowing water

Trade name	Common name
Copper sulphate	Copper sulphate (various forms)
Magnacide H	Acrolein
Hydrothol 191	Endothall (mono N,N,-dimethyl alkylamine) salt
Xylene	Xylene
Ortho Diquat	Diquat

management to induce changes in weed morphology (via heterophylly), which may facilitate herbicide action; (4) establishment of beneficial competitive species, such as small spikerush, *Eleocharis acicularis* (Anderson, 1986*b*). All such approaches must take into account the fate of the water as well as the water use patterns and physical limitations of the systems (see McNabb and Anderson 1985 for more detailed examples of treatment decision schemes).

The recent development of the triploid grass carp has led to several on-going research and demonstration projects in many western states, including California, Washington, Oregon, Colorado, and New Mexico. This herbivorous, and presumptively sterile, fish has performed very well in both static and flowing systems and holds great promise for further applications, as a single-use tool or in fully-integrated management schemes, in lakes, ponds, and canal systems. The major caveat to the enthusiasm regarding the fish is its overall non-selectivity and limitations on our ability to control its movement in flowing systems. While clear feeding preferences are exhibited by the grass carp when several macrophytes are available, eventually less-preferred plants are consumed, as the abundance of preferred plants diminishes. Economical methods for 'herding' and overwintering the fish must be found if it is to be of practical use in the seasonally-dewatered irrigation channels in the West.

Canada

Although there is some use of herbicides for control of plants in irrigation districts (Burland and Catling 1986), various mechanical methods are relied upon for most Canadian sites (e.g. Armour *et al.*1980). A good example is the response in British Columbia to increasing *M. spicatum* populations (MacKenzie, Oldham, and Powrie 1979). Five distinct mechanical methods have been used over the past 10 years: harvesting, rototilling (Fig. 19a.5a, b), shallow water cultivation, diver-operated dredging, and benthic barriers (Ministry of Environment and Parks, 1986). None provides very long-term control, but these approaches reflect a reluctance to use herbicides; in fact, the use of 2,4-D for control of *M. spicatum* in the Okanagan Valley was discontinued after 1981. Costs (Canadian dollars) of these methods range from \$125–\$400 per ha for shallow water tillage to \$1 200 per ha for harvesting and \$26 000 per ha for some types of bottom barriers (Ministry of Environment and Parks 1986). These costs do not include capital outlays for machinery, which can be nearly \$100 000. In addition to Eurasian water milfoil, several other lakes are managed to control excessive growth of plants including *Brasenia, Ceratophyllum, Nuphar*, and *Chara*. These problems too are dealt with via mechanical methods which are expensive and generally labour-intensive.

FIG. 19a.5. (a) Eurasian water milfoil is often removed mechanically in Canada using a 'rotovator'. Note the long floats in the foreground: these support a screen which is put in place to retain the many fragments provided by rotovation. (b) Close view of hydraulically operated rotavator 'business' end: note the rotating tines and the fragments of Eurasian water milfoil.

SPECIAL CASES: *HYDRILLA VERTICILLATA* AND *EICHHORNIA CRASSIPES*

Hydrilla verticillata

Since its introduction into the south-eastern United States in the early 1960s, *H. verticillata* has progressively spread into Texas, California, and Arizona, and these states constitute its current distribution in the West. Table 19a.7 summarizes the incidence of dioecious *H. verticillata* in these states and points out

Table 19a.7. Occurrence of dioecious *Hydrilla verticillata* in the western United States[a]

State	Year	Infested area (ha)
Arizona	1982	<2 (under eradication)
California	1975	None reported
	1976	12 (Lake Ellis, now eradicated)
	1977	200–250 (Increase due to infestation and spread in Imperial Irrigation District where it is currently under successful management with triploid grass carp)
	1984–1986	250–320 (Additional sites in Santa Rosa and Shasta counties; Santa Rosa site eradicated in 1985)
	1987–1988	320–330 (Additional 10 sites in Calaveras county and a 0.8 ha lake in San Francisco)
Texas	1976[b]	~500
	1984	4900
	1986	>24000

[a] Modified from Anderson (1986*a*). Note that these are maximum areas. Weed coverages at several sites have been reduced, especially in California.
[b] From Tyndall (1982).

that the California introductions are the most recent. Fig. 19a.6 shows the distribution and size of infestations in California by County. Note that the most northern site (Shasta county) is located adjacent to the Sacramento River, a major source of water for northern California domestic and agricultural uses and also a major contributor to the Sacramento Delta. This infestation is the most immediate and current threat to the extensive water transport systems in the state since water from the Sacramento Delta is exported to central and southern California via the California Aqueduct and the Delta Mendota Canal.

As evidence that not all hydrilla in California has been located (or that it may still be spreading) eleven 'new' infestations were discovered in 1988. All except one (McLaren Park, San Francisco) are in Calaveras County on fairly remote tracts of land where these small ponds are used for fishing and some irrigation. Although the Calaveras sites are 50 to 60 kilometres from the Sacramento Delta, there is potential for movement into the expansive Delta waterways via a small, seasonally flowing tributary. As in the Shasta County sites, these infestations,

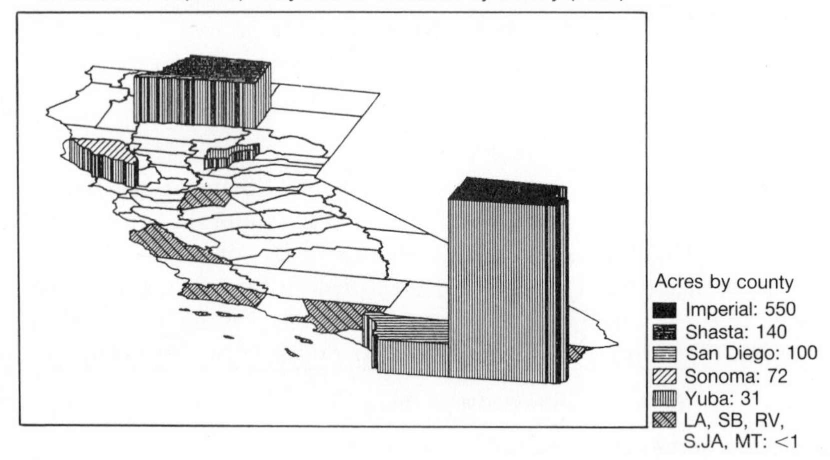

FIG. 19.6. Distribution of *Hydrilla verticillata* in California showing the greatest areal coverage reported at any one time prior to 1987. Note that much of the acreage shown has now been reduced and that infestations in Shasta county and Imperial county are the only remaining large populations (from Anderson 1986a). In 1988, additional hydrilla infestations were discovered in northern California (10 sites in Calaveras county; one lake in San Francisco: N. Dechoretz, pers. comm.).

FIG. 19a.7. Aerial view of a hydrilla infested area of Sheldon Reservoir in southern California (1984). Fragmentation from upstream populations caused this reservoir to become 80 per cent covered with hydrilla within 6 months of its construction. A major canal system is seen to the left.

once found, were immediately treated with chelated copper (Komeen™) to remove above-sediment growth. Subsequent treatments with Fluridone (Sonar™) will be made to prevent regrowth from turions.

The impacts of *H. verticillata* infestations (Figs 19a.7, 19a.8) are not unlike those caused by excessive growth of many other rooted submerged aquatic plants: interference with water transport in canals, lost recreational revenues, higher costs of management and wildlife habitat impairment. A unique feature of the response in California to *Hydrilla* introduction was its classification as a 'Type A' pest. This designation dictated that the plant was to be eradicated whenever possible and that its sale, intrastate movement and importation is illegal. As a consequence, California has expended roughly \$1.3 million annually since 1978 on various eradication efforts on several sites (Anderson 1986*a*).

FIG. 19a.8. A partially dewatered main canal in the Imperial Irrigation District in southern California (1979). *Hydrilla* occupied 50 to 75 per cent of the canal volume when water was flowing. This irrigation system was stocked with triploid grasscarp in 1986 and 1987 which has resulted in >90 per cent clearance of *Hydrilla*.

Results of these eradication programmes have been good (Fig. 19a.9) and only two major infestations remain: Redding (Shasta County) and the Imperial Irrigation District (IID) in the southern-most part of the state. The Redding sites will probably be eradicated by 1989 but the IID site is extensive and in rapidly flowing water. There, several thousand triploid grass carp were introduced in 1985–1987 and early results are extremely promising with >90 per cent

FIG. 19a.9. *Hydrilla* eradication has a high priority in California. Here the complete excavation and renovation of a 70 hectare recreational and flood control lake is being done in order to remove all plants and tubers. At a cost of about US $1 000 000 this lake was restored from 1985 to 1986 and is now in use again.

reduction in plant populations reported (Stocker 1987). It remains to be seen if the continued presence of the fish will eventually deplete the existing *Hydrilla* turion bank, and bring about at least near eradication. It should be noted too that, given the propensity for *H. verticillata* to spread, the expenses incurred with California's assertive eradication programme probably have saved many millions of dollars over the long-term since most of the potential sources for spread have been removed relatively quickly.

Finally, to complete the picture for the western distribution of *H. verticillata*, recently documented infestations in an irrigation system in the Mexicali Valley, Mexico (near the United States border at Calexico) should be mentioned (Stocker 1987). Surveys in 1987 revealed about 38 km of earthen canals infested with *Hydrilla*. An introduction of triploid grass carp is planned for *c.* 62 km in 1988–1989. This irrigation system is immediately south of the extensive infestation in the Imperial Irrigation District, though there is no direct waterway connection to the IID infestations.

Eichhornia crassipes

The problem infestations of water hyacinth in the western United States are primarily in Texas and in the Sacramento Delta of California (Fig. 19a.10a,b).

FIG. 19a.10. (a) Water hyacinth has become a menace in many regions of the Sacramento Delta where it blocks marinas and interferes with boat traffic. This picture shows a dense stand encroaching on an irrigation pump. Since 1984, the massive growths have been kept under control with the application of 2,4-D in a carefully monitored management programme.
(b) Before the herbicide programme was instituted, several hundred truckloads of water hyacinth were hauled away daily from the conveyer system on the Delta, a process costing hundreds of thousands of dollars annually. With the herbicide programme, costs have been cut by at least 90 per cent.

According to Tyndall (1982), this exotic species, which was noted as a problem in the 1930s, occurred on about 6 500 ha in Texas in 1981, and *c.* 7 000 ha in 1984 (Guerra 1984). In California, although *E. crassipes* was noted by the early 1900s, problems did not arise until the late 1940s and then again in 1980 (Thomas and Anderson 1984). The total area, even recently, has been small (*c.* 200 ha) compared to Texas or Florida (1985 Florida estimates: 494 000 ha according to Schardt 1986). However, the nature of the Sacramento Delta site with narrow sloughs, tidal movement, and large pumping stations made this infestation a serious problem. At the US Bureau of Reclamation pump works, tons of the plants were removed daily and had to be hauled away, a job costing several hundred thousand dollars annually. This expensive activity has not been required for the past 3 years since 2,4-D applications began.

In both Texas and California, foliar application of the herbicide 2,4-D has been used very effectively to reduce the growth and spread of the water hyacinth. About 4 300 ha of water hyacinth were treated with herbicides in Texas in 1987, and perhaps twice that area may need treatments in 1988 (Texas Parks and Wildlife Service, pers. comm.)

In addition, biological control agents (*Neochetina eichhorniae, N. bruchi, Sameodes albiguttalis*) have been released in California, but have not proved effective in the Sacramento Delta as yet. The key to management in this site has been the early spring applications of 2,4-D to prevent the formation of massive and mobile mats of the plants which would otherwise block commercial navigation, recreational uses, and the vital pumping system on which Central and Southern California depends. This strategy is coupled with regular monitoring for residues of the herbicide at strategic points in the Delta. In the four years of this programme, no 2,4-D residue problems have arisen.

THE FUTURE

The intensive urbanization, and expanded agricultural and industrial activities in the western United States over the past 30–40 years have placed greater and greater demands on the water resources of this region. These demands will intensify since the 1985 population for the seventeenth western states, about 71 million, is projected to increase to over 90 million by the year 2000 (USBR 1985).

Because general concerns for environmental protection are ever more intensely focused on water as a vital yet mobile resource, aquatic plant management strategies must become more tailored, sophisticated, efficient, and cost-effective. The public's concerns about groundwater contamination from industrial and agricultural chemicals, including herbicides, place more urgency on the development of non-chemical aquatic weed management approaches.

Experience has shown that key components of aquatic weed management in this age must include multi-agency co-ordination where public waters are concerned, and a highly effective public awareness and education programme. Managers in Canada have been especially thoughtful and diligent in their public education and quarantine efforts, and the western US states are increasing their activities here too. With these ingredients, and continued basic and applied research on the nuisance aquatic plants, it should be possible to reduce the economic and health impacts of aquatic weeds greatly in the next decade.

ACKNOWLEDGEMENTS

The following people provided various information and resources and their help is greatly appreciated: Mr Carl Tennis, Mr Greg Crossman (US Bureau of Reclamation); Mr William Mitchell (California Dept. of Water Resources); Mr Winn Winkyaw (Salt River Project, Arizona); Mr Nathan Dechoretz (California Dept. of Food and Agriculture); Dr Peter Newroth (Ministry of Environment, Vancouver BC), Ms Sherry Williams (USDA/ARS, Davis, Ca.).

(b) Aquatic weed problems and management in the eastern United States

K. K. Steward

The total area of surface water in the eastern states of the USA, in which aquatic weed problems may occur, is approximately 86 937 km^2 (8.7 million ha, Table 19b.1). This is an area about the size of Austria, or slightly larger than the state of Maine in the USA. The geographical boundaries of this area are 24° to 47° N, and 67° to 92° W. There is a wide diversity of climatic and habitat conditions. The southern states with warmer temperatures, longer growing seasons, and generally hard, fertile water, tend to have more serious aquatic weed problems than do the northern states. As would be expected, aquatic plant management programmes have been established longer and are more active in the southern states. Table 19b.2 lists the aquatic plants which may create problems in the eastern USA. Each species is identified by a two-character symbol when reported as a problem (Table 19b.4). Table 19b.3 lists aquatic herbicides used to control various problem species and are also identified by a two character symbol when used on a particular problem species (Table 19b.5).

The largest proportion of the activities to manage aquatic plants is performed by public agencies of federal, state, county, and municipal governments. The US Army Corps of Engineers (CE) and the Tennessee Valley Authority (TVA) are

Table 19b.1. Surface water area for the eastern United States

North-east		South-east	
State	Area (ha)	State	Area (ha)
Connecticut	53 450	Alabama	333 720
Delaware	34 830	Florida	1 418 310
Indiana	119 880	Georgia	371 385
Kentucky	246 240	Louisiana	1 080 135
Maine	617 220	Mississippi	293 625
Maryland	198 450	North Carolina	1 109 700
Massachusetts	115 020	South Carolina	302 535
Michigan	447 525	Tennessee	317 520
New Hampshire	91 935		
New Jersey	98 415		
New York	564 165		
Ohio	151 470		
Pennsylvania	176 580		
Rhode Island	46 170		
Vermont	100 035		
Virginia	345 060		
West Virginia	60 345		

Table 19b.2. Aquatic plants of the eastern and north-central United States

Common name	Scientific name
Alligator weed	*Alternanthera philoxeroides*
Lemon bacopa	*Bacopa caroliniana*
Watershield (ws)	*Brasenia schreberi*
Fanwort (fw) cabomba (cb)	*Cabomba caroliniana*
Sedge	*Carex* spp.
Coontail (ct)	*Ceratophyllum demersum*
Muskgrass (mg)	*Chara* spp.
Sawgrass	*Cladium jamaicense*
Colocasia	*Colocasia esculentum*
Brazilian elodea	*Egeria densa*
Water hyacinth	*Eichhornia crassipes*
Spikerush	*Eleocharis* spp.
Canadian waterweed (cw)	*Elodea canadensis*
Water stargrass	*Heteranthera dubia*
Hydrilla (hy)	*Hydrilla verticillata*
Water pennywort	*Hydrocotyle* spp.
Hygrophila	*Hygrophila polysperma*
Trompetilla	*Hymenachne amplexicaulis*
Duckweed	*Lemna* spp.
Frog's bit	*Limnobium spongia*
Ambulia	*Limnophila sessiliflora*
Water primrose	*Ludwigia uruguayensis*
Water milfoil (wm)	*Myriophyllum* spp.
Parrot feather	*M. aquaticum*
American water milfoil	*M. exalbescens*
Variable-leaf milfoil	*M. heterophyllum*
Southern milfoil	*M. laxum*
Green parrots feather	*M. pinnatum*
Eurasian water milfoil	*M. spicatum*

Table 19b.2. (*contd*)

Common name	Scientific name
Naiad (nd)	*Najas* spp.
Southern naiad	*N. guadalupensis*
Marine naiad	*N. marina*
Brittle naiad (bn)	*N. minor*
American lotus (al)	*Nelumbo lutea*
Stonewort (sw)	*Nitella* spp.
Spatterdock (sd)	*Nuphar luteum*
Water lily (wl)	*Nymphaea* spp.
Fragrant waterlily	*N. odorata*
Floating heart	*Nymphoides aquatica*
Maidencane	*Panicum hemitomon*
Paragrass	*P. purpurascense*
Torpedo grass	*P. repens*
Water paspalum	*Paspalum fluitans*
Napier grass	*Pennisetum purpureum*
Giant reed (gr)	*Phragmites australis*
Water lettuce	*Pistia stratiotes*
Giant smartweed	*Polygonum coccineum*
Smartweed	*P. densiflorum*
Pickerel weed	*Pontederia lanceolata*
Pondweed (pw)	*Potamogeton* spp.
Curlyleaf pondweed	*P. crispus*
Illinois pondweed	*P. illinoensis*
Floating pondweed	*P. natans*
American pondweed	*P. nodosus*
Sago pondweed	*P. pectinatus*
Small pondweed	*P. pusillus*
Widgeon grass	*Ruppia maritima*
Strap leaf sag	*Sagittaria kurziana*
Arrowhead	*Sagittaria lancifolia*
Floating fern	*Salvinia rotundifolia*
Angel's hair	*Scirpus confervoides*
Sphagnum moss (sm)	*Sphagnum* spp.
Duckweed	*Spirodela* spp.
Water chestnut (wc)	*Trapa natans*
Cattails	*Typha* spp.
Bladderwort (bw)	*Utricularia* spp.
Eel grass (eg)	*Vallisneria americana*
Horned pondweed	*Zannichellia palustris*
Giant cutgrass	*Zizaniopsis miliacea*
Filamentous algae (fa)	several

Note: two-letter species codes are given for major weed species.

the only federal agencies actively engaged in aquatic weed control programmes in the regions covered in this report. The CE has been a dominant force since enactment of the Rivers and Harbours Act of 1899. Initially, this federal legislation authorized removal of water hyacinth from navigable waters of the states of Louisiana and Florida. The scope of the CE programme was expanded by subsequent legislation in 1902, 1905, 1912, 1916, 1958, and 1965 and at the present time the CE is authorized to expend $15 million annually to provide for

Table 19b.3. Common names of aquatic herbicides listed in the text

Code	Herbicide
ax	2,4-D
Bn	Banvel (dicamba)
Bn7	Banvel 720 (dicamba-dimethylammonium)
ci	inorganic copper
co	organic copper
db	dichlobenil
dn	diuron
dp	dalapon
dq	diquat
ea	endothall-amine
ek	endothall-potassium
fl	fluridone
gl	glyphosate
im	imazapyr
sz	simazine

'control and progressive eradication of noxious aquatic plant growth'. The 1965 legislation which established the Aquatic Plant Control (APC) Program, provides for establishment of co-operative programmes with other federal agencies and states. Seventy per cent of the cost of control operations of co-operative programmes will be provided by the CE APC Program (50 per cent as of 1 October 1987). This requires formal agreement from co-operators to hold the US government free from claim which may occur from control operations. This legislation also provides for continuing research for development of the most effective and economical control measures. The cost of this research is fully borne by the APC Program. The research programme is administered and located at the Waterways Experiment Station, Vicksburg, Mississippi. Several states participate in the APC cost sharing programme (Alabama, Florida, Georgia, Louisiana, Maryland, North Carolina, South Carolina, Virginia, Vermont) as well as those conducting management programmes supported by state funds (Alabama, Florida, Georgia, Louisiana, North Carolina, South Carolina, Connecticut, Delaware, Indiana, Kentucky, Maryland, Massachusetts, Michigan, New Jersey, New York, Ohio, Pennsylvania, Rhode Island, Vermont, Virginia, West Virginia). Jubinsky (pers. comm.) estimated expenditure on aquatic weed control in Florida in 1987 to be nearly US $23 million: US $5.9 million through the CE APC Program; US $4.8 million of state and local funds; and US $12 million by the private sector.

A difficult-to-estimate proportion of aquatic plant control is carried out privately by land developers, golf courses, farmers, homeowners' associations, and, occasionally, individual property owners. The use of herbicides is regulated by federal and state agencies. Permits are required by most states before application of these materials to water is allowed. Some states require that pesticide application to public water only be done by trained and certified applicators. This requirement further restricts the activities of individuals and shifts the responsibility to public agencies and commercial applicators.

Documentation of aquatic weed problems and management activities is poor nationwide since management programmes throughout the country function independently and have no organized system of information exchange. There is no national programme other than that of the CE.

Florida has the best organized state programme (Burkhalter 1975) and consequently the best records. Florida's government has designated their Department of Natural Resources, Bureau of Aquatic Plant Management as the lead agency responsible for aquatic plant control. They conduct a spraying programme through contracts to local agencies, a permit programme for control activities, and conduct surveys of state waters for aquatic plant infestation. They adminster co-operative cost sharing programmes with the CE and with municipal and county agencies. They also, through the permit programme, regulate the importation, transportation, and culture of exotic aquatic plant species. A list of prohibited species is maintained and revised as needed. The law is enforced and violators are subject to fines as well as loss of permits. This pioneering legislation has been in place since 1969 (regulation) or 1970 (control) and has been serving as models for use by other states and countries. The Federal Noxious Weed Act of 1974 was passed by the US Congress in January 1975 and provides '... for the control and eradication of noxious weeds, and the regulation of the movement in interstate or foreign commerce of noxious weeds and potential carriers thereof, and for other purposes.' This legislation was recently reviewed by committees of the US Departments of Agriculture and Interior and recommendations are being made to improve the function of the law. In Figs 19b.1–9 aquatic weed infestations are shown in various areas in Florida.

The most recent surveys of aquatic weed problems and management costs in the eastern United States were reported in 1982. Funds for these surveys were provided by the US Environmental Protection Agency, Office of Environmental Processes and Effects Research. The results of these surveys are contained in two reports (Aurand 1982; Trudeau 1982). The data in Tables 19b.4 and 19b.5 were compiled from these reports as well as from responses from water resource managers of several states.

Table 19b.4. Aquatic weed problems and management in the north-eastern and north-central states of the USA

State	Weeds (for codes see Table 19b.2)	Impairment	Management	Management costs	Notes
Connecticut	wm, ct, sd, bw, pw	Recreation, aesthetic	Chemical, mechanical, drawdown	\$40 000–50 000/yr	American and Eurasian water milfoil are present. Management costs are mainly for chemical control. Herbicides used to treat 10050 acres (4070 ha) in 1980 were co, ci, dq, ek, sz, ax. Use of grass carp is illegal.
Delaware	fw, hy, sd, wm, bw	Fishing, swimming	Chemical, drawdown	\$10 000/yr	
Indiana	mg, cw, wm, pw, ct, bn	Recreation, aesthetic, drainage, irrigation	Chemical, mechanical, physical	?	The predominant control technique in the state is chemical. Three types of physical management techniques are used: benthic screens, dyes to reduce light penetration and drawdown. Grass carp are used though illegal. Sago and curlyleaf pondweeds are in the state.
Kentucky	pw, bn, wm, ct, sd, wl	Fishing, aesthetic, farm ponds, irrigation	Chemical, drawdown	?	
Maine	no problems	None	Mechanical, physical	?	
Maryland	wm, hy, cw, wl, pw, bw, fa	Recreation, fishing	Mechanical, chemical	?	*Hydrilla* is a serious problem in the Potomac River. Eurasian water milfoil is present. White waterlilies are present.
Massachusetts	wm, pw, fw, wc, bw, ct, cw	Recreation, aesthetic	Chemical, mechanical, drawdown	\$520/ha	Eurasian, variable-leaf and American water milfoil present. Pondweeds present: *P. crispus, amplifolius, robbinsii, richardsonii,* and *pectinatus.* White and yellow water-lilies present.
Michigan	wm, pw, ct, cw, eg	Swimming, boating, fishing	Chemical, mechanical, drawdown	?	Eurasian and American water milfoil present. Curlyleaf and Sago pondweed present.
New Hampshire	bw, wm, fw, pw	Aesthetic, recreation	Chemical, mechanical, drawdown	?	Variable-leaf water milfoil present.

New Jersey	wm, wl, sd, pw, cw, bw, fa, na, cb, sm	Recreation, drainage	Chemical, mechanical	$160/ha	Fragrant water-lily present. Curlyleaf pondweed present. Sphagnum resistant to chemicals. Drainage in cranberry bogs affected by weeds.
New York	wm, pw, wc	Recreation	Chemical, mechanical, physical	?	Eurasian and American water milfoil present. Curlyleaf pondweed present. Water-chestnut estimated to infest 2025 ha in 1980. Dredging and harvesting are used as are benthic screens and drawdown.
Ohio	wm, pw, nd, mg, ct, wl, sd, al	Recreation, farm ponds, irrigation	Chemical, drawdown	?	Leafy pondweed, variable-leaf and eurasian water milfoil and white water-lily present.
Pennsylvania	pw, cw, wm, nd, bw, mg, sd, ws	Recreation, aesthetic	Chemical, drawdown	?	Curlyleaf pondweed present.
Rhode Island	wm, bw, pw, fw, wl, sd, ws	Recreation, aesthetic	Chemical	?	Pondweeds: *P. richardsonii, P. nodosus, P. amplifolius,* and white water-lily present.
Vermont	wc, wm, pw, cw, ct	Recreation, minor navigation	Mechanical, physical, chemical	?	Eurasian water milfoil and curlyleaf pondweed present. Management involves harvesting, benthic barriers, drawdown and limited chemical use.
Virginia	hy, pw, cw, wm, ws, ct, wl	Recreation, farm ponds	Drawdown, chemical, grass carp	?	Curlyleaf and sago pondweeds, Eurasian water milfoil and white water-lily present.
West Virginia	pw, wl, nd, wm, cw, mg	Recreation, farm ponds	Chemical, drawdown	?	

Table 19b.5. Aquatic weed problems and management in the south-eastern states of the USA

State	Weed	Weed area (ha)	Chemical use and cost	Biological control and cost	Date reported
Alabama	Algae (TVA)	101	ci		Sep 86
	American lotus (TVA)	444	ax		Sep 86
	Eurasian water milfoil (TVA)	5014	ax		Sep 86
	Hydrilla (TVA)	533	dq, ek, fl		Sep 86
	Mixed milfoil, naids (TVA)	1188	dq, ek		Sep 86
	Naiads (TVA)	1258	dq, ek		Sep 86
	Pondweeds (TVA)	77	ea		Sep 86
	Water hyacinth				Dec 80
	Other (TVA)	458	ax, dq, ek, fl, gl		
Florida	Alligator weed	1007	Bn, Bn7, dq, ax, gl	*Agasicles, Vogtia, Amynothrips*	Dec 84
	Cattails	12327	dp, gl		Dec 84
	Eurasian water milfoil	874	ax		Dec 84
	Filamentous algae	4530			Dec 84
	Fragrant water-lily	3748	db, gl		Dec 84
	Spatterdock	3839	gl, db		Dec 84
	Torpedo grass	11809			Dec 84
	Great bulrush	4696			Dec 84
	Water hyacinth	2344	ax, $148/ha		Dec 85
	Illinois pondweed	5932			Dec 84
	Southern naiad	2021			Dec 84
	Giant reed	2179			Dec 84
	Hydrilla	21253	dq + co, $1015/ha		Dec 85
	Maidencane	3707			Dec 84
	Water lettuce	2976	dq		Dec 84
Georgia	Brazilian elodea	8	dq		Dec 80
	Eurasian water milfoil	3240	ax		Dec 80
	Hydrilla	1012			Dec 79
	Illinois pondweed	1053			Dec 80
	Water hyacinth	375			Dec 79
Kentucky	Eurasian water milfoil (TVA)	16	ax		Sep 86
	Naiads (TVA)	648	dq, ek		Sep 86
	Mixed milfoil, naiad (TVA)	125	dq, ek		Sep 86
	American lotus (TVA)	4	dq, ek		Sep 86
	Mixed (TVA)	43			Sep 86

Louisiana	Water hyacinth	84 361	ax, $136/ha		Jan 86
	Alligator weed	60 750	Bn, Bn7, ax, dq, gl	*Agasicles, Vogtia*	Dec 82
Mississippi	Water hyacinth	247			Dec 76
North Carolina	Alligator weed	1 023	gl, im		Feb 86
	Bladderwort	0	dq, fl	Sterile grass carp	Feb 86
	Brazilian elodea	500	db, dq, fl		Feb 86
	Brittle naiad	1 000	dq, ek, fl		Feb 86
	Duckweeds	0	fl, sz	Sterile grass carp	Feb 86
	Eurasian water milfoil	0	ax, dq	Sterile grass carp	Feb 86
	Hydrilla	1 000	dq, ek, fl	Sterile grass carp	Feb 86
	Pondweed	0	dq, ek, fl	Sterile grass carp	Feb 86
	Water primrose	500	gl	Sterile grass carp	Feb 86
South Carolina	Alligator weed	125	gl, $198/ha	*Agasicles,* flea beetle $20/ha	Jan 86
	Brazilian elodea	2 662	dq, $252–341/ha		Jan 86
	Brittle naiad	684	ax, dq, $252–417/ha	Triploid grass carp $254/ha	Jan 86
	Water milfoil	324	ax, dq, $252–417/ha	Triploid grass carp $254/ha	Jan 86
	Giant reed	1 417	gl, $575/ha		Jan 86
	Hydrilla	376	co, dq, ek, $341–518/ha		Jan 86
	Water hyacinth	247	ax, dq, $123–160/ha	*Neochetina*	Jan 86
	Water primrose	1 255	gl, $198–252/ha		Jan 86
Tennessee	Eurasian water milfoil (TVA)	448	ax		Sep 86
	Naiads (TVA)	1 962	dq, ek, fl		Sep 86
	Mixed milfoil, naiads (TVA)	133	dq, ek, fl		Sep 86
	American lotus	19	ax		Sep 86
	Algae (TVA)	7	ci		
	Other	959	dq, ek, fl, ax, gl		

Note: TVA = Tennessee Valley Authority region; for chemical codes see Table 19b.3.

FIG. 19b.1. Aerial view of water hyacinth invasion of a large South Florida irrigation-drainage canal. (Photo: K. K. Steward.)

FIG. 19b.2. Water hyacinth infestation of a small south Florida drainage canal. (Photo: K. K. Steward.)

FIG. 19b.3. Water hyacinth mat developing over *Hydrilla* in a south Florida water storage area. (Photo: K. K. Steward.)

FIG. 19b.4. Mixed growth of white and yellow water-lily (*Nymphaea* and *Nuphar*) in southwest Florida's Lake Trafford. (Photo: K. K. Steward.)

FIG. 19b.5. Water-lilies and cattail in background, Lake Trafford, Florida. (Photo: K. K. Steward.)

FIG. 19b.6. *Hydrilla* growth in a large south Florida flood control canal (Broward County). (Photo: K. K. Steward.)

FIG. 19b.7. *Hydrilla* growth in a medium to small south Florida flood control canal (Dade County). (Photo: K. K. Steward.)

FIG. 19b.8. *Hydrilla* growth in a south Florida golf course pond surrounded by private residences. (Photo: K. K. Steward.)

FIG. 19b.9. Extensive *Hydrilla* growth (developed since 1983) in the Potomac estuary (large patches fringing main channel), south of Washington DC. Access to Bella Haven Marina (lower right) is impeded by weed grow. Outbound motor scars in the main weed bed are visible. (Photo: R. S. Hammerschlag.)

ACKNOWLEDGEMENTS

I am indebted to the following colleagues who responded to my request for data on aquatic weeds from their particular region: Leon Bates, Tennessee Valley Authority, Muscle Shoals, AL; Steven de Kozlowski, South Carolina Water Resources Commission, Columbia, SC; Mr Larry Hartmann, Louisiana Dept. of Fish and Wildlife, Baton Rouge, LA; Kenneth Langeland, Dept. of Weed

Science, North Carolina State University, Raleigh, NC (currently—Center for Aquatic Plants, University of Florida, Gainesville, FL); Larry Nall, Florida Dept. of Natural Resources, Tallahassee, FL; Donald Riemer, Soils and Crops Dept., Cook College, Rutgers University, New Brunswick, NJ.

20

Aquatic weed problems and management in South and Central America

O. A. Fernández, D. L. Sutton, V. H. Lallana, M. R. Sabbatini, and J. H. Irigoyen

INTRODUCTION

South and Central America, with a rapidly growing human population currently around 400 million, extends from latitude 20°N to 56°S and is 8 000 km in length and 5 000 km at its widest point. The altitude ranges from sea level to 7 000 m in the highest mountains. Over such a vast expanse of territory the variation in vertical and horizontal bio-geographical zones is, therefore, of extraordinary magnitude, ranging from tropical rain forests to deserts such as that of Puna de Atacama, temperate zones like the Pampas, and the cold sub-antarctic regions. An enormous number of aquatic plant species occurs in natural bodies of water with the majority of these being native and generally not regarded as troublesome. However, when they start to invade bodies of water constructed, or used, by man, they become not only troublesome but also at times harmful.

The last 30 years have witnessed a huge increase in the number of aquatic environments created by man in South America. The need to increase food and energy production and to provide supplies of water for urban development in rapidly growing population centres, has led to the construction of a wide range of standing and flowing waters, from farm ponds and irrigation complexes to man-made reservoirs covering thousands of hectares.

With the growing utilization and management of water for human needs, concern over aquatic weeds has assumed greater importance as reservoirs and dams on the continent are experiencing serious problems. Among the worst affected systems are irrigation zones, where invading aquatic plants interfere not only with systems for the collection and storage of water, but also with irrigation and drainage channel networks.

Many areas of South and Central America are now beginning to experience a problem already encountered in the other parts of the world: dense infestations of aquatic plants threaten the economy and welfare of human communities by disrupting irrigation, fouling water supplies, impeding transportation, adversely affecting the fishing industry, and interfering with the functioning of hydroelectric power stations. Recreational uses of water, and the aesthetic value of aquatic environments are also impaired.

The tropical and subtropical climates of a considerable part of the continent provide favourable conditions for the rapid growth of aquatic plants to nuisance proportions. The problem is aggravated by the fact that elementary information on the biology and ecology of these plants is often lacking at the local level, as is knowledge of adequate aquatic weed control methods.

Compared with other areas of the world, published information on aquatic weeds and their management in South and Central America is quite scarce. Relevant information can be gleaned from certain studies of taxonomy, geographical distribution, and biology of aquatic macrophytes in South and Central America (Table 20.1). Tur (1977) developed an extensive bibliography but by

Table 20.1. Studies on aquatic macrophytes in South and Central America giving taxonomic identification, geographical distribution, biology and related data relevant to the aquatic weeds discussed in this chapter.

Aspect of study	References
1. Taxonomic studies	Pérez-Moreau 1938; Lourteig 1951, 1956; Castellanos 1952, 1958; Barros, 1953, 1960; Hoehne 1955; di Fulvio 1956, 1961; Burkart 1957; de la Sota 1962*a, b, c,* 1963, 1964; St. John 1963, 1964, 1965; Cabrera 1964; Foster 1965; Crespo and Pérez-Moreau 1967; Gamerro 1968; Rataj 1970, 1972, 1975, 1978; Dawson 1973; Agostini 1974; Cáceres 1978; Ancibor 1979; Ramírez and Stegmeier 1982; Tur 1982; Marta 1983; Forno 1983.
2. Geographical distribution	Hoehne 1955; St. John 1963, 1964, 1965; Hunziker 1966; Lombardo 1970; Armitage and Fassett 1971; Pieterse and Schulten 1976; Pisano 1976; Neiff and Marchessi 1978; Forno and Harley 1979; Borhidi, Muñiz, and Del Risco 1979; Ramírez, Romero, and Riveros 1979, 1980; Ramírez, Godoy and Hauenstein 1981; Alburqueque 1981; Ramírez and Beck 1981; Lallana, Marta and Sabattini 1981; Ramírez and Stegmeier 1982; Forno 1983; Rangel and Aguirre 1983; Irgang, Pedralli, and Waetcher 1984; Beck 1984; Pedralli, Irgang, and Pereira 1985; Tell 1985; Cruz and Delgado 1986.
3. Biology of aquatic macrophytes	Cabrera 1964; Gamerro 1968; Mitchell and Tur 1975; Rossi and Tur 1976; Tur and Rossi 1976; Neiff and Marchessi 1978; Borhidi, Muñiz, and Del Risco 1979; Rangel and Aguirre 1983; Bravo Velázquez, and Balslev 1985.

Table 20.2. Distribution by country[a], habitat, and ranking importance of major aquatic weeds of South and Central America

Taxon	Habitat						Status
	Rivers	Streams	Lakes and ponds	Marshes and fens	Irrigation and drainage canals	Reservoirs	Overall[b] ranking as weed in country
Eichhornia crassipes	ARG	ARG	ARG		ARG	ARG	C
			BOL	BOL			F
	BRA	BRA	BRA	BRA		BRA	P
			CHI	CHI			C
	COL	COL	COL	COL	COL	COL	P
	ECU	ECU		ECU	ECU	ECU	P
	GUA		GUA	GUA			C
			PAN	PAN		PAN	P
			PER				C
	PUR	PUR	PUR				C
	VEN	VEN	VEN	VEN	VEN	VEN	P
	CUB		CUB			CUB	S
Pistia stratiotes			BOL	BOL			F
	BRA	BRA	BRA	BRA		BRA	C
	ARG	ARG	ARG	ARG	ARG		C
	COL	COL	COL		COL		C
	ECU			ECU	ECU		C
	GUA		GUA				C
				PAN			X
	PER		PER				X
			VEN	VEN	VEN	VEN	C
	HON		HON	HON	HON	HON	X
Elodea spp.[c]			ARG		ARG		X
	BRA		BRA				X
					CHI	CHI	X
	CUB		CUB			CUB	X
			PER			PUR	C

Eichhornia azurea				BOL			F
	BRA	BRA	BRA			BRA	C
	COL	COL	COL	COL	COL		C
	VEN	VEN	VEN		VEN	VEN	C
Typha spp.[c]		URU	URU		URU		C
		PER			PER	PER	P
			ARG		ARG	ARG	P
			BOL	BOL			F
		BRA	BRA	BRA			C
			CHI	CHI			C
	COL	COL	COL		COL		C
			CUB	CUB		CUB	C
		VEN	VEN	VEN			C
		ECU	ECU	ECU			P
		PUR			PUR		C
Chara spp.[c]					ARG	ARG	P
	ECU		ECU		ECU		P
			PER	PER			C
					VEN		C
Potamogeton spp.[c]	ARG		ARG		ARG	ARG	P
			BOL				F
Myriophyllum spp.			ARG	ARG	ARG		F
			BRA	BRA			F
			CHI				X
Schoenoplectus tatora			ARG	ARG	ARG		C
			CHI	CHI			C
			PER				C
Lemna spp.[c]		ARG	ARG	ARG	ARG		X
			BOL	BOL			X
		BRA	BRA	BRA		BRA	X
			PER	PER			F
			CHI				X

Table 20.2. (*contd*)

Taxon	Habitat						Status
	Rivers	Streams	Lakes and ponds	Marshes and fens	Irrigation and drainage canals	Reservoirs	Overall[b] ranking as weed in country
Salvinia spp.[c]		ARG	ARG	ARG	ARG	ARG	X
		BRA	BRA	BRA			X
		ECU	ECU	ECU	ECU		C
			VEN		VEN	VEN	C
Paspalum repens	ARG	ARG	ARG	ARG	ARG		C
	VEN	VEN	VEN		VEN		C
	BRA	BRA	BRA				C
Azolla spp.[c]			ARG	ARG	ARG		F
			BOL	BOL			F
		BRA	BRA	BRA			F
			CHI				X
Najas spp.[c]						ARG	F
					COL		F
		ECU					F

[a] Countries: ARG, Argentina; BOL, Bolivia; BRA, Brasil; COL, Colombia; CUB, Cuba; CHI, Chile; ECU, Ecuador; GUA, Guatemala; HON, Honduras; PAN, Panamá; PER, Perú; PUR, Puerto Rico; URU, Uruguay; VEN, Venezuela.

[b] Ranking of species in terms of overall nuisance value as aquatic weeds in countries listed: (after Holm *et al.* 1979): S = serious weed; P = principal weed; C = common weed; X = present as weed, importance unknown; F = further information needed to confirm species behaves as a weed.

[c] Species reported to cause weed problems within the genera listed include: *Elodea: E. canadensis, E. potamogeton; Potamogeton: P. striatus, P. pectinatus, P. linguatus. Chara: C. contraria. Typha: T. angustifolia, T. domingensis, T. latifolia; Myriophyllum: M. brasiliense, M. aquaticum, M. elatinoides; Lemna: L. minor, L. gibba; Salvinia: S. auriculata, S. radula, S. rotundifolia, S. herzogii; Azolla: A. caroliniana, A. filiculoides; Najas: N. indica, N. marina.*

no means claimed it to be exhaustive, comprising almost 300 citations of South American studies on vascular hydrophytes.

To obtain data specifically on aquatic weeds, a questionnaire survey was undertaken in 1986 of institutions throughout South and Central America connected with the utilization and management of water (e.g. irrigation districts, hydroelectricity complexes, universities, experimental stations), requesting information on aquatic weed types and their habitat, the problems they cause and their management and control. Of 400 questionnaires sent out, more than 80 replies dealt specifically with aquatic weed problems at the regional and local level. Data from the survey have been used extensively in this chapter.

GEOGRAPHICAL DISTRIBUTION OF AQUATIC WEEDS IN SOUTH AND CENTRAL AMERICA

Survey data showing the distribution by country of the most frequently mentioned genera are summarized in Table 20.2. The habitats in which individual species are known to exert a damaging influence, and an estimation of their overall ranking as aquatic weeds in each country, are also included. About 50 per cent of the returned questionnaires reported the presence of *Eichhornia crassipes*.

Studies which provided additional information to complement and corroborate the survey data on aquatic weed distribution and status in South and Central America include: Sosa González (1956), Vera (1970), Liederman and Grassi (1970), Sierra *et al.* (1970), Hearne and Pasco (1972), Agostini (1974), Dorna (1974*a*, *b*), Zambrano (1974), Marzocca (1976), Casamiquela (1977), Gangstad (1977*b*), Petetin and Molinari (1977), Toscani (1981, 1983), Lorenzi (1982), Toro, Briones, and Pinoargote (1982), CEDEGE (1983), San Martín and Ramírez (1983), Toscani, Pizzolo, and Maradei (1983), Méndez (1984), and Bristow *et al.* (undated).

Other aquatic weed genera not listed in Table 20.2 but known to occur in various South and Central America countries include: *Juncus*, *Polygonum*, *Limnobium*, *Spirodela*, *Hydrocotyle*, *Nitella*, *Panicum*, *Eleocharis*, *Hydrocleis*, *Nymphaea*, *Heteranthera*, *Alternanthera*, and *Cladophora*.

Aquatic weed species which were cited only once or twice in the survey, or were found in the literature, are grouped by country as follows:

Argentina: *Sagittaria montevidensis*, *Alternanthera philoxeroides*, *Polygonum punctatum*, *Hydrocotyle bonariensis*, *Zannichellia palustris*, *Panicum elephantipes*, *Phragmites australis*, *Ceratophyllum demersum*, *Echinodorus grandiflorus*, *Ludwigia peploides*, *L. uruguayensis*.

Brazil: *Echinodorus grandiflorus*, *Alternanthera philoxeroides*, *Nymphoides humboldtiana*, *Spirodela polyrrhiza*, *Utricularia gibba*, *Nymphaea ampla*, *Heteranthera limosa, H. reniformis, Pontederia cordata, Hydrocotyle umbellata, Sagittaria montevidensis, Montrichardia linifera.*

Chile: *Alisma plantago-aquatica, Egeria densa, Zannichellia palustris, Caltha sagittata, Ranunculus aquatilis, Hydrocotyle ranunculoides, Nymphaea alba, Jussiaea repens, Juncus procerus, J. bufonius.*

Cuba: *Brachiaria mutica, Mimosa pigra.*

Guatemala: *Nymphaea ampla.*

Panama: *Heteranthera reniformis.*

Venezuela: *Thalia geniculata.*

The importance of these species is difficult to assess, and the extent of problems caused by them is unknown.

The geographical distribution of seven major aquatic weeds in South and Central America is shown in Fig. 20.1. Each dot represents one or more citation in the literature or the survey, indicating a troublesome aquatic plant. Of the aquatic plants considered, submerged plants appear to cause the greatest problems in the temperate and cooler latitudes of the continent, whereas in tropical and subtropical regions it is the floating species which are the most troublesome.

Rice, one of the most prominent cereals in South America, is seriously endangered by the presence of aquatic weeds. Estimates of the fall in rice production due to the presence of aquatic weeds in the littoral area of Argentina range from 35 to 70 per cent. Similar figures have been reported in the rice-growing areas of Brazil and Peru (Victoria Filho and Carvalho 1981; Cerna Bazán and Díaz 1982). The principal weeds occurring in this crop varied by region (Zambrano 1974; Aranha and Pio 1981, 1982; Cerna Bazán and Díaz 1982; Lockett 1983*a, b*; Ormeño 1983; San Martín and Ramírez 1983) but species of the *Echinochloa* and *Cyperus* genera are almost universally present as weeds in South American ricefields.

In many parts of the continent algae are a frequent cause of trouble in irrigation and drainage canals, and ponds, forming a scum on the surface of the water, or growing as entangled mats. Algal blooms pose serious problems in reservoirs supplying water to urban communities. These blooms pollute the water and obstruct its circulation by blocking filters and pipelines. Emiliani and Rodríguez (1974) listed 31 genera and species of algae in Argentina which form blooms, some of which (e.g. *Volvox aureus, Synedra ulna, Melosira* spp.) affect the supply of water to large towns. A taxonomic study of the phytoplankton of the Paso Piedras reservoir, supplying Bahía Blanca, was given by Sala and Intartagio (1985). In Brasilia (Brazil) reservoirs infested with blue-green algae (*Microcystis aeruginosa, Anabaenopsis raciborskii*) cause considerable problems in water destined for human consumption, the functioning of hydroelectric complexes and recreational purposes (I. Guimaraes Altafin Masini, pers. comm.). In the hydroelectric complex El Nihuil (province of Mendoza, Argentina) the submerged aquatic plants *Myriophyllum aquaticum, Potamogeton pectinatus*, and *Chara* spp. cause major problems for hydroelectric power production (Toscani 1983).

FIG. 20.1. Geographical distribution of aquatic weeds in South and Central America.

STATUS OF AQUATIC WEED CONTROL

An increasing awareness of the serious problems caused by aquatic plants has led to the adoption of weed control measures in all countries of South and Central America. The methods employed vary as much as the environments in which they are applied. Primitive, empirical, and simple methods of control exist side by side with some planned management projects based on experimentation and more elaborate techniques (Fernández *et al.* 1987*b*).

Physical methods, including manual and power-operated devices, are intensively used. In some regions, where labour is plentiful, manual methods are economical for small reservoirs and canals. In water distribution and drainage systems, bank-based hydraulic excavators, rakes, and grab-buckets, as well as heavy chains pulled by tractors, are widely used against submerged and emergent vegetation. Boats equipped with submersible cutting blades are employed in some cases; a second operation is then required to remove the floating foliage. The rapid regrowth of aquatic plants may require the use of control measures one or more times during the year, depending on the region and the type of weed involved. Mechanical methods are also frequently used against floating weeds such as *Eichhornia crassipes* and *Pistia stratiotes*. Whenever possible, watercourses and reservoirs are drained so the weeds can be removed by physical means or simply left to dry out and die. Perennation of the underground reproductive structures usually brings about a rapid recovery of the aquatic weed population once the water level is raised.

Mechanical methods are costly and time-consuming when used for aquatic weed problems in large watersheds and irrigation areas, hence the progressive increase in the use of herbicides. Depending on the circumstances, the use of chemicals to manage aquatic weeds is generally easier and less expensive than mechanical methods. Herbicides are often considered an immediate solution for heavy and long-established weed infestations. Herbicides recommended for aquatic weed control include: 2,4-D, paraquat, diquat, aminotriazole, glyphosate, acrolein, and copper sulphate. The effect of several herbicides on *Potamogeton striatus* and *Chara contraria* was tested in the laboratory by Irigoyen (1981*a*). As in the case of mechanical control, herbicides provide only temporary clearance and the treatment has to be repeated as required.

The complexity of using chemical products in aquatic environments and the mounting concern regarding safety, means that one of the most important factors limiting their more intensive use is the lack of personnel trained in their application and with knowledge of the potential environmental hazards involved. There is a critical need to improve the level of knowledge of chemical methods, by research and small-scale preliminary field trials, in areas where aquatic weeds pose problems.

The problem of weed control in rice fields is particularly complex because of the mixture of aquatic and terrestrial plants. Herbicides are now the main method of weed control in rice field systems in South and Central America (Victoria Filho and Carvalho 1981; Cerna Bazán and Díaz 1982; Lockett 1983*a*; Marchesini 1985).

Copper sulphate is the most commonly used chemical to control algal blooms in ponds and reservoirs (Emiliani and Rodríguez 1974; Altafín Masini pers. comm.).

Biological control provides another option for control of a number of aquatic weed problems but this method has not been properly investigated. The manatee (*Trichechus manatus*), listed as an endangered species, has been suggested as a potential agent for aquatic weed control in tropical regions. In Guyana it has been used for many years to keep the Georgetown Botanical Garden free of aquatic weeds, and has also been introduced into some irrigation canals, reservoirs, and drainage systems of this country for the purpose of aquatic weed control (Allsopp 1969; National Academy of Sciences 1976; De León Hernández and Zaballa 1985). The manatee is apparently an efficient way of keeping limited, confined areas clear of aquatic weeds. However, estimates of its rate of food consumption (Etheridge *et al.* 1985) show that it would be impractical to use this herbivore as the only means of controlling the more troublesome aquatic weeds, with their high productivity over large areas. Preliminary trials to control aquatic weeds using grass carp (*Ctenopharyngodon idella*) have been initiated in Argentina (Toscani 1983; Crouzel and Cordo 1984).

BIOLOGY AND MANAGEMENT OF SOME MAJOR AQUATIC WEED TAXA IN SOUTH AND CENTRAL AMERICA

Potamogeton spp. (pondweeds)

Pondweeds occur throughout most of South America (Fig. 20.1; Sculthorpe 1967). More than fifteen species were listed by Tur (1982) in Argentina, extending in this country (and in Chile) as far south as the cold Patagonian latitudes (Pisano 1976; Casamiquela 1977).

In the temperate and cold regions of South America pondweeds grow profusely in irrigation and drainage systems. *Potamogeton* spp. are present in almost all irrigation complexes throughout Argentina covering an area of approximately 1.5 million ha. The usual problem is one of massive infestations of *Potamogeton* spp. obstructing the free flow of water. The resulting malfunctioning of drainage canals causes impaired land drainage, and problems of salinity in the soil of cultivated land. *Potamogeton* spp. are also troublesome in hydroelectric systems (Toscani 1983).

A few years ago a multi-disciplinary programme began in Argentina, aimed at developing a rationalized system of submerged weed control in the irrigation districts of the central and southern regions of the country (Fernández *et al.* 1987*a*, *b*). The study site was located in the lower valley of the Colorado river (62°37′ W, 39°23′ S), where the main problems are caused by *Potamogeton striatus* (Tur 1982) and *Chara contraria*. Until recently only mechanical control methods were available, and this proved to be a costly and slow way of dealing with the large areas involved. The project has served to expand knowledge on the biology and ecology of these two species, and has given rise to recommendations related to their management by herbicides, and appropriate methods of chemical application. Some of the findings (Fernández *et al.* 1987*a*, *b*) on the biology and control of *P. striatus* are summarized here. Those related to *Chara contraria* are given in the next section.

In the southern regions of South America growth of *P. striatus* occurs in spring and summer, when the stem can attain a length of up to 3 m following the direction of the current. Maximum shoot biomass of about 500 g (dry wt.)m^{-2} was recorded in February. During the autumn and winter, the plants remain at vegetative rest, and at this time the underground biomass represents about 80 per cent of the total plant biomass. The species reproduces sexually by means of drupaceous fruits dispersed by water. Flowering occurs between early spring to the beginning of autumn, though it reaches its peak in mid to late summer. The plant has a vigorous system of vegetative propagation by rhizomes and tubers. The formation of turions has not been observed, though the detachment of small shoots capable of developing into new plant was recorded. Plants grown from tubers under semi-controlled conditions in the canals showed a capacity to develop more than 14 m of rhizomes, 17 new tubers, and more than 200 fruits during the first year of growth. The main period for tuber production extended from late spring to the middle of summer, before the period of maximum flower activity (Irigoyen 1981*b*). Seedlings grown from tubers showed a high tolerance of pH fluctuations.

Traditional methods for the control of *P. striatus* are mechanical. Hydraulic grab-buckets (1.05 m in width), which extract vegetation together with some of the sediment, work at a rate of 40 m of canal length per hour. The buckets are frequently replaced by 3 m wide rakes, which eliminate 60 or 70 per cent of the existing vegetation at a time, and are capable of covering around 100 m per hour. Although the use of rakes is rapid as well as economical and leads to the swift removal of blocking plant growths from water, the results are short-lived because of the fast regrowth of the remaining plants and propagules. Mechanical maintenance of the canals has to be performed yearly and it is sometimes even necessary to repeat the operation during the same season.

Chemical treatment with acrolein provides a useful alternative to mechanical control, and can also be used in conjunction with the latter. The best results for the control of *P. striatus* were obtained with the use of a low concentration of acrolein applied over long injection times. The optimum concentration proved to be 2–5 mg l^{-1} maintained over 24 hours. This resulted in a reduction of biomass of 30–50 per cent, for a distance of 7 to 10 km downstream from the point of application. Since acrolein is a contact herbicide, the reservoir of underground propagules in the canal soil is not affected. Under favourable conditions, regrowth of the plants after treatment can reverse the clearance process within a matter of months, so that the canals come to resemble untreated ones. Successive applications of acrolein at regular intervals during the growth season were sufficient to ensure the normal functioning of the canals throughout the whole year. Once the plant population has successfully been reduced to satisfactory levels, the canals can usually be maintained by the application of lower concentrations of herbicide at less frequent intervals.

It was observed that a single chemical treatment early in the growing season disrupted the formation of vegetative reproductive structures. Summer applications had a similar effect on the period of maximum fruit production.

Chara spp. (stoneworts)

Representatives of the Characeae are distributed throughout the South and Central American continent (Fig. 20.1; Wood and Imahori 1959). Tell (1985) reported that 27 species, including 15 species of *Chara* occur in Argentina. Bicudo (1972) identified 22 species of *Chara* in Brazil. In central Argentina a study carried out on 11 Characeae species included a taxonomic study of *C. contraria* (Cáceres 1978). As mentioned in the previous section, *C. contraria* has been identified as one of the more important weeds in irrigation districts of temperate South America and has been the subject of several studies, particularly with respect to its biology and management (Fernández *et al.* 1987*a, b*). The results of these studies show that this plant is a strong competitor, frequently forming extensive monospecific stands anchored to the bottom of the canals, and extending vertically up towards the surface with a stem length of 50–60 cm. The thalli form a compact interwoven mass, which offers a high flow resistance for water movement in the channel. In many cases, particularly in secondary canals, this densely packed structure almost reaches the surface of the water, seriously reducing flow and impairing the functioning of the canals. During the annual growth cycle, the plants show two distinct patterns of development; the first characterized by a rapid increase in plant growth during spring, attaining a maximum biomass and growth rate of 1 600 g (dry wt.) m^{-2}, and 10.2 g (dry wt.) m^{-2} day^{-1} respectively, in the early summer. The second phase is one of a slow but continual loss of biomass, to a winter minimum of 600 g (dry wt.) m^{-2} (Sabbatini, Irigoyen, and Fernández 1986).

Chara contraria reproduces sexually by oospore formation, and vegetatively by fragmenting thalli segments which root easily in the sediment of the canals. The formation of oospores begins in mid-spring and continues until well into the summer. In autumn, the oospores are liberated and join the oospore bank in the sediments (Sabbatini, Irigoyen, and Fernández 1986). Germination rates of 44 to 85 per cent have been recorded for oospores harvested directly from the plants throughout the phenological cycle and placed immediately *in vitro* to germinate (Sabbatini *et al.* 1983). Oospores can germinate after remaining for at least 1 year in hostile environments (e.g. in air or in water under wide by ranging temperature conditions) (Sabbatini, Fernández, and Argüello 1986).

Mechanical methods as for *P. striatus* are widely used to control of this alga. Chemical methods (acrolein, copper sulphate pentahydrate: CSP) have also been tested (Fernández *et al.* 1987*a, b*). *C. contraria* is highly susceptible to acrolein even at low concentrations (1–2 mg l^{-1}) or exposure periods as short as 3 hours. The effect of a single treatment can extend for up to 10 km downstream from the point of injection.

Due to its low cost and easy availability, the use of CSP is a viable alternative to acrolein and mechanical control of *C. contraria*. A simple method was employed to assure a reasonable control of the concentration and injection time into the canal water (Fernández *et al.* 1987*b*). A 50 litre polyethylene drum regularly perforated with 5 mm diameter holes and containing 50 kg CSP was submerged in the water. The number of open orifices regulates the amount of water entering into the drum and flowing out to the external medium in the form of a copper solution. A gradual liberation of the salt is obtained. By measuring the amount of copper ion in the water it was possible to determine that 2 to 5 mg l^{-1} was a satisfactory amount for reducing the troublesome effect of the algae over distances of 500 to 1000 m, depending on the type of canal and the time of year. Placing several drums at regular intervals enhances the efficiency of the treatment. One single application of CSP during the summer was sufficient to impede the formation of *C. contraria* oospores for the rest of the growth period. A significant advantage of using CSP instead of acrolein, apart from cost, is that it does not require the use of any specialized equipment.

Neither acrolein nor CSP can prevent regrowth of the alga and it is, therefore, necessary to repeat the treatment. Nevertheless, once the canals are reasonably free of the algae, the density of the plant population can be maintained through the application of lower concentrations of the chemicals at less frequent intervals and at the appropriate time.

Research is currently under way on a promising Argentinian species of the apple snail, *Ampullaria canaliculata*, in connection with its possible introduction as a biological control against this alga (Cazzaniga 1981, 1983). Four years of work have shown that the snail feeds voraciously on this plant. Attempts to control submerged weeds in reservoirs by using the grass carp (*Ctenopharyngodon idella*), showed that *Chara* spp. are a preferred food for this fish (Toscani 1983).

3. *Myriophyllum aquaticum* (Parrot-feather)

Parrot-feather was first named in 1829 by Cambessedes (Muhlberg 1982). A nomenclature change combined *Myriophyllum proserpinacoides* and *Myriophyllum brasiliense* under its present name of *Myriophyllum aquaticum*. The status of the species as an aquatic weed has recently been reviewed by Sutton (1985). *M.*

aquaticum is a dioecious plant, indigenous to South America. In a comprehensive study of *Myriophyllum* species, Orchard (1981) found only a few male flowers, and only two plants with immature fruits, on parrot-feather specimens collected from the lowlands of central and western South America. However, female flowers were found on plants collected from a much wider area in South America. These were primarily from low altitudes but some specimens were collected from areas as high as 3 250 m in Peru and 1 900 m in Brazil.

Little is known of the factors which regulate growth of parrot-feather in its native habitat; although herbivory may play an important role. In Argentina, the native flea beetle, *Lysathia flavipes* was found to cause moderate damage to *M. aquaticum* under field and laboratory conditions (Cordo and Deloach, 1982*a*). A native curculionid weevil, *Listronotus marginicollis*, was also reported as host specific to *M. aquaticum* in Argentina (Cordo and Deloach 1982*b*).

Parrot-feather may form profuse growths under suitable conditions. Perhaps because of the presence of natural regulating agents, it is less of a problem in South America than in various other parts of the world to which it has been introduced by human intervention. Colonization by seed is rare. Vegetative propagation seems to be the main means of spread once an initial introduction has successfully established.

The presence of emergent stems is a characteristic identifying feature of parrot-feather (Godfrey and Wooten 1981). These stems, generally with few branches, grow up to 30 cm or more above the surface of moist, muddy areas, or they may be attached to the bank and extend out several metres over the surface of the water. In shallow bodies of water, it is quite common to have submerged and emergent leaves on the same stem; however, under these conditions, the emergent portion tends to lie on the surface, or extend only a few centimetres above it. Roots grow from nodes of the stem within one week of submergence of the lower portion of excised apical stem sections (Sutton *et al.* 1969).

M. aquaticum appears to prefer a warm climate and muddy banks or shallow water, in which it grows in discrete patches, primarily in emergent form. Eutrophic water (e.g. water bodies enriched by sewage effluent) may favour the development of nuisance growths of the weed. Sutton (1968) found that emergent parrot-feather plants cultured for 6 weeks in glasshouse conditions in 10 per cent Hoaglands nutrient solution (Hoagland and Arnon 1950), had a lower dry weight biomass than plants grown in $\geqslant$50 per cent Hoaglands solution, suggesting a strong growth response to high nutrient availability. Few data are available on the growth and nutrient uptake of *M. aquaticum* under field conditions.

Barko and Smart (1981) found that, under controlled environment conditions, submerged *M. aquaticum* plants readily absorbed nitrogen and phosphorus from three different sediments, and concentrated these nutrients in the shoots at levels well above those required for growth.

There is some experimental evidence to support the observation that invading

M. aquaticum may colonize coastal or other saline-influenced habitats, to the exclusion of pre-existing macrophyte species. Haller, Sutton, and Barlowe (1974) found that seawater at salinity $\geqslant 10.0$ parts per thousand (‰S) was toxic to the emergent form of *M. aquaticum*, under greenhouse conditions. However, root growth was stimulated at 0.8–3.3 ‰S. Since all concentrations of seawater reduced transpiration of the plants, compared to fresh water controls, the increased root growth may represent an attempt to overcome a water deficit, by increasing the surface area of the roots.

Control of *M. aquaticum* may be achieved with low-volatility ester of 2,4-D at 4.4–8.9 kg ha^{-1}, sprayed onto the emergent foliage (Blackburn and Weldon 1963). The granular formulation of 2,4-D was noted by Braddock (1966) to control parrot-feather for periods $\geqslant 12$ months. Experimental studies using radio-labelled 2,4-D (Sutton and Bingham 1970) suggested that the herbicide is translocated symplastically in *M. aquaticum* and is, therefore, more effective when applied to young, actively-growing plants than against native plants.

Copper at a concentration of 1.0 mg l^{-1} or higher in the root zone was found to reduce growth of parrot-feather under greenhouse conditions (Sutton and Blackburn 1971). Under field conditions, the use of copper sulphate would probably not be practical as it might be difficult to maintain this concentration for the period of time necessary to kill the plants.

There have been a few studies of biological control of *M. aquaticum*. As stated above, two South American insects, *Lysathia flavipes* and *Listronotus marginicollis*, cause some damage to parrot-feather, in its native habitat. Habeck and Wilkerson (1980) found under laboratory conditions that parrot-feather was a natural host for larvae of the flea beetle *Lysathia ludoviciana*, indigenous to the southern United States and the Caribbean region. However, the scarcity of this flea beetle on *M. aquaticum* in field sites suggests that the beetle may have additional preferred hosts. Newly-hatched larvae of the moth *Parapoynx allionealis* preferred parrot-feather leaves to those of *Sagittaria stagorum, Eleocharis vivipara*, and duckweed, *Lemna* spp. (Habeck 1974). The first stage larvae mined the leaves of parrot-feather but the potential extent of damage is not known.

A *Rhizoctonia solani* isolate (RhEa) collected from diseased anchoring water hyacinth (*Eichhornia azurea*) plants in Panama was phytotoxic only to the tips of submerged parrot-feather which protruded above the surface of the water (Joyner and Freeman 1973). In northern California, an isolate of *Pythium carolinianum*, was collected from diseased emergent parrot-feather plants, by Bernhardt and Duniway (1984). When mycelium of this isolate was inserted into the stems of parrot-feather cuttings, the fungi girdled the stems and caused collapse of the plants. Additional testing under field conditions showed that growth of parrot-feather plants inoculated with this isolate was lower than uninoculated controls. These studies showed that various fungi may have potential for biological control of parrot-feather.

Little is known about the efficiency of mechanical clearance of nuisance growths of *M. aquaticum*. Since the plant fragments readily, and individual fragments can regenerate shoots and roots, rapid regrowth may occur, from fragments and any whole plant surviving clearance (Jacot-Guillarmod 1977). Research is needed to develop effective mechanical control techniques for use against parrot-feather, with the emphasis on methods which minimize the likelihood of reinfestation from surviving plant fragments.

Eichhornia crassipes (water hyacinth) and other floating weeds

The floating macrophyte vegetation of South America is closely associated with the network of rivers in the tropical and subtropical climates (Fig. 20.1). Many of these are floodplain rivers subjected to periodic seasonal water fluctuations. Vast areas of adjacent plains are flooded during the periods of high water. Numerous inland lakes are left behind after the river has receded. The aquatic habitats and the flora of the floodplain system of the Rivers Orinoco, Amazon, and Paraná have been the subject of several studies (Burkart 1957; Schulz 1961; Drago 1976; Armitage and Fasset 1971; Neiff 1979, 1981, 1986; Junk 1980*a*, 1984; Lallana, Marta, and Sabattini 1981; Rangel and Aguirre 1983; Sabattini, Lallana, and Marta 1983; Beck 1984; Melack 1984; Bravo Velázquez and Balslev 1985; Franceschi, Prado, and Lewis 1985; Sánchez and Vásquez 1986).

Aquatic macrophytes constitute a significant portion of the plant biomass of such amphibious systems (Pérez del Viso, Tur, and Montovani 1968; Neiff 1975; Bayo *et al.* 1981; Sabattini, Lallana, and Marta 1983). Floating macrophytes play an important role in the functioning of the natural ecosystems of the flowing and standing waters of these regions (Junk 1970, 1980*b*; Bonetto 1971, 1975; Lallana, Sabattini, and Lallana 1987), by providing habitat and food for other organisms (Dioni 1967; Poi de Neiff 1977; Poi de Neiff and Neiff 1977; Sazima and Zamprogno 1985).

Of all the aquatic plants listed in the 1986 survey, *Eichhornia crassipes* is the most important in terms of distribution, biomass, and the problems it causes to human activities. It is native to tropical South America (Castellanos 1958; Calderón García 1971; Zambrano 1978; Holm *et al.* 1979), and its geographical distribution covers an area stretching from Central America to the Rio de La Plata, Argentina (35° S). Similar in appearance is *Eichhornia azurea*. The abundance of this plant in the upper Paraná river is attributed to the allochthonous contribution of the Paraguay river, which has its source in a huge flat basin known as El Pantanal, with numerous static or slow-flowing reservoirs, favouring the growth of this and other floating species (Neiff and Bonetto 1983). Besides the two *Eichhornia* spp., *Pistia stratiotes*, *Azolla caroliniana*, and *Salvinia* spp. are important components of the floating weed flora of South and Central America. These latter three plants are also native to the area.

Relatively little information is available on the biology of these floating weeds. *E. crassipes, E. azurea*, and *Pistia stratiotes* reproduce for the most part asexually (Penfound and Earle 1948; Sculthorpe 1967; Lallana, 1980, 1981). However, their ability to reproduce sexually should not be overlooked since this is a potential means of invasion into new areas, a means of reinfestation where clearance has been undertaken, and is important to the formation of new genotypes. The production of disseminules of sexual origin by these plants in the middle Paraná river was studied by Lallana and Marta (1980, 1981) and Lallana (1984*a, b*, 1987). These studies showed that a high proportion of the flowers and fruits of *E. crassipes* are destroyed or damaged by insects, snails, and other organisms. Only 65 per cent of the fruits analysed contained seeds, and the number of seeds per square metre of vegetation ranged from 400 to 3 400, depending on the sampling site and time of year. More than 70 per cent of *E. azurea* fruits contained seeds. Between 13 to 60 per cent of the fruits of *Pistia stratiotes* had seeds with an average of 2 300 seeds m^{-2} of vegetation. The biology of *P. stratiotes* in the Amazon area was studied by Da Silva (1981). The dispersion of the disseminules of these plants is facilitated by the action of birds, water, and the plants themselves, which carry disseminules along in flowing waters, adhering to their roots.

In certain areas of the continent excessive populations of *E. crassipes*, frequently accompanied by other floating weeds, cause severe problems. Navigation is made hazardous by the solid mass of floating material. Water resources held in reservoirs, lakes, canals, and streams can be difficult or even impossible to manage properly. Many of the artificial water reservoirs of Central American countries are particularly badly affected by *E. crassipes*. (Gangstad 1977, 1980; De León Hernández and Zaballa 1985). The problem of floating aquatic weeds is a cause of serious concern in some of the large hydroelectric projects in South America (Botry and Otaegui 1979; Micotz 1981; CEDEGE 1983; Neiff and Bonetto 1983; IPPC 1984).

In Venezuela water hyacinth is troublesome in certain tributaries of the Orinoco River, where water flow has been altered as a result of dike construction for flood control purposes (Pieterse and Schulten 1976: Fig. 20.2). Moreover, in affluents of the Gulf of Maracaibo (caños), water hyacinth (in association with *Paspalum repens, Pistia stratiotes*, and *Ludwigia natans*) frequently forms blocking mats of vegetation, extending over many kilometres (Pieterse and Schulten 1976: Fig. 20.3).

In Argentina, when the water of the Paraná river recedes after flooding the adjacent territories, it carries back with it the floating vegetation of inland standing waters. This mass of plants forms what is locally known as 'camalotales', floating islands made up of densely packed vegetation moving downstream (Schulz 1961; Tur 1972; Lallana 1982). The size of these drifting islands varies from 1 to 200 m, though it is not uncommon to finds some as large as 1 500 m. Lallana (1982)

FIG. 20.2. A dense cover of water hyacinth (*Eichhornia crassipes*) in the Tucupita river, in the Orinoco Delta in Venezuela. (Photo: A. H. Pieterse.)

FIG. 20.3. Blockage of the caño Caiman, a waterway which connects the Laguna del Zulia and the Gulf of Maracaibo in Venezuela, by *Ludwigia natans* and *Eichhornia crassipes* (water hyacinth). (Photo: A. H. Pieterse.)

indicated that more than 4 ha of camalotales pass daily downstream from the city of Paraná (Entre Ríos, Argentina) at certain times of the year. *E. crassipes* is always the predominant component of these islands, together with *E. azurea, P. stratiotes,* and *Salvinia herzogii*. Other species also commonly carried along are *Pontederia rotundifolia, Paspalum repens,* and *Polygonum* spp. (Lallana 1982). Frequently, a rich fauna made up of insects, snakes, snails, and other animals is also transported in the floating islands.

The mass of the camalotales at several tonnes (fresh weight) per ha (Lallana 1982), combined with the velocity of the river current is capable of inflicting damage on any structures blocking their downstream advance (Schulz 1961). During the extraordinary flood of 1966, these drifting masses of vegetation endangered the stability of a bridge over the Santa Fe river (Santa Fe, Argentina), and it was necessary to use explosives to release the pressure exerted by the vegetation caught against it. Under similar circumstances part of the bridge was destroyed in 1983.

Navigational traffic on the Paraná river and its tributaries, ranging from small boats to trans-oceanic vessels, is hindered during the high water periods by these drifting masses of vegetation. The recreational use of beaches on the river banks is periodically impaired by the plants deposited there, which have to be cleared by the local authorities.

The delta formed by the lower Paraná and Uruguay rivers where they join to form the La Plata river, is used intensively for forestry purposes. Exploitation of the land is carried out on the basis of a network of about 5 000 km of drainage canals most of which have become infested with floating species such as *E. crassipes, E. azurea, P. stratiotes, Azolla* spp., and *Lemna* spp. Other accompanying plants are emergents (e.g. *Alternanthera philoxeroides, Limnobium laevigatum, Typha* spp., *Scirpus* spp., *Myriophyllum aquaticum*) or marginal weeds such as *Panicum elephantipes* and *Paspalum repens* (Toscani, Pizzolo, and Maradei 1983). This type of vegetation jeopardizes forestry development by hindering drainage of the fields, blocking the pump outlets, obstructing canals and ditches, preventing the passage of boats, and ultimately leading to the closure of the waterways. Manual control is used in the narrower canals. For those of 5–15 m width mechanical buckets and rakers are employed. In recent years herbicides have proved satisfactory for controlling both floating weeds and other aquatic plants present. The herbicides most frequently used are 2,4-D, diquat, paraquat, bromacil, dalapon, atrazine, diuron, glyphosate, and mixtures of these (Liederman and Figueiredo 1967; Toscani, Pizzolo, and Maradei 1983, Toscani 1984).

The subject of biological control of floating weeds appears not to have been dealt with systematically in South America. Permanent control of *E. crassipes* by the weevil *Neochetina bruchi* was reported by Deloach and Cordo (1983) for the Los Sauces reservoir in La Rioja, Argentina. Their results indicate that this insect can successfully reduce the plant population under certain conditions. While it is generally accepted that a better response can be expected from the biological control of introduced weeds rather than from native species, the possibility that at least some of the native weed populations of floating macrophytes may be regulated by their natural enemies seems reasonable. South America, and particularly Argentina, has until now played the role of an exporter of organisms for use in biological control programmes in other countries (Bennett 1970, 1975; Silveira-Guido 1971; Deloach and Cordo 1974, 1978, 1981, 1982; Silveira-Guido and Perkins 1975; Deloach, Deloach, and Cordo 1976, 1979; Deloach *et al.* 1980; Poi de Neiff, Neiff and Bonetto 1977; Cordo *et al.* 1978; Cordo, Deloach, and Ferrer 1984; Crouzel and Cordo 1984; Forno and Bourne 1984). Given an adequate level of research within the continent, biological control may hold a real prospect of solving some of the native weed problems of South and Central America.

APPENDIX A
Aquatic weeds

Freshwater macrophyte and algal species named in text as causing nuisance vegetation problems.
Authorities are also given for non-nuisance aquatic plant species mentioned in text.

Life/growth form classification of vegetation type: S = submerged; FF = free floating; RF = rooted, floating leaves; E = emergent; A = algae; * = rice weed; † = non-nuisance species

Achyranthes philoxeroides (Mart.) Standley E/RF
(= *Alternanthera philoxeroides* (Mart.) Griseb.)
Acorus L. E
Acorus calamus L. E
Aldrovanda vesiculosa L. S
Alisma L. E
Alisma canaliculatum A. Br. & Bouche *E
Alisma gramineum Lejeune E
Alisma lanceolatum With. *E
Alisma plantago-aquatica L. E
Alternanthera Forsk. E/RF
Alternanthera philoxeroides (Mart.) Griseb. E/RF
(= *Achyranthes philoxeroides* (Mart.) Standley)
Alternanthera sessilis (L.) DC. *E
Ammania coccinea Rottb. *E
Anabaena Bory A
Anabaenopsis (Wolozsynska) Miller A
Anacharis alsinastrum Bab. S
(= *Elodea canadensis* Michx.)
Anacharis chilensis Planch. S
(= *Elodea potamogeton* (Bert.) Espin.)
Anacystis Menegh. A
(= *Synechococcus* Nageli)
Aneilema keisak Hassk. *E
(= *Murdannia keisak* (Hassk.) Hand.-Mazz.
Aphanizomenon Morren A
Aphanizomenon flos-aquae (Lyngbye) Brebisson A
Apium L. E
Aponogeton L. S/RF
Azolla Lamarck FF
Azolla caroliniana Willdenow FF
Azolla filiculoides Lam. *FF

Bacopa Aublet *S/RF
Bacopa caroliniana (Walt.) Robinson S/RF
Bacopa rotundifolia (Michx.) Wettst. *S/RF
Berula erecta (Hudosn) Coville E
Blyxa aubertii Rich. *S
Blyxa japonica (Miq.) Aschers & Guerke *S
Brachiaria Griseb. E/RF
Brachiaria mutica (Forsk.) Stapf E/RF
(= *Panicum purpurescens* Raddi)
Brasenia schreberi J. F. Gmelin RF
Butomus L. E
Butomus umbellatus L. E

Cabomba Aublet S/RF
Cabomba caroliniana A. Gray S/RF
Calla palustris L. E
Callitriche L. S
Callitriche stagnalis Scop. S
Callitriche truncata Guss. S

Callitriche verna L. S
Caltha sagittata Cav. E
Canna L. E
Canna flaccida Rosc. E
Cardamine L. E
Carex L. E
Carex acuta L. E
Carex riparia Curtis E
Ceratophyllum L. S
Ceratophyllum demersum L. S
Ceratophyllum echinatum A. Gray S
(= *Ceratophyllum muricatum* Chamisso)
Ceratophyllum muricatum Chamisso S
(= *Ceratophyllum echinatum* A. Gray)
Ceratopteris Brongn. *E/FF
Ceratopteris thalictroides (L.) Brongn. *E
Chaetomorpha linum (Muell.) Kuetz. A
Chara L. A
Chara contraria A. Braun ex Kutz. A
Chara corallina Klein ex Willdenow A
Chara fragilis Desv. A
(= *Chara globularis* Thuill.)
Chara globularis Thuill. A
(= *Chara fragilis* Desv.)
Chara vulgaris L. A
Chlamydomonas reinhardii Dangeard A
Cicuta virosa L. E
Cladium jamaicense Crantz E
Cladophora Agardh. A
Cladophora glomerata (L.) Kutzing A
Coelosphaerium Nägeli A
Colocasia Schott E
Colocasia esculentum (L.) Schott E
Cosmarium Corda A
Crassula helmsii Kirk S/E
Cryptocoryne Fischer ex Wydler E
Cyperus L. E
Cyperus difformis L. *E
Cyperus esculentus L. *E
Cyperus flavescens (non L.) Benth. E
Cyperus nitidus Boeck. E
Cyperus papyrus L. E
Cyperus procerus Rottb. *E
Cyperus serotinus Rottb. *E
Cyrtosperma chamissonis Merrill E

Damasonium minus (R.Br.) Buch. *E
Diplachne fusca (L.) Beauv. *E
Dopatrium junceum Buch.-Ham. ex Benth. *S/E
Dulichium arundinaceum (L.) Britt. E

Echinochloa colona Link *E
(= *Echinochloa colonum* (L.) Link.)
Echinochloa colonum (L.) Link *E
(= *Echinochloa colona* Link
Echinochloa crus-galli (L.) Beauv. *E
Echinochloa pyramidalis Hitchc. & Chase E
Echinochloa stagnina (Retz.) Beauv. E
Echinodorus Rich. ex Engelm. S/RF/E
Echinodorus grandiflorus (Cham. & Schlecht.) M. Micheli E
Eclipta alba (L.) Hassk. S/E
(= *Eclipta prostrata* (L.) L.)
Eclipta prostrata (L.) L. S/E
(= *Eclipta alba* (L.) Hassk.)
Egeria densa Planchon S
(= *Elodea densa* (Planch.) Casp.)
Eichhornia Kunth FF/RF
Eichhornia azurea (Sw.) Kunth RF
Eichhornia crassipes (Mart.) Solms. FF
Elatine L. S/E
Elatine americana Arm. *S/E
(= *Elatine triandra* Schkuhr)
Elatine triandra Schkuhr *S/E
(= *Elatine americana* Arm.)
Eleocharis R.Br. S/E
Eleocharis acicularis (L.) R. & S. *S
Eleocharis acutangula (Roxb.) Schutt. *E
Eleocharis coloradoensis (Britt.) Gilly† S
Eleocharis dulcis (Burm. f) Henschel *E
Eleocharis kuroguwai Ohwi *
Eleocharis plantaginea R.Br. E
Eleocharis vivipara Link. E
Elodea Michx. S
Elodea canadensis Michx. S
(= *Anacharis alsinastrum* Bab.)
Elodea nuttallii (Planch.) St. John S
Elodea potamogeton (Bert.) Espin. S
(= *Anacharis chilensis* Planch.)

Enhalus acoroides (L.f.) Rich. ex Steud. S
Enteromorpha intestinalis (L.) Link A
Equisetum fluviatile L. E
Euryale ferox Salisb. RF

Fimbristylis Vahl. *E
Fimbristylis miliacea (L.) Vahl. *E
Fontinalis (Dill.) Hedwig. S
Fontinalis antipyretica L. S

Gloeocystis Nägeli A
Gloeotrichia J. Agardh. A
Glyceria declinata Bréb. RF
Glyceria fluitans (L.) R. Br. RF
Glyceria maxima (Hartman) Holmberg E

Heteranthera Ruiz & Pavon S/RF
Heteranthera dubia (Jacq.) MacMillan F
Heteranthera limosa (Schwartz.) Willd. *E
Heteranthera reniformis Ruiz & Pavon E/RF
Hydrilla L. C. Rich. S
Hydrilla verticillata (L.f.) Royle S
Hydrocharis morsus-ranae L. E/RF
Hydrocleis L. C. Rich. RF
Hydrocotyle L. RF/E
Hydrocotyle bonariensis Lamarck E
Hydrocotyle ranunculoides L. fil. RF
Hydrocotyle umbellata L. E/RF
Hydrodictyon reticulatum (L.) Lagerheim A
Hydrolythrum wallichii Hook S
(= *Rotala myriophylloides* Welw. ex Hiern)
Hygrophila R. Br. E
(= *Nomaphila* Blume)
(= *Synnema* Bentham)
Hygrophila auriculata (Schumach.) Heine E
Hygrophila polysperma (Roxb.) T. Anderson E
Hygroryza aristata (Retz.) Nees RF
Hymenachne amplexicaulis (non Nees) Monod RF/E

Ipomoea L. RF
Ipomoea aquatica Forsk. *RF
(= *Ipomoea reptans* Poir.)
Ipomoea crassicaulis (Benth.) B. L. Robinson RF
Ipomoea reptans Poir. *RF
(= *Ipomoea aquatica* Forsk.)
Isachne globosa (Thunb.) O. Ktze. *E
Isoetes lacustris L. S

Juncus L. E/S
Juncus bufonius L. E
Juncus bulbosus L. S
Juncus procerus E. Mey E
Jussiaea repens L. RF
(= *Ludwigia adscendens* (L.) Hara)
Jussiaea suffruticosa L. E
Jussiaea uruguayensis Cambess. E
(= *Ludwigia uruguayensis* (Cambess.) Hara

Lagarosiphon Harvey S
Lagarosiphon major (Ridley) Moss S
Lagenandra Dalzell F
Leersia hexandra Swartz *E
Lemna L. FF
Lemna aequinoctialis Welw. ex Hegelm FF
(= *Lemna perpusilla* Torr.)
Lemna gibba L. FF
Lemna minor L. FF
Lemna minor L. agg. FF
(= *Lemna minor* L. + *Lemna gibba* L.)
Lemna minuscula Herter FF
Lemna perpusilla Torr. FF
(= *Lemna aequinoctalis* Welw. ex Hegelm.
Lemna trisulca L. FF/S
Leptochloa chinensis (L.) Nees *S/E
Limnanthemum indicum L. Griseb. *RF
(= *Nymphoides indica* (L.) O. Kuntze)
Limnobium laevigatum (Humb. & Bond. ex Willd.) Heine RF/FF
Limnobium spongia (Bosc.) Rich. ex Steud. FF/RF

Limnocharis flava (L.) Buch. E
Limnophila R. Br. *S/E
Limnophila ceratophylloides (Hiern) Skan. S/E
Limnophila conferta Benth. *E
Limnophila sessiliflora Blume S/E
Lobelia chinensis Lour. E
Ludwigia adscendens (L.) Hara RF
(= *Jussiaea repens* L.)
Ludwigia natans Elliot RF
Ludwigia peploides (HBK) Raven *RF
Ludwigia repens Forst S/E/FF
Ludwigia uruguayensis (Cambess.) Hara E
(= *Jussiaea uruguayensis* Cambess.)
Lyngbya C. A. Agardh A

Marsilea quadrifolia L. *RF
Megalodonta beckii (Torrey) Greene S/E
Melosira Agardh A
Mentha aquatica L. E
Menyanthes L. E
Menyanthes trifoliata L. E/RF
Micranthemum Michx. E
Microcystis Kutzing A
Microcystis aeruginosa Kutzing A
Microspora Thuret A
Mimosa pigra L. E
Mimulus orbicularis Benth. *E
Monochoria Presl *E
Monochoria hastata (L.) Solms E
Monochoria korsakowii Regel & Maack *E
Monochoria vaginalis (Burm. f.) Kunth *E
Montrichardia linifera Schott E
Murdannia keisak (Hassk.) Hand.-Mazz. *E
(= *Aneilema keisak* Hassk.)
Myriophyllum L. S/E
Myriophyllum aquaticum (Vell.) Verd. S/E
(= *Myriophyllum brasiliense* Cambess.)
(= *Myriophyllum proserpinacoides* Gillies)
Myriophyllum brasiliense Cambess. S/E
(= *Myriophyllum aquaticum* (Vell.) Verd.)
Myriophyllum elatinoides Gaudich. S
Myriophyllum exalbescens Fern. S
(= *Myriophyllum spicatum* var. *exalbescens* (Fernald) Jepson)
(= (?) *Myriophyllum sibiricum* Komarov)
Myriophyllum heterophyllum Michx. S
Myriophyllum laxum Shuttl. ex Chapm. S
Myriophyllum pinnatum Britton, Stern & Poggenb. S
Myriophyllum propinquum A. Cunn. S
Myriophyllum proserpinacoides Gillies S/E
(= *Myriophyllum aquaticum* (Vell.) Verd.)
Myriophyllum salsugineum A. E. Orchard S
Myriophyllum sibiricum Komarov
(= (?) *Myriophyllum exalbescens* Fern.)
Myriophyllum spicatum L. S
Myriophyllum spicatum var. *exalbescens* (Fernald) Jepson S
(= *Myriophyllum exalbescens* Fern.)
Myriophyllum verticillatum L. S

Najas L. S
Najas flexilis (Wild) R. & S. S
Najas gracillima (A. Braun) Magnus *S
Najas guadalupensis (Spreng.) Magnus S
Najas indica Cham. S
Najas marina L. S
Najas minor All. S
Najas pectinata (Parl.) Magnus S
Nasturtium amphibium (L.) R. Br. E
Nasturtium officinale R. Br. E
(= *Rorippa nasturtium-aquaticum* (L.) Hayek)
Navicula Bory A
Nelumbo Adanson RF
Nelumbo lutea (Willd.) Persoon RF
Nelumbo nucifera Gaertn. RF
Nitella (C. A. Agardh) Leonhardi *A
Nitella hookeri A. Br. A
Nitella tasmanica F. Müll. ex A. Br. A
Nodularia Mertens A
Nomaphila Blume E
(= *Hygrophila* R. Br.)

Nostoc Vaucher. A
Nuphar J. E. Smith RF
Nuphar advena Ait. F. RF
(= *Nuphar luteum* Sibth. & Small)
Nuphar lutea (L.) Sm. RF
Nuphar luteum Sibth. & Small RF
(= *Nuphar advena* (Ait.) Ait. F.)
Nuphar polysepalum Engelm. RF
Nymphaea L. RF
Nymphaea alba L. RF
Nymphaea ampla (Salisb.) DC RF
Nymphaea caerulea Savigny RF
Nymphaea guineensis Schum. & Thonn. RF
Nymphaea lotus L., non Hook. f. & Thoms. RF
Nymphaea mexicana Zuccarini RF
Nymphaea nouchali Burm. fil. RF
(= *Nymphaea stellata* Willd.)
Nymphaea odorata Ait. RF
Nymphaea stellata Willd. RF
(= *Nymphaea nouchali* Burm. fil.)
Nymphoides Hill RF
Nymphoides aquatica (Walt.) Kuntze RF
Nymphoides cristata (Roxb.) O. Kuntze *RF
(= *Nymphoides hydrophylla* (Lour.) O. Kuntze)
Nymphoides humboldtiana (Kunth) Hoehne RF
(= *Nymphoides indica* (L.) O. Kuntze)
Nymphoides hydrophylla (Lour.) O. Kuntze *RF
(= *Nymphoides cristata* (Roxb. O. Kuntze)
Nymphoides indica (L.) O. Kuntze *RF
(= *Limnanthemum indicum* L. Griseb.)
(= *Nymphoides humboldtiana* (Kunth) Hoehne)
Nymphoides peltata (S. G. Gmel.) O. Kuntze RF

Oedogonium Link A
Oenanthe L. E
Oenanthe javanica DC *E
Opephora Petit A
Oryza L. E
Oryza perennis Moench. E
Oryza rufipogon Griff. *E
Oryza sativa L.† E
Oscillatoria Vauch. A
Ottelia alismoides (L.) Pers. *S
(= *Ottelia japonica* Miguel)
Ottelia japonica Miguel *S
(= *Ottelia alismoides* (L.) Pers.)
Ottelia kunenensis Gürke S

Pandanus tectorius Soland. ex Balf. f.† E
Panicum L. E/RF
Panicum elephantipes Nees E
Panicum hemitomon Schult. E
Panicum purpurascens Raddi E
(= *Brachiaria mutica* (Forssk.) Stapf.)
Panicum repens L. *E
Paspalum distichum L. *E
Paspalum fluitans (Ell.) Kunth. RF
Paspalum repens Berg. E
Paspalum vaginatum Sw. *E
Pennisetum purpureum Schumach. E
Pentapetes phoenica L. *
Phalaris arundinacea L. E
Phragmites Adanson E
Phragmites australis (Cav.) Trin. ex Steudel E
(= *Pragmites communis* Trin.)
Phragmites communis Trin. E
(= *Phragmites australis* (Cav.) Trin. ex Steudel)
Phragmites karka (Retz.) Trin. ex Steudel E
Pistia L. FF
Pistia stratiotes L. *FF
Pithophora Wittrock A
Polygonum L. E/RF
Polygonum amphibium L. RF
Polygonum coccineum Muhl. S/RF/E
Polygonum densiflorum Meisn. E
Polygonum punctatum Ell. E
Pontederia L. S/RF/E
Pontederia cordata L. E

Pontederia lanceolata Nutt. E
Pontederia rotundifolia L. E
Potamogeton L. S/RF
Potamogeton amplifolius Tuckerman S
Potamogeton cheesemanii A. Benn. S
Potamogeton crispus L. S
Potamogeton distinctus A. Benn. *S
Potamogeton foliosus Raf. S
Potamogeton gramineus L. S
Potamogeton illinoensis Morong S
Potamogeton indicus Roxb. non Roth. *S/RF
(= *Potamogeton nodosus* Porr)
Potamogeton linguatus Hagstr. S/RF
Potamogeton lucens L. S
Potamogeton malaianus Miq. S
Potamogeton natans L. RF/S
Potamogeton nodosus Poir. *S/RF
(= *Potamogeton indicus* Roxb. non Roth.)
Potamogeton ochreatus Raoul. S
Potamogeton panormitanus Biv. S
(= *Potamogeton pusillus* L.)
Potamogeton pectinatus L. S
var. *interruptus* S
Potamogeton perfoliatus L. S
Potamogeton praelongus Wulfen S
Potamogeton pusillus L. S
(= *Potamogeton panormitanus* Biv.)
Potamogeton richardsonii (Benn.) Rydb. S
Potamogeton robbinsii Oakes S
Potamogeton schweinfurthii A. Benn. S/RF
Potamogeton striatus Ruiz & Pavon S
Potamogeton thunbergii Cham. & Schlecht. S/RF
Potamogeton tricarinatus F. Muell. & A. Benn., ex A. Benn. S/RF
Potamogeton trichoides Cham. & Schlecht. S
Potamogeton vaginatus Turczaninow S
Potamogeton × *nitens* Weber S
Potamogeton zosteriformis Fernald S

Ranunculus L. S/RF/E
Ranunculus aquatilis L. S
Ranunculus circinatus Sibth. S
Ranunculus fluitans Lam. S
Ranunculus hederaceus L. S/RF
Ranunculus lingua L. E
Ranunculus penicillatus (Dumort.) Bab. S
var. *calcareus* (Butcher) C. D. K. Cook S
Ranunculus repens L. E
Ranunculus trichophyllus Chaix S
Rhizoclonium Kützing A
Rhynchospora corymbosa (L.) Britton *E
Rorippa Scop. E
Rorippa nasturtium-aquaticum (L.) Hayek E
(= *Nasturtium officinale* R. Br.)
Rotala indica (Willd.) Koehne *RF/E
Rotala myriophylloides Welw, ex Hiern S
(= *Hydrolythrum wallichii* Hook.)
Rumex crispus L. *E
Ruppia L. S
Ruppia drepanensis Tineo S
Ruppia maritima L. S
Ruppia megacarpa Mason S

Sacciolepis Nash *E
Sagittaria L. S/RF/E
Sagittaria kurziana Glück E
(= *Sagittaria subulata* (L.) Buchen. var. *kurziana* (Glück) Bogin)
Sagittaria lancifolia L. E
Sagittaria latifolia Willdenow E
Sagittaria montevidensis Cham. & Schlecht. *E ssp. *calycina* (Engelm.) Bogin *E
Sagittaria pygmaea Miq. *E
Sagittaria stagorum Small S/RF/E
Sagittaria subulata (L.) Buchen. var. *kurziana* (Glück) Bogin E
(= *Sagittaria kurziana* Glück)
Sagittaria trifolia L. *S/RF/E
Salvinia Seguier FF
Salvinia auriculata Aubl. FF
Salvinia biloba Raddi FF
Salvinia cucullata Roxb. *FF
Salvinia herzogii de la Sota FF

Salvinia molesta D. S. Mitchell *FF
Salvinia natans (L.) All. FF
Salvinia radula Baker FF
Salvinia rotundifolia Willdenow FF
Samolus L. E
Schinus terebinthifolius Raddi E
Schoenoplectus (Reichenb. fil.) Palla E
(= *Scirpus* L.)
Schoenoplectus lacustris (L.) Palla E
(= *Scirpus lacustris* L.)
Schoenoplectus tatora (Nees & Meyen) Kunth. E
(= *Scirpus californicus* (C. A. Meyer Steud)
Scirpus L. E
(= *Schoenoplectus* (Reichenb. fil.) Palla)
Scirpus californicus (C. A. Meyer) Steud. E
(= *Schoenoplectus tatora* (Nees & Meyen) Kunth.
Scirpus confervoides Poir. E
Scirpus cubensis Poepp. & Kuenth. E
Scirpus grossus L. f. *E
Scirpus hotarui Ohwi *E
Scirpus lacustris L. E
(= *Schoenoplectus lacustris* (L.) Palla)
Scirpus mucronatus L. *E
Scirpus tabernaemontani C. C. Gmelin E
Selenastrum Reinsch. A
Sium latifolium L. E
Sparganium emersum Rehmann RF
Sparganium erectum L. E
Sphaeranthus indicus L. *E
Sphagnum L. S/E
Sphenoclea zeylanica Gaertn. *E
Spirodela Schleiden FF
Spirodela polyrrhiza (L.) Schleiden FF
Spirogyra Link A
Spirulina Turpin A
Stratiotes L. FF/S
Stratiotes aloides L. FF/S
Synedra ulna (Nitzsch.) Ehrenberg A
Synnema Bentham E
(= *Hygrophila* R. Br.)

Thalia geniculata L. E
Trapa L. S/RF
Trapa bispinosa Roxb. S/RF
Trapa incisa Sieb. & Zucc. S/RF
Trapa natans L. S/RF
Typha L. E
Typha angustata Bory & Chaub. E
(= (?) *Typha domingensis* Pers.)
Typha angustifolia L. E
Typha domingensis Pers. E
(= (?) *Typha angustata* Bory & Chaub.
Typha elephantina Roxb. E
Typha javanica Schnizl. E
Typha latifolia L. E
Typha orientalis Presl. E

Utricularia L. S
Utricularia flexuosa Vahl. *S
Utricularia gibba L. S
Utricularia vulgaris L. S

Vallisneria L. S
Vallisneria americana Michx. S
Vallisneria spiralis L. S
Vaucheria DC. A
Vaucheria dichotoma (L.) C. A. Agardh. A
Veronica L. E
Veronica beccabunga L. E
Victoria amazonica (Poppig) Sowerby RF
Volvox aureus Ehrenberg A
Vossia cuspidata (Roxb.) Griff. E

Wiesneria schweinfurthii Hooker fil. S/RF
Wolffia Horkel. ex Schleiden FF
Wolffia arrhiza (L.) Wimm. FF
Wolffia microscopica (Griff.) Kurz. FF
Wolffiella Hegelm. FF

Zannichellia palustris L. S
Zannichellia pedunculata Rchb. S
Zannichellia peltata Bertol. S
Zizania L. E
Zizania aquatica L. E
Zizaniopsis miliacea (Michx.) Doell. & Aschers. E
Zostera capricorni Aschers. S

Acknowledgement

We thank Prof. C. D. K. Cook for reviewing the list of plant species in Appendix A.

APPENDIX B

Biological control agents used against aquatic weeds

B1. PATHOGENS

Species	Subdivision; Class; Order; Family	Target weeds
Acremonium curvulum W. Gams.	Deuteromycotina; Hyphomycetes; Moniliales; Moniliaceae	*Myriophyllum spicatum*
Acremonium zonatum (Saw.) Gams.	Deuteromycotina; Hyphomycetes; Moniliales; Moniliaceae	*Eichhornia* spp.
Alternaria alternantherae Holcomb & Antonopoulos	Deuteromycotina; Hyphomycetes; Moniliales; Dematiaceae	*Alternanthera* spp.
Alternaria alternata (Fr.) Keissler	Deuteromycotina; Hyphomycetes; Moniliales; Dematiaceae	*M. spicatum; Eichhornia crassipes*
Alternaria eichhorniae Nag Raj & Ponnappa	Deuteromycotina; Hyphomycetes; Moniliales; Dematiaceae	*E. crassipes*
Aphanomyces euteiches Drechs.	Mastigomycotina; Oomycetes; Saprolegniales; Saprolegniaceae	*M. spicatum*
Articulospora tetracladia Ingold	Deuteromycotina; Hyphomycetes; Moniliales; Moniliaceae	*M. spicatum*
Bipolaris oryzae (B. de H.) Shoemaker	Deuteromycotina; Hyphomycetes; Moniliales; Dematiaceae	*E. crassipes*
Botrytis spp.	Deuteromycotina; Hyphomycetes; Moniliales; Moniliaceae	*M. spicatum*
Burgoa sp.	Deuteromycotina; Hyphomycetes; Moniliales; Mycelia sterilia	Potamogetonaceae
Cercospora piaropi Tharp	Deuteromycotina; Hyphomycetes; Moniliales; Moniliaceae	*E. crassipes*
Cercospora rodmanii Conway	Deuteromycotina; Hyphomycetes; Moniliales; Moniliaceae	*E. crassipes*
Colletrotrichum gloeosporioides (Penz.) Sacc.	Deuteromycotina; Coelomycetes; Melanconiales; Malanconiaceae	*M. spicatum*

Dactylella microaquatica Ingold	Deuteromycotina; Hyphomycetes; Moniliales; Moniliaceae	*M. spicatum*
Flagellospora stricta Ingold	Deuteromycotina; Hyphomycetes; Moniliales; Moniliaceae	*M. spicatum*
Fusarium acuminatum El. & Ev.	Deuteromycotina; Hyphomycetes; Moniliales; Moniliaceae	*M. spicatum*
Fusarium lateritium Nees	Deuteromycotina; Hyphomycetes; Moniliales; Moniliaceae	Potamogetonaceae
Fusarium oxysporum Schlecht.	Deuteromycotina; Hyphomycetes; Moniliales; Moniliaceae	*M. spicatum*
Fusarium poae (Peck) (Wr.	Deuteromycotina; Hyphomycetes; Moniliales; Moniliaceae	*M. spicatum*
Fusarium roseum Link	Deuteromycotina; Hyphomycetes; Moniliales; Moniliaceae	*Hydrilla verticillata; Stratiotes aloides*
Fusarium roseum (Link) emend. Snyd. & Hans. 'Avenaceum'	Deuteromycotina; Hyphomycetes; Moniliales; Moniliaceae	*M. spicatum*
Fusarium roseum (Link) emend. Snyd. & Hans. 'Culmorum'	Deuteromycotina; Hyphomycetes; Moniliales; Moniliaceae	*H. verticillata; S. aloides*
Fusarium solani (Mart.) Appl. & Wr.	Deuteromycotina; Hyphomycetes; Moniliales; Moniliaceae	*M. spicatum*
Fusarium sporotrichioides Sherb.	Deuteromycotina; Hyphomycetes; Moniliales; Moniliaceae	*M. spicatum*
Fusidium sp.	Deuteromycotina; Hyphomycetes; Moniliales; Moniliales	Potamogetonaceae
Marasmiellus inoderma (Berk.) Sing.	Basidiomycotina; Hymenomycetes; Agaricales; Agaricaceae	*E. crassipes*
Myrothecium roridum Tode ex Fr.	Deuteromycotina; Hyphomycetes; Moniliales; Moniliaceae	*E. crassipes; Salvinia* spp.
Papalospora sp.	Deuteromycotina; Hyphomycetes; Moniliales; Mycelia sterilia	Potamogetonaceae

B1. PATHOGENS (*contd*)

Species	Subdivision; Class; Order; Family	Target weeds
Penicillium spp.	Deuteromycotina; Hyphomycetes; Moniliales; Moniliaceae	*H. verticillata; M. spicatum*
Phytophthora spp.	Mastigomycotina; Oomycetes; Peronosporales; Pythiaceae	Not specified
Pythium carolinianum Matthews (not a valid species?)	Mastigomycotina; Oomycetes; Peronosporales; Pythiaceae	*Myriophyllum brasiliense*
Rhizoctonia solani Kuehn	Deuteromycotina; Hyphomycetes; Moniliales; Mycelia sterilia	*E. crassipes*
(= *Corticium solani* [Prill. & Delacr.] Bond. & Galz.)	Basidiomycotina; Hymenomycetes; Aphyllophorales; Corticiaceae	
(= *Aquathanatephorus pendulus* Tu & Kimbrough)	Basidiomycotina; Hymenomycetes; Aphyllophorales; Corticiaceae	
Sclerotium hydrophyllum Sacc.	Deuteromycotina; Hyphomycetes; Moniliales; Mycelia sterilia	*Myriophyllum* spp.
Sclerotium rolfsii (Sacc.) Curzi	Deuteromycotina; Hyphomycetes; Moniliales; Mycelia sterilia	*H. verticillata; Pistia stratiotes*
Uredo eichhorniae Gonz.-Frag. & Ciferri	Basidiomycotina; Teliomycetes; Uredinales	*E. crassipes*
Uredo nitidula Arth.	Basidiomycotina; Teliomycetes; Uredinales	*Alternanthera philoxeroides*

B2. INVERTEBRATES

1. Insects

Species	Order; Family	Target weeds
Acentropus niveus (Olivier)	Lepidoptera; Pyralidae	*Myriophyllum spicatum*
(= *Acentria nivea* Olivier)		
Acigona infusella (Walker)	Lepidoptera; Pyralidae	*Eichhornia crassipes*
Agasicles hygrophila Selman & Vogt	Coleoptera; Chrysomelidae	*Alternanthera philoxeroides*
Amynothrips andersoni O'Neill	Thysanoptera; Palaeothripidae	*Alternanthera philoxeroides*
(= *Bellura densa* (Walker))		
Arzama densa (Walker)	Lepidoptera; Noctuidae	*Eichhornia crassipes*
Bagous geniculatus Hustache	Coleoptera; Curculionidae	*Hydrilla verticillata*
Bagous sp. nr. *limosus*	Coleoptera; Curculionidae	*Hydrilla verticillata*
Bagous sp. nr. *lutulosus*	Coleoptera; Curculionidae	*Hydrilla verticillata*
(= *Bagous affinis* Hustache)		
Bagous vicinus Hustache	Coleoptera; Curculionidae	*Myriophyllum spicatum*
Cyrtobagous salviniae Calder & Sands	Coleoptera; Curculionidae	*Salvinia molesta*
Disonycha argentinensis Jacoby	Coleoptera; Chrysomelidae	*Alternanthera philoxeroides*
Epipsammia pectinicornis Hampson	Lepidoptera; Noctuidae	*Pistia stratiotes*
(= *Namangana pectinicornis* Hampson)		
Gesonula punctifrons Stal	Orthoptera; Acrididae	*Eichhornia crassipes*
Hydrellia pakistanae Deonier	Diptera; Ephydridae	*Hydrilla verticillata*
Listronotus marginicollis (Hustache)	Coleoptera; Curculionidae	*Myriophyllum spicatum*
Litodactylus leucogaster (Marsham)	Coleoptera; Curculionidae	*Myriophyllum spicatum*
Lysathia flavipes (Boheman)	Coleoptera; Chrysomelidae	*Myriophyllum aquaticum*
Lysathia ludoviciana (Fall)	Coleoptera; Chrysomelidae	*Myriophyllum aquaticum*
Neochetina bruchi Hustache	Coleoptera; Curculionidae	*Eichhornia crassipes*

B2. INVERTEBRATES (*contd*)

1. Insects (*contd*)

Species	Order; Family	Target weeds
Neochetina eichhorniae Warner	Coleoptera; Curculionidae	*Eichhornia crassipes*
Neohydronomous pulchellus Hustache	Coleoptera; Curculionidae	*Pistia stratiotes*
Nymphula responsalis Walker	Lepidoptera; Pyralidae	*Salvinia molesta*
Nymphula tenebralis Lower	Lepidoptera; Pyralidae	*Pistia stratiotes*
Parapoynx allionealis Walker	Lepidoptera; Pyralidae	*Myriophyllum aquaticum*
Parapoynx diminutalis Snellen	Lepidoptera; Pyralidae	*Hydrilla verticillata*
Paulinia acuminata DeGeer	Orthoptera; Pauliniidae	*Salvinia molesta*
Phytobius sp.	Coleoptera; Curculionidae	*Myriophyllum spicatum*
Proxenus hennia Swinhoe	Lepidoptera; Noctuidae	*Pistia stratiotes*
Samea multiplicalis Guenée	Lepidoptera; Pyralidae	*Salvinia molesta*
Sameodes albiguttalis (Warren)	Lepidoptera; Pyralidae	*Eichhornia crassipes*
Vogtia malloi Pastrana	Lepidoptera; Pyralidae	*Alternanthera philoxeroides*

2. Mites

Orthagolumna terebrantis Wallwork	Acarina; Galumnidae	*Eichhornia crassipes*

3. Molluscs

Ampullaria canaliculata Lam.	Gastropoda; Ampullariidae	Non-specific grazer
Marisa cornuarietis L.	Gastropoda; Ampullariidae	Semi-specific grazer
Pomacea australis d'Orbigny	Gastropoda;	Semi-specific grazer

B3. OTHER ORGANISMS

1. Plants

Species	Order; Family	Target weeds
Casuarina L.	Verticillatae; Casuarinaceae	Non-specific shade plant
Eleocharis acicularis R.	Cyperales; Cyperaceae	Competitor plant against submerged weeds
Eleocharis coloradoensis (Britt.) Gilly	Cyperales; Cyperaceae	Competitor plant against submerged weeds
Salvinia spp.	Hydropteridales; Salviniaceae	Shade plant against submerged weeds

2. Fish

Species	Order; Family	Target weeds
Aristichthys nobilis Rich	Ostariophysi; Cyprinidae	Non-specific grazer
Ctenopharyngodon idella Val.	Ostariophysi; Cyprinidae	Semi-specific grazer
Hypophthalmichthys molitrix Val.	Ostariophysi; Cyprinidae	Phytoplankton-grazer
Metynnis roosevelti Eigenmann	Ostariophysi; Characidae	Non-specific grazer
Mylossoma argenteum E. Ahl.	Ostariophysi; Characidae	Non-specific grazer
Tilapia melanopleura Dumeril (= *T. rendalli*)	Percomorphi; Cichlidae	Non-specific grazer
Tilapia zillii (Gervais)	Percomorphi; Cichlidae	Non-specific grazer

3. Mammals

Species	Order; Family	Target weeds
Hydrochoerus hydrochaeris L.	Rodentia; Hydrochoeridae	Non-specific grazer
Myocaster coypus Mol.	Rodentia; Capromyidae	Non-specific grazer
Trichechus manatus L.	Sirenia; Trichechidae	Non-specific grazer

ACKNOWLEDGEMENTS

We thank Prof. R. Charudattan and Dr K. L. S. Harley for respectively reviewing the lists of species in Appendices B1 and B2.

APPENDIX C

Herbicides used in fresh waters

Nomenclature: IUPAC (Pesticides Manual, 7th edn).
Formulation data refer to aquatic-use formulations only.

Common name	Chemical name
acrolein	acrylaldehyde (= 2-propenal)
ametryne	*N*-ethyl-*N'*-isopropyl-6-methylthio-1,3,5-triazine-2,4-diyldiamine
6-aminotoluic acid	6-aminotoluic acid
aminotriazole (= amitrole)	1*H*-1,2,4-triazol-3-ylamine
ammonia	NH_3
asulam	methylsulphanilylcarbamate
atrazine	6-chloro-*N*-ethyl-N'isopropyl-1,3,5-triazine-2,4-diyldiamine
bensulfuron methyl	methyl-2 (4,6-dimethoxypirimidin 2-yl)aminosulphonylmethylbenzoate
bentazone (= 'Basagran')	3-isopropyl-(1*H*)-2,1,3-benzothiadiazin-4(3*H*)-one 2,2-dioxide
bromacil	5-bromo-3-*sec*-butyl-6-methyluracil
copper (formulations include: copper sulphate pentahydrate salt, copper EDA, copper TEA)	Cu
cyanatryn	2-(4-ethylamino-6-methylthio-1,3,5-triazin-2-ylamino)-2-methylpropionitrile
2,4-D (formulations include: sodium, dimethylamine (DMA), butoxyethanol ester (BEE), and PGBE)	(2,4-dichlorophenoxy)acetic acid
dalapon	2,2-dichloropropionic acid

Common name	Chemical name
dicamba (= banvel)	3,6-dichloro-*o*-anisic acid
dichlobenil	2,6-dichlorobenzonitrile
diquat (formations include: dibromide salt and alginate gel)	9,10-dihydro-8*a*,10*a*-diazoniaphenanthrene ion
diuron	3-(3,4-dichlorophenyl)-1,1-dimethylurea
endothall (formulations include: disodium, dipotassium, and mono (*n*,*n*-dimethylalkylamine) salts)	7-oxabicyclo [2.2.1] heptane-2,3-dicarboxylic acid
fenac (= chlorfenac) (formulated usually as sodium salt)	(2,3,6-trichlorophenyl)acetic acid
fluridone	1-methyl-3-phenyl-5(α,α,α-trifluoro-*m*-tolyl)-4-pyridone
glyphosate (formulated as mono (isopropylammonium) salt)	*N*-(phosphonomethyl)glycine
hexazinone	3-cyclohexyl-6-dimethylamino-1-methyl-1,3,5-triazine-2,4(1*H*,3*H*)-dione
hydrogen peroxide	H_2O_2
imazapyr	2-(4-isopropyl-4-methyl-5-oxo-2-imidazolin-2-yl)nicotinic acid
imidazoline	ethyleneurea
MCPP	(±)-2-(4-chloro-*o*-tolyoxy)propionic acid
methoxone (= MCPA) (= phenoxylene-30)	2-methyl-4-chlorophenoxyaceto-*o*-chloroanilide
molinate	*S*-ethyl azepane-1-carbothioate
monuron	3-(4-chlorophenyl)-1,1-dimethylurea
paraquat (formulated as dichloride salt)	1,1″-dimethyl-4,4″-bipyridinium dichloride
silvex (= fenoprop) (= 2,4,5-TP)	2-(2,4,5-trichlorophenoxy)propionic acid

Common name	Chemical name
simazine	6-chloro-*N*,*N*′-diethyl-1,3,5-triazine-2,4-diyldiamine
sodium arsenite	$NaAsO_2$
sulfometuron	2-3-(4,6-dimethylpyrimidin-2-yl) ureidosulphonylbenzoic acid
2,4,5-T	(2,4,5-trichlorophenoxy)acetic acid
TCA	trichloroacetic acid
terbutryne	*N*-*t*-butyl-*N*′-ethyl-6-methylthio-1,3,5-triazine-2,4-diylamine
thiobencarb	*S*-4-chlorobenzyl diethyl (thiocarbamate)

REFERENCES

Abdalla, A. A. and Abdel Hafeez, A. T. (1969). Some aspects of utilization of water hyacinth (*Eichhornia crassipes*). *Pest Articles and News Summaries*, **15**, (2), 204–7.

Abdusamadov, A. S. (1986). Biology of white amur, *Ctenopharyngodon idella*, silver carp, *Hypophthalmichthys molitrix*, and bighead, *Aristichthys nobilis* acclimatized in the Terek Region of the Caspian Basin. *Journal of Ichthyology*, **26**, 41–9.

Achmad, S. (1971). Problems and control of aquatic weeds in Indonesian open water. In *Tropical weeds: some problems, biology and control* (ed. M. Soerjani), pp. 107–13. Biotrop, Bogor.

Ackers, P., White, W. R., Perkins, J. A., and Harrison, A. J. (1978). *Weirs and flumes for flow measurement*. John Wiley & Sons, Winchester. pp. 327.

Adam, N. (1979). The floating islands of Rawa Pening and their socio-economic aspects. *Biotrop Bulletin*, **11**, 209–18.

Adamek, Z. and Sanh, T. D. (1981). The food of grass carp fry (*Ctenopharyngodon idella*) in southern Moravian fingerling ponds. *Folia Zoologica*, **30**, 263–70.

Adams, M. S., McCracken, M. D., and Schmidt, K. (1971). *Preliminary results of measurements of primary productivity of aquatic macrophytes and periphyton in Lake Wingra during 1970–71*. US IBP-EDFB Memo Report 71–54.

Adams, M. S., Scarpace, F. L., Scherz, J. P., and Woelkerling, W. J. (1977). *Assessment of aquatic environment by remote sensing*. University of Wisconsin, IES Report No. 84. University of Wisconsin, Madison.

Adams, M. S., Titus, J. E., and McCracken, M. D. (1974). Depth distribution of photosynthetic activity in a *Myriophyllum spicatum* community in Lake Wingra, Wisconsin. *Limnology and Oceanology*, **19**, 377–89.

Agaronian, A. G., Aslanian, A. M., and Gevorkian, K. H. A. (1980*a*). [Influence of dalapon on reed (*Phragmites communis*) biomass and its renewal under chemical land-reclamation.] In Armenian, English summary. *Biologicheskii Zhurnal Armenia*, **33**, 317–20.

Agaronian, A. G., Bazmanova, N. V., Altunian, M. G., and Arutiunian, K. A. (1980*b*). [Migration and detoxification of dalapon in soil and rhizomes of *Phragmites communis*.] In Armenian, English summary. *Biologicheskii Zhurnal Armenia*, **33**, 337–40.

Agostini, G. (1974). El género *Heteranthera* (Pontederiaceae) en Venezuela. *Acta Bótanica Venezuelica*, **9**, 295–301.

Agusti, S., Duarte, C., and Guimaraes, T. (1984). *Influencia da carpa espelhada* (Cyprinus carpio *L., f.* specularis *Lac.*) *no desenvolvimento das infestantes aquaticas em valas de drenagem*. Paper H-4/84, INIP Sem. Aquacult. Centro Botanico Aplicato Agricultua, Univ. Tecnica Lisbõa.

Ahling, B. B. and Jernelöv, A. (1971). Weed control with grass carp in Lake Ösbysjön. *Vatten*, **27**, 253.

Ahmad, N. (1953). Control of submerged vegetation in fish ponds. *Journal of Agriculture, Pakistan*, **4**, 13–16.

Ahmad, N. (1968). Review of research work done by the Directorate of Fisheries, West Pakistan. *Journal of Agriculture, Pakistan,* **19**, 557–72.

Ahmed, S. A., Ito, M., and Kunikazu, U. (1980). Water quality as affected by water hyacinth decomposition after 2,4-D and ametryne application. *Weed Research, Japan,* **25**, (Supp.), 83–4.

Ahmed, S. A., Ito, M., and Ueki, K. (1981). Water quality as affected by water hyacinth decomposition after 2,4-D and ametryne application. *Weed Research, Japan,* **25**, 286–93.

Ahmed, S. A., Ito, M., and Ueki, K. (1982). Water quality as affected by water hyacinth decomposition after cutting or 2,4-D application. *Weed Research, Japan,* **27**, 34–9.

Ahti, T. Hämet-Ahti, L., and Pettersson, B. (1973). Flora of the inundated Wadi Halfa reach of the Nile, Sudanese Nubia, with notes on adjacent areas. *Annales Botanicae Fennici,* **10**, 131–62.

AIBS (1978). Aquatic hazard evaluation of chemical pesticides. *BioScience,* **28** (9), 600–1.

Aiken, S. (1976). Turion formation in watermilfoil, *Myriophyllum farwelli. Michigan Botanist,* **15**, 99–102.

Aiken, S. G., Newroth, P. R., and Wile, I. (1979). The biology of Canadian weeds. 34. *Myriophyllum spicatum* L. *Canadian Journal of Plant Sciences,* **59**, 201–15.

Alburqueque, B. W. P. de (1981). Plantas forrageiras da Amazonia. I. Aquáticas flutuantes livres. *Acta Amazónica,* **11**. 457–71.

Aliyev, D. S. (1976*a*). Biological method for preventing the overgrowth and deformation of canals of hydro-engineering and reclamation systems. *Voprosi Ichtiologii,* **16**, 247–51.

Aliyev, D. S. (1976*b*). The role of phytophagous fishes in the reconstruction of commercial fish fauna and the biological improvement of water. *Journal of Ichthyology,* **16 (2)**, 216–29.

Alikunhi, K. H. and Sukumaran, K. K. (1965). Preliminary observations on Chinese carps in India. *Proceedings of the Indian Academy of Sciences Section B,* **60**, pp. 171–89.

Allen, S. K. and Wattendorf, R. J. (1987). Triploid grass carp: status and management implications. *Fisheries,* **12**, 20–4.

Allsopp, W. H. L. (1960). The manatee–ecology and use for weed control. *Nature* (*London*), **188**, 762.

Allsop, W. H. L. (1961). Putting manatees to work. *New Scientist,* **12**, 263.

Allsopp, W. H. L. (1969). Aquatic weed control by manatees—its prospects and problems. In *Man-made lakes* (ed. L. E. Obeng), pp. 344–51. Ghana University Press, Accra.

Almazan, G. and Boyd, C. E. (1978). Effects of N levels on rates of O_2 consumption during decay of aquatic plants. *Aquatic Botany,* **5**, 119–26.

Almkvist, B. (1975). The influence of flight altitude and type of film on photo-interpretation of aquatic macrophytes. *Svensk Botanisk Tidskrift,* **69**, 181–7.

Altinayar, G. and Onursal, N. F. (1982). [Investigation of species and distribution of aquatic weeds in the irrigation and drainage systems of Turkey.] In Turkish. *Bitki Kornma Bulteni,* **22**, 120–41.

Ancibor, E. (1979). Systematic anatomy of vegetative organs of the Hydrocharitaceae. *Botanical Journal of the Linnaean Society,* **78**, 237–66.

Anderson, L. W. J. (1978). *Light requirement of efficacy of fluridone in preventing growth of American pondweed* (Potamogeton nodosus *Poir.*) *and sago pondweed* (P. pectinatus *L.*). Report in Abstracts 1978 Meeting Weed Science Society America.

Anderson. L. W. J. (1981*a*). Control of aquatic weeds with hexazinone. *Journal of Aquatic Plant Management,* **19,** 9–14.

Anderson, L. W. J. (1981*b*). Effect of light on the phytotoxicity of fluridone in American pondweed (*Potamogeton nodosus*) and sago pondweed (*P. pectinatus*). *Weed Science,* **29,** 723–8.

Anderson, L. W. J. (1983). Translocation of Oust[R] (DPX 5648) in hydrilla and American pondweed, using a silicon sealant. *Western Society Weed Science Research Programme Reports 1983,* pp. 294–7.

Anderson, L. W. J. (ed.) (1985). *Proceedings 1st International Symposium on Watermilfoil* (Myriophyllum spicatum) *and related Haloragaceae species; Vancouver, British Columbia, Canada.* Aquatic Plant Management Society, Vicksburg, Mississippi. pp. 223.

Anderson, L. W. J. (1986*a*). *Exotic pest profile:* Hydrilla verticillata. Report to California Department of Food and Agriculture, USDA Agricultural Research Service, University of California, Davis, California. pp. 32.

Anderson, L. W. J. (1986*b*). Recent developments and future trends in aquatic weed management. *Proceedings EWRS/AAB 7th Symposium on Aquatic Weeds 1986,* pp. 9–16.

Anderson, T. M. (1981). Spray raft simplifies aquatic weed control. *Tropical Pest Management,* **27,** 278–9.

Andrews, D. S., Webb, D. H., and Bates, A. L. (1984). The use of aerial remote sensing in quantifying submersed aquatic macrophytes. *Special Technical Publications ASTM* STP 843, 92–9. American Society for Testing Materials.

Andrews, F. W. (1945). Water plants in the Gezira canals. *Annals of Applied Biology,* **32,** 1–14.

Andrews, J. H. and Hecht, E. P. (1981). Evidence for pathogenicity of *Fusarium sporotrichioides* to Eurasian watermilfoil *Myriophyllum spicatum. Canadian Journal of Botany,* **59,** 1069–77.

Andrews, J. H., Hecht, E. P., and Bashirian, S. (1982). Association between the fungus *Acremonium curvulum* and Eurasian watermilfoil *Myriophyllum spicatum. Canadian Journal of Botany,* **60,** 1216–21.

Angerelli, N. P. D. and Beirne, N. P. (1982). Mortality of introduced mosquito larvae in natural and artificial ponds containing aquatic vegetation. *Protection Ecology,* **4,** 381–6.

Anon. (1971). *Herbicide monitoring in irrigation systems—acrolein residues in irrigation water.* Progress Report 1st Quarter Fiscal Year, 1971. Federal Water Quality Administration, US Bureau of Reclamation. pp. 7.

Anon. (1980). New aquatic herbicide controls *Hydrilla. Agrichemical Age,* December **1980,** 46–7.

Anon. (1982). *Ecological studies for conservation of shore birds in Songkhla lake. Vol. 1.* Office of the National Environment Board, Bangkok.

Anon. (1985*a*). *Biological control of* Salvinia molesta *in the Eastern Caprivi.* Leaflet: Water Quality Division, Department of Water Affairs, South West Africa.

Anon. (1985*b*). *Ultrasound technique may eradicate milfoil*. Aquatic Plant Management Society Newsletter 20.

Anon. (1986). *Salvinia* in Sri Lanka. *Aquaphyte*, **6** (1), 3.

Antalfi, A. and Tölg, I. (1971). *Graskarpfen, pflanzenfressende Fische*. Donau Verlag, Günzburg. pp. 207.

Antanyniene, G. and Trainauskaite, I. (1985). [Production—biochemical characteristics of water plants in Lake Druksial in 1982.] In Russian. *Lietuvos TSR Mokslu Akademija. Darbai B Serija*, **1 (89)**, 68–75.

Aranha, C. and Pio, R. M. (1981). Plantas invasoras da cultura do arroz (*Oryza sativa* L.) no estado de São Paulo. 1. Dicotiledoneas. *Planta Daninha*, **4**, 33–57.

Aranha, C. and Pio, R. M. (1982). Plantas invasoras da cultura do arroz (*Oryza sativa* L.) no estado de São Paulo. 2. Monocotiledoneas. *Planta Daninha*, **5**, 65–81.

Arber, A. (1920). *Water plants—a study of aquatic angiosperms*. Cambridge University Press, Cambridge. pp. 436.

Armitage, K. B. and Fassett, N. C. (1971). Aquatic plants of El Salvador. *Archiv für Hydrobiologie*, **69**, 234–55.

Armour, G. D., Hanna, R. S., Walters, I. P., and Maxnuk, M. D. (1980). *Studies on aquatic macrophytes. XIV. Summary of mechanical aquatic plant management, Okanagan Valley, 1978*. Report, Ministry of Environment, Inventory & Engineering Branch, British Columbia. pp. 36.

Arsenović, M., Dimitrijević, M., and Konstantinović, B. (1986). Control of aquatic weeds with herbicides in drainage system 'Dunav–Tisa–Dunav' in Yugoslavia. *Proceedings EWRS/AAB 7th Symposium on Aquatic Weeds 1986*, pp. 31–5.

Arsenović, M., Stanković, A., Dimitrijević, M., and Konstantinović, B. (1982). Investigations into the control of submerged vegetation in an irrigation system by specific herbicides. *Proceedings EWRS 6th Symposium on Aquatic Weeds 1982*, pp. 177–84.

Arthington, A. H. and Mitchell, D. S. (1986). Aquatic invading species. In *Ecology of biological invasions: an Australian perspective* (ed. R. H. Groves and J. J. Burdon), pp. 34–53. Australia Academy of Science, Canberra.

Arthur, J. C. (1920). New species of Uredineae XII. *Torrey Botanical Club Bulletin*, **47**, 465–80.

Arts, W. B. M. and van Wijk, A. (1978). New developments in the mechanical control of water weeds. *Proceedings EWRS 5th Symposium on Aquatic Weeds 1978*, pp. 195–201.

Arts, W. B. M. and van Wijk, A. (1982). New implements for weeding out the vegetation from watercourses. *Proceedings EWRS 6th Symposium on Aquatic Weeds 1982*, pp. 159–63.

Asensio, D. (1985). Uptake of bicarbonate by waterhyacinth (*Eichhornia crassipes* (Mart.) Solms). *Abstract of paper presented at International Symposium on Aquatic Macrophytes, Silkeborg, Denmark 1985*.

Ashton, F. M., Bissell, S. R., Di Tomaso, J. M., and Wach, M. J. (1984). *Research on biological control of aquatic weeds with competitive species of spikerush* (Eleocharis *spp.*) Final report, Cooperative Agreement No. 58-9AHZ-9-437, USDA and University of California. pp. 187.

Ashton, F. M. and Bissell, S. R. (1987). Influence of temperature and light on dwarf spikerush and slender spikerush growth. *Journal of Aquatic Plant Management*, **25**, 4–7.

Ashton, F. M., Bissell, S. R., Di Tomaso, J. M., and Wach, M. J. (1984). *Research on biological control of aquatic weeds with competitive species of spikerush* (Eleocharis *spp.*), Final Report, Cooperative Agreement No. 58-9AHZ-9-437, USDA and University of California. pp. 187.

Ashton, P. J., Scott, W. E., and Steyn, D. J. (1981). The chemical control of the water hyacinth (*Eichhornia crassipes* (Mart.) Solms). *Progress in Water Technology*, **12**, 865–82.

Assemat, L., Morishima, K., and Oka, H. I., (1981). Neighbour effects between rice (*Oryza sativa* L.) and barnyard grass (*Echinochloa crusgalli* Beauv.) strains. II. Some experiments on the mechanisms of interaction between plants. *Acta Oecologica*, **2**, 63–78.

Aston, H. I. (1973). *Aquatic plants of Australia*. Melbourne University Press, Melbourne. pp. 368.

Aston, H. I. (1977). *Supplement to aquatic plants of Australia*. Melbourne University Press, Melbourne.

Augsten, H. and Jungnickel, F. (1983). Control of turion formation by light, phosphate and sugar in *Spirodela polyrhiza* (L.) Schleiden. *Wissenschaftliche Zeitschrift der Ernst-Moritz-Arndt Universität Greitswald, Math. Naturwissenschaft Reihe*, **32**, 64–6.

Aurand, D. (1982). *Nuisance aquatic plants and aquatic plant management programs in the United States, Volume 2. South-eastern Region*. Report: The Mitre Corporation, Metrek Division, McLean, Virginia, USA.

Aurand, D. (1983). *Nuisance aquatic plants and aquatic plant management programs in the United States, Volume 4. Northwestern Region.*, Report: The MITRE Corporation, No. MTR-82W47-04. p. 157.

Aurand, D., Tyndall, R. W., and Trudeau, P. (1982). The role of integrated pest management principles in aquatic plant management programs. *Proceedings Conference on Strategies for Aquatic Weed Management* (ed. T. L. Chesnut), pp. 2–12. University of Florida, Gainesville.

Australian Water Resources Council (1985). *Guidelines for the use of herbicides in or near water*. Australian Government Publishing Service, Canberra.

Badar-ud-Din, A. A. (1978). *Control of aquatic weeds*. Project No. FG-Pa-271 Grant No. PK-ARS-72. Institute of Chemistry, University of Punjab, Lahore, Pakistan.

Badger, M. R., Kaplan, A., and Berry, J. A. (1978). A mechanism for concentrating CO_2 in *Chlamydonas reinhardii* and *Anabaena variabilis*, and its role in photosynthetic CO_2 fixation. *Carnegie Institute Yearbook*, **77**, 251–61.

Baer, R. G. and Quimby, J. P. C. (1981). Laboratory rearing and life history of *Arzama densa*, a potential native biological control agent against water hyacinth. *Journal of Aquatic Plant Management*, **19**, 22–6.

Bagnall, L. O. (1980*a*). *Intermediate technology screw presses for dewatering aquatic plants*. American Society of Agricultural Engineers, Paper No. 80-50-44, pp. 15.

Bagnall, L. O. (1980*b*). Bulk mechanical properties of hydrilla. *Journal of Aquatic Plant Management*, **18**, 23–6.

Bagnall, L. O. (1981). *Aquatic plant harvesting and harvesters*. American Society of Agricultural Engineers, Paper No. 81-50-19. pp. 6.

Bailey, W. M. (1975). Operational experiences with the white amur in weed control programs. *Proceedings of a Symposium on Water Quality and Management through Biological Control, Gainesville 1975,* pp. 75–8.

Bailey-Watts, A. E. and Duncan, P. (1981). A review of macrophyte studies. In *The ecology of Scotland's largest lochs* (ed. P. S. Maitland). *Monographiae Biologicae,* **44,** 119–34.

Bain, J. T. and Proctor, M. C. F. (1980). The requirement of aquatic bryophytes for free CO_2 as an inorganic carbon source: some experimental evidence. *New Phytologist,* **86,** 393–400.

Baird, D. D., Baker, G. E., Brown, H. F., and Urrutia, V. M. (1983). Aquatic weed control with glyphosate in south Florida. *Proceedings Southern Weed Society 36th Annual Meeting,* 430–3.

Baird, D. D., Upchurch, R. P., Homesley, W. B., and Franz, J. E. (1971). Introduction of a new broadspectrum post-emergence herbicide class with utility for herbaceous perennial weed control. *Proceedings North Central Weed Control Conference,* **26,** pp. 64–8.

Balasooriya, I., Gunasekera, S. A., Hettiarachchi, S., and Gunasekera, I. J. (1984). Biology of water hyacinth: fungi associated with water hyacinth in Sri Lanka. *Proceedings International Conference on Water Hyacinth* (ed. G. Thyagarajan), pp. 304–17. UNEP, Nairobi, Kenya.

Balciunas, J. K. (1982). *Insects and other macro-invertebrates associated with Eurasian watermilfoil in the United States.* Technical Report A-82-5, US Army Engineer Waterways Experiment Station, Vicksburg, Mississippi. pp. 87.

Balciunas, J. K. (1983). Overseas searches for insects on hydrilla in Southeast Asia and Australia. In *Proceedings 17th Annual Meeting Aquatic Plant Control Research Program.* pp. 104–114. US Army Engineer Waterways Experiment Station, Vicksburg, Mississippi.

Balciunas, J. K. and Habeck, D. H. (1981). Recent range of a hydrilla-damaging moth, *Parapoynx diminutalis* (Lepidoptera: Pyralidae). *Florida Entomologist,* **64,** 195–6.

Baldwin, B. C., Bray, M. F., and Geoghegan, M. J. (1966). Microbial decomposition of paraquat. *Biochemical Journal,* **101,** 15.

Baloch, G. M. and Sana-Ullah (1973). Insects and other organisms associated with *Hydrilla verticillata* (L.f.) Royle (Hydrocharitaceae). *Proceedings in Pakistan 3rd International Symposium Biological Control of Weeds, CIBC Miscellaneous Publication,* **8,** pp. 61–6.

Baloch, G. M., Sana-Ullah, and Ghani, M. A. (1980). Some promising insects for biological control of *Hydrilla verticillata* in Pakistan. *Tropical Pest Management,* **26,** 194–200.

Balyan, R. S., Bhan, V. M., Kamboj, R. K., and Singh, S. P. (1981). Translocation of foliage-applied glyphosate into the rhizomes of cattail. *Proceedings 8th Asian–Pacific Weed Science Society Conference,* pp. 341–5.

Banerjee, S. and Mitra, E. (1954). Preliminary observations on the use of copper sulphate to control submerged aquatic weeds in alkaline waters. *Indian Journal of Fisheries,* **1,** 204–16.

Barko, J. W. (1982). Influence of potassium source (sediment vs. open water) and sediment composition on the growth and nutrition of a submerged freshwater macrophyte (*Hydrilla verticillata* (L.f.) Royle). *Aquatic Botany,* **12,** 157–72.

Barko, J. W. (1983). The growth of *Myriophyllum spicatum* L. in relation to selected characteristics of sediment and solution. *Aquatic Botany*, **15**, 91–103.

Barko, J. W. and Smart, R. M. (1979). The nutritional ecology of *Cyperus esculentus*, an emergent aquatic plant, grown on different sediments. *Aquatic Botany*, **6**, 13–28.

Barko, J. W. and Smart, R. M. (1980). Mobilisation of sediment phosphorus by submerged freshwater macrophytes. *Freshwater Biology*, **10**, 229–38.

Barko, J. W. and Smart, R. M. (1981*a*). Comparative influences of light and temperature on the growth of selected submerged freshwater macrophytes. *Ecological Monographs*, **51**, 219–35.

Barko, J. W. and Smart, R. M. (1981*b*). Sediment-based nutrition of submerged macrophytes. *Aquatic Botany*, **10**, 339–52.

Barko, J. W. and Smart, R. M. (1983). Effects of organic matter additions to sediment on the growth of aquatic plants. *Journal of Ecology*, **71**, 161–76.

Barko, J. W., Smart, R. M., Hardin, D. G., and Matthews, M. S. (1980). *Growth and metabolism of three introduced submerged plant species in relation to the influence of temperature and light*. Technical Report A-80-1, US Army Engineer Waterways Experiment Station, Vicksburg, Mississippi.

Barltrop, J., Martin, B. B., and Martin, D. F. (1982). Response of *Hydrilla* to selected dyes. *Journal of Environmental Science and Health*, **A17 (5)**, 725–35.

Barnes, H. H. (1967). *Roughness characteristics of natural channels*. US Geological Survey Water Supply Paper No. 1849, Washington DC. pp. 214.

Barrett, P. R. F. (1974*a*). A spraying rig for the experimental application of herbicides to the floating leaves of water plants. *Weed Research*, **14**, 313–15.

Barrett, P. R. F. (1974*b*). The effect of spraying large plots of *Nuphar lutea* L. with glyphosate. *Proceedings 12th British Weed Control Conference*, **1**, 229–32.

Barrett, P. R. F. (1976). The effect of dalapon and glyphosate on *Glyceria maxima* (Hartm.) Holmberg. *Proceedings 1976 British Crop Protection Conference—Weeds*, pp. 79–82.

Barrett, P. R. F. (1978*a*). Some studies on the use of alginate for the placement and controlled release of diquat on submerged plants. *Pesticide Science*, **9**, 425–33.

Barrett, P. R. F. (1978*b*). Aquatic weed control: necessity and methods. *Fisheries Management*, **9**, 93–101.

Barrett, P. R. F. (1981*a*). A comparison of two formulations of diquat for weed control in rivers. *Proceedings Association of Applied Biologists Conference: Aquatic Weeds and their Control 1981*, pp. 183–8. AAB, Wellesbourne, UK.

Barrett, P. R. F. (1981*b*). Diquat and sodium alginate for weed control in rivers. *Journal of Aquatic Plant Management*, **19**, 51–2.

Barrett, P. R. F. (1981*c*). Aquatic herbicides in Great Britain, recent changes and possible future development. *Proceedings Association of Applied Biologists Conference: Aquatic Weeds and their Control 1981*, pp. 95–103. AAB, Wellesbourne, UK.

Barrett, P. R. F. (1983). Localized control of weeds in fisheries. *Proceedings 3rd Freshwater Fisheries Conference 1983*, pp. 95–105. University of Liverpool, Liverpool.

Barrett, P. R. F. (1985). Efficacy of glyphosate in the control of aquatic weeds. In *The herbicide glyphosate* (ed. E. Grossbard and D. Atkinson), pp. 365–74. Butterworths, London.

Barrett, P. R. F. and Logan, P. (1982). The localized control of submerged aquatic weeds in lakes with diquat–alginate. *Proceedings EWRS 6th Symposium on Aquatic Weeds 1982*, pp. 193–9.

Barrett, P. R. F. and Murphy, K. J. (1982). The use of diquat–alginate for weed control in flowing waters. *Proceedings EWRS 6th Symposium on Aquatic Weeds 1982*, pp. 200–8.

Barrett, P. R. F. and Robson, T. O. (1971). The effect of time of application on the susceptibility of some emergent water plants to dalapon. *Proceedings EWRC 3rd Symposium on Aquatic Weeds 1971*, pp. 197–205.

Barrett, P. R. F. and Robson, T. O. (1974). Further studies on the seasonal changes in the susceptibility of some emergent plants to dalapon. *Proceedings 12th British Weed Control Conference*, **1**, pp. 249–53.

Barrett, S. C. H. (1980). Sexual reproduction in *Eichhornia crassipes* (water hyacinth). I. Fertility of clones from diverse regions. *Journal of Applied Ecology*, **17**, 101–12.

Barrett, S. C. H. and Seaman, D. E. (1980). The weed flora of Californian rice fields. *Aquatic Botany*, **9**, 351–76.

Barros, M. (1953). Las Juncóceas de la Argentina, Chile y Uruguay. *Darwiniana*, **10**, 279–460.

Barros, M. (1960). Las Ciperáceas del Estado de Santa Catalina. *Sellowia*, **12**, 181–450.

Bartels, P. G. and Watson, C. W. (1978). Inhibition of carotenoid synthesis by fluridone and norflurazon. *Weed Science*, **26**, 198–203.

Barthelmes, D. (1985). Ziele und bisherige Ergebnisse der Seenbewirtschaftung mit Silber- und Marmorkarpfen, Teil I; II. *Zeitschrift für die Binnenfisherei der DDR*, **32**, 224–33 and 261–6.

Bartley, T. R. and Hattrup, A. R. (1975). *Acrolein residues in irrigation water and effects on rainbow trout*. Report REC-GRC-75-8, Bureau of Reclamation, Denver, Engineering & Research Center. pp. 19.

Basiouny, F. M., Haller, W., and Garrard, L. A. (1978). Survival of hydrilla (*Hydrilla verticillata*) plants and propagules after removal from the aquatic habitat. *Weed Science*, **26**, 502–4.

Batra, S. W. T. (1977). Bionomics of the aquatic moth *Acentropus niveus* (Oliver), a potential biological control agent for Eurasian watermilfoil and hydrilla. *Journal of the New York Entomological Society*, **85**, 143–52.

Baudo, R., Galanti, B., Giulizzoni, P., Merlini, L., and Varini, P. G. (1981). Relationships between heavy metals and aquatic organisms in Lake Mezzola hydrographic system (Northern Italy). 5. Net photosynthesis of the submerged macrophytes *Potamogeton crispus* L. and *Potamogeton perfoliatus* L. *Memorie dell' Istituto Italiano di Idrobiologie*, **39**, 227–42.

Baudo, R. and Varini, P. G. (1976). Copper, manganese and chromium concentration in five macrophytes from the delta of River Toce (northern Italy). *Memorie dell' Istituto Italiano di Idrobiologie*, **33**, 305–24.

Bayley, I. A. E. and Williams, W. D. (1973). *Inland waters and their ecology*. Longman, Melbourne.

Bayley, S. (1970). The ecology and disease of Eurasian watermilfoil (*Myriophyllum spicatum* L.) in the Chesapeake Bay. Ph. D. Thesis, Johns Hopkins University, Baltimore. pp. 190.

Bayley, S., Rabin, H., and Southwick, C. H. (1968). Recent decline in the distribution and abundance of Eurasian milfoil in Chesapeake Bay. *Chesapeake Science,* **9,** 173–81.

Bayley, S., Stotts, V. D., Springer, P. F., and Steenis, J. (1978). Changes in submerged aquatic macrophyte populations at the head of Chesapeake Bay 1958–75. *Estuaries,* **1,** 73–84.

Bayo, V., Lallana, V., Lerenzatti, E., and Marta, M. C. (1981). Evaluación cuantitativa de la vegetación acuática en islas del valle aluvial del Rio Paraná Medio. *Ecologia,* **6,** 67–72.

Beach, M. L., Lazor, R. L., and Burkhalter, A. I. (1978). Some aspects of the environmental impact of the white amur (*C. idella* (Val.)) in Florida, and its use for aquatic weed control. *Proceedings IV International Symposium on Biological Control of Weeds,* pp. 269–89.

Beadle, L. C. (1981). *The inland waters of tropical Africa: an introduction to tropical limnology,* (2nd edn), pp. 475. Longman, London.

Beadle, L. C., Long, S. P., Imbamba, S. K., Hall, D. O., and Olembo, R. J. (1985). *Photosynthesis in relation to plant production in terrestrial environments.* Tycooly Publishing, Oxford. pp. 156.

Bean, G. A., Fusco, M., and Klarman, W. L. (1973). Studies on the 'Lake Venice' disease of Eurasian milfoil in the Chesapeake Bay. *Chesapeake Science,* **14,** 279–80.

Beard, T. D. (1969). *Impact of an overwinter drawdown on the aquatic vegetation in Murphy Flowage, Wisconsin.* Department of Natural Resources Research Report 43, Madison, Wisconsin. pp. 16.

Beardall, J. and Raven, J. A. (1981). Transport of inorganic carbon and the CO_2 concentrating mechanism in *Chlorella emersonii* (Chlorophyceae). *Journal of Phycology,* **17,** 134–41.

Bebawi, F. F. and Mohamed, B. F. (1984). Effects of 2,4-D on mature and juvenile plants of water hyacinth (*Eichhornia crassipes* (Mart.) Solms.). *Hydrobiologia,* **110,** 91–3.

Beck, S. G. (1984). Communidades vegetales de las sabanas inundadizas en el NE de Bolivia. *Phytocoenología,* **12,** 321–50.

Beckett, R. (1989). The application of field-flow fractionation techniques to the characterisation of complex environmental samples. *Environmental Technology Letters,* (in press).

Beer, S. and Eshel, A. (1983). Photosynthesis of *Ulva* spp. II. Utilization of HCO^-_3 when submerged. *Journal of Experimental Marine Ecology,* **70,** 99–106.

Beer, S., Spencer, W., and Bowes, G. (1985). Photosynthesis and growth of the filamentous blue-green alga *Lyngbya birgei* in relation to its environment. *Journal of Aquatic Plant Management,* **24,** 61–5.

Behl, N. K., Pahuja, S. S., and Moolani, M. K. (1973). Herbicidal control of water hyacinth (*Eichhornia crassipes*). 1. Effect of 2,4-D and amitrole alone and in combination. *Proceedings 4th Asian–Pacific Weed Science Society Conference,* pp. 316–21.

Beitinger, T. L. and Freeman, L. (1983). Behavioural avoidance and selection responses of fishes to chemicals. *Residue Reviews,* **90,** 35–55.

Beitz, H., Ackermann, H., Schmidt, H., and Horing, H. (1982). [Use of plant protection chemicals and plant growth regulators in protection zone II of drinking water catchment areas.] In German, English summary. *Nachrichtenblatt für den Pflanzenschutz in der DDR,* **36,** (1), 14–19.

Belokon, G. S. (1971). The development of plant growth along canals in the Southern Ukraine. *Hydrobiological Journal,* **7 (4),** 33–9.

Beltman, B. (1979). Research in ditches and canals in the province of Utrecht (the Netherlands) (summary). *Hydrobiological Bulletin,* **13 (1),** 29.

Beltman, B. (1984). Management of ditches. The effect of cleaning of ditches on the water coenoses. *Verhandlungen der Internationalen Vereinigung für theoretische und angewandte Limnologie,* **22,** 2022–8.

Bennett, F. D. (1970). Insects attaching water hyacinth in the West Indies, British Honduras and the USA. *Hyacinth Control Journal,* **8,** 10–13.

Bennett, F. D. (1974). Biological control. In *Aquatic vegetation and its use and control* (ed. D. S. Mitchell), pp. 85–98. UNESCO, Paris.

Bennett, F. D. (1975). Insects and plant pathogens for the control of *Salvinia* and *Pistia*. *Proceedings of a Symposium on Water Quality Management through Biological Control* (ed. P. L. Brezonik and J. L. Fox), pp. 28–35. Report No. ENV-07-75-1. Department of Environmental Engineering Science, University of Florida, Gainesville.

Bennett, F. D. (1984). Biological control of aquatic weeds. In *Proceedings International Conference on Water Hyacinth* (ed. G. Thyagarajan), pp. 14–40. UNEP, Nairobi.

Benton, A. R. (1980). Remote sensing of aquatic plants. In *Weed control methods for public health application* (ed. E. O. Gangstad), pp. 45–6. CRC Press, Boca Raton, Florida.

Benton, A. R., Clark, C. A., and Snell, W. W. (1980). Monitoring aquatic plants in reservoirs. *Transportation Engineering,* **106,** 453.

Benton, A. R., James, W. P., and Rouse, J. W. (1978). Evapotranspiration from water hyacinth *Eichhornia crassipes* (Mart.) Solms in Texas reservoirs. *Water Resources Bulletin,* **14,** 919–30.

Benton, A. R. and Newman, R. M. (1976). Colour aerial photography for aquatic plant monitoring. *Journal of Aquatic Plant Management,* **14,** 14–16.

Bernard, J. M. (ed.) (1988). *Carex*. In *Proceedings 2nd INTECOL Wetlands Conference, Czechoslovakia. Aquatic Botany,* **30,** pp. 168.

Bernatowicz, S. (1965). Effects of mowing on the occurrences of macrophytes in Dgal Maly Lake. *Acta Hydrobiologia,* **7,** 71–82.

Bernatowicz, S. (1966). The effects of shading on the growth of macrophytes in lakes. *Ekologiya Polska,* **14,** 607–16.

Bernhardt, E. A. and Duniway, J. M. (1982). Endemic diseases of aquatic weeds in California. *Phytopathology,* **72,** 986.

Bernhardt, E. A. and Duniway, J. M. (1984). Root and stem rot of parrot-feather (*Myriophyllum brasiliense*) caused by *Pythium carolinianum. Plant Disease,* **68,** 999–1003.

Berry, C. R. (1984). Toxicity of the herbicides diquat and endothall to goldfish. *Environmental Pollution Series A,* **34,** 251–8.

Bertram, G. C. L. and Bertram, C. K. R. (1966). Sea cows could be useful. *Sea Frontiers,* **12,** 210–17.

Beshir, M. E. and Gadir, L. A. (1975). Aquatic environment in the Sudan with special reference to the Gezira canalization system. In *Aquatic weeds in the Sudan with special reference to water hyacinth* (ed. M. Obeid), pp. 10–30. National Academy of Sciences, Washington DC.

Beshir, M. O. and Bennett, F. D. (1984). Biological control of water hyacinth on the White Nile, Sudan. *Proceedings VI International Symposium on Biological Control of Weeds* (ed. E. S. Delfosse), pp. 491–6. Agriculture Canada.

Best, E. P. H. (1981). A preliminary model for growth of *Ceratophyllum demersum* L. *Verhandlungen der Internationalen Vereinigung für theoretische und angewandte Limnologie*, **21**, 1484–91.

Best, E. P. H. (1982*a*). The aquatic macrophytes of Lake Vechten. Species composition, spatial distribution and production. *Hydrobiologia*, **95**, 65–77.

Best, E. P. H. (1982*b*). Growth modelling in aquatic macrophytes. In *Studies on aquatic vascular plants* (ed. J. J. Symoens, S. S. Hooper, and P. Compère), pp. 102–11. Royal Botanical Society of Belgium, Brussels.

Best, E. P. H. (1986) Photosynthetic characteristics of the submerged macrophyte *Ceratophyllum demersum. Physiologia Plantarum*, **68**, 502–10.

Best, E. P. H. and Dassen, J. H. A. (1987). Biomass, stand area, primary production characteristics and oxygen regime of the *Ceratophyllum demersum* L. population in Lake Vechten, the Netherlands. *Archiv für Hydrobiologie Supplement*, **76**, 347–67.

Best, E. P. H. and Meulemans, J. T. (1979). Photosynthesis in relation to growth and dormancy in *Ceratophyllum demersum. Aquatic Botany*, **6**, 53–67.

Best, E. P. H. and Visser, H. W. C. (1987). Seasonal growth of the submerged macrophyte *Ceratophyllum demersum* L. in mesotrophic Lake Vechten in relation to isolation, temperature and reserve carbohydrates. *Hydrobiologia*, **148**, 231–43.

Best, E. P. H., de Vries, D., and Reins, A. (1984). The macrophytes in the Loosdrecht Lakes: a story of their decline in the course of eutrophication. *Verhandlungen der Internationalen Vereinigung für theoretische und angewandte Limnologie*, **22**, 868–75.

Best, E. P. H. and van der Wittenboer, J. P. (1978). Effects of paraquat on growth and photosynthesis of *C. demersum* and *E. canadensis. Proceedings EWRS 5th Symposium on Aquatic Weeds, 1978*, pp. 157–62.

Best, E. P. H., van der Zweerde, W., and Zwietering, F. W. (1986). A management approach to the Netherlands' water courses using models on macrophyte growth, aquatic weed control and hydrology. *Proceedings EWRS/AAB 7th Symposium on Aquatic Weeds 1986*, pp. 37–42.

Best, M. D. and Mantai, K. E. (1978). Growth of *Myriophyllum*: sediment or lake water as the source of nitrogen and phosphorus. *Ecology*, **59**, 1075–80.

Best, R. G., Wehde, M. E., and Linder, R. L. (1981). Spectral reflectance of hydrophytes. *Remote Sensing Environment*, **11**, 27–35.

Beyers, R. J. (1966). The pattern of photosynthesis and respiration in laboratory microecosystems. In *Primary productivity in aquatic environments*, pp. 61–74. University of California Press, California.

Bicudo, R. M. T. (1972). O genero Chara (Charophyceae) no Brasil; Tesis Doctoral. Universit. São Paulo. pp. 230.

Birch, E. (1976). Trout stream weeds: how to cut and how not to, for fish and fly. *Salmon and Trout Magazine*, **206**, 49–53.

Birch, J. B. and Cooley, J. L. (1983). Regrowth of giant cutgrass (*Zizaniopsis miliacea*) following cutting. *Aquatic Botany*, **15**, 105–11.

Birmingham, B. C. and Colman B. (1983). Potential phytotoxicity of diquat accumulated by aquatic plants and sediments. *Water, Air and Soil Pollution,* **19**, 123–31.

Bishop, N. I. (1962). Inhibition of the oxygen-evolving system of photosynthesis by amino-triazines. *Biochimica et Biophysica Acta,* **57**, 186–9.

Björk, S. (1968). Metodik och forskningsproblem vid sjörestaurering. *Vatten,* **1**.

Björkqvist, I. (1967/68). Studies in *Alisma* (2 parts). *Opera Botanica* 17/19, pp. 266.

Black, C. C. (1973). Photosynthetic carbon fixation in relation to net CO_2 uptake. *Annual Review of Plant Physiology,* **24**, 253–86.

Blackburn, R. D. (1974). Chemical control of ditchbank and aquatic weeds in Australian irrigation systems. *Proceedings 7th Asian–Pacific Weed Science Conference,* pp. 219–22.

Blackburn, R. D. and Andres, L. A. (1968). The snail, the mermaid and the flea beetle. *United States Department of Agriculture Yearbook of Agriculture*, pp. 229–34.

Blackburn, R. D., Taylor, T. M. and Sutton, D. L. (1971). Temperature tolerance and necessary stocking rates of *Marisa cornuarietis* L. for aquatic weed control. *Proceedings EWRC 3rd Symposium on Aquatic Weeds 1971*, pp. 79–86.

Blackburn, R. D. and Durden, W. C. (1975). *Integrated control of alligator weed.* Aquatic Plant Control Program Technical Report 10, A3-A4. US Army Engineer Waterways Experiment Station, Vicksburg, Mississippi.

Blackburn, R. D., Gangstad, E. O., Yeo, R. R., Dechoretz, N., and Frank, P. A. (1976). *Aquatic use pattern for diquat for control of* Egeria *and* Hydrilla. Technical Report 13. US Army Engineer Waterways Experiment Station, Vicksburg, Mississippi. pp. 157.

Blackburn, R. D. and Taylor, J. B. (1976). Aquazine—a promising algicide for use in south-eastern waters. *Proceedings 29th Annual Meeting Southern Weed Science Society,* pp. 365–73.

Blackburn, R. D. and Weldon, L. W. (1963). Suggested control measures of common aquatic weeds of Florida. *Hyacinth Control Journal,* **2**, 2–5.

Blackman, G. E. (1960). Responses to environmental factors by plants in the vegetative phase. In *Growth of living systems* (ed. M. X. Yarrow), pp. 525–56. Basic Books, New York.

Blake, G., Dubois, J., and Gerbeaux, P. (1986). Distributional changes of populations of macrophytes in an Alpine lake (Lac d'Aiguebelette, Savoie, France). *Proceedings EWRS/AAB 7th Symposium on Aquatic Weeds, 1986,* pp. 43–50.

Blanchard, J. L. (1970). Mechanical and herbicidal lake weed management. *Journal of Aquatic Plant Management,* **8**, 36–7.

Blaszyk, P. (1977). Erhebung über die Anwendung von Herbiziden in und an Entwasserungsgraben in Weser–Ems–Gebiet. *Nachrichtenblatt des Deutschen Pflanzen-Schutzdienstes,* **29**, 146–8.

Bloemendaal, F. H. J. L. and Roelofs, J. G. M. (1986). *Waterkwaliteit en waterplanten.* Report: Laboratory of Aquatic Ecology, Catholic University Nijmegen, The Netherlands.

Blümel, F. (1974). Der derzeitige-Stand de Grabenerhaltung in Österreich. *Proceedings EWRC 4th Symposium on Aquatic Weeds 1974,* pp. 77–83.

Bohl, M. (1971). Die teichwirtschaftliche Bedeutung der Wasserpflanzen und die Möglichkeiten ihrer Bekämpfung unter besonderer Berücksichtigung der chinesischen pflanzenfressenden Fische. *Wasser-und Abwasser-Forschung,* **3**, 82–9.

Bohl, M. and Wagner, H. (1981). Umweltverhalten von Kupfersulfat als Herbizidin Forellenteichen. In *Beitraege zur Fischtoxicologie und-Parasitologie* (ed. H.-H. Klinke and W. Ahne), pp. 127–40. Gustav-Fischer Verlag, Stuttgart.

Bohl, M., Wagner, H., and Hoffman, R. (1982). Long-term investigations with periodic applications of copper sulphate as a herbicide in trout ponds. *Proceedings EWRS 6th Symposium on Aquatic Weeds 1982*, pp. 244–54.

Bolas, P. M. and Lund, J. W. G. (1974). Some factors affecting the growth of *Cladophora glomerata* in the Kentish Stour. *Water Treatment and Examination*, **23**, 25–49.

Bole, J. B. and Allan, J. R. (1978). Uptake of phosphorus from sediment by aquatic plants, *Myriophyllum spicatum* and *Hydrilla verticillata*. *Water Research* **12**, 353–8.

Bolier, G., van der Maas, H. L., and Bootsma, R. (1973). The toxicity of the herbicide dichlobenil to goldfish (*Carassius auratus*). *Mededelingen van de Faculteit Landbouwwetenschappen Rijksuniversteit Gent*, **38**, 733–40.

Bolman, J. (1977). Opgang en neergang van de Brede waterpest (*Elodea canadensis* Michx.) *Natura* (*Amsterdam*), **74**, 191–7.

Bond, W. J. and Roberts, M. G. (1978). The colonization of Cabora Bassa, Moçambique, a new man-made lake, by floating aquatic macrophytes. *Hydrobiologia*, **60**, 243–59.

Bonetto, A. A. (1971). Dinámica de las principales communidades bióticas en aguas de la cuenca del Plata y problemas relacionados. In *Recicentes adelantos en Biologia*, pp. 186–93. Mejia and Moguilevsky, Buenos Aires.

Bonetto, A. A. (1975). Hydrologic regime of the Paraná River and its influence on ecosystems. In *Coupling land and water systems* (ed. A. D. Hasler), pp. 175–97. Springer-Verlag, New York.

Bonham, A. J. (1978). A role for flood retarding vegetation in urban stormwater management and creek preservation. In *Hydrology Symposium, Canberra 1978*. Institution of Engineers, Australia, pp. 174–5.

Bonham, A. J. (1980). *Bank protection using emergent plants against boat wash in rivers and canals*. Report No. IT 206, Hydraulics Research Station, Wallingford. pp. 12.

Bonnewell, V., Koukkari, W. L., and Pratt, D. C. (1983). Light, oxygen, and temperature requirements for *Typha latifolia* seed germination. *Canadian Journal of Botany*, **61**, 1330–6.

Boorman, L. A., and Fuller, R. M. (1981). The changing status of reedswamps in the Norfolk Broads. *Journal of Applied Ecology*, **18**, 241–69.

Boorman, L. A., Sheail, J., and Fuller, R. M. (1978). *The* Phragmites *die-back problem*. Institute of Terrestrial Ecology, I.T.E. Project No. 539, Progress Report. pp. 48.

Borhidi, A., Muñiz, O., and Del Risco, E. (1979). Classificación fitocoenológica de la vegetación de Cuba. *Acta Botanica Academiae Scientiarum Hungaricae*, **25**, 263–301.

Born, S. M., Wirth, T. L., Brick, E., and Peterson, J. O. (1973). *Restoring the recreation potential of small impoundments: the Marion Millpond experience*. Bulletin No. 71, Wisconsin Department of Natural Resource Technology, Madison, Wisconsin. pp. 20.

Bose, J. C. (1924). *The physiology of photosynthesis*. Longmans, Green and Co., London.

Boston, H. L. and Adams, M. S. (1983). Evidence of Crassulacean acid metabolism in two North American isoetids. *Aquatic Botany*, **15**, 381–6.

Boston, H. L. and Adams, M. S. (1986). The contribution of Crassulacean acid metabolism to the annual productivity of two aquatic vascular plants. *Oecologia* (*Berlin*), **68**, 615–22.

Boston, H. L. and Perkins, M. A. (1982). Water column impacts of macrophyte decomposition below fibreglass screens. *Aquatic Botany*, **14**, 15–27.

Botry, L. and Otaegui, A. (1979). *Malezas acuáticas. Vigilancia y control en el embalse Salto Grande.* Sante Fé (Argentina), Gerencia de Salud, Ecologia y Desarollo Regional CTMSG 11 STH/37–79. pp. 10.

Boughey, A. S. (1963). The explosive development of a floating weed vegetation on Lake Kariba. *Adansonia*, **3**, 49–61.

Bouquet, H. (1977). Grass carp in the Netherlands. *Proceedings 8th British Coarse Fish Conference Liverpool, 1977*, pp. 108–11. University of Liverpool, Liverpool.

Bowes, G. (1985). Pathways of CO_2 fixation by aquatic organisms. In *Inorganic carbon uptake by aquatic photosynthetic organisms* (ed. W. J. Lucas and J. A. Berry), pp. 187–210. American Society for Plant Physiology, Rockville, Maryland.

Bowes, G. (1987). Aquatic plant photosynthesis: strategies that enhance carbon gain. In *Plant life in aquatic and amphibious habitats* (ed. R. M. M. Crawford), pp. 76–96. British Ecological Society Publication, No. 5. Blackwell, Oxford.

Bowes, G. and Beer, S. (1987). Physiological processes: photosynthesis. In *Aquatic plants for water treatment and resource recovery* (ed. K. R. Reddy and W. H. Smith), pp. 311–35. Magnolia Publishing, Orlando, Florida.

Bowes, G., Holaday, A. S., and Haller, W. T. (1979). Seasonal variation in the biomass, tuber density and photosynthetic metabolism of *Hydrilla* in three Florida lakes. *Journal of Aquatic Plant Management*, **17**, 61–5.

Bowes, G., Holaday, A. S., Van, T. K., and Haller, W. T. (1978). Photosynthetic and photorespiratory carbon metabolism in aquatic plants. In *Proceedings 4th International Congress on Photosynthesis 1977* (ed. D. O. Hall, J. Coombs and T. W. Goodwin). The Biochemical Society, London. pp. 289–98.

Bowes, G. and Reiskind, J. (1987). Inorganic carbon concentrating systems from an environmental perspective. In *Progress in photosynthesis research*, **4**, (ed. J. Biggens), pp. 345–52. Martinus Nijhoff Publishers, the Netherlands.

Bowes, G. and Salvucci, M. E. (1984). *Hydrilla:* inducible C_4-type photosynthesis without Kranz anatomy. *Proceedings 6th International Congress on Photosynthesis III*, (ed. C. Sybesma), pp. 892–32. Martinus Nijhoff/Dr W. Junk Publishers, Boston.

Bowes, G., Van, T. K., Garrard, L. A., and Haller, W. T. (1977). Adaptation to low light levels by *Hydrilla. Journal of Aquatic Plant Management*, **15**, 32–35.

Bowker, D. W. and Duffield, A. N. (1981). Population growth of *Lemna gibba* L. in the Gwent levels drainage channels. *Proceedings Association of Applied Biologists Conference: Aquatic Weeds and their Control 1981*, pp. 67–76. AAB, Wellesbourne, UK.

Bowmer, K. H. (1975). Discussion of Proceedings 'Environmental aspects of aquatic plant control by Bartley, T. R. and Gangstad, E. O.' *Paper 10810, Journal of Irrigation and Drainage Division ASCE*, **101**, (IR3), 230–2.

Bowmer, K. H. (1979). Management of aquatic weeds in Australian irrigation systems. *Proceedings 7th Asian–Pacific Weed Science Society Conference 1979*, pp. 219–22.

Bowmer, K. H. (1982*a*). Adsorption characteristics of seston in irrigation water: implications for the use of aquatic herbicides. *Australian Journal of Marine and Freshwater Research*, **33**, 443–58.

Bowmer, K. H. (1982*b*). Aggregates of particulate matter and aufwuchs on *Elodea canadensis* in irrigation water and inactivation of diquat. *Australian Journal of Marine and Freshwater Research*, **33**, 589–93.

Bowmer, K. H. (1982*c*). Residues of glyphosate in irrigation water. *Pesticide Science*, **33**, 623–38.

Bowmer, K. H. (1986). Rapid biological assay and limitations in macrophyte ecotoxicology: a review. *Australian Journal of Marine and Freshwater Research*, **37**, 297–308.

Bowmer, K. H. (1987*a*). Residues of dalapon and TCA in sediments and irrigation water. *Pesticide Science*, **18**, 1–13.

Bowmer, K. H. (1987*b*). Herbicides in water. In *Progress in pesticide biochemistry and toxicology, Volume 6. Herbicides* (ed. D. H. Hutson and T. K. Roberts), pp. 272–355. Wiley, Chichester.

Bowmer, K. H. and Adeney, J. A. (1978). Residues of diuron and phytotoxic degradation products in aquatic situations. II. Diuron in irrigation water. *Pesticide Science*, **9**, 354–64.

Bowmer, K. H. and Higgins, M. L. (1976). Some aspects of persistence and fate of acrolein herbicide in water. *Archives Environmental Contamination Toxicology*, **5**, 87–96.

Bowmer, K. H., Lang, A. R. G., Higgins, M. L., Pillay, A. R., and Tchan, Y. T. (1974). Loss of acrolein from water by volatilization and degradation. *Weed Research*, **14**, 325–8.

Bowmer, K. H., Mitchell, D. S., and Short, D. L. (1984). Biology of *Elodea canadensis* Michx. and its management in Australian irrigation systems. *Aquatic Botany*, **18**, 231–8.

Bowmer, K. H., O'Loughlin, E. M., Shaw, K., and Sainty, G. R. (1976). Residues of dichlobenil in irrigation water. *Journal of Environmental Quality*, **5**, 315–19.

Bowmer, K. H. and Sainty, G. R. (1977). Management of aquatic plants with acrolein. *Journal of Aquatic Plant Management*, **15**, 40–6.

Bowmer, K. H. and Sainty, G. R. (1978). Herbicide residues in the Murrumbidgee and Coleambally Irrigation Areas, Australia. *Proceedings EWRS 5th Symposium on Aquatic Weeds, 1978*, pp. 163–70.

Bowmer, K. H., Sainty, G. R., Smith, G., and Shaw, K. (1979). Management of *Elodea* in Australian irrigation systems. *Journal of Aquatic Plant Management*, **17**, 4–12.

Bowmer, K. H., Shaw, K., and Adeney, J. A. (1985). Management of submerged aquatic macrophytes with terbutryne: availance, phytotoxicity and persistence. *Weed Research*, **25**, 449–59.

Bowmer, K. H. and Smith, G. H. (1984). Herbicides for injection into flowing water: acrolein and endothal-amine. *Weed Research*, **24**, 201–11.

Bownik, L. J. (1970). The periphyton of the submerged macrophytes of Mikolajskie Lake. *Ekologiya Polska*, **18**, 503–20.

Boyd, C. E. (1968). Freshwater plants: a potential source of protein. *Economic Botany*, **22**, 359–68.

Boyd, C. E. (1970*a*). Vascular aquatic plants for mineral nutrient removal from polluted waters. *Economic Botany*, **24**, 95–103.

Boyd, C. E. (1970*b*). Losses of mineral nutrients during the decomposition of *Typha latifolia*. *Archiv für Hydrobiologie*, **66**, 511–17.

Boyd, C. E. (1974). Utilization of aquatic plants. In *Aquatic vegetation and its use and control* (ed. D. S. Mitchell), pp. 107–15. UNESCO, Paris.

Boyd, C. E. (1978). Chemical composition of wetland plants. In *Freshwater wetlands*, pp. 395–468. Academic Press, New York.

Boyd, C. E. (1987). Evapotranspiration/evaporation (E/E_o) ratios for aquatic plants. *Journal of Aquatic Plant Management*, **25**, 1–3.

Boyd, C. E. and Noor, M. H. M. (1982). Aquashade treatment of channel catfish ponds. *North American Journal of Fish Management*, **2**, 193–6.

Boylen, C. W. and Sheldon, R. B. (1976). Submerged macrophytes: growth under winter ice cover. *Science*, **194**, 841–2.

Braakhekke, W. G. (1980). On co-existence: a causal approach to diversity and stability in grassland vegetation. *Verslagen van Landbouwkundige Onderzoekingen*, **902**, 1–164.

Brabben, T. E. (1986). Monitoring the effects of aquatic weeds in Egyptian irrigation systems. *Proceedings EWRS/AAB 7th Symposium on Aquatic Weeds 1986*, p. 51.

Braddock, W. B. (1966). Weed control problems in the East Volusia Mosquito Control District. *Hyacinth Control Journal*, **5**, 31.

Braendle, R. and Crawford, R. M. M. (1987). Rhizome tolerance and habitat specialization in wetland plants. In *Plant life in aquatic and amphibious habitats*. Special Publication British Ecological Society, No. 5. (ed. R. M. M. Crawford), pp. 397–410. Blackwell Scientific Publications, Oxford.

Bravo Velásquez, E. and Balslev, H. (1985). *Dinámica de las plantas vasculares de dos ciénagas tropicales en Ecuador*. University of Aarhus, Botanical Institute, Contribution 67. pp. 50.

Breedlove, B. W. and Dennis, W. M. (1984). The use of small-format aerial photography in aquatic macrophyton sampling. *Special Technical Publications, ASTM*, STP 843, 100–11. American Society for Testing Materials.

Breen, C. M., Lillig, C., Oliver, J., and Furness, H. D. (1976). A quantitative sampler for aquatic macrophytes. *Journal of the Limnological Society of South Africa*, **2**, 59–62.

Brezny, O. (1970). *Draft Final Technical Report on Aquatic Weed Control in Chambal Command Area, Rajasthan*. pp. 68.

Brezny, O. and Mehta, I. (1970). Aquatic weeds and their distribution in Chambal Command Area in Kota. *Indian Journal of Weed Science*, **2**, 70–80.

Brian, R. C., Homer, R. F., Stubbs, J., and Jones, R. L. (1958). A new herbicide 1:1′ ethylene-2:2′ dipyridylium dibromide. *Nature* (*London*), **181**, 446–7.

Bristow, J. M., Cardenas, J., Fulleron, T. M., and Sierra, F. (undated). *Malezas acuáticas*. Bogota, Instituto Colombia Agrop e International Plant Protection Center Document 681. pp. 116.

Bristow, J. M. and Whitcombe, W. (1971). The role of roots in the nutrition of aquatic vascular plants. *American Journal of Botany*, **58**, 8–13.

British Agrochemicals Association (1976). *Technical data on herbicides cleared for use in or near water*. British Agrochemical Association, London. pp. 14.

British Waterways Board (1981). *Vegetation Control Manual 1981*. British Waterways Board, London. pp. 56.

Britton, R. H. (1974). Factors affecting the distribution and productivity of emergent vegetation at Loch Leven, Kinross. *Proceedings Royal Society of Edinburgh, B*, **74**, 209–19.

Brock, T. C. M. (1985). Remarks on the distribution and survival biology of the white, yellow and fringed waterlily: an introduction. In *Ecological studies on nymphaeid water plants*. M.Sc. Thesis, Catholic University, Nijmegen. pp. 204.

Brønstad, J. O. and Friestad, H. O. (1985). Behaviour of glyphosate in the aquatic environment. In *The herbicide glyphosate* (ed. E. Grossbard and D. Atkinson), pp. 200–5. Butterworths, London.

Brooker, M. P. (1974). The risk of deoxygenation of water in herbicide application for aquatic weed control. *Journal of the Institution of Water Engineers*, **28**, 206–10.

Brooker, M. P. (1976*a*). The ecological effects of the use of dalapon and 2,4-D for drainage channel management. I. Flora and chemistry. *Archiv für Hydrobiologie*, **78**, 396–412.

Brooker, M. P. (1976*b*). The ecological effects of the use of dalapon and 2,4-D for drainage channel management. II. Fauna. *Archiv für Hydrobiologie*, **78**, 507–25.

Brooker, M. P. (1979). The life cycle and growth of *Sialis lutaria* L. (Megaloptera) in a drainage channel under different methods. *Ecological Entomology*, **4**, 111–17.

Brooker, M. P. and Baird, J. H. (1974). Cost evaluation of watercourse management in Essex. *Surveyor*, **144**, 34–7.

Brooker, M. P. and Edwards, R. W. (1973*a*). Effects of the herbicide paraquat on the ecology of a reservoir. I. Botanical and chemical aspects. *Freshwater Biology*, **3**, 157–76.

Brooker, M. P. and Edwards, R. W. (1973*b*). Effects of the herbicide paraquat on the ecology of a reservoir. II. Community metabolism. *Freshwater Biology*, **3**, 383–92.

Brooker, M. P. and Edwards, R. W. (1973*c*). The use of a herbicide in a fisheries reservoir. *Journal of the Institute of Fisheries Management*, **4**, 102–8.

Brooker, M. P. and Edwards, R. W. (1974). Effects of the herbicide paraquat on the ecology of a reservoir. III. Fauna and general discussion. *Freshwater Biology*, **4**, 311–35.

Brooker, M. P. and Edwards, R. W. (1975). Review paper: aquatic herbicides and the control of water weeds. *Water Research*, **9**, 1–15.

Brooker, M. P., Morris, D. L., and Wilson, C. J. (1978). Plant-flow relationships in the River Wye catchment. *Proceedings EWRS 5th Symposium on Aquatic Weeds 1978*, pp. 63–70.

Brookes, A. (1981). *Waterways and wetlands. A practical conservation handbook*. British Trust for Conservation Volunteers, Reading.

Brookes, A. (1984). *Recommendations bearing on the sinnosity of Danish stream channels*. National Agency of Environmental Protection, Freshwater Laboratory. Silkeborg, Denmark. Technical Report No. 6. pp. 130.

Brown, A. W. A. (1978). Herbicides in water. In *Ecology of pesticides* (ed. A. W. A. Brown), pp. 426–31. Wiley, New York.

Brown, A. W. A. and Deom, J. O. (1973). Summary: health aspects of man-made lakes. In *Man-made lakes: their problems and environmental effects* (ed. W. C. Ackermann, C. F. White, and E. B. Worthington), pp. 755–64. American Geophysical Union, Washington DC.

Brown, B. T. and Rattigen, B. M. (1979). Toxicity of soluble copper and other metal ions to *Elodea canadensis*. *Environmental Pollution*, **20**, 302–14.

Brown, C. L. (1984). Improved above-substrate sampler for macrophytes and phytomacrofauna. *Progressive Fish Culture*, **46**, 142–4.

Brown, J. L. and Spencer, N. R. (1973). *Vogtia malloi*, a newly introduced phycitine moth (Lepidoptera: Pyralidae) to control alligator weed, *Environmental Entomology*, **2**, 519–23.

Brown, J. M. A., Dromgoole, F. I., Towsey, M. W., and Browse, J. (1974). Photosynthesis and photorespiration in aquatic macrophytes. In *Mechanisms of regulation of plant growth* (ed. R. L. Bieleski, A. R. Ferguson, and M. M. Cresswell). *Royal Society of New Zealand Bulletin*, **12**, 243–9.

Brown, V. M. (1968). The calculation of the acute toxicity of mixtures of poisons to rainbow trout. *Water Research*, **2**, 723–33.

Brownell, R. L., Ralls, K., and Reeves, R. R. (1981). Report of the West Indian manatee workshop. In *The West Indian manatee in Florida* (ed. R. L. Brownell and K. Ralls), pp. 3–16. Proceedings of a Workshop in Orlando, Florida. Florida Department of Natural Resources, Tallahassee.

Browse, J. A., Brown, J. M. A., and Dromgoole, F. I. (1980). Malate synthesis and metabolism during photosynthesis in *Egeria densa* Planch. *Aquatic Botany*, **8**, 295–305.

Browse, J. A., Dromgoole, F. J., and Brown, J. M. A. (1977). Photosynthesis in the aquatic macrophyte *Egeria densa*. I. $^{14}CO_2$ fixation at natural CO_2 concentrations. *Australian Journal of Plant Physiology*, **6**, 169–76.

Browse, J. A., Dromgoole, F. I., and Brown, J. M. A. (1979). Photosynthesis in the aquatic macrophyte *Egeria densa*. III. Gas exchange studies. *Australian Journal of Plant Physiology*, **6**, 499–512.

Bruhn, H. D., Livermore, D. F., and Aboaba, F. O. (1971). Processing characteristics of aquatic macrophytes as related to mechanical harvesting. *Transactions American Society of Agricultural Engineers*, **14**, 1004–8.

Bruns, V. F. (1973). *Studies on the control of reed canarygrass along irrigation systems.* United States Department of Agriculture Research Service, Report ARS-W-3. pp. 17.

Bruwer, C. A. (1979). *The economic impact of eutrophication in South Africa*. Technical Report No. TR94, South Africa Department of Water Affairs, Pretoria.

Bryan, K. and Hellawell, J. M. (1980). *The use of herbicides in water supply catchments.* Severn–Trent Water Authority Biologists Group Report No. 1, S.T.W.A. Birmingham, UK.

Bryant, C. B. (1974). Aquatic weed harvesting costs and equipment—1972. *Hyacinth Control Journal*, **12**, 53–5.

Bucha, H. C. and Todd, C. W. (1981). 3-(*p*-chlorophenyl)-1, 1-dimethylurea—a new herbicide. *Science* (*New York*), **114**, 493–4.

Buckingham, G. R. and Bennett, C. A. (1981). Laboratory biology and behaviour of *Litodactylus leucogaster*, a Ceutorhychine weevil that feeds on watermilfoils. *Annals Entomological Society America*, **74**, 451–8.

Buckingham, G. R., Boucias, D. and Theriot, R. F. (1983). Reintroduction of the alligator flea beetle (*Agasicles hygrophyla* Selman and Vogt) into the United States from Argentina. *Journal of Aquatic Plant Management*, **21**, 101–2.

Buckingham, G. R. and Ross, B. M. (1981). Notes on the biology and host specificity of *Acentria nivea* (*Acentropus niveus*). *Journal of Aquatic Plant Management*, **19**, 32–6.

Buckmaster, D. E. (1980). Grass carp for Australia: a fisheries viewpoint. *Proceedings Vth International Symposium on Biological Control of Weeds, Brisbane 1979*, pp. 273–80.

Bulcke, R. A. J. (1978). Biology of *Vaucheria dichotoma* Agardh. *Proceedings EWRS 5th Symposium on Aquatic Weeds*, pp. 107–15.

Bulthuis, D. A. (1984). Control of the seagrass *Heterozostera tasmanica* by benthic screens. *Journal of Aquatic Plant Management*, **22**, 41–2.

Burdich, G. E., Dean, H. J., and Harris, E. J. (1964). Toxicity of aqualin to fingerling brown trout and bluegills. *New York Fish and Game Journal*, **11**, 106–14.

Burian, K. and Sieghardt, H. (1979). The primary producers of the *Phragmites* belt, their energy utilization and water balance. In *Neusiedlersee: the limnology of a shallow lake in Central Europe* (ed. H. Loffler). *Monographiae Biologicae*, **37**, 273–9.

Burkhalter, A. P. (1975). The state of Florida aquatic weed control program. *Proceedings Symposium on Water Quality Management through Biological Control* (ed. P. L. Brezonik and J. L. Fox), pp. 15–17. Report No. ENV-07-75-1. Department of Environmental Engineering and Science, University of Florida, Gainesville, Florida.

Burkhard, N. and Guth, J. A. (1976). Photodegredation of atrazine, atraton and ametryne in aqueous solution with acetone as a photosensitizer. *Pesticide Science*, **7**, 65–71.

Burkart, A. (1957). Ojeada sinóptica sobre la vegetación del Delta del Río Paraná. *Darwiniana*, **11**, 457–561.

Burland, R. and Catling, P. (1986). *Vascular aquatic weeds in Alberta*. Aquatic Plant Management Newsletter No. 22.

Burton, G. J. (1960). Studies on the bionomics of mosquito vectors which transmit filariasis in India. II. The role of water hyacinth (*Eichhornia speciosa* Kunth) as an important host plant in the life cycle of *Mansonia uniformis* (Theobald) with notes on the differentiation of late embryonic and newly hatched stages of *Mansonia uniformis* (*Theobald*) and *Mansonia annulifera* (Theobald). *Indian Journal of Malariology*, **14**, 81–105.

Buryi, V. S., Goshka, A. T., Kudevich, S. N., Sannikov, G. P., and Gubareva, K. P. (1973). [Residues in environment of herbicides used for cleaning canals]. In Russian. *Kimiya v Sels'kom Khozyaistve*, **10**, 688–94.

Butler, P. A. (1963). Commercial fisheries investigations. In *Pesticide Wildlife Studies 1962*. US Fisheries and Wildlife Service Circular **167**, pp. 11–25.

Buttery, B. and Lambert, J. M. (1965). Competition between *Glyceria maxima* and *Phragmites communis* in the region of Surlingham Broad. I. Competition mechanism. *Journal of Ecology*, **53**, 163–82.

Cabrera, A. L. (1964). *Las plantas acuáticas*. Buenos Aires, Editorial Universitaria. pp. 55.

Cáceres, E. J. (1978). Carófitas del Centro de Argentina. *Boletin de la Academia Nacional de Ciencias en Córdoba*, **52**, 315–72.

Caffrey, J. (1982). Notes on hydrophytes in selected waters of Galway (H17 and H15) and Monaghan (H32). *Irish Naturalists Journal*, **20**, 509–60.

Caffrey, J. M. (1986). The impact of peat siltation on macrophyte communities in the River Suck: an Irish coarse fishery. *Proceedings EWRS/AAB 7th Symposium on Aquatic Weeds, 1986*, pp. 53–60.

Calder, A. A. and Sands, D. P. A. (1985). A new Brazilian *Cyrtobagous* Hustache (Coleoptera, Curculionidae) introduced into Australia to control salvinia. *Journal of the Australian Entomological Society*, **24**, 57–64.

Calderbank, A. (1968). The bipyridilium herbicides. *Advances in Pest Control Research,* **8,** 127–235.

Calderon Garcia, A. (1971). Estudio del lirio acuático *Eichhornia crassipes* (Mart.) Solms en la lago de Patzcuaro, Michoacan, Mexico. *Publ. Comis. Forestal del Estudo Morellia, Michoacan, Mexico,* pp. 5–41.

Callahan, J. L. and Osborne, J. A. (1983). Comparison of the grass carp and hybrid grass carp. *Aquatics,* **5,** 10–15.

Cancek, M. (1969). On aquatic macrophytes developing in the main land network of the Danube–Tissa–Danube system. *Ann. Sci. Novi Sad Univ. Pojoprio vredni Fakultet,* **13,** 121–32.

Canellos, G. (1981). *Aquatic plants and mechanical methods for their control.* Report: MTR-81W55 to Environmental Protection Agency. Mitre Corporation, Metrek Division, McLean, Virginia, USA. pp. 140.

Canfield, D. E. (1983). Impact of integrated aquatic weed management on water quality in a citrus grove. *Journal of Aquatic Plant Management,* **21,** 69–73.

Cardarelli, N. F. (1980). Control of snail-borne parasitic diseases. In *Weed control methods for public health applications.* CRC Press, Boca Raton, Florida.

Carignan, R. and Kalff, J. (1980). Phosphorus sources for aquatic weeds: water or sediments? *Science* (*New York*), **207,** 987–9.

Carlsson, R. (1980). Södra Vätterloygdens flora. *Svensk Botanisk Tidsskrift,* **73,** 417–22.

Carmichael, W. W. and Bent, P. E. (1980). *The practical directory to toxic blue-green algae* (*Cyanobacteria*). Wright State University, Dayton, Ohio.

Carpenter, S. R. and Adams, M. S. (1978). Macrophyte control by harvesting and herbicides: implications for phosphorus cycling in Lake Wingra. *Journal of Aquatic Plant Management,* **16,** 20–3.

Carpenter, S. R. and Gasith, A. (1978). Mechanical cutting of submersed macrophytes: immediate effects on littoral water chemistry and metabolism. *Water Research,* **12,** 55–7.

Carpenter, S. R. and Greenlee, J. K. (1981). Lake deoxygenation after herbicide use: a simulation model analysis. *Aquatic Botany,* **11,** 173–86.

Carter, C. C. and Hestand, R. S. (1977*a*). The effects of selected herbicides on phytoplankton and sulfur bacteria populations. *Journal of Aquatic Plant Management,* **15,** 47–56.

Carter, C. C. and Hestand, R. S. (1977*b*). Relationship of regrowth of aquatic macrophytes after treatment with herbicides to water quality and phytoplankton populations. *Journal of Aquatic Plant Management,* **15,** 65–9.

Carter, G. S. (1955). *The papyrus swamps of Uganda.* W. Heffer & Sons, Cambridge.

Cary, P. R. and Weerts, P. G. J. (1981). Growth of *Salvinia molesta* as affected by nutrition and water temperature. In *Proceedings 5th International Symposium on Biological Control of Weeds* (ed. E. S. Delfosse), pp. 35–46. Commonwealth Scientific and Industrial Research Organisation, Australia, Melbourne.

Cary, P. R. and Weerts, P. G. J. (1983*a*). Growth of *Salvinia molesta* as affected by water temperature and nutrition. I. Effects of nitrogen level and nitrogen compounds. *Aquatic Botany,* **16,** 163–72.

Cary, P. R. and Weerts, P. G. J. (1983*b*). Growth of *Salvinia molesta* as affected by water temperature and nutrition. II. Effects of phosphorus level. *Aquatic Botany,* **17,** 61–70.

Cary, P. R. and Weerts, P. G. J. (1984*a*). Growth of *Salvinia molesta* as affected by water temperature and nutrition. III. Nitrogen–phosphorus interactions and effect of pH. *Aquatic Botany,* **19**, 171–82.

Cary, P. R. and Weerts, P. G. J. (1984*b*). Growth and nutrient composition of *Typha orientalis* as affected by water temperature and nitrogen and phosphorus supply. *Aquatic Botany,* **19**, 105–18.

Casamiquela, O. M. (1977). *Organización de la Intendencia de Riego de IDEVI.* Viedma, Río Negro, 90–109.

Caseley, J. C. and Coupland, D. (1985). Environmental and plant factors affecting glyphosate uptake, movement, and activity. In *The herbicide glyphosate* (ed. E. Grossbard and D. Atkinson), pp. 92–123. Butterworths, London.

Cassani, J. R. and Caton, W. E. (1985). Effects of chemical and biological weed control on the ecology of a South Florida pond. *Journal of Aquatic Plant Management,* **23**, 51–8.

Cassie, V. (1966). Effects of spraying on phytoplankton in Lake Rotorua. *Proceedings Rotorua Seminar on Water Weeds, University Extension Service, University of Auckland, New Zealand, 1966,* pp. 31–40.

Castellanos, A. (1952). Revisión de las Pontederiaceae argentinas. *Lilloa,* **25**, 585–94.

Castellanos, A. (1958). Las Pontederiaceae de Brasil. *Arquivos do Jardim Botânico, Rio de Janeiro,* **16**, 147–237.

Cave, T. G. (1981). Current weed control problems in land drainage channels. *Proceedings Association of Applied Biologists Aquatic Weeds and their Control 1981,* pp. 5–14. AAB, Wellesbourne, UK.

Cazzaniga, N. J. (1981). Evaluación preliminar de un gasterópodo para el control de malezas acuáticas. In *2da. Reunión sobre malezas subacuáticas en los canales de desagüe de CORFO,* pp. 131–65. Publicacion C.I.C., La Plata, Argentina.

Cazzaniga, N. J. (1983). Apple-snails eating *Chara. Aquaphyte,* **3 (2)**, 1–4.

CEDEGE (1983). Diagnóstico del grado de infestación de malezas acuáticas flotantes en el sistema hidrográfico del embalse Daule-Peripa y proyecto para su control. In *Plan de Conservación Ambiental del Proyecto de Proposito Multiple 'Jaime Roldos Aguilera'.* Comis. de Estudios para el Desarallo del Rio Guayas. Unidad Estudios Agrologicos, Departamento Técnico, pp. 259.

Cedeno-Maldonado, A. and Liu, L. C. (1976). The effects of two substituted urea and two s-triazine type herbicides on the photosynthesis of *Lemna perpusilla* Torr. *Journal of the Agricultural University, Puerto Rico,* **60**, 369–74.

Center, T. D. and Spencer, N. R. (1981). The phenology and growth of water hyacinth (*Eichhornia crassipes*) in a eutrophic north-central Florida lake. *Aquatic Botany,* **10**, 1–32.

Center, T. D., Steward, K. K., and Bruner, M. C. (1982). Growth of water hyacinth (*Eichhornia crassipes*) with *Neochetina eichhornia* (Coleoptera: Curculionidae) and a growth retardant. *Weed Science,* **30**, 453–7.

Cerna Bazán, L. A. and Díaz, C. J. (1982). Control quimico de malezas en arroz (*Oryza sativa* L. cv. INTI) de trasplante con herbicidas granulados. *Turrialba,* **32**, 111–17.

Cernohous, F. (1980). *Callitriche hermaphroditica* v Ceskoslovensku. *Preslia,* **52**, 203–8.

Chabwela, H. N. and Siwela, A. A. (1986). The vegetative structure of the Kafue Flats North Bank, after the construction of the dams. *Proceedings EWRS/AAB 7th Symposium on Aquatic Weeds 1986*, pp. 61–72.

Chachu, R. E. O. (1979). The vascular flora of Lake Kainji. In *Proceedings International Conference Kainji Lake and River Basins Development in Africa*, Kainji Lake Research Institute, New Bussa, Nigeria, Volume 2, pp. 479–487.

Chadwick, M. J. and Obeid, M. (1966). A comparative study of the growth of *Eichhornia crassipes* Solms and *Pistia stratiotes* L. in water-culture. *Journal of Ecology*, **45**, 563–75.

Chakravarty, A. K. (1957). Weed flora of paddy fields of West Bengal. *Indian Agriculturist*, **1**, 19–26.

Chalmers, M. I. (1968). Report to World Food Programme on a visit to Sudan. In *Animal Production WFP. Mission report on animal nutrition in Sudan. Study on the use of water hyacinth in ruminant animal feeding and also as a means of weed control.* FAO, Rome.

Chambers, P. A., Spence, D. H. N., and Weeks, D. C. (1985). Photocontrol of turion formation by *Potamogeton crispus* L. in the laboratory and natural water. *New Phytologist*, **99**, 183–94.

Chan, K. Y. and Leung, S. C. (1986). Effects of paraquat and glyphosate on growth, respiration and enzyme activity of aquatic bacteria. *Bulletin of Environmental Contamination and Toxicology*, **36**, 52–9.

Chancellor, R. J. (1960*a*). Results of some preliminary experiments on the use of pelleted herbicides for controlling submerged water plants. *Proceedings 5th British Weed Control Conference*, **2**, pp. 559–67.

Chancellor, R. J. (1960*b*). Experiments on the chemical control of emergent water plants. *Proceedings 5th British Weed Control Conference*, **2**, pp. 569–78.

Chancellor, R. J. (1981). The manipulation of weed behaviour for control purposes. *Philosophical Transactions Royal Society London*, **B295**, pp. 103–10.

Chandler, J. H. (1975). Is *Elodea nuttallii* in the Fens? *Watsonia*, **10**, 464.

Chang, S.-M., Yang, C.-C., and Sung, S.-C. (1977). The cultivation and the nutritional value of Lemnaceae. *Bulletin of the Institute of Chemistry, Academia Sinica*, **24**, 19–30.

Chapman, V. J., Brown, J. M. A., Hill, C. F., and Carr, J. L. (1974). Biology of excessive weed growth in the hydroelectric lakes of the Waikato River, New Zealand. *Hydrobiologia*, **44**, 349–63.

Charudattan, R. (1973). Pathogenicity of fungi and bacteria from India to hydrilla and water hyacinth. *Hyacinth Control Journal*, **11**, 44–8.

Charudattan, R. (1982). Regulation of microbial weed control agents. In *Biological control of weeds with plant pathogens* (ed. R. Charudattan and H. L. Walker), pp. 175–88. John Wiley, New York.

Charudattan, R. (1983). Biological control of aquatic weeds. *Phytopathology*, **73**, 775.

Charudattan, R. (1984). Role of *Cercospora rodmanii* and other pathogens in the biological and integrated controls of water hyacinth. *Proceedings International Conference on Water Hyacinth* (ed. G. Thyagarajan), pp. 834–59. UNEP, Nairobi.

Charudattan, R. (1985). The use of natural and genetically altered strains of pathogens for weed control. In *Biological control in agricultural IPM systems* (ed. M. A. Hoy and D. C. Herzog), pp. 347–72. Academic Press, Orlando.

Charudattan, R. (1986). Integrated control of water hyacinth (*Eichhornia crassipes*) with a pathogen, insects and herbicides. *Weed Science*, **34**, (Supplement 1), 26–30.

Charudattan, R. and Conway, K. E. (1975). Comparison of *Uredo eichhorniae* the water hyacinth rust, and *Uromyces pontederiae*. *Mycologia*, **67**, 653–7.

Charudattan, R., Conway, K. E., and Freeman, T. E. (1975). A blight of water hyacinth *Eichhornia crassipes* caused by *Bipolaris stenospila* (*Helminthosporium stenospilum*). *Proceedings American Phytopathological Society* **2**, 65.

Charudattan, R., Freeman, T. E., Cullen, R. E., and Hofmeister, F. M. (1984). *Evaluation of* Fusarium roseum *'Culmorum' as a biological control agent for* Hydrilla verticillata. Technical Report A-84-5 prepared by University of Florida, Gainesville, Florida for US Army Engineer Waterways Experiment Station, Vicksburg, Mississippi, pp. 30.

Charudattan, R., Linda, S. B., Kluepfel, M., and Osman, Y. A. (1985). Biocontrol efficacy of *Cercospora rodmanii* on water hyacinth. *Phytopathology*, **75**, 1263–7.

Charudattan, R. and McKinney, D. E. (1978). A Dutch isolate of *Fusarium roseum* 'Culmorum' may control *Hydrilla verticillata* in Florida. *Proceedings EWRS 5th Symposium on Aquatic Weeds 1978*, pp. 219–24.

Charudattan, R., McKinney, D. E., Cordo, H. A., and Silveira-Guido, A. (1976). *Uredo eichhorniae*, a potential biocontrol agent for water hyacinth. *Proceedings IV International Symposium on Biological Control of Weeds*, pp. 210–13.

Charudattan, R., McKinney, D. E., and Hepting, K. T. (1981). Production storage, germination and infectivity of uredo spores of *Uredo eichhorniae* and *Uromyces pontederiae*. *Phytopathology*, **71**, 1203–7.

Charudattan, R., Perkins, B. D., and Littell, R. C. (1978). The effects of fungi and bacteria on the decline of arthropod-damaged water hyacinth in Florida. *Weed Science*, **26**, 101–7.

Charudattan, R. and Rao, K. V. (1982). Bostrycin and 4-deoxy-bostrycin: two non-specific phytotoxins produced by *Alternaria eichhorniae*. *Applied Environmental Microbiology*, **43**, 846–9.

Charudattan, R. and Walker, H. L. (1982). *Biological control of weeds with plant pathogens*. John Wiley, New York.

Charyev, R. (1980). [The grass carp and plant succession in reservoirs]. In Russian. *Ekologiya*, **11 (4)**, 93–4.

Charyev, R. (1984). Some consequences of the introduction and acclimatization of grass carp, *Ctenopharyngodon idella* (Cyprinidae) in the Kara Kum Canal. *Journal of Ichthyology*, **24**, 1–8.

Chaudhuri, H. and Ram, K. J. (1975). Control of aquatic weed by moth larvae. *Nature*, (*London*), **253**, 40–1.

Chaudhuri, H., Murty, D. S., Rey, R. K., and Reddy, P. V. G. K. (1976). Role of Chinese grass carp, *Ctenopharyngodon idella* (Val.) in biological control of noxious aquatic weeds in India: a review. In *Aquatic weeds in south-east Asia*, (ed. C. K. Varshney and J. Rzóska), pp. 315–22. W. Junk, The Hague.

Chokder, A. H. (1967). Biological control of aquatic vegetation. *Agriculture Pakistan*, **18**, 225–9.

Chokder, A. H. (1968). Further investigations on control of aquatic vegetation in fisheries. *Agriculture Pakistan*, **19**, 101–18.

Chokder, A. H. and Begum, A. (1965). Control of aquatic vegetation in fisheries. *Agriculture Pakistan*, **16**, 235–47.

Chomchalow, N. and Pongapangan, S. (1976). Aquatic weeds in Thailand: occurrence, problems and existing and proposed control measures. In *Aquatic weeds in south-east Asia* (ed. C. R. Varshney and J. Rzóska), pp. 43–50. W. Junk, The Hague.

Chou, Chang-Hung, Lee, M. L., and Oka, H. I. (1984). Possible allelopathic interactions between *Oryza perennis* and *Leersia hexandra*. *Botanical Bulletin Academia Sinica*, **25**, 1–19.

Chow, C. Y., Thevasagayam, E. S., and Wambeek, E. G. (1955). Control of *Salvinia*—a host of *Mansonia* mosquitoes. *Bulletin WHO*, **12**, 365–9.

Chow, Ven Te (1981). *Open-channel hydraulics*, (2nd edn). McGraw Hill, New York. pp. 680.

Chynoweth, D. P. (1987). Biomass conversion options. In *Aquatic plants for water treatment and resource recovery*, pp. 621–42. Magnolia Press, Orlando, Florida.

Clare, P. and Edwards, R. W. (1983). The macro-invertebrate fauna of the drainage channels of the Gwent Levels, S. Wales. *Freshwater Biology*, **13**, 205–27.

Clark, R. P. (1982). What is IAWM? *Aquatics*, **4 (1)**, 8, 18.

Clason, E. W. (1953). Notes on the Potamogetons of the Zuidlaren Lake and its adjacent waters. *Acta Botanica Neerlandica*, **1**, 489–96.

Clayton, J. S. (1977). A new approach for the selective control of aquatic weeds. *Proceedings 30th New Zealand Weed and Pest Control Conference 1977*, pp. 141–4.

Clayton, J. S. (1982). Effects of fluctuation in water level and growth of *Lagarosiphon major* on the aquatic vascular plants in Lake Rotorua from 1973 to 1980. *New Zealand Journal of Marine and Freshwater Research*, **16**, 89–94.

Clayton, J. S. (1986). Review of diquat use in New Zealand for submerged weed control. *Proceedings EWRS/AAB 7th Symposium on Aquatic Weeds 1986*, pp. 73–9.

Clayton, J. S. and Tanner, C. C. (1982). PP100: an alternative formulation of diquat for control of submerged aquatic weeds. *Proceedings 35th New Zealand Weed and Pest Control Conference* (ed. M. J. Hartley), pp. 261–4.

Coates, D. and Redding-Coates, T. A. (1981). Ecological problems associated with irrigation canals in the Sudan, with particular references to the spread of bilharziasis, malaria and aquatic weeds and the ameliorative rate of fishes. *International Journal of Environmental Studies*, **16**, 207–12.

Coats, G. E., Funderburk, H. H., Lawrence, J. M., and Davis, D. E. (1964). Persistence of diquat and paraquat in pools and ponds. *Proceedings 17th Southern Weed Control Conference*, **17**, pp. 308–20.

Coble, T. A. and Vance, B. D. (1987). Seed germination in *Myriophyllum spicatum* L. *Journal of Aquatic Plant Management*, **25**, 8–10.

Coffey, B. T. (1975). Submerged weed control by lake lowering. *Proceedings 28th New Zealand Weed and Pest Control Conference 1975*, pp. 143–52.

Coffey, B. T. (1983). Survey and management requirements for plants in New Zealand waterways. *Proceedings Aquatic Weeds Seminar, Massey, New Zealand 1983*, pp. 7–11.

Coffey, B. T. (1988). Aquatic macrophyte management in New Zealand. Plenary Address: XXIII S.I.L. Congress, Hamilton, New Zealand 1987. *Verhandlungen der Internationalen Vereinigung für theoretische und angewandte Limnologie*, **23**(4), 42–7.

Coffey, B. T. and McNabb, C. D. (1974). Eurasian water-milfoil in Michigan. *The Michigan Botanist*, **13**, 159–65.

Colle, D. E., Shireman, J. V., and Rottman, D. W. (1978). Food selection by grass carp fingerlings in a vegetated pond. *Transactions American Fisheries Society*, **107**, 149–52.

Collett, L. C., Collina, A. J., Gibbs, P. J., and West, R. J. (1981). Shallow dredging as a strategy for the control of sublittoral macrophytes: a case study in Tuggerah Lakes, New South Wales. *Australian Journal of Marine and Freshwater Research*, **32**, 563–72.

Collins, C. D., Park, R. A., and Boylen, C. W. (1985). *A mathematical model of submerged aquatic plants*. Miscellaneous Paper A-85-2, prepared by Rensselaer Polytechnic Institute, Troy, New York, for US Army Engineer Waterways Experiment Station, Vicksburg, Mississippi. pp. 21.

Collins, C. D., Sheldon, R. D., and Boylen, C. W. (1987). Littoral zone macrophyte community structure distribution and association of species along physical gradients in Lake George, New York, USA. *Aquatic Botany*, **29**, 177–94.

Comes, R. D., Bruns, V. F., and Kelley, A. D. (1976). Residues and persistence of glyphosate in irrigation water. *Weed Science*, **24**, 47–50.

Comes, R. D., Marquis, L. Y., and Kelley, A. D. (1981). Response of seedlings of 3 perennial grasses to dalapon, amitrole, and glyphosate. *Weed Science*, **29**, 619–21.

Conway, K. E. (1976*a*). *Cercospora rodmanii*, a new pathogen of water hyacinth with biological control potential. *Canadian Journal of Botany*, **54**, 1079–83.

Conway, K. E. (1976*b*). Evaluation of *Cercospora rodmanii* as a biological control of water hyacinths. *Phytopathology*, **66**, 914–17.

Conway, K. E. and Cullen, R. E. (1978). The effect of *Cercospora rodmanii*, a biological control for water hyacinth, on the fish *Gambusia affinis*. *Mycopathologia*, **66**, 113–16.

Conway, K. E. and Freeman, T. E. (1977). Host specificity of *Cercospora rodmanii*, a potential biological control of water hyacinth. *Plant Disease Report*, **61**, 262–6.

Conway, K. E., Freeman, T. E., and Charudattan, R. (1978). Development of *Cercospora rodmanii* as a biological control for *Eichhornia crassipes*. *Proceedings EWRS 5th Symposium on Aquatic Weeds 1978*, pp. 225–30.

Conyers, D. L. and Cooke, G. D. (1982). A comparative study of harvesting and herbicides for control of nuisance plants. *Ohio Journal of Science*, **82**, 93.

Cook, C. D. K. (1968). The vegetation of the Kainji Reservoir site in Northern Nigeria. *Vegetatio*, **15**, 225–43.

Cook, C. D. K. (1976). *Salvinia* in Kerala, S. India and its control. In *Aquatic weeds in south east Asia* (ed. C. K. Varshney and J. Rzoska), pp. 241–3. W. Junk, The Hague.

Cook, C. D. K. (1983). Aquatic plants endemic to Europe and the Mediterranean. *Bot. Jahrb. Syst.*, **103**, 539–83.

Cook, C. D. K. (1985). Range extensions of aquatic vascular plant species. *Journal of Aquatic Plant Management*, **23**, 1–6.

Cook, C. D. K. (1987*a*). Dispersion in aquatic and amphibian vascular plants. In *Plant life in aquatic and amphibious habitats*. Special Publication British Ecological Society, No. 5. (ed. R. M. M. Crawford), pp. 179–90. Blackwell Scientific Publication, Oxford.

Cook, C. D. K. (1987*b*). Vegetative growth and genetic mobility in some aquatic weeds. In *Differentiation patterns in higher plants* (ed. K. M. Urbanska), pp. 217–25. Academic Press, London.

Cook, C. D. K. and Gut, B. J. (1971). *Salvinia* in the State of Kerala, India. *Pest Articles and News Summaries,* **17 (4)**, 438–47.

Cook, C. D. K., Gut, B. J., Rix, E. M., Schneller, J., and Seitz, M. (1974). *Water plants of the world.* W. Junk, The Hague. pp. 561.

Cook, C. D. K. and Lüönd, R. (1982). A revision of the genus *Hydrilla* (Hydrocharitaceae). *Aquatic Botany,* **13**, 485–504.

Cook, C. D. K. and Urmi-König, K. (1984). A revision of the genus *Egeria* (Hydrocharitaceae). *Aquatic Botany,* **19**, 73–96.

Cook, C. D. K. and Urmi-König, K. (1985). A revision of the genus *Elodea* Hydrocharitaceae). *Aquatic Botany,* **21**, 111–56.

Cooke, A. S. and Newbold, C. (1986). III. *Freshwater herbicides.* In *The use of herbicides on nature reserves* (ed. A. S. Cooke). Focus on Nature Conservation Series Report No. 145, Chief Scientist Directorate, Nature Conservancy Council, Peterborough, UK, pp. 65–74.

Cooke, G. D. (1980). Lake-level drawdown as a macrophyte control technique. *Water Resources Bulletin,* **16**, 317–22.

Cooke, G. D. and Gorman, M. E. (1980). Effectiveness of DuPont Typar sheeting in controlling macrophyte regrowth after overwinter drawdown. *Water Resources Bulletin,* **16**, 353–5.

Cooley, T. N., Dooris, P. M., and Martin, D. F. (1980). Aeration as a tool to improve water quality and reduce the growth of *Hydrilla. Water Research,* **14**, 485–9.

Coordinatiecommissie Onkruidonderzoek (1984). *Report on terminology in weed science in the Dutch language area,* (2nd edn). Report: Coordinatiecommissie Onkruidonderzoek, NRLO, The Netherlands.

Cope, O. B., McRaren, J. P., and Eller, L. (1969). Effects of dichlobenil on two fish pond environments. *Weed Science,* **17**, 158–65.

Cope, O. B., Wood, E. M., and Wallen, G. H. (1970). Some chronic effects of 2,4-D on the bluegill (*Lepomis macrochirus*). *Transactions of the American Fisheries Society,* **99**, 1–12.

Corbet, P. S., Longfield, C., and Moore, N. W. (1960). *Dragonflies.* Collins, London.

Corbus, F. G. (1982). Aquatic weed control with endothall in a Salt River Project canal. *Journal of Aquatic Plant Management,* **20**, 1–3.

Cordo, H. A. and Deloach, C. J. (1982*a*). The flea beetle *Lysathia flavipes* that attacks *Ludwigia* (water primrose) and *Myriophyllum* (parrot feather) in Argentina. *Coleoptera Bulletin,* **36**, 298–301.

Cordo, H. A. and Deloach, C. J. (1982*b*). Weevils *Listronotus marginicollis* and *L. cinnamomeus* that feed on *Limnobium* and *Myriophyllum* in Argentina. *Coleoptera Bulletin,* **36**, 302–8.

Cordo, H. A., Deloach, C. J., and Ferrer, M. (1984). Biology and larval host range of flea beetle *Dysonicha argentinensis* (Coleoptera: Chrysomelidae) on alligator weed in Argentina. *Annals Entomological Society America,* **22**, 134–41.

Cordo, H. A., Deloach, C. J., Runnacles, J., and Ferrer, R. (1978). *Argentinorhynchus bruchi,* a weevil from *Pista stratiotes* in Argentina: biological studies. *Environmental Entomology,* **7**, 329–33.

Corillion, R. (1957). Les Charophycées de France et d'Europe occidentale. *Bulletin de la Société Science Bretagne,* **32,** 1–499.

Corwin, D. L. and Farmer, W. J. (1984). Nonsingle-valued adsorption-desorption of bromacil and diquat by freshwater sediments. *Environmental Science and Technology,* **18,** 507–14.

Cosby, B. J., Hornberger, G. M., and Kelly, M. G. (1984). Identification of photosynthesis-light models for aquatic systems. II. Application to a macrophyte-dominated stream. *Ecological Modelling,* **23,** 25–51.

Cottam, G. and Nichols, S. (1970). *Changes in the water environment resulting from aquatic plant control.* Technical Report OWRR B-019-WIS, University of Wisconsin Water Research Center, Madison. pp. 27.

Couch, R. W. and Gangstad, G. O. (1974). The response of duckweed to CO_2-laser radiation. *Hyacinth Control Journal,* **12,** 25–6.

Couch, R. W. and Nelson, E. N. (1982). Effects of 2,4-D on non-target species in Kerr Reservoir. *Journal of Aquatic Plant Management,* **20,** 8–13.

Couch, R. W. and Nelson, E. N. (1985). *Myriophyllum spicatum* in North America. In *1st International Symposium on Water milfoil* (Myriophyllum spicatum) *and related Haloragaceae species* (ed. L. W. J. Anderson), pp. 8–18. Aquatic Plant Management Society, Vicksburg.

Coulson, J. R. (1977). *Biological control of alligator weed 1959–1972. A review and evaluation.* Technical Bulletin US Department of Agriculture No. 1547. pp. 98.

Courtney, A. D. (1974). The control of weeds in drainage channels and minor watercourses. *Agriculture in Northern Ireland,* **49,** 41–4.

Cragg, B. A. (1980). The role of micro-organisms in the deoxygenation of water treated with herbicides. Ph.D. thesis. UWIST, University of Wales, Cardiff. pp. 265.

Cragg, B. A. and Fry, J. C. (1984). The use of microcosms to simulate field experiments to determine the effects of herbicides on aquatic bacteria. *Journal of General Microbiology,* **130,** 2309–16.

Crawford, R. M. M. (ed.) (1987). *Plant life in aquatic and amphibious habitats,* Special Publication British Ecological Society No. 5. Blackwell Scientific Publications, Oxford. pp. 452.

Crawford, S. A. (1981). Successional events following simazine application. *Hydrobiologia,* **77,** 217–23.

Crespo, S. and Perez-Moreau, R. L. (1967). Revisión dél género *Typha* en Argentina. *Darwiniana,* **14,** 413–29.

Crisp, D. T. and Gledhill, T. (1970). A quantitative description of the recovery of the bottom fauna in a muddy reach of a mill stream in Southern England after draining and dredging. *Archiv für Hydrobiologie,* **64,** 502–41.

Croll, B. T. (1986). The effects of the agricultural use of herbicides on freshwaters. In *Effects of land use on freshwaters* (ed. J. F. de L. G. Solbé), pp. 201–9. Water Research Centre/Ellis Horwood, Chichester.

Crosby, D. G. and Tucker, R. K. (1966). Toxicity of aquatic herbicides to *Daphnia magna. Science (New York),* **154,** 289–91.

Crossland, N. O. and Adams, W. M. (1963). The use of acrolein for the control of aquatic weeds and snails in an irrigation system in Tanganyika. *Pest Articles C,* **9,** 221–4.

Crossland, N. O. and Elgar, K. E. (1974). Evaluation of the herbicide WL 63611 in aquatic systems. *Proceedings EWRC 4th Symposium on Aquatic Weeds,* 58–68.

Crouzel, Z. S. and Cordo, H. A. (1984). Control biológico de malezas y su applicación en la Républica Argentina. *Malezas,* **12,** 74–82. Asociacion Argentina Control Malezas, Buenos Aires.

Crowell, T. E. and Steenis, J. H. (1968). Status of Eurasian watermilfoil in the Currituck Sound area, 1967. *Abstracts Weed Science Society of America 1968,* 57.

Cruz, G. A. and Delgado, R. (1986). Distribución de los macrófitos en el lago Yojoa, Honduras. *Revista de Biologia Tropical,* **34,** 141–9.

Cullen, P. and Maher, W. (1987). Nutrient cycling in lakes. In *Proceedings 9th Australian Symposium on Analytical Chemistry,* Volume 2, pp. 905–8.

Cullen, P. and O'Loughlin, E. M. (1982). Non-point sources of pollution. In *Prediction in water quality* (ed. E. M. O'Loughlin and P. Cullen), pp. 437–53. Australian Academy of Science, Canberra.

Cullen, P. and Rosich, R. S. (1979). Effect of rural and urban sources of phosphorus on Lake Burley Griffin. *Progress in Water Technology,* **11,** 219–30.

Culpepper, M. M. and Decell, J. L. (1978). *Field evaluation of the Aqua-Trio system.* In Technical Report A-78-3. Mechanical Harvesting of Aquatic Plants (2 vols). US Army Engineer Waterways Experiment Station, Vicksburg, Mississippi. pp. 438–49.

Cummings, J. G. (1977). Status of EPA regulations of aquatic herbicides. *Proceedings Research Planning Conference Aquatic Plant Control Program 1976, Florida,* pp. 16–18.

Cure, V. (1974). [Improvement of ecosystem in macrophyte-overrun Frăsinet pond after stocking with *C. idella* (Val.)] In German. *Archiv für Hydrobiologie,* **44,** 338–51.

Cure, V., Snaider, A., and Chiosila, I. (1970). [Macrophytes from Frăsinet pond (Ilfov)—their influence on ecosystem two years after the introduction of *C. idella*] In Rumanian. *Buletinul Institutului de Cercetări si Projectari Piscicole,* **29,** 5–27.

Custer, P. E., Halverson, F. D., Malone, J., and Chong, C. V. (1978). The white amur as a biological control agent of aquatic weeds in the Panama Canal. *Fisheries,* **3,** 2–9.

Czopek, M. (1963). Studies on the external factors inducing the formation of turions in *Spirodela polyrrhiza* (L.) Schleiden. *Acta Societatis Botanicarum Poloniae,* **32,** 199–211.

Dabydeen, S. and Leavitt, J. R. C. (1981). Adsorption and effect of simazine and atrazine on *Elodea canadensis. Bulletin of Environmental Contamination Toxicology,* **26,** 385.

Dalrymple, R. L. (1971). Experiences with diuron for aquatic weed control. *Proceedings 24th Annual Meeting Southern Weed Science Society,* pp. 333.

Dandy, J. E. (1937). The genus *Potamogeton* L. in tropical Africa. *Journal of the Linnaean Society,* **50,** 507–40.

Danell, K. (1979). Reduction of aquatic vegetation following the colonisation of a Northern Swedish lake by the muskrat, *Ondatra zibethica. Oecologia,* **38,** 101–6.

Daniel, T. C. (1972). Evaluation of diquat and endothall for the control of watermilfoil (*Myriophyllum exalbescens*) and the effect of weed-kill on the nitrogen and phosphorus status of a water body. Ph.D. thesis, University of Wisconsin, Madison, Wisconsin.

Dardeau, E. A. (1983). *Aerial survey techniques to map and monitor aquatic plant populations—four case studies.* Technical Report A-83-1, US Army Engineer Waterways Experiment Station, Vicksburg, Mississippi. pp. 42.

Dardeau, E. A. and Hogg, E. A. (1983). *Inventory and assessment of aquatic plant management methodologies*. Technical Report A-83-2, US Army Engineer Waterways Experiment Station, Vicksburg, Mississippi.

Das, R. R. (1969). A study of reproduction in *Eichhornia crassipes* (Mart.) Solms. *Tropical Botany*, **10**, 195–8.

Da Silva, C. J. (1981). Oservacoes sobre a biología reproductiva de *Pistia stratiotes* L. (Araceae). *Acta Amazónica*, **11**, 487–504.

Dassanayake, W. L. P. and Chow, C. Y. (1954). The control of *Pistia stratiotes* in Ceylon by means of herbicides. *Annals Tropical Medicine Parasitology*, **48**, 129–34.

Datta, P. C. and Maiti, R. K. (1963). Paddy field weeds of Midnapore district. *Indian Agriculturalist*, **7**, 147–66.

Datta, S. C. and Biswas, K. K. (1970). Germination pattern and seedling morphology of *Pistia stratiotes* L. *Phyton* (*Argentina*), **27**(2), 157–61.

Datta, S. C. and Biswas, K. K. (1977). Autecological studies on weeds of West Bengal. VII. *Nymphaea nouchali* Burm. F. *Bulletin Botanical Society Bengal*, **31**, 129–40.

Davies, H. R. J. (1959). Effects of the water hyacinth (*Eichhornia crassipes*) in the Nile Valley. *Nature, London*, **184**, 1085–6.

Davies, P. J. and Seaman, D. E. (1964). Physiological effects of diquat on submersed aquatic weeds. *Abstracts 1964 Meeting Weed Science Society America*, 100.

Dawood, K. E., Farooq, M., Dazo, B. C., Miguel, L. C., and Unrau, G. O. (1965). Herbicide trials in the snail habitats of the Egypt-49 Project Area. *Bulletin WHO*, **32**, 269–87.

Dawson, F. H. (1973). Notes on the production of stream bryophytes in the High Pyrenees (France). *Annals of Limnology*, **9**, 231–40.

Dawson, F. H. (1976). The annual production of the aquatic macrophyte *Ranunculus penicillatus* var. *calcareus* (R. W. Butcher) C. D. K. Cook. *Aquatic Botany*, **2**, 51–73.

Dawson, F. H. (1978*a*). The seasonal effects of aquatic plant growth on the flow of water in a stream. *Proceedings EWRS 5th symposium on Aquatic Weeds 1978*, pp. 71–8.

Dawson, F. H. (1978*b*). Aquatic plant management in semi-natural streams: the role of marginal vegetation. *Journal of Environmental Management*, **6**, 231–21.

Dawson, F. H. (1981). The reduction of light as a technique for the control of aquatic plants—an assessment. *Proceedings Association of Applied Biologists Conference: Aquatic Weeds and their Control 1981*, pp. 157–64. AAB, Wellesbourne, UK.

Dawson, F. H. (1985). Light reduction techniques for aquatic plant management. *NALMS Symposium on Applied Lake and Watershed Management, Lake Geneva 1985*, pp. 1–32.

Dawson, F. H., Castellano, E., and Ladle, M. (1978). Concept of species succession in relation to river vegetation and management. *Verhandlungen der Internationalen Vereinigung für theoretische und angewandte Limnologie*, **20**, 1429–34.

Dawson, F. H. and Charlton, F. G. (1988). *A bibliography on the hydraulic resistance or roughness of vegetated watercourses*. Freshwater Biological Association Occasional Publication 25. FBA, Windermere, UK.

Dawson, F. G. and Hallows, H. B. (1983). Practical applications of shading material for aquatic plant control. *Aquatic Botany*, **17**, 301–8.

Dawson, F. H. and Haslam, S. M. (1983). The management of river vegetation with particular reference to shading effects of marginal vegetation. *Landscape Planning,* **10,** 147–69.

Dawson, F. H. and Kern-Hansen, U. (1978). Aquatic weed management in natural streams: the effect of shade by the marginal vegetation. *Verdhandlungen der Internationalen Vereinigung für theoretische und angewandte Limnologie,* **20,** 1451–6.

Dawson, F. H. and Kern-Hansen, U. (1979). The effect of natural and artificial shade on the macrophytes of lowland streams, and the use of shade as a management technique. *Internationale Revue der gesamten Hydrobiologie,* **64 (4),** 437–55.

Dawson, F. H. and Robinson, W. N. (1985). Submerged macrophytes and the hydraulic roughness of a lowland chalkstream. *Verhandlungen der Internationalen Vereinigung für theoretische und angewandte Limnologie,* **22,** 1944–8.

Dawson, G. (1973). Flora Argentina. Lentibulariáceas. *Revista del Museo de La Plata. Sección Botánica,* **13,** 1–59.

Dazo, B. C., Hairston, N. G., and Dawood, J. K. (1966). The ecology of *Bulinus truncatus* and *Biomphalaria alexandrina* and its implications for the control of bilharziasis in Egypt-49 project area. *Bulletin WHO,* **35,** 339–56.

Dearden, P. (1982). Comparative risk assessment associated with the growth of Eurasian water milfoil, (*Myriophyllum spicatum* L.) in the Okanagan valley, British Columbia, Canada. *Proceedings EWRS 6th Symposium on Aquatic Weeds 1982,* pp. 113–22.

DeBusk, T. A., Ryther, J. H., and Williams, L. D. (1983). Evapotranspiration of *Eichhornia crassipes* (Mart.) Solms. and *Lemna minor* L. in Central Florida: relation to canopy structure and season. *Aquatic Botany,* **16,** 31–9.

Dechoretz, N. (1980). Control of submerged aquatic weeds in irrigation canals with fluridone. *Proceedings Western Society Weed Science,* **33,** pp. 133.

Dechoretz, N. and Anderson, L. W. J. (1983). Growth of American pondweed in soil treated with DPX 5648 (Oust[R]). *Western Society Weed Science Research Progress Reports 1983,* pp. 287–8.

De Groot, S. J. (1985). Introductions of non-indigenous fish species for release and culture in the Netherlands: a review. *Aquaculture,* **46,** 237–57.

Degroote, D. and Kennedy, R. A. (1977). Photosynthesis in *Elodea canadensis. Plant Physiology* (*Lancaster*), **59,** 1133–5.

Deighton, K. A. (1984). Developments in long-reach excavators. *Public Works Congress 1984, Birmingham,* pp. 1–15.

de Lange, L. and van Zon, J. C. J. (1978). Evaluation of the botanical response of different methods of aquatic weed control, based on the structure and floristic composition of the macrophyte vegetation. *Proceedings EWRS 5th Symposium on Aquatic Weeds 1978,* pp. 279–86.

de la Sota, E. R. (1962*a*). Contribución al conocimiento de las Salviniaceae neotropicales. I. *Salvinia oblongifolia* Martius. *Darwiniana,* **12,** 465–98.

de la Sota, E. R. (1962*b*). Contribución al conocimiento de las Salviniaceae neotropicales. II. *Salvinia auriculata* Aublet. *Darwiniana,* **12,** 499–513.

de la Sota, E. R. (1962*c*). Contribución al conocimiento de las Salviniaceae neotropicales. III. *Salvinia herzogii* sp. nov. *Darwiniana,* **12,** 514–20.

de la Sota, E. R. (1963). Contribución al conocimiento de las Salviniaceae neotropicales. IV. Datos morfoanatómicos sobre *Salvinia rotundifolia* Willdenow y *Salvinia herzogii* de la Sota. *Darwiniana,* **12**, 612–23.

de la Sota, E. R. (1964). Contribución al conocimiento de las Salviniaceae neotropicales. V. *Salvinia spruceri* Kuhn. *Darwinia,* **13**, 529–36.

de Leon Hernandez, G. and Zaballa, M. P. (1985). Vegetación acuática. Características generales y recommendaciones para su control. *Voluntad Hidráulica,* **23**, 2–9.

Del Fosse, E. S., Perkins, B. D., and Steward, K. K. (1976). A new US record for *Parapoynx diminutalis* (Lepidoptera: Pyralidae) a possible biological control agent for *Hydrilla verticillata. The Florida Entomologist,* **59 (1)**, 19–20.

Del Fosse, E. S., Sutton, D. L., and Perkins, B. D. (1976). Combination of the mottled water hyacinth weevil and the white amur for biological control of water hyacinth. *Journal of Aquatic Plant Management,* **14**, 64–7.

DeLoach, C. J. and Cordo, H. A. (1974). Biological control of aquatic weeds. *Malezas,* 3, 76–82. Asociacion Argentina Control Malezas, Buenos Aires.

DeLoach, C. J. and Cordo, H. A. (1976). Ecological studies of *Neochetina bruchi* and *N. eichhorniae* on water hyacinth in Argentina. *Journal of Aquatic Plant Management,* **14**, 53–9.

DeLoach, C. J. and Cordo, H. A. (1978). Life history and ecology of the moth *Sameodes albiguttalis,* a candidate for biological control of water hyacinth. *Environmental Entomology,* **7**, 309–21.

DeLoach, C. J. and Cordo, H. A. (1981). Biology and host range of the weevil *Neochetina affinis* which feeds on Pontederiaceae in Argentina. *Annals Entomological Society America,* **74**, 14–19.

DeLoach, C. J. and Cordo, H. A. (1982). Natural enemies of *Neochetina bruchi* and *N. eichhorniae*, two weevils from water hyacinth in Argentina. *Annals Entomological Society America,* **75**, 115–18.

DeLoach, C. J. and Cordo, H. A. (1983). Control of water hyacinth by *Neochetina bruchi* (Coleoptera: Curculionidae: Bagoini) in Argentina. *Environmental Entomology,* **12**, 9–23.

DeLoach, C. J., DeLoach, A. D., and Cordo, H. A. (1976). *Neohydronomous pulchellus,* a weevil attacking *Pistia stratiotes* in South America: biology and host specificity. *Annals Entomological Society America,* **69**, 830–4.

DeLoach, C. J., DeLoach, D. J., and Cordo, H. A. (1979). Observations on the biology of the moth *Samea multiplicalis*, on water lettuce in Argentina. *Journal of Aquatic Plant Management,* **17**, 42–4.

DeLoach, C. J., Cordo, H. A., Ferrer, R., and Runnacles, J. (1980). *Acigona infusella,* a potential biological control agent for water hyacinth: observations in Argentina (with descriptions of 2 new species of *Apanteles* by L. de Santis). *Annals Entomological Society America,* **73**, 138–46.

den Hartog, C. (1983). Synecological classification of aquatic plant communities. *Colloques phytosociologiques 10, vegetations aquatiques, Lille 1981,* 171–82.

den Hartog, C. and Segal, S. (1963). A new classification of the water-plant communities. *Acta Botanica Neerlandica,* **13**. 367–93.

Dennis, W. M. (1984). Aquatic macrophyton sampling: an overview. *Special Technical Publications ASTM STP 843,* 2–6. American Society for Testing Materials.

Dennis, W. M. and Isom, B. G. (ed.) (1984) Ecological assessment of macrophyton: collection, use and meaning of data, *Special Technical Publications ASTM STP 843*, pp. 122. American Society for Testing Materials.

Denny, P. (1972). Sites of nutrient absorption in aquatic macrophytes. *Journal of Ecology*, **60**, 819–29.

Denny, P. (1973). Lakes of South West Uganda. II. Vegetation studies on Lake Bunyoni. *Freshwater Biology*, **3**, 123–30.

Denny, P. (1984). Permanent swamp vegetation of the Upper Nile. *Hydrobiologia*, **110**, 79–90.

Denny, P. (ed.) (1985). *The ecology and management of African wetland vegetation*. W. Junk, The Hague. pp. 344.

Denny, P., Orr, P. T., and Erskine, D. J. C. (1983). Potentiometric measurements of carbon dioxide flux of submerged aquatic macrophytes in pH-statted natural waters. *Freshwater Biology*, **13**, 507–19.

Denoyelles, F. and Kettle, D. (1980). *Herbicides in Kansas waters—evaluations of the effects of agricultural runoff and aquatic weed control on aquatic food chains*. Final Report of Office of Water Research & Technology, US Department of the Interior, Washington DC. pp. 40.

Department of the Environment (1975). *The Waterways of the British Waterways Board. A study of operating and maintenance costs. Volume 1*. Peter Fraenkel & Partners, London & Glasgow, HMSO London.

de Silva, S. S. and Weerakoon, D. E. M. (1981). Growth, food intake and evacuation rates of grass carp, *Ctenopharyngodon idella* fry. *Aquaculture*, **25**, 67–76.

Desougi, L. A. (1979). A survey of aquatic macrophytes in Gezira irrigation canals. *Weed research in the Sudan: Volume 1. Proceedings of a Symposium* (ed. M. E. Beshir & W. Koch), pp. 146–52. Berichte aus dem Fachgebeit Herbologie, Universität Hohenheim, FRG.

Dethioux, M. (1982). Données sur l'écologie de *Ranunculus penicillatus* (Dum.) Bab. et *R. fluitans* Lam. en Belgique. In *Studies on aquatic vascular plants* (ed. J. J. Symoens, S. S. Hooper, and P. Compère), pp. 187–91. Royal Botanical Society of Belgium, Brussels.

Devlin, R. M., Saras, C. N., and Kisiel, M. J. (1978). Influence of fluridone on chlorophyll production. *Abstracts Weed Science Society America 1978*, 3.

de Vries, P. J. R. (1987). Ervaringen met de graskarper bij waterschappen. *Waterschapsbelangen*, **72**, 384–9.

de Wit, C. T. (1960). On competition. *Verslagen van Landbouwkundige Onderzoekingen*, **66**, 1–82.

de Wit, H. C. D. (1971). *Aquarienpflanzen*. Verlag Eugen Ulmer, Stuttgart. pp. 351.

Dhahiyat, Y., Lankester, K., Pieterse, A. H., Siregar, H., Sutisna, T., and Vink, J. A. C. (1982). A study on *Hydrilla verticillata* (L. f.) Royle in Lake Curug (Indonesia). *Proceedings EWRS 6th Symposium on Aquatic Weeds 1982*, pp. 63–76.

Dial, N. A. and Bauer, C. A. (1984). Teratogenic and lethal effects of paraquat on developing frog embryos (*Rana pipiens*). *Bulletin Environmental Contamination Toxicology*, **33**, 592–7.

Diem, J. R. and Davies, D. E. (1974). Effect of 2,4-D on ametryn toxicity. *Weeds*, **22**, 285–92.

di Fulvio, T. E. (1956). *Observaciones morfológicas y taxonómicas de las especies de* Azolla *del centro de Argentina*. Revista Ciencias Exactas Fis. Naturales (Córdoba, Argentina).

di Fulvio, T. E. (1961). Sobre el episporio de las especies americanas de *Azolla* con especial referencia a *A. mexicana* Presl. *Kurtziana*, **1**, 299–302.

Dill, W. A. (1961). Some notes on the use of the manatee (*Trichechus*) for the control of aquatic weeds. *Fisheries Biology Paper No. 13*, Food and Agriculture Organization, United Nations, Rome.

Dimitrov, M. (1984). Intensive polyculture of common carp (*Cyprinus carpio*) and herbivorous fish (silver carp: *Hypophthalmichthys molitrix* and grass carp: *Ctenopharyngodon idella*). *Aquaculture*, **38**, 241–53.

Dinka, M. (1986). Accumulation and distribution of elements in cattail species (*Typha latifolia* L., *T. angustifolia* L.) and reed (*Phragmites australis* (Cav.) Trin. ex. Steudel) living in Lake Balaton. *Proceedings EWRS/AAB 7th Symposium on Aquatic Weeds 1986*, pp. 81–6.

Dioni, W. (1967). Investigación preliminar de la estructura básica de las asociaiones de la micro y meso fauna de las raíces de las plantas flotantes. *Acta Zoologica Lilloana*. **23**, 111–37.

Doarks, C. (1980). *Botanical survey of marsh dykes in Broadland*. Report BA RS2. Broads Authority, Norwich, UK.

Doarks, C. (1984). *A study of marsh dykes in Broadland*. Report BA RS9. Broads Authority, Norwich, UK. pp. 148.

Dolberg, F., Saadullah, M., and Haque, M. (1981). A short review of the feeding values of water plants. *Tropical Animal Production*, **6**, 322–6.

Dooris, P. M. and Martin, D. F. (1980). Growth inhibition of *Hydrilla verticillata* by selected lake sediment extracts. *Water Resource Bulletin*, **16**, 112–17.

Dorna, J. M. (1974*a*). Control de malezas en canales y zanjas de desagüe. *Malezas*, **3**, 99–102.

Dorna, J. M. (1974*b*). Malezas acuáticas. Su control. *Malezas*, **3**, 83–98.

Dove, R. and Wallis, M. (1981). *Studies on aquatic macrophytes Part. XXXIV*. APD Bulletin 17, 1980 Aquatic Plant Quarantine Project, Ministry of Environment, British Columbia. pp. 31.

Dowidar, A. E. and Robson, T. O. (1971). Preliminary studies on the biology and control of *Vaucheria dichotoma* from British waters. *Proceedings EWRC 3rd Symposium on Aquatic Weeds 1971*, pp. 275–83.

Dowidar, A. R. and Robson, T. O. (1972). Studies on the biology and control of *Vaucheria dichotoma* found in freshwaters in Britain. *Weed Research*, **12**, 221–8.

Downton, W. J. S. (1975). The occurrence of C_4 photosynthesis among plants. *Photosynthetica*, **9**, 96–105.

Drago, E (1976). Origen y clasificación de ambientes léniticos en llanuras aluviales. *Revista Asociacion Ciencias Naturales Litoral*, **7**, 123–37.

Drew, M. C. (1983). Plant injury and adaptation to oxygen deficiency in the root environment: a review. *Plant and Soil*, **75**, 179–99.

Drijver, C. A. and Marchand, M. (1985). *Taming the floods—environmental aspects of floodplain development in Africa.* Report: University of Leiden, the Netherlands. pp. 208.

Driscoll, R. J. (1975). *Factors affecting the status of aquatic macrophytes in drainage dykes in Broadland.* Report, Nature Conservancy Council, Norwich, UK.

Druijff, A. H. (1973). The chain scythe, a simple tool for controlling aquatic weeds in irrigation canals. *Pest Articles and News Summaries,* **19**, 216–18.

Druijff, A. H. (1979). Manual and mechanical control of aquatic weeds in watercourses. *Weed Research in the Sudan: Volume 1. Proceedings of a Symposium* (ed. M. E. Beshir and W. Koch) pp. 139–45. Berichte aus dem Fachgebeit Herbologie Universität Hohenheim, FRG.

Dubbers, F. A. A., Ghoneim, S., Siemelink, M. E., El Gharably, Z., Pieterse, A. H., and Blom, J. E. (1981). Aquatic weed control in irrigation and drainage canals in Egypt by means of grass carp (*Ctenopharyngodon idella*). *Proceedings Vth International Symposium on Biological Control of Weeds, Brisbane 1980,* pp. 261–71. Commonwealth Scientific and Industrial Research Organisation, Australia.

Dubois, J. P., Blake, G., Gerbeaux, P., and Jensen, S. (1984). Methodology for the study of aquatic vegetation in the French Alpine lakes. *Verhandlungen der Internationalen Vereinigung für theoretische und angewandte Limnologie,* **22**, 1036–9.

Dubyna, D. V. and Chornaya, G. A. (1984). [Species of the *Potamogeton* L. genus in the water flora of the Seversky-Donets River valley]. In Russian; English summary. *Ukrainski Botanichnyi Zhurnal,* **41 (4)**, 22–8.

Duffield, A. N. (1981). The impact of *Lemna* on the oxygen resources of channels of potential value as fisheries. *Proceedings Association of Applied Biologists Conference: Aquatic Weeds and their Control 1981,* pp. 257–64. AAB Wellesbourne, UK.

Duffield, A. N. and Edwards, R. W. (1981). Predicting the distribution of *Lemna* spp. in a complex system of drainage channels. *Proceedings Association of Applied Biologists Conference: Aquatic Weeds and their Control 1981,* pp. 59–65. AAB, Wellesbourne, UK.

Dunk, W. P. (1954). A report on mechanical and chemical methods for controlling weed growth in Victorian irrigation and drainage systems. *Congress Irrigation Drainage,* **2**, 21–32.

Dunn, R. W. (1976). Aspects of weed control in water supply reservoir with amenity uses. *Proceedings Symposium on Aquatic Herbicides: British Crop Protection Council Monograph No. 16,* pp. 14–18.

Dupont, P. and Visset, L. (1968). L'envahissement de la vallée de l'Erdre et de quelques canaux de Loire–Atlantique par *Elodea densa*. *Bulletin de la Société Scientifique de Bretagne,* **43**, 285–7.

Durden, W. C., Blackburn, R. D., and Gangstad, E. O. (1975). Control program for alligator weed in the south-eastern states. *Hyacinth Control Journal,* **13**, 27–30.

du Rietz, G. E., Hannerz, A. G., Lohammer, G., Santesson, R., and Waern, M. (1939). Zur Kenntnis der Vegetation des Sees Takern. *Acta Phytogeographica Suecica,* **12**, 1–65.

Dutartre, A. (1980). [Preliminary studies of an aquatic weed infesting lakes along the Aquitanian coast. *Lagarosiphon major* (Ridley) Moss. (Hydrochariteae)]. In French. *Proceedings 6th International Colloquium on Weed Ecology, Biology and Systematics,* pp. 141–57. COLUMA-EWRS, Montpellier.

Dutartre, A. (1986). Aquatic plants introduced in freshwater lakes and ponds of Aquitaine (France): dispersion and ecology of *Lagarosiphon major* and *Ludwigia peploides*. *Proceedings EWRS/AAB 7th International Symposium on Aquatic Weeds 1986*, pp. 93–8.

Dutartre, A. and Capdevielle, P. (1982). [Present distribution of vascular aquatic plants introduced into South-West France]. In French. In *Studies on aquatic vascular plants* (ed. J. J. Symoens, S. S. Hooper, and P. Compère), pp. 390–93. Royal Botanical Society of Belgium, Brussels.

Dutartre, A. and Dubois, J. P. (1986). Biological control of Eurasian watermilfoil (*Myriophyllum spicatum* L.) using waterfowl, La Jarnelle pond (France), *Proceedings EWRS/AAB 7th Symposium Aquatic Weeds 1986*, pp. 99–104.

Dutartre, A. and Gross, F. (1982). Evolution des végétaux aquatiques dans les cours d'eaux recalibrés (examples pris dans le sud-ouest de la France). In *Studies on aquatic vascular plants* (ed. J. J. Symoens, S. S. Hooper, and P. Compère), pp. 394–7. Royal Botanical Society of Belgium, Brussels.

Dvořak, J. (1970). Horizontal zonation of macrovegetation, water properties and macrofauna in a littoral stand of *Glyceria aquatica* in a South Bohemian pond. *Hydrobiologia*, **35**, 17–20.

DVWK (1984). *Okologische Aspekte bei Ausbau und Unterhaltung von Fliessgewässern.* Merkblätter zur Wasserwirtschaft H204, Deutscher Verband für Wasserwirtschaft und Kulturban, Paul Parey, Hamburg and Berlin, pp. 188.

DVWK (1987). *Erfahrungen bei Ausbau und Unterhaltung von Fliessgewässern I. Verfahren und Kosten bei der naturnahen Gestaltung und Unterhaltung. II. Auswirkungen von Massnahmen der Gewässerunterhaltung auf Gewässerlebensgemeinschaften.* Schriftenreihe des Deutschen Verbandes für Wasserwirtschaft und Kulturbau, H79. Paul Parey, Hamburg and Berlin, pp. 276.

Dykyjova, D., Veber, D., and Priban, K. (1967). Photosynthetic production and growth of reedswamp macrophytes under controlled conditions of mineral nutrition. *Annual Report Laboratory of Experimental Algology and Department of Applied Algology*, pp. 153–9. Czechoslovak Academy of Science, Trebon, Czechoslovakia.

Easley, J. F. and Shirley, R. L. (1974). Nutrient elements for livestock in aquatic plants. *Hyacinth Control Journal*, **12**, 82–5.

Eaton, J. W. and Freeman, J. (1982). Ten years experience of weed control in the Leeds and Liverpool Canal. *Proceedings EWRS 6th Symposium on Aquatic Weeds 1982*, pp. 96–104.

Eaton, J. W., Murphy, K. J., and Hyde, T. M. (1981). Comparative trials of herbicidal and mechanical control of aquatic weeds in canals. *Proceedings Association of Applied Biologists Conference: Aquatic Weeds and their Control 1981*, pp. 105–16. AAB, Wellesbourne, UK.

Edie, H. and Ho, B. (1969). *Ipomoea aquatica* as a vegetable crop in Hong Kong. *Economic Botany*, **23**, 32–6.

Edwards, D. and Thomas, P. A. (1977). The *Salvinia molesta* problem in the northern Botswana and eastern Caprivi area. *Proceedings 2nd National Weeds Conference of South Africa, Stellenbosch 1977*, pp. 221–37. A. A. Balkema, Cape Town.

Edwards, D. J. and Hine, P. M. (1974). Introduction, preliminary handling and diseases of grass carp in New Zealand. *New Zealand Journal of Marine and Freshwater Research*, **8**, 441–54.

Edwards, D. J. and Moore, E. (1975). Control of water weeds by grass carp in a drainage ditch in New Zealand. *New Zealand Journal of Marine and Freshwater Research,* **9**, 283–92.

Edwards, G. E. and Hueber, S. C. (1981). The C_4 pathway. In *Photosynthesis,* Vol. 8, (ed. M. D. Hatch and N. K. Boardman), pp. 238–78. Academic Press, New York.

Edwards, R. W. and Brown, M. W. (1960). Aerial photographic method for studying distribution of aquatic macrophytes in shallow waters. *Journal of Ecology,* **48**, 161–70.

Edwards, R. W. and Owens, M. (1962). The effects of plants on river conditions. IV. The oxygen balance of a chalk stream. *Journal of Ecology*, **50**, 207–20.

Edwards, R. W., Owens, M., and Gibbs, J. W. (1961). Estimates of surface aeration in two streams. *Journal of the Institute of Water Engineers,* **15**, 395–405.

Eelman, W. and van der Ploeg, D. T. E. (1979). *Potamogeton coloratus* Hornem. opnieuw in Nederland gevonden. *Gorteria*, **9**, 325–30.

Eighmy, T. T., Jahnke, L. S., and Fagerberg, W. R. (1987). Evidence for bicarbonate active transport in *Elodea nuttallii.* In *Progress in photosynthesis research,* Vol. 4 (ed. J. Biggens), pp. 335–56. Martinus Nijhoff Publishers, the Netherlands.

Elamson, R. (1977). Clarosan IG for control of submerged vegetation. *Weeds & Weed Control: 18th Swedish Weed Conference, Uppsala,* Part 1, pp. 1–3.

El Gharably, Z., Khattab, A. F., and Dubbers, F. A. A. (1982). Experience with grass carps for the control of aquatic weeds in irrigation canals in Egypt. *Proceedings 2nd International Symposium on Herbivorous Fish 1982,* pp. 17–26. EWRS, Wageningen, The Netherlands.

El Gharably, Z., Tolba, A., Pieterse, A. H., and Druijff, A. H. (1978). Preliminary experiments with grass carp for the control of aquatic weeds in Egypt. *Proceedings EWRS 5th Symposium on Aquatic Weeds 1978,* pp. 369–73.

El Gindy, H. I. (1962). Ecology of snail vectors in bilharziasis. *Proceedings Ist International Symposium on Bilharziasis,* pp. 305–18. Cairo Government Printers.

El Hadidi, M. N. (1965). *Potamogeton trichoides* Cham. & Schlecht in Egypt. *Candollea,* **20**, 159–66.

El Hadidi, M. N. (1968). *Vallisneria spiralis* L. in Egypt. *Candollea,* **23/1**, 51–8.

El Hadidi, M. N. (1976). The riverain flora in Nubia. In *The Nile, biology of an ancient river,* (ed. J. Rzoska), pp. 87–91. W. Junk, The Hague.

Ellis, P. A. and Camper, N. D. (1982). Aerobic degradation of diuron by aquatic micro-organisms. *Journal of Environmental Science Health,* **17**, 277-89.

Elliston, R. A. and Steward, K. K. (1972). The response of Eurasian watermilfoil to various concentrations and exposure periods of 2,4-D. *Hyacinth Control Journal,* **10**, 38–40.

El Sammani, M. O. (1984). *Jonglei Canal: dynamics of planned change in the Twic area.* Monograph Series, Development Studies and Research Centre, University of Khartoum (Sudan), No. 8. pp. 191.

El Sayed, J. K. A., Tolba, A., and Druijff, A. H. (1978). Evaluation of some machines for mechanical control of aquatic weeds in Egypt. *Proceedings EWRS 5th Symposium on Aquatic Weeds, 1978,* pp. 359–67.

El Tigani, K. B. (1979). Water hyacinth control: organization, strategy and cost of large-scale control operations. *Weed Research in the Sudan, Volume 1: Proceedings of a*

Symposium, (ed. M. E. Beshir and W. Koch), pp. 123–9. Berichte aus dem Fachgebeit Herbologie, Universität Hohenheim.

Emerson, S. and Broecker, W. (1973). Gas-exchange rates in a small lake as determined by the radon method. *Journal of the Fisheries Research Board, Canada*, **30**, 1475–84.

Emiliani, F. and Rodríguez, R. C. (1974). Control de floraciones algales (I nota). *Revista Asociacion Ciencias Naturales Litoral*, **5**, 99–126.

Engel, S. (1984). Evaluating stationary blankets and removable screens for macrophyte control in lakes. *Journal of Aquatic Plant Management*, **22**, 43–8.

Engel, S. and Nichols, S. (1984). Lake sediment alteration for macrophyte control. *Journal of Aquatic Plant Management*, **22**, 38–41.

Engelhardt, D. (1974). Die rechtliche Situation beim Einsatz von Herbiziden zur Krautbekämpfung an und in Gewässern in der BRD. *Proceedings EWRC 4th Symposium Aquatic Weeds, 1974*, pp. 260–6.

Entz, B.A.G. (1976). Lake Nasser and Lake Nubia. In *The Nile, biology of an ancient river* (ed. J. Rzoska), pp. 271–91. W. Junk, The Hague.

Entz, B.A.G. (1980). Ecological aspects of Lake Nasser–Nubia. The first decade of its existence, with special reference to the development of insect populations and the land and water vegetation. In *The Nile and its environment* (ed. M. Khassass and S. I. Ghabbour). *Water Supply Management*. **4 (1–2)**, 67–72.

Erixon, G. (1979). Environment and aquatic vegetation of a riverside lagoon in northern Sweden. *Aquatic Botany*, **6**, 95–110.

Esler, H. J. (1966). Control of water chestnuts by machine in Maryland 1964–5. *Proceedings Northeast Weed Control Conference*, **20**, 682–7.

Etheridge, K., Rathbun, G. B., Powell, J. A., and Kochman, H. I. (1985). Consumption of aquatic plants by the West Indian manatees. *Journal of Aquatic Plant Management*, **23**, 21–5.

European Weed Research Council (1964). *Proceedings of the EWRC 1st Symposium on Aquatic Weeds, La Rochelle, France*. EWRC, Wageningen, The Netherlands.

European Weed Research Council (1967). *Proceedings of the EWRC 2nd Symposium on Aquatic Weeds, Oldenburg, West Germany*. pp. 197. EWRC, Wageningen, the Netherlands.

European Weed Research Council (1971). *Proceedings of the EWRC 3rd Symposium on Aquatic Weeds, Oxford, UK*. pp. 318. EWRC, Wageningen, the Netherlands.

European Weed Research Council (1974). *Proceedings of the EWRC 4th Symposium on Aquatic Weeds, Vienna, Austria*. pp. 278 EWRC, Wageningen, the Netherlands.

European Weed Research Society (1978). *Proceedings of the EWRS 5th Symposium on Aquatic Weeds, Amsterdam, the Netherlands*. pp. 427. EWRS, Wageningen, the Netherlands.

European Weed Research Society (1982). *Proceedings of the EWRS 6th Symposium on Aquatic Weeds, Novi Sad, Yugoslavia*. pp 305. EWRS, Wageningen, the Netherlands.

European Weed Research Society (1986). *Proceedings of the EWRS/AAB 7th Symposium on Aquatic Weeds, Loughborough, UK*. pp. 427. EWRS, Wageningen, the Netherlands.

Evans, D. M. (1978). Aquatic weed control with the isopropylamine salt of *N*-phosphonomethyl glycine. *Proceedings EWRS 5th Symposium on Aquatic Weeds, 1978*, pp. 171–8.

Evans, D. M. (1982). *Phragmites* control with glyphosate through selective equipment. *Proceedings EWRS 6th Symposium on Aquatic Weeds, 1982,* pp. 209–11.

Evans, W. F. and Gallagher, J. E. (1969). Performance of fenac for aquatic weed control. *Proceedings North Central Weed Control Conference 24,* pp. 85.

Ewel, K. C., Braat, L., and Stevens, M. L. (1975). Use of models for evaluating aquatic weed control strategies. *Hyacinth Control Journal,* **13**, 34–9.

Ewel, K. C. and Fontaine, T. D. (1982). Effect of white amur (*Ctenopharyngodon idella*) on a Florida lake: a model. *Ecological Modelling,* **16**, 251–73.

Ewel, K. C. and Fontaine, T. D. (1983). Structure and function of a warm monomictic lake. *Ecological Modelling,* **19**, 139–61.

Faculteit der Wiskunde en Natuurwetenschappen (1983). *Proceedings of the International Symposium on Aquatic Macrophytes, Nijmegen, 18–23 September 1983,* pp. 326. University of Nijmegen, The Netherlands.

Faegri, K. (1982). The *Myriophyllum spicatum* group in North Europe. *Taxon,* **31**, 467–71.

Fair, P. and Meeke, L. (1983). Seasonal variations in the pattern of photosynthesis and possible adaptive response to varying light flux regimes in *Ceratophyllum demersum* L. *Aquatic Botany,* **15**, 81–90.

Falter, C. M., Leonard, J., Naskali, R., Rabe, F. O., and Bobisud, H. (1974). *Aquatic macrophytes of the Columbia and Snake River drainages* (*United States*). Report: US Army Corps of Engineers, No. OACW 68–72–C–0269. pp. 275.

Fanst, S. D. and Zarins, A. (1969). Interactions of diquat and paraquat with clay minerals and carbon in aqueous solutions. *Residue Reviews,* **29**, 151–70.

FAO (1968). Copper sulfate – mud pellets for weed control. *Food & Agriculture Organization Fish Culture Bulletin,* **1**, 6.

Fasken, G. B. (1963). *Guide to selecting roughness coefficient 'n' values for channels.* US Department Agriculture, Soil Conservation Service. pp. 33.

Fayed, A. A. (1985). The distribution of *Myriophyllum spicatum* L. in the inland waters of Egypt. *Folia geobotanica et Phytotoxonomica,* **20**, 197–9.

Feichtinger, F. (1974). Mechanische Grabenerhaltung in Österreich. *Proceedings EWRS 4th Symposium on Aquatic Weeds 1974,* pp. 84–6.

Ferguson, F. F. (1971). Some current controls for aquatic weeds inimical to public health. *East African Medical Journal,* **48**, 456–9.

Ferguson, F. F. (1980). Aquatic weeds and man's well-being. In *Weed control methods for public health applications,* pp. 3–15. CRC Press, Boca Raton, Florida.

Fernandes, J. D., Guerreiro, A. R., Vasconcelos, T., and Moreira, I. (1978). Essais de lutte contre les plantes aquatiques au Portugal. *Proceedings EWRS 5th Symposium on Aquatic Weeds 1978,* pp. 189–94.

Fernández, O. A., Irigoyen, J. H., Sabbatini, M. R., and Brevedan, R. E. (1987*a*). Aquatic plant management in drainage canals of southern Argentina. *Journal of Aquatic Plant Management,* **25**, 65–7.

Fernández, O.A., Irigoyen, J. H., Sabbatini, M. R., and Svachka, O. (1987*b*). Recomendaciones para el control de *Potamogeton striatus* y *Chara contraria* en distritos de riego. *ASAM* (*Asociacione Argentina Control Malezas*) *Malezas,* **15**, (in press).

Figueiredo, J., Duarte, C., Moreira, I., and Agusti, S. (1984). *As infestantes aquaticas nos sistemas de irrigaçao e drenagem do Ribatejo*. Paper H-3/84, Centro de Botanico Aplicada a Agricultura da Universitie Tecnica de Lisboa, pp. 11.

Finlayson, C. M. (1980). Aspects of the hydrobiology of the Lake Moondarra–Leichhardt River water supply system, Mount Isa, Queensland. Dissertation, James Cook University of North Queensland. pp. 448.

Finlayson, C. M., Farrell, T. P., and Griffiths, D. J. (1984). Studies of the hydrobiology of a tropical lake in Northwestern Queensland. III. Growth, chemical composition and potential for harvesting of the aquatic vegetation. *Australian Journal of Marine and Freshwater Research,* **35**, 525–36.

Finlayson, C. M. and Mitchell, D. S. (1982/83). Management of salvinia (*Salvinia molesta*) in Australia. *Australian Weeds,* **2 (2)**, 71–6.

Finlayson, C. M., Roberts, J., Chick, A. J., and Sale, P. J. M. (1983). The biology of Australian weeds. II. *Typha domingensis* Pers. and *Typha orientalis* Presl. *Journal of the Australian Institute Agricultural Science,* **49 (1)**, 3–10.

Fischer, Z. and Lyakhnovich, V. P. (1973). Biology and bioenergetics of grass carp (*Ctenopharyngodon idella* Val.). *Polskie Archiwum Hydrobiologie,* **20**, 521–57.

Fish, G. R. (1966). Some effects of the destruction of aquatic weeds in Lake Rotoiti, New Zealand. *Weed Research,* **6**, 350–8.

Fjerdingstad, E., Fjerdingstad, E., and Kemp, K. (1979). Further studies in Danish *Lobelia-Isoetes* lakes. *Archiv für Hydrobiologie,* **85**, 72–97.

Flock, H., Klug, K., and Canvin, D. T. (1979). Effect of carbon dioxide and temperature on photosynthetic CO_2 uptake and photorespiratory CO_2 evolution in sunflower leaves. *Planta,* **145**, 219–23.

Florida Aquaculture Review Council (1981). *Florida Aquaculture Plan*. Florida Department of Agriculture and Consumer Services. pp. 83.

Folmar, L. C. (1976). Overt avoidance reaction of rainbow trout fry in nine herbicides. *Bulletin of Environmental Contamination Toxicology,* **15**, 509–12.

Folmar, L. C. (1978). Avoidance chamber responses of mayfly nymphs exposed to eight herbicides. *Bulletin of Environmental Contamination Toxicology,* **19**, 312–18.

Folmar, L. C., Sanders, H. O., and Julin, A. M. (1979). Toxicity of the herbicide glyphosate and several of its formulations to fish and aquatic invertebrates. *Archives of Environmental Contamination Toxicology,* **8**, 269–78.

Foret, J. A. (1974). An integrated program for alligator weed control in rice irrigation canals. *Proceedings Southern Weed Science Society 1974,* pp. 282.

Foret, J. A., Spencer, M. R., and Gangstad, E. O. (1974). *Towards integrated control of alligator weed. Aquatic use patterns for 2,4-D dimethylamine and integrated control.* Aquatic Plant Control Program TR. 7, pp. H3–H16. US Army Engineer Waterways Experiment Station, Vicksburg, Mississippi.

Forno, I. W. (1983). Native distribution of the *Salvinia auriculata* complex and keys to species identification. *Aquatic Botany,* **17**, 71–83.

Forno, I. W. (1985). How quickly can insects control salvinia in the tropics? *Proceedings 10th Asian–Pacific Weed Science Society Conference, Chaing-mai, Thailand 1985,* pp. 271–6.

Forno, I. W. (1987). Biological control of the floating fern *Salvinia molesta* in north-eastern Australia: plant herbivore interactions. *Bulletin of Entomological Research*, **77**, 9–17.

Forno, I. W. and Bourne, A. S. (1984). Studies in South America of arthropods of the *Salvinia auriculata* species complex of floating ferns and their effect on *S. molesta*. *Bulletin Entomological Research*, **74**, 609–21.

Forno, I. W. and Harley, K. L. S. (1979). The occurrence of *Salvinia molesta* in Brazil. *Aquatic Botany*, **6**, 185–7.

Forno, I. W. and Wright, A. D. (1981). The biology of Australian weeds. 5. *Eichhornia crassipes* (Mart.) Solms. *Journal of the Australian Institute of Agricultural Science*, **47**, 21–8.

Forsberg, C. (1965*a*). Sterile germination of oospores of *Chara* and seeds of *Najas maxima*. *Physiologia Plantarum*, **18**, 128–37.

Forsberg, C. (1965*b*). Environmental conditions of Swedish charophytes. *Symbolae Botanicae Uppsalienses*, **18**, 1–67.

Forsberg, C. and Ryding, S. (1980). Eutrophication parameters and trophic state indices in 30 Swedish waste-receiving lakes. *Archiv für Hydrobiologie*, **80**, 189–207.

Foster, R. C. (1965). Studies in the flora of Bolivia. III. Cyperaceae. Part 1. *Rhodora*, **67**, 97–139.

Fowler, M. C. (1977). Laboratory trials of a new triazine herbicide (DPX 3674) on various species of macrophytes and algae. *Weed Research*, **17**, 191–5.

Fowler, M. C. (1982). Experiments on food conversion ratios and growth rates of grass carp (*Ctenopharyngodon idella* Val.) in England. *Proceedings 2nd International Symposium on Herbivorous Fish 1982*, pp. 107–10. EWRS, Wageningen, The Netherlands.

Fowler, M. C. (1985*a*). The present state of grass carp (*Ctenopharyngodon idella* Val.) for the control of aquatic weeds in England and Wales. *Proceedings VIth International Symposium on Biological Control of Weeds, Vancouver*, pp. 537–42.

Fowler, M. C. (1985*b*). The results of introducing grass carp, *Ctenopharyngodon idella* Val., into small lakes. *Aquaculture and Fisheries Management*, **16**, 189–201.

Fowler, M. C. (1986). *Survey of the problems caused by filamentous algae in freshwater in Great Britain*. Report: AFRC Long Ashton Research Station, Weed Research Division, Aquatic Weeds Group, Oxford. pp. 11.

Fowler, M. C. and Barrett, P. R. F. (1986). Preliminary studies on the potential of hydrogen peroxide as an algicide on filamentous species. *Proceedings EWRS/AAB 7th Symposium on Aquatic Weeds 1986*, pp. 113–18.

Fowler, M. C. and Robson, T. D. (1974). Studies on the volatilization of dichlobenil from water. *Proceedings EWRC 4th Symposium on Aquatic Weeds 1974*, pp. 180–4.

Fowler, M. C. and Robson, T. O. (1978). The effects of the food preferences and stocking rates of grass carp (*Ctenopharyngodon idella* Val.) on mixed plant communities. *Aquatic Botany*, **5**, 261–76.

Fox, A. M. (1987). The efficacy and ecological impact of the management of submerged vegetation in flowing water. Ph.D thesis, University of Glasgow. pp. 373.

Fox, A. M., Murphy, K. J., and Westlake, D. F. (1986). Effects of diquat alginate and cutting on the submerged macrophyte community of a *Ranunculus* stream in Northern England. *Proceedings EWRS/AAB 7th Symposium on Aquatic Weeds 1986*, pp. 105–12.

Foy, C. L. (1969). The chlorinated aliphatic acids. In *Degradation of herbicides,* (ed. P. C. Kearney and D. D. Kaufman), pp. 207–53. Marcel Dekker, New York.

Frake, A. (1979). The Hampshire Avon Survey. *Proceedings 1st British Freshwater Fisheries Conference 1979,* pp. 100–5. University of Liverpool, Liverpool, UK.

Franceschi, E. A., Prado, D. E., and Lewis, J. P. (1985). Communidades herbáceas y otros. In *Communidades vegetales y mapas de vegetación. Reserva 'El Rico' e Islas aledañas,* pp. 13–22. Prov. Santa Fé (Argentina) Servicio Publico Universidad Nacional Rosario.

Frank, P. A. (1972). Herbicidal residues in aquatic environments. In *Fate of organic pesticides in the aquatic environment: Advances in Chemistry Series,* **111**, pp. 135–48.

Frank, P. A. and Comes, R. D. (1967). Herbicidal residues in pond water and hydrosoil. *Weeds,* **15**, 210–13.

Frank, P. A. and Dechoretz, N. (1980). Allelopathy in dwarf spikerush (*Eleocharis coloradoensis*). *Weed Science*, **28**, 499–505.

Frankland, B., Bartley, M. R., and Spence, D. H. N. (1987). Germination under water. In *Plant life in aquatic and amphibious habitats,* Special Publication British Ecological Society, No 5, (ed. R. M. M. Crawford), pp. 167–77. Blackwell Scientific Publications, Oxford.

Freeman, T. E. (1977). Biological control of aquatic weeds using plant pathogens. *Aquatic Botany,* **3**, 175–84.

Freeman, T. E. and Charudattan, R. (1974). *Cercospora piaropi* on water hyacinth in Florida. *Plant Disease Reports,* **58**, 277–8.

Freeman, T. E. and Charudattan, R. (1984). *Occurrence of* Cercospora rodmanii *Conway, a biocontrol agent for water hyacinth.* Technical Bulletin 842, Florida Agricultural Experimental Station, University of Florida, Gainesville.

Freeman, T. E., Charudattan, R., Cullen, R. E., and Addor, E. E. (1982). *Rhizoctonia* blight on water hyacinth in the United States. *Plant Disease,* **66**, 861–2.

Freidel, J. W. (1978). Populationsdynamik der Wasserhyazinthe (*Eichhornia crassipes* (Mart.) Solms) under besonderer Berücksichtigung der Sudanesischen Befallsgebietes. Ph.D. thesis, Universität Hohenheim. pp. 145.

Freidel, J. W. and Beshir, M. E. (1979). On the dynamics of populations and distribution of water hyacinth in the White Nile, Sudan. *Weed research in the Sudan: Volume 1. Proceedings of a Symposium,* (ed. M. E. Beshir and W. Koch). pp. 94–105. Berichte aus dem Fachgebiet Herbologie, Universität Hohenheim.

Fricke, G. and Steubing, L. (1984). [Distribution of macrophytes in hard water of the Eder reservoir influents]. In German; English summary. *Archiv für Hydrobiologie,* **107**, 361–72.

Fritz-Sheridan, R. P. (1982). Impact of the herbicide Magnacide-H (2-propenal) on algae. *Bulletin Environmental Contamination Toxicology,* **28**, 245–9.

Fry, J. C., Brooker, M. P., and Thomas, P. L. (1973) Changes in the microbial population of a reservoir treated with the herbicide paraquat. *Water Research,* **7**, 395–407.

Fry, J. G. and Humphrey, N. C. B. (1978). The effect of paraquat-induced death of aquatic plants on the heterotrophic activity of freshwater bacteria. *Proceedings 1978 British Crop Protection Conference—Weeds,* **2**, pp. 595–602.

Funderburk, H. H. (1969). Diquat and paraquat. In *Degradation of herbicides* (ed. P. C. Kearney and D. D. Kaufman), pp. 283–98. Marcel Dekker, New York.

Funderburk, H. H. and Lawrence, J. M. (1963). Absorption and translocation of radioactive herbicides in submerged and emerged aquatic weeds. *Weed Research* 3, 304–11.

Gaddis, C. W. and Kissel, C. L. (1982). Acrolein in irrigation waterways. *Proceedings EWRS 6th Symposium on Aquatic Weeds 1982*, pp. 164–70.

Galbraith, J. C. and Hayward, A. C. (1984). *The potential of indigenous micro-organisms in the biological control of water hyacinth in Australia*. Final Report, Australian Water Resource Council, Research Project No. 80/132. Department of Resources & Energy, Canberra.

Gambrell, R. P., Taylor, B. A., Reddy, K. S., and Patrick, W. H. (1984). *Fate of selected toxic compounds under controlled redox potential and pH conditions in soil and sediment–water systems*. Project Summary, Environmental Protection Agency, USA EPA-600/53-84-018. pp. 4.

Gamerro, J. C. (1968). Observaciones sobre la biologiá floral y morfologiá de la Potamogetonacea *Ruppia cirrhosa* (Petag.) Grande. *Darwiniana*, **14**, 575–608.

Gangstad, E. O. (1971). Aquatic plant control program. *Hyacinth Control Journal*, **9**, 46–8.

Gangstad, E. O. (1975a). *Integrated control of alligator weed and water hyacinth in Texas*. Aquatic Plant Control Program. Report TR 9, pp. 1–9. US Army Engineer Waterways Experiment Station, Vicksburg, Mississippi.

Gangstad, E. O. (1975*b*). *Integrated program for alligator weed management*. Aquatic Plant Control Program. Report TR 10, pp. 1–8. US Army Engineer Waterways Experiment Station, Vicksburg, Mississippi.

Gangstad, E. O. (1976). Biological control operations in alligator weed. *Journal of Aquatic Plant Management*, **14**, 50–3.

Gangstad, E. O. (1977*a*). *Research and development studies on the use of the butoxyethanol ester of 2,4-D granular herbicide (2,4-D BEE) for control of Eurasian water milfoil*, M. spicatum. Report, Office of Chief of Engineers (Army), Washington DC. pp. 162.

Gangstad, E. O. (1977*b*). Aquatic weed problems in Puerto Rico. *Journal of Aquatic Plant Management*, **15**, 3–5.

Gangstad, E. O. (1980). Aquatic weed problems in the Panama Canal. In *Weed control methods for public health applications* (ed. E. O. Gangstad), pp. 27–36. CRC Press, Boca Raton, Florida.

Gangstad, E. O. (1981). The role of silvex herbicide residues in the environment. *Proceedings Slow Release Society*, pp. 3–5.

Gangstad, E. O. (1982*a*). Benefit/cost analysis of silvex cancellation. *Journal of Aquatic Plant Management*, **20**, 45–9.

Gangstad, E. O. (1982*b*). Dissipation of 2,4-D residues in large reservoirs. *Journal of Aquatic Plant Management*, **20**, 13–16.

Gangstad, E. O. (1983). Benefit/risk analysis of silvex cancellation. *Journal of Aquatic Plant Management*, **21**, 65–9.

Gangstad, E. O. (1984). Aquatic use pattern for silvex cancellation. *Journal of Aquatic Plant Management*, **22**, 78–80.

Gangstad, E. O. (1986). *Freshwater vegetation management*. Thomas Publications, Fresno, California.

Garg, S. P. (1968). *Studies on aquatic weed control in U.P.* Technical Memorandum 39-RR(B1). Uttar Pradesh Irrigation Research Institute, Roorkee. pp. 32.

Garrard, L. A. and Van, T. K. (1982). General characteristics of freshwater vascular plants. In *CRC Handbook of biosolar resources 1, Part 2, Basic principles* (ed. A. Mitsui and C. C. Black), pp. 77–85. CRC Press, Boca Raton, Florida.

Gass, R. F., Deesin, T., Surathin, K., Vutiles, S., Sucharit, S., and Harinsuta, C. (1983). Studies on oviposition characteristics of *Mansonia mansonioides* mosquitoes in southern Thailand. *Annals of Tropical Medicine Parasitology,* 77, (6), 605–14.

Gast, A., Grob, H., and Fankhauser, E. (1965). New selective triazines in cereals. *Proceedings EWRC 2nd Symposium on New Herbicides,* pp. 297–303.

Gast, A., Knüsli, E., and Gysin, H. (1956). Uber Pflanzenwachstumstegulatoren über weitere phytotoxische Triazine. *Experientia,* **12**, 146–8.

Gates, R. J. (1974). Submerged control of aquatics by use of the Bifluid-invert system. *Proceedings 27th Annual Meeting Southern Weed Science Society,* 310–13.

Gaudet, J. J. (1973). Growth of a floating aquatic weed, *Salvinia*, under standard conditions. *Hydrobiologia,* **41**, 77–106.

Gaudet, J. J. (1975). Mineral concentration in papyrus in various African swamps. *Journal of Ecology,* **63**, 483–91.

Gaudet, J. J. (1976*a*). Nutrient relationships in the detritus of a tropical swamp. *Archiv für Hydrobiologie*, **78**, 213–39.

Gaudet, J. J. (1976*b*). *Salvinia* infestation on Lake Naivasha in East Africa (Kenya). In *Aquatic weeds in south east Asia* (ed. C. K. Varshney and J. Rzoska), pp. 193–209. W. Junk, The Hague.

Gaudet, J. J. (1977). Uptake, accumulation and loss of nutrients by papyrus in tropical swamps. *Ecology,* **58**, 415–22.

Gaudet, J. J. (1979*a*). Seasonal changes in nutrients in a tropical swamp: North Swamp, Lake Naivasha, Kenya. *Journal of Ecology*, **67**, 953–81.

Gaudet, J. J. (1979*b*). Weeds in African man-made lakes. *Pest Articles and News Summaries*, **25**, 279–86.

Gay, P. A. (1958). *Eichhornia crassipes* in the Nile of the Sudan. *Nature* (*London*), **182**, 538–9.

Gay, P. A. (1960). Ecological studies of *Eichhornia crassipes* Solms. in the Sudan. 1. Analysis of spread in the Nile. *Journal of Ecology*, **48**, 183–91.

Gay, P. A. and Berry, L. (1959). The water hyacinth: a new problem on the Nile. *Geographical Journal*, **125**, 89–91.

Gehu, J. M. (ed.) (1983). *Les vegetations aquatiques et amphibies.* Colloques phytosociologiques, No. 10, Lille 1981. pp. 520.

Gentile, J. H. (1971). Blue-green and green algal toxins. In *Microbial toxins* (ed, S. J. Ajl), p. 27. Academic Press, New York.

George, M. (1976). Mechanical methods of weed control in watercourses—an ecologist's view. *Proceedings Symposium on Aquatic Herbicides, British Crop Protection Council, Monograph No. 16*, pp. 91–9.

George, T. T. (1982). The Chinese grass carp, *Ctenopharyngodon idella,* its biology, introduction and control of aquatic macrophytes, and breeding in the Sudan. *Aquaculture*, **27**, 317–27.

Gerbeaux, P., and Ward, J. (1986). The disappearance of macrophytes and its importance in the management of shallow lakes in New Zealand. *Proceedings EWRS/AAB 7th Symposium on Aquatic Weeds 1986*, pp. 119–24.

Gerking, S. D. (1948). Destruction of aquatic plants by 2,4-D. *Journal of Wildlife Management*, **12**, 221–7.

Gerritsen, A. (1979). Van vaarweg tot recreatievooziening, het Apeldoorns Kanaal, meerdere functies, meervoudig gebruik. *Recreatie*, **17 (2)**, 27–39.

Gessner, F. (1955). *Hydrobotanik I. Energiehaushalt*. Hochschulbücher für Biologie 3. VEB Deutscher Verlag der Wissenschaften. Berlin. pp. 517.

Gessner, F. (1959). *Hydrobotanik II. Stoffhaushalt*. Hochschulbücher für Biologie B, VEB Deutscher Verlag der Wissenschaften. Berlin. pp. 701.

Getsinger, K., Davis, G. J., and Brinson, M. M. (1982). Changes in a *Myriophyllum spicatum* L. community following 2,4-D treatment. *Journal of Aquatic Plant Management*, **20**, 4–8.

Getsinger, K. D. and Dillon, C. R. (1984). Quiescence, growth and senescence of *Egeria densa* in Lake Marion. *Aquatic Botany*, **20**, 329–38.

Getsinger, K. D., and Westerdahl, H. E. (1984). *Field evaluation of Garlon 3A (triclopyr) and 14-ACE-B (2,4-D BEE) for the control of Eurasian watermilfoil.* Aquatic Plant Control Research Program, Miscellaneous Paper A-84-5. US Army Engineer Waterways Experiment Station, Vicksburg, Mississippi. pp. 12.

Gewertz, D. B. (1983). *Sepik River Societies.* Yale University Press, New Haven, USA.

Ghani, M. A. (1965). *Biologies and host plant ranges of insects that attack noxious weeds common to Pakistan and United States.* Annual Report Commonwealth Institute of Biological Control, London. pp. 30.

Gholson, A. K. (1982). Aquatic Weeds in Lake Seminole. In *Weed control methods for recreation facilities management*, pp. 175–86. CRC Press, Boca Raton, Florida.

Gibbons, E. J. (1975). *The flora of Lincolnshire.* Lincolnshire Naturalists' Union, Lincoln, UK.

Gilderhus, P. A. (1966). Some effects of sublethal concentrations of sodium arsenite on bluegills and the aquatic environment. *Transactions American Fish Society*, **95**, 289–96.

Gilderhus, P. A. (1967). Effects of diquat on bluegills and their food organisms. *Progressive Fish Culturist*, **29**, 67–74.

Giulizzoni, P., Bonomi, G., Galanti, G., and Ruggiu, D. (1984). Basic trophic status and recent development of some Italian lakes as revealed by plant pigments and other chemical components in sediment cores. *Memorie dell' Istituto Italiano di Idrobiologie*, **40**, 79–98.

Gloser, J. (1977). Characteristics of CO_2 exchange in *Phragmites communis* Trin. derived from measurements *in situ*. *Photosynthetica*, **7**, 139–47.

Gloser, J. (1978). Net photosynthesis and dark respiration of reed estimated by gas-exchange measurements. In *Pond littoral ecosystems* (ed. D. Dykyjova and J. Květ), pp. 227–34. Springer-Verlag, New York.

Godfrey, R. K. and Wooten, J. W.. (1981). *Aquatic and wetland plants of the south-eastern United States: Dicotyledons.* University of Georgia Press, Athens, Georgia. pp. 933.

Godwin, H. (1923). Dispersal of pond floras. *Journal of Ecology*, **11**, 160–4.

Goldman, C. R. (1972). The role of minor nutrients in limiting the productivity of aquatic ecosystems. In *Nutrients and eutrophication: the limiting nutrient controversy* (ed. E. Likens) *Special Symposium of the American Society for Limnology and Oceanography*, **1**, 21–38.

Goldsby, T. L., Bates, A. L., and Stanley, R. A. (1978). Effect of water level fluctuation and herbicide on Eurasian watermilfoil in Melton Hill Reservoir. *Journal of Aquatic Plant Management*, **16**, 34–8.

Goldsby, T. L. and Sanders, D. R. (1977). Effects of consecutive water fluctuations on the submerged vegetation of Black Lake, Louisana. *Journal of Aquatic Plant Management*, **15**, 23–8.

Gomez, L. D. (1984). *Las plantas acuaticas y anfibas de Costa Rica y Centroamerica. 1.* Liliopsida. Editorial Universal Estatal a Distancia, San Jose, Costa Rica. pp. 430.

Good, R. E., Whigham, D. F., and Simpson, R. L. (1978). *Freshwater methods.* Academic Press, London. pp. 378.

Gopal, B. (1982). Biological and ecological considerations in aquatic weed management. *Proceedings EWRS 6th Symposium on Aquatic Weeds 1982*, pp. 152–8.

Gopal, B. (1984). Utilization of water hyacinth as a new resource or for its control? Some environmental considerations. In *Proceedings International Conference on Water Hyacinth* (ed. G. Thyagarajan), pp. 193–211. UNEP, Nairobi.

Gopal, B. (1987). *Water hyacinth.* Aquatic Plant Studies 1. Elsevier, Amsterdam. pp. 471.

Gopal, B. and Sharma, K. P. (1979). Aquatic weed control versus utilization. *Economic Botany*, **30**, 340–6.

Gopal, B. and Sharma, K. P. (1981). *Water hyacinth* (Eichhornia crassipes): *the most troublesome weed of the world.* Hindosia Publishers, Delhi, India. pp: 128.

Gopal, B., Trivedy, R. K., and Goel, P. K. (1984). Influence of water hyacinth cover on the physico-chemical characteristics of water and phytoplankton in a reservoir near Jaipur (India). *Internationale Revue der Gesamten Hydrobiologie*, **69**, 859–65.

Gophen, H. (1982). Unusually dense watermilfoil (*Myriophyllum spicatum L.*) vegetation in the southern basin of Lake Kinneret (Israel) in 1979. *Aquatic Botany*, **13**, 307–15.

Gorbach, E. I. (1961). Age composition, growth and age of onset sexual maturity of the white, *Ctenopharyngodon idella*, and black, *Mylopharyngodon piceus*, amurs in the Amur River basin. *Journal of Ichthyology*, **1**, 119–26.

Gorbach, E. I. (1972). Fecundity of the grass carp (*Ctenopharyngodon idella* Val.) in the Amur basin. *Journal of Ichthyology*, **12**, 616–25.

Gorden, R. W., Waite, S. W., Tazik, P., and Wiley, M. J. (1982). The effects of simazine and endothall in aquatic macrophyte control. *Proceedings EWRS 6th Symposium on Aquatic Weeds*, pp. 287–96.

Gossett, D. R. and Norriss, W. E. (1971). Relationship between nutrient availability and content of nitrogen and phosphorus in tissues of the aquatic macrophyte, *Eichhornia crassipes* (Mart.) Solms. *Hydrobiologia*, **38**, 15–28.

Goulder, R. (1969). Interactions between the rates of production of a freshwater macrophyte and phytoplankton in a pond. *Oikos*, **20**, 300–9.

Goulding, R. (1981). Toxicity assessment for the acceptance of aquatic herbicides. *Proceedings Association of Applied Biologists Conference: Aquatic Weeds and their Control 1981*, pp. 287–93. AAB, Wellesbourne, UK.

Grace, J. B. (1983). Autotoxic inhibition of seed germination by *Typha latifolia*: an evaluation. *Oecologia*, **59**, 366–9.

Grace, J. B. and Wetzel, R. G. (1978). The production biology of Eurasian watermilfoil (*Myriophyllum spicatum* L.): a review. *Journal of Aquatic Plant Management*, **16**, 1–11.

Grace, J. B. and Wetzel, R. G. (1981*a*). Effects of size and growth rate in vegetative reproduction in *Typha*. *Oecologia* (*Berlin*), **50**, 158–61.

Grace, J. B. and Wetzel, R. G. (1981*b*). Phenotypic and genotypic components of growth and reproduction in *Typha latifolia*: experimental studies in marshes of differing successional maturity. *Ecology*, **62**, 789–801.

Grace, J. B. and Wetzel, R. G. (1981*c*). Habitat partitioning and competitive displacement in cattails (*Typha*): experimental field studies. *American Naturalist*, **118**, 463–74.

Grace, J. B. and Wetzel, R. G. (1982). Variations in growth and reproduction within populations of two rhizomatous plant species: *Typha latifolia* and *Typha angustifolia*. *Oecologia*, **53**, 258–63.

Graham, W. A. E. (1976). *Aquatic weeds. Observations on their growth and control in New Zealand*. State Rivers & Water Supply Commission, Victoria, Australia. pp. 154.

Grahn, O. (1977). Macrophyte succession in Swedish lakes caused by deposition of airborne acid substrates. *Water, Air and Soil Pollution*, **7**, 295–306.

Grahn, O. (1985). Macrophyte biomass and production in Lake Gardsjön—an acidified clear water lake in south west Sweden. *Ecological Bulletins NFR* (*Naturveterisk Forskningsradet*), **37**, 203–12.

Grainger, J. (1947). Nutrition and flowering of water plants. *Journal of Ecology*, **35**, 49–64.

Graneli, W. (1977). [Measurement of sediment oxygen uptake in the laboratory using undistributed sediment cores]. In Swedish. *Vatten*, **3**, 1–15.

Greco E. C. (1951). 2,4-D eradicates water hyacinth. *Oil Gas Journal*, **49**, 85.

Grillas, P. and Duncan, P. (1986). On the distribution and abundance of submerged macrophytes in temporary marshes in the Camargue (S. France). *Proceedings EWRS/AAB 7th Symposium on Aquatic Weeds 1986*, pp. 133–41.

Grime, J. P. (1979). *Plant strategies and vegetation processes*. Wiley, New York. pp. 222.

Grinwald, M. E. (1968). Harvesting aquatic vegetation. *Hyacinth Control Journal*, **7**, 31–2.

Grossbard, E. and Atkinson, D. (1985). *The herbicide glyphosate*. Butterworths, London. pp. 490.

Grover, R. and Smith, A. E. (1974). Adsorption studies with the acid and dimethylamine forms of 2,4-D and dicamba. *Canadian Journal of Soil Science*, **54**, 179–86.

Guerlesquin, M. and Podlejski, V. (1980). Characées et végétaux submerges et flottants associés dans quelques milieux Camarguais *Naturalia monspeliensia, series Botanica*, **36**, 1–20.

Guerra, L. V. (1974). Integrated controls on noxious aquatic plants. In *Proceedings Research Planning Conference on Integrated Systems of Aquatic Plant Control, 1973*, pp. 85–87. US Army Engineer Waterways Experiment Station, Vicksburg, Mississippi, pp. 174.

Guerra, L. V. (1977). *Hydrilla* in Texas. *Proceedings Southern Weed Science Society*, **30**, 335.

Guerra, L. V. (1984). Status of water hyacinths in Texas. *Aquatics*, **6** (**1**), 21.

Guerreiro, A. R. (1976). O jacinto aquático (*Eichhornia crassipes* (Mart.) Solms.) em Portugal. *II Simposia Nacionale Herbologia*, **1**, 1–18.

Guillory, V. and Gasaway, R. D. (1978). Zoogeography of grass carp in the United States. *Transactions American Fisheries Society*, **107**, 105–12.

Gunasekera, I. J., Balasooriya, I., and Gunasekara, S. A. (1983). Some leaf surface fungi of water hyacinth in Sri Lanka. *Journal of Aquatic Plant Management*, **21**, 99–100.

Gunatilaka, A., Broeshart, H., and Araujo de Oliveira, M. E. (1983). Isotopically exchangeable P-pool in a reed marsh in comparison to phosphorus yield of reed (*Phragmites australis*) in summer. *Proceedings International Symposium on Aquatic Macrophytes, Nijmegen 1983*, pp. 84–8.

Guppy, R. H. (1967). A comparison of various methods for the control of Florida elodea in Orange County, Florida. *Journal of Aquatic Plant Management*, **6**, 24–5.

Gupta, O. P. (1979). *Aquatic weeds: their menace and control. A textbook and manual.* Today & Tomorrows Printing Publishers, New Delhi.

Gurmukh, S.G., Gray, E. C., and Seckler, D. (1971). The California Water Plan and its critics: a brief review. In *California water: a study in resource management*, (ed. D. Seckler). University of California Press, Berkeley, California. pp. 348.

Gustafson, T. D. and Adams, M. S. (1973*a*). Remote sensing of *Myriophyllum spicatum* L. in a shallow, eutrophic lake. *Remote Sensing and Water Resource Management*, **17**, 387–91.

Gustafson, T. D. and Adams, M. S. (1973*b*). *The remote sensing of aquatic macrophytes.* University of Wisconsin, Institute of Environmental Studies, Remote Sensing Report No. 24.

Gutierrez, M., Gracen, W. E., and Edwards, G. G. (1974). Biochemical and cytological relationships in C_4 plants. *Planta*, **119**, 279–300.

Haag, K. H. (1984). Behavioural and physiological response of water hyacinth weevils (*Neochetina eichhorniae* and *Neochetina bruchi*) to herbicide application. *Proceedings Annual Meeting Entomological Society of America, San Antonio, Texas 1984*, pp. 9–13.

Haag, K. H. (1985). Does herbicide application affect water hyacinth weevils? *Aquatics*, **7**, 13–15.

Haag, K. H., Glenn, M. S., and Jordan, J. C. (1988). Selective patterns of herbicide application for improved biological control of water hyacinth. *Journal of Aquatic Plant Management*, **26**, 17–19.

Haag, R. W. (1979). The ecological significance of dormancy in some rooted aquatic plants. *Journal of Ecology*, **67**, 727–38.

Haag, R. W. (1983). Emergence of seedlings of aquatic macrophytes from lake sediments. *Canadian Journal of Botany*, **61**, 148–56.

Habeck, D. H. (1974). Caterpillars of *Parapoynx* in relation to aquatic plants in Florida. *Hyacinth Control Journal*, **12**, 15–18.

Habeck, D. H. (1983). The potential of *Parapoynx stratiota* L. as a biological control agent for Eurasian watermilfoil. *Journal of Aquatic Plant Management*, **21**, 26–9.

Habeck, D. H. and Wilkerson, R. (1980). The life cycle of *Lysathia ludoviciana* (Fall) (Coleoptera: Chrysomelidae) on parrot feather, *Myriophyllum aquaticum* (Velloso) Verde. *Coleoptera Bulletin*, **34**, 167–70.

Habib-ur-Rahman, Mushtaque, M., Baloch, G. M., and Ghani, M. A. (1969). Preliminary observations on the biological control of watermilfoil (*Myriophyllum* spp., Haloragidaceae). *Technical Bulletin CIBC*, **11**, 165–71. Commonwealth Institute of Biological Control, London.

Haddow, B. C., Stovell, F. R., and Payne, D. H. (1974). Field trials with cyanatryn (WL63611) for the control of aquatic weeds. *Proceedings 12th British Weed Control Conference*, **1**, pp. 239–48.

Hall, B. L. (1969). Cost and effectiveness of control of weeds in secondary canals in Dade County, Florida. *Journal of Aquatic Plant Management*, **8**, 34–5.

Hall, J. B. and Okali, D. V. V. (1974). Phenology and productivity of *Pistia stratiotes* L. on the Volta Lake, Ghana. *Journal of Applied Ecology*, **11**, 709–25.

Hall, J. B., Laing, E., Hossain, M., and Lawson, G. W. (1969). Observations on aquatic weeds in the Volta Basin. In *Man-made lakes: the Accra Symposium*, (ed. L. E. Obeng), pp. 331–43. Ghana Universities Press, Accra.

Hall, J. F., Westerdahl, H. E., Hoeppel, R. E., and Williams, L. (1982). *The 2,4-D threshhold concentrations for control of Eurasian watermilfoil and sago pondweed.* Technical Report A-82-6, US Army Engineer Waterways Experiment Station, Vicksburg, Mississippi.

Hall, J. F., Westerdahl, H. E., and Stewart, T. J. (1984). *Growth response of* Myriophyllum spicatum *and* Hydrilla verticillata *when exposed to continuous, low concentrations of fluridone.* Technical Report A-84-1, US Army Engineer Waterways Experiment Station, Vicksburg, Mississippi. pp. 32.

Haller, W. T. (1976). *Hydrilla, a new and rapidly spreading aquatic weed problem.* Circular S-245, Florida Agricultural Experiment Station, IFAS, University of Florida, pp. 1–13.

Haller, W. T. (1982). Hydrilla goes to Washington. *Aquatics*, **4**, 6–7.

Haller, W. T., Miller, J. L., and Garrard L. A. (1976). Seasonal production and germination of hydrilla vegetative propagules. *Journal of Aquatic Plant Management*, **14**, 26–9.

Haller, W. T., Shireman, J. V., and Durant, D. F. (1980). Fish harvest resulting from mechanical control of hydrilla. *Transactions of the American Fish Society 109*, pp. 517–20.

Haller, W. T. and Sutton, D. L. (1973*a*). Factors affecting the uptake of endothall – ^{14}C by hydrilla. Weed Science, **21**, 446–8.

Haller, W. T. and Sutton, D. L. (1973*b*). Effect of pH and high phosphorus concentrations on growth of water hyacinth. *Hyacinth Control Journal*, **11**, 59–61.

Haller, W. T. and Sutton, D. L. (1975). Community structure and competition between *Hydrilla* and *Vallisneria*. *Hyacinth Control Journal*, **13**, 48–50.

Haller, W. T., Sutton, D. L., and Barlow, W. C. (1974). Effects of salinity on growth of several aquatic macrophytes. *Ecology*, **55**, 891–4.

Haller, W. T. and Tag el Seed, M. (1979). *Study of water hyacinths showing possible resistance to 2,4-D chemical control programs.* Final Report, Miscellaneous Paper A-79-8. US Army Engineer Waterways Experiment Station, Vicksburg, Mississippi. pp. 14.

Ham, S. F., Wright, J. F., and Berrie, A. D. (1982). The effect of cutting on the growth and recession of the freshwater macrophyte *Ranunculus penicillatus* (Dumort.) Bab. var. *calcareus* (R. W. Butcher) C. D. K. Cook. *Journal of Environmental Management*, **15**, 263–71.

Hambric, R. N. (1968). Control of duckweed and azolla with diuron. *Proceedings 8th Annual Meeting Weed Science Society of America*, pp. 65.

Hambric, R. N. (1969). *Testing diuron as an economical herbicide in recreational lakes.* Research Report, Texas Parks and Wildlife Department, 321.

Hamdoun, A. M. and El Tigani, K. B. (1977). Weed control problems in Sudan. *Pest Abstracts & News Summaries*, **23**, 190–4.

Hamel, C. and Bhéreur, P. (1982). Méthode d'interprétation de l'évolution spatiale et temporelle des hydrophytes vasculaires. In *Studies on Aquatic Vascular Plants* (ed. J. J. Symoens, S. S. Hooper, and P. Compère), pp. 274–303. Royal Botanical Society of Belgium, Brussels.

Hanafiah, A. R., Sisombat, L., and Sathal, H. (1979). Weeds in irrigated lowland rice of different soil types in Bogor district. *Biotrop Bulletin*, **11**, 131–8.

Hanbury, R. G. (1986). Conservation on canals: a review of the present status and management of British navigable canals with particular reference to aquatic plants. *Proceedings EWRS/AAB 7th Symposium on Aquatic Weeds, 1986,* pp. 143–50.

Hanbury, R. G., Murphy, K. J., and Eaton, J. W. (1981). The ecological effects of 2-methylthio triazine herbicides used for aquatic weed control in navigable canals. II. Effects on macro-invertebrate fauna, and general discussion. *Archiv für Hydrobiologie*, **91**, 408–26.

Hanley, S. (1981). The effect of glyphosate on *Scirpus maritimus*. *Proceedings Association of Applied Biologists Conference: Aquatic Weeds and their Control 1981*, pp. 199–200. AAB, Wellesbourne, UK.

Hanley, S. (1982). A technique for surveying aquatic plant populations using an echo-sounder. *Proceedings EWRS 6th Symposium on Aquatic Weeds 1982*, pp. 171–6.

Haque, M. M. and Rahman, M. A. (1976). Effect of water hyacinth and sediments on the flow in irrigation and drainage channels. *Bangladesh Journal of Agricultural Science*, **3 (2)**, 192–6.

Harberg, M. C. and Modde, T. (1985). Feeding behaviour, food consumption, growth and survival of hybrid grass carp in two South Dakota ponds. *North American Journal of Fisheries Management*, **5**, 457–64.

Harbott, B. J. and Rey, C. J. (1981). The implications of long-term aquatic herbicide application: problems associated with environmental assessment. *Proceedings Association of Applied Biologists Conference: Aquatic Weeds and their Control 1981*, pp. 219–31. AAB, Wellesbourne, UK.

Harding, J. P. C. (1979). *River macrophytes of the Mersey and Ribble basins, summer 1978.* Report: TS-BS-79-1 North West Water Authority, Warrington, UK.

Hardjosoewarno, S., Pudjoarinto, A., Taka, A., and Soenarta, A. (1972). *Floating islands of Rawa Pening from the viewpoint of vegetation succession.* Report: Gajah Mada University, Indonesia.

Harley, K. L. S., Forno, I. W., Kassulke, R. C., and Sands, D. P. A. (1984). Biological control of water lettuce. *Journal of Aquatic Plant Management*, **22**, 101–2.

Harley, K. L. S. and Mitchell, D. S. (1981). The biology of Australian weeds: *Salvinia molesta* D. S. Mitchell. *Journal of the Australian Institute of Agricultural Science*, **47**, 21–8.

Harley, K. L. S. and Wright, A. D. (1984). Implementing a program for biological control of water hyacinth, *Eichhornia crassipes*. *Proceedings International Conference on Water Hyacinth* (ed. G. Thyagarajan), pp. 58–69. UNEP, Nairobi.

Harp, B. L. and Campbell, R. S. (1964). Effects of the herbicide silvex on benthos of a farm pond. *Journal of Wildlife Management*, **28**, 307–17.

Harris, B. B. and Silvey, J. K. B. (1940). Limnological investigations of Texas reservoir lakes. *Ecological Monographs*, **10**, 111–43.

Harris, C. I. and Warren, B. F. (1964). Adsorption and desorption of herbicides by soil. *Weeds*, **12**, 120–6.

Harris, F. W. (1985). *Evaluation of polymers for controlled delivery of herbicides for aquatic plant control.* Technical Report A-85-2, prepared by Wright State University, Dayton, Ohio for US Army Engineer Waterways Experiment Station, Vicksburg, Mississippi. pp. 85.

Harris, F. W. and Talukder, M. A. (1982). Development of polymeric controlled-release herbicide systems. *Proceedings 16th Annual Meeting Aquatic Plant Control Research Planning Operations Review*. Miscellaneous Paper A-82-3, pp. 68–77. US Army Engineer Waterways Experiment Station, Vicksburg, Mississippi.

Hart, B. T. (1982). Uptake of trace metals by sediment and suspended particulates: a review. *Hydrobiologia*, **91**, 299–313.

Hartley, G. S. and Graham-Bryce, I. J. (1980). *Physical principles of pesticide behaviour*, Volume 1. Academic Press, London.

Hartman, P., Faina, R., and Máchová, J. (1984). The effect of the Midstream herbicide on fishpond ecosystem [Vliv herbicidního přípavku Midstream na rybniční ekosystém.] In English; Czech summary. *Práce Vyskumnehu Ustavu Ryb. Hydrobiol. Vodňany*, **13**, 21–33.

Hartman, W. A. and Martin, D. B. (1984). Effect of suspended bentonite clay on the acute toxicity of glyphosate to *Daphnia pulex* and *Lemna minor*. *Bulletin Environmental Contamination Toxicology*, **33**, 355–61.

Haslam, S. M. (1968). The biology of reed (*Phragmites communis*) in relation to its control. *Proceedings 9th British Weed Control Conference 1968*, 392–7.

Haslam, S. M. (1969). *The reed.* Monograph 1, Norfolk Reed Growers Association, Norwich. pp. 42.

Haslam, S. M. (1972). Biological flora of the British Isles. *Phragmites communis* Trin. *Journal of Ecology*, **60**, 585–610.

Haslam, S. M. (1973*a*). The management of British wetlands. II. Conservation. *Journal of Environmental Management*, **1**, 345–61.

Haslam, S. M. (1973*b*). Some aspects of the life history and autecology of *Phragmites communis* Trin. A review. *Polskie Archiwum Hydrobiologie*, **20**, 79–100.

Haslam, S. M. (1978). *River plants.* Cambridge University Press, Cambridge. pp. 396.

Haslam, S. M. (1979). Infra-red colour photography and *Phragmites communis* Trin. *Polskie Archiwum Hydrobiologie*, **26**, 65–72.

Haslam, S. M. (1982). Major factors determining the distribution of macrophytic vegetation in the watercourses of the European Economic Community. *Proceedings EWRS 6th Symposium on Aquatic Weeds 1982*, pp. 105–12.

Haslam, B. M. (1986). Causes of changes in river vegetation giving rise to complaints. *Proceedings EWRS/AAB 7th Symposium on Aquatic Weeds 1986*, pp. 151–6.

Haslam, S. M. (1987). *River plants of Western Europe*. Cambridge University Press, Cambridge. pp. 512.

Hasler, A. D. and Jones, E. (1949). Demonstration of the antagonistic action of large aquatic plants on algae and rotifers. *Ecology*, **30**, 359–64.

Hattersley, P. W. and Watson, L. (1976). C_4 grasses: an anatomical criterion for distinguishing between NADP-ME species and PCK or NAD-ME species. *Australian Journal of Botany*, **14**, 297–308.

Hattingh, E. R. (1962). *Report on investigations into the control of* Salvinia auriculata *on Lake Kariba. Part I.* Union Weedkiller Services, Johannesburg, South Africa.

Hattingh, E. R. (1963). *Report on investigations into the control of* Salvinia auriculata *on Lake Kariba. Part II. Field trials and recommendations.* Union Weedkiller Services, Johannesburg, South Africa.

Hauser, W. J., Legner, E. F., Medved, R. A., and Platt, S. (1976). *Tilapia*—a management tool. *Fisheries*, **1**, (6), 24.

Hawxby, K., Tubea, B., Ownby, J., and Basler, E. (1977). Effects of various classes of herbicide on four species of algae. *Pesticide Biochemistry Physiology*, **7**, 203–9.

Hayslip, H. F. and Zettler, F. W. (1973). Past and current research on the diseases of Eurasian watermilfoil (*M. spicatum*). *Hyacinth Control Journal*, **11**, 38–40.

Hearne, J. S. and Pasco, R. A. (1972). Aquatic weed control trials in Gatun Lake, Panama Canal. *Hyacinth Control Journal*, **10**, 33–35.

Heckman, C. W. (1980). Rice field ecology in northeastern Thailand. *Monographiae Biologicae*, **34**. W. Junk, The Hague.

Heckman, C. W. (1982). Ecological functions of aquatic tracheophytes in lentic ecosystems. A comparative study of artificial urban water bodies in Hamburg. *Internationale Revue der gesamtem Hydrobiologie*, **67**, (2), 187–207.

Hejny, S. (1960). *Ökologische Charakteristik der Wasser- und Sumpfpflanzen in den Slowakischen Tiefebenen.* Czechoslovak Academy of Science, Bratislava. pp. 487.

Hejny, S. (1978). Management aspects of fishpond drainage. In *Pond littoral ecosystems: ecological studies 28*, (ed. D. Dykyjova and J. Květ), pp. 399–403. Springer-Verlag, Berlin.

Hejny, S., Květ, J., and Dykyjova, D. (1981). Survey of biomass and net production of higher plant communities in fishponds. *Folia Geobotanica et Phytotoxonomica, Praha*, **16**, 73–94.

Helder, R. J. and van Harmelen, M. J. (1982). Carbon assimilation patterns in the submerged leaves of the aquatic angiosperm *Vallisneria spiralis* L. *Acta Botanica Neerlandica*, **31**, 281–95.

Hellawell, J. M. (1978). *Biological surveillance of rivers.* Natural Environment Research Council/Water Research Centre, Stevenage.

Henderson, F. M. (1966). *Open channel flow.* Macmillan, London. pp. 522.

Henderson, S. (1980). The use of grass carp for control of aquatic vegetation in Arkansas. *Proceedings Vth International Symposium on Biological Control of Weeds, Brisbane 1979*, pp. 287–9.

Henley, D. E. (1982). A threshold decision model and decision matrix for integrated control of aquatic weeds. *Proceedings Conference on Strategies for Aquatic Weed Management* (ed. T. C. Chesnut), pp. 51–78. University of Florida, Gainesville, Florida.

Henshaw, G. G., Coult, D. A., and Boulter, D. (1961). Cytochrome-c oxidase, the terminal oxidase of *Iris pseudacorus* L. *Nature* (*London*), **197**, 579.

Hermens, L. C. M. (1978). Grüne Bäche in Limburg. *Proceedings EWRS 5th Symposium on Aquatic Weeds 1978*, pp. 212–18.

Hertzman, T. (1985). Hornborgasjön-restaureringsplaner och aktuella biotopforsök. *Var Fagelvarld* (*Supplement*), **10**, 77–86.

Hertzman, T. (1986). Restoration of Lake Hornborga—results from development of methods of wetland management. *Proceedings EWRS/AAB 7th Symposium on Aquatic Weeds 1986*, pp. 157–62.

Heslop-Harrison, Y. (1955*a*). Biological flora of the British Isles *Nuphar* Sm. genus; *N. lutea* (L.) Sm.; *N. pumila* (Timm.) D.C.: *N.* × *intermedia* Ledeb. *Journal of Ecology*, **43**, 342–64.

Heslop-Harrison, Y. (1955*b*). Biological flora of the British Isles: *Nymphaea* L. em. Sm.: *N. alba* L. *Journal of Ecology*, **43**, 719–34.

Hestand, R. S. and Carter, C. C. (1975). Succession of aquatic vegetation in Lake Ocklawaha, two growing seasons following a winter drawdown. *Hyacinth Control Journal*, **13**, 43–7.

Hestand, R. S. and Carter, C. C. (1977). Succession of various aquatic plants after treatment with four herbicides. *Journal of Aquatic Plant Management*, **15**, 60–4.

Hestand, R. S., May, B. E., Schultz, D. P., and Walker, C. R. (1973). Ecological implications of water levels on plant growth in a shallow water reservoir. *Hyacinth Control Journal*, **11**, 54–8.

Hettiarachchi, S., Gunasekera, S. A., and Balasooriya, I. (1983). Leaf spot diseases of water hyacinth in Sri Lanka. *Journal of Aquatic Plant Management*, **21**, 62–5.

Heuss, K. (1971). Der Einfluss von Herbiziden auf aquatische Biozönosen und deren physiologische Leistungen. *Proceedings EWRC 3rd Symposium on Aquatic Weeds 1971*, pp. 139–46.

Heuss, K. (1972). [Effects of some herbicides on limnic plankton communities]. In German. *Schriftenreihe des Vereins für Wasser-Boden-und Lufthygiene*, **37**, 221–9.

Hickling, C. F. (1965). Biological control of aquatic vegetation. *Pest Articles and News Summaries*, **11**, 237–44.

Hickling, C. F. (1966). On feeding processes in *C. idella*. *Journal and Zoological Proceedings of the Zoological Society of London*, **148**, 408–19.

Hickling, C. F. (1967). On the biology of a herbivorous fish, the white amur *Ctenopharyngodon idella* Val. *Proceedings Royal Society Edinburgh*, **70**, pp. 62–81.

Hildebrand, E. M. (1946). Herbicidal action of 2,4-dichlorophenoxy acetic acid on the water hyacinth *Eichhornia crassipes*. *Science* (*New York*), **103**, 447–9.

Hildebrand, L. D., Sullivan, D. S., and Sullivan, T. P. (1980). Effects of Roundup[R] herbicide on populations of *Daphnia magna* in a forest pond. *Bulletin Environmental Contamination Toxicology*, **25**, 353–7.

Hiley, P. D., Wright, J. F., and Berrie, A. D. (1981). A new sampler for stream benthos, epiphytic macrofauna and aquatic macrophytes. *Freshwater Biology*, **11**, 79–85.

Hill, B. H. (1987). *Typha* productivity in a Texas pond: implications for energy and nutrient dynamics in freshwater wetlands. *Aquatic Botany*, **27**, 385–94.

Hillebrand, D. (1950). Verkrautung und Abfluss. *Deutsche Gewasserkundliche Jahrbuch*, **2**, 1–30.

Hilsenhoff, W. L. (1966). Effect of diquat on aquatic insects and related animals. *Journal of Economic Entomology*, **59**, 1520–1.

Hiltibran, R. C. (1962). Duration of toxicity of endothal in water. *Weeds*, **10**, 17–19.

Hiltibran, R. C. (1965). The effect of diquat in aquatic plants in central Illinois. *Weeds*, **13**, 71–2.

Hiltibran, R. C. (1967). Effects of some herbicides on fertilized fish eggs and fry. *Transactions American Fish Society*, **96**, 414–16.

Hiltibran, R. C. (1980). 1980 Illinois weed macrophyte control research report. *Research Report: North Central Weed Control Conference*, **37**, 13–15.

Hiltibran, R. C. (1981). Uptake and translocation of ^{14}C-labelled glyphosate by aquatic plants. *Abstracts & Proceedings Weed Science Society of America*, **114**, pp. 53–4.

Hiltibran, R. C. (1984). Long-term continuous control of submerged aquatic macrophytes. *Abstracts of the Weed Science Society of America Meeting 1984*, p. 41.

Hiltibran, R. C., Underwood, D. L., and Fickle, J. (1972). Fate of diquat in the environment. *Research Report Water Resources Centre, University of Illinois, Urbana-Champaign*, **52**, 1–45.

Hilton, J. (1984). Airborne remote sensing for freshwater and estuarine monitoring. *Water Research*, **18**, 1195–223.

Hindmarsh, A. C. (1975). *GEARB: solutions of ordinary differential equations having banded Jacobian L-NL*. Report UCID-30059, Rev. 1.

Hitchcock, A. E., Zimmerman, P. W., Kirkpatrick, H., and Earle, T. T. (1949). Water hyacinth: its growth, reproduction and practical control by 2,4-D. *Contributions Boyce Thompson Institute*, **15**, 363–401.

Hitchcock, A. E., Zimmerman, P. W., Kirkpatrick, H., and Earle, T. T. (1950). Growth and reproduction of water hyacinth and alligator weed and their control by means of 2,4-D. *Contributions Boyce Thompson Institute*, **16**, 91–130.

Ho, Y. B. (1979). Shoot development and production studies of *Phragmites australis* (Cav.) Trin. ex Steudel in Scottish lochs. *Hydrobiologia*, **64**, 215–22.

Hoagland, D. R. and Arnon, D. I. (1950). *The water-culture method for growing plants without soil* (Revised edition). California Agricultural Experimental Station, Berkeley, Circular 347, pp. 32.

Hocking, P. J., Finlayson, C. M., and Chick, A. J. (1983). The biology of Australian weeds. *Phragmites australis* (Cav.) Trin. ex Steud. *Journal of the Australian Institute of Agricultural Science*, **49**, 123–32.

Hodge, W. H. (1956). Chinese water chestnut or matai—a paddy crop of China. *Economic Botany*, **10**, 49–65.

Hodgkin, E. P., Black, R. E., Birch, P. B., and Hillman, K. (1985). *The Peel Harvey estuarine system proposals: for management*. Department of Conservation and Environment, Western Australia. Report No. 14.

Hodgson, L. M. and Linda, S. B. (1984). Response of periphyton and phytoplankton to chemical control of hydrilla in artificial pools. *Journal of Aquatic Plant Management*, **22**, 48–52.

Hoehne, F. C. (1955). *Plantas aquáticas*. Publicacione Serie D, Secret. Agric. Instituto Botanico, São Paulo. pp. 168.

Hoeppel, R. E. and Westerdahl, H. E. (1982). Evaluation of controlled-release 2,4-D formulations in Lake Seminole, Georgia. *Proceedings 16th Annual Meeting Aquatic Plant Control Research Planning and Operations Review*. Miscellaneous Paper A-82-3, pp. 78–86. US Army Engineer Waterways Experiment Station, Vicksburg, Mississippi.

Hoeppel, R. E. and Westerdahl, H. E. (1983). Dissipation of 2,4-D DMA and BEE from water, mud and fish at Lake Seminole, Georgia. *Water Resources Bulletin*, **19 (2)**, 197–204.

Hofstede, A. E. (1960). [Control of excess aquatic vegetation]. In Dutch. *Visserij Nieuws*, **13 (1–2)**, 2–6.

Holaday, A. S. and Bowes, G. (1980). C_4 acid metabolism and dark CO_2 fixation in a submerged aquatic macrophyte (*Hydrilla verticillata*). *Plant Physiology*, **65**, 331–5.

Holaday, A. S., Salvucci, M. E., and Bowes, G. (1983). Variable photosynthesis/photorespiration ratios in *Hydrilla* and other submerged aquatic macrophyte species. *Canadian Journal of Botany*, **61**, 229–36.

Holcik, J. (1976). On the occurrence of far eastern plantivorous fishes in the Danube River with regard to the possibility of their natural reproduction. *Vestnik Ceskoslovenske Spolecnosti Zoologicke*, **40**, 88–103.

Holcomb, G. E. (1978). *Alternaria alternantherae* from alligator weed also is pathogenic on ornamental Amaranthaceae species. *Phytopathology*, **68**, 265–6.

Holcomb, C. E. and Antonopoulos, A. A. (1976). *Alternaria alternantherae*: a new species found on alligatorweed. *Mycologia*, **68**, 1125–9.

Holden, A. V. (1972). Effects of pesticides on life in fresh waters. *Proceedings Royal Society* (*B*), **180**, 383.

Holden, A. V. (1976). The relative importance of agricultural fertilizers as a source of nitrogen and phosphorus in Loch Leven. In *Agriculture and Water Quality*, Technical Bulletin Ministry of Agriculture, Fisheries and Food, No. 32, pp. 306–14. HMSO, London.

Holm, L. G, Pancho, J. V., Herberger, J., and Plucknett, D. L. (1979). *A geographical atlas of world weeds*. John Wiley, New York. pp. 391.

Holm, L. G., Plucknett, D. L., Pancho, J. V., and Herberger, J. P. (1977). *The world's worst weeds: Distribution and biology*. University Press of Hawaii, Honolulu. pp. 609.

Holm, L ,G., Weldon, L. W., and Blackburn, R. D. (1969). Aquatic weeds. *Science* (*New York*), **166**, 699–709.

Holmberg, D. J. and Lee, G. F. (1976). Effects and persistence of endothall in the aquatic environment. *Journal of Water Pollution Control Federation*, **48**, 2738–46.

Holmes, N. (1983). *Typing British rivers according to their flora*. Focus on Nature Conservation Series, Report No. 4, Chief Scientist Directorate, Nature Conservancy Council, Peterborough, UK.

Holz, W. (1963). [Chemical weed control in ditches: trials on the control of submerged plants]. In German. *5. Deutsche Arbeitsbesprechung über Fragen der Unkrautbiologie und-bekämpfung, Hohenheim*. pp. 4.

Holzner, W. (1974). Die Vegetation der Gewässer Österreichs. *Proceedings EWRS 4th Symposium on Aquatic Weeds 1974*, pp. 12–13.

Hong-Jun, Hu, and Xue-Ming, Ni (1986). The aquatic flora of the lakes of the middle and lower reaches of the Yangtze River. *Proceedings EWRS/AAB 7th Symposium on Aquatic Weeds 1986*, pp. 169–73.

Hoogerkamp, M. and Rozenboom, G. (1978). The management of vegetation on the slopes of waterways in the Netherlands. *Proceedings EWRS 5th Symposium on Aquatic Weeds 1978*, pp. 203–11.

Hook, S. M. (1977). Control of waterchestnut in New York State. *Proceedings Research Planning Conference, Aquatic Plant Control, 1976, Florida*, pp. 55–9.

Hooper, F. F. and Cook, A. B. (1957). Chemical control of submerged water weeds with sodium arsenite. *Pamphlet, Fisheries Department, Michigan Department of Conservation, No. 16.*

Hoq, A. (1955). Weed flora of paddy fields and its control in eastern Uttar Pradesh. *Science Culture*, **27**, 277.

Hörmann, W. D., Tounayre, J. C., and Ëgli, H. (1979). Triazine herbicide residues in central European streams. *Pesticide Monitoring Journal*, **13**, 128–31.

Horn, C. N. (1983). The annual growth cycle of *Heteranthera dubia* in Ohio. *Michigan Botanist*, **23**, 29–34.

Horowitz, M. (1966). Breakdown of endothal in soil. *Weed Research*, **6**, 168–71.

Hotchkiss, N. (1972). *Common marsh, underwater and floating-leaved plants of the United States and Canada.* Dover Publications, New York. pp. 124.

Hough, R. A. (1979). Photosynthesis, respiration and organic carbon release in *Elodea canadensis* Michx. *Aquatic Botany*, **7**, 1–11.

Howard-Williams, C. (1978). Growth and production of aquatic macrophytes in a south temperate saline lake. *Verhandlungen der Internationalen Vereinigung für theoretische und angewandte Limnologie*, **20**, 1153–8.

Howard-Williams, C. (1979). Distribution, biomass and role of aquatic macrophytes in Lake Sibaya. In *Lake Sibaya*, (ed. J. Illies) *Monographiae Biologicae*, **36**, 88–107.

Howard-Williams, C. (1981). Studies on the ability of a *Potamogeton pectinatus* community to remove dissolved nitrogen and phosphorus compounds from lake water. *Journal of Applied Ecology*, **18**, 619–37.

Howard-Williams, C., Clayton, J. S., Coffey, B. J., and Johnstone, I. M. (1987). Macrophyte invasions. In *Inland Waters of New Zealand*, (ed. A. B. Viner). Science Information Publishing Centre, Department of Scientific and Industrial Research, Wellington, New Zealand.

Howard-Williams, C., Davies, J., and Vincent, W. F. (1986). Horizontal and vertical variability in the distribution of aquatic macrophytes in Lake Waikaremoana. *New Zealand Journal of Marine and Freshwater Research*, **20**, 55–65.

Howard-Williams, C. and Longman, T. G. (1976). A quantitative sampler for submerged aquatic macrophytes. *Journal of the Limnological Society of South Africa*, **2**, 31–3.

Howard-Williams, C. and Vincent, W. F. (1983). Plants of the littoral zone. In *Lake Taupo: ecology of a New Zealand lake* (ed. D. J. Forsyth and C. Howard-Williams), pp. 73–84. Science Information Publishing Centre, Department of Scientific and Industrial Research, Wellington, New Zealand.

Howard-Williams, C. and Walker, B. H. (1974). The vegetation of a tropical African lake: classification and ordination of the vegetation of Lake Chilwa (Malawi). *Journal of Ecology*, **62**, 831–54.

Hübschman, J. H. (1967). Effects on copper on the crayfish *Oronectes rusticus* (Girard). I. Acute toxicity. *Crustaceana*, **12**, 33–42.

Huebert, D. B. and Gorham, P. R. (1983). Biphasic mineral nutrition of the submerged aquatic macrophyte *Potamogeton pectinatus* L. *Aquatic Botany*, **16**, 269–74.

Hughes, H. R. (1976). *Research into aquatic weeds in New Zealand waterways: a review*. Information Series, Department of Scientific and Industrial Research, Wellington.

Hughes, H. R. and Meeklah, F. A. (1977). Control of *Lagarosiphon* in Lake Wanaka: *Proceedings 30th New Zealand Weed & Pest Control Conference 1977*, pp. 135–40.

Hughes, J. S. and Davis, J. T. (1962). Comparative toxicity to bluegill sunfish of granular and liquid herbicides. *Proceedings Conference South East Association of Game & Fish Commissioners*, **16**, 319–23.

Hunziker, A. T. (1966). Sobre la presencia de *Eichhornia crassipes* en el Dique Los Sauces (Provincia de la Rioja). *Kurtziana*, **3**, 235.

Hürlimann, H. (1951). Zur Lebensgeschichte des Schilfs an den Ufen der Schweizer seen. *Beitrage zur Geobotanischen Landesaufnahme der Schweiz*, **30**, 1–232.

Husak, S. (1973). Destructive control of stands of *Phragmites communis* and *Typha angustifolia* and its effects on shoot production followed for three seasons. In *Littoral of the Nesyt fishpond* (ed. J. Květ), pp. 89–91. Studie C.S.A.V. 15.

Husak, S. (1978). Control of reed and reedmace stands by cutting. In *Pond littoral ecosystems: ecological studies 28*, (ed. D. Dykyjova and J. Květ), pp. 404–8. Springer-Verlag, Berlin.

Husak, S., Květ, J., and Plasencia Fraga, J. M. (1986). Experiments with mechanical control of *Typha* spp. stands. *Proceedings EWRS/AAB 7th Symposium on Aquatic Weeds 1986*, pp. 175–81.

Husak, S. and Otahelova, H. (1982). Contributions to the biology of *Stratiotes aloides* in Eastern Slovakia. *Proceedings EWRS 6th Symposium on Aquatic Weeds 1982*, pp. 297–305.

Hutchinson, G. E. (1970). The chemical ecology of three species of *Myriophyllum* (Angiospermae: Haloragaceae). *Limnology and Oceanography*, **15**, 1–5.

Hutchinson, G. E. (1975). *A treatise on limnology. III. Limnological Botany*. John Wiley, New York. pp. 660.

Hutto, T. D. (1981). Simulation modeling of mechanical control systems. In *Proceedings 16th Annual Meeting Aquatic Control Research Planning & Operations Review*, pp. 33–51. US Army Engineer Waterways Experiment Station, Vicksburg, Mississippi.

Hyde, T. M. H. (1977). Water supply for waterways. In *Proceedings 24th International Navigation Conference, Leningrad*. PIANC, Brussels.

Hydraulics Research Ltd. (1985). *The hydraulic roughness of vegetation in open channels*. Report IT 281. Hydraulics Research Ltd., Wallingford, Oxford.

Ikusima, I. (1965). Ecological studies on the productivity of aquatic plant communities. I. Measurement of photosynthetic activity. *Botanical Magazine, Tokyo*, **78**, 202–11.

Ikusima, I. (1983). Human impact on aquatic macrophytes. In *Man's impact on vegetation*, (ed. W. Holzner, M. J. A. Werger, and M. Numata), pp. 69–75. W. Junk, The Hague.

Ikusima, I. (1984). Aquatic macrophytes. In *Lake Biwa*, (ed. S. Horie), pp. 303–11. W. Junk, The Hague.

ILACO (1978). *Aquatic weed control project*. ILACO (International Land Development Consultants) Final Report. pp. 123.

ILACO (1982). *Grass carp project. Economical analysis of the use of grass carps for aquatic weed control*. ILACO (International Land Development Consultants) Technical Note No. 6. pp. 47.

Imahori, K. (1954). *Ecology, phytogeography and taxonomy of the Japanese Charophyta*. Kanazawa, Japan. pp. 234.

Imevbore, A. M. A., Odu, E. A., and Adebona, A. C. (1986). Growth of a tropical aquatic weed, *Salvinia nymphellula* Desv. in natural conditions. *Proceedings EWRS/AAB 7th Symposium on Aquatic Weeds 1986*, pp. 183–8.

Imhof, G. and Burian, K. (1972). Energy flow systems in a wetland ecosystem (Reed belt of L. Neusiedler See). *Special Publication, Austrian Academy of Science, T.B.P.*, pp. 1–15. Springer-Verlag, Vienna.

IPPC (International Plant Protection Center) (1984). Aquatic weed survey in Ecuador. *Aquaphyte*, **4**, 1.

Irgang, B. E., Pedralli, G., and Waetcher, J. L. (1984). Macrófitos aquáticos da Estaçao Ecol do Taim, Río Grande do Sul, Brasil, *Roessiéria, Instituto de Pesquisas de Recursos Naturais Renováveis 'Ataliba Paz'*, **6**, 395–404.

Irigoyen, J. H. (1981*a*). Evaluación de herbicidas en laboratorio para el control de *Potamogeton pectinatus* y *Chara* sp. In *II. Renunión sobre malezas subacuáticas en los canales de desague de CORFO, La Plata (Argentina)*. pp. 113–30. Publicaciones CIC.

Irigoyen, J. H. (1981*b*). Crecimiento y desarrollo de *Potamogeton pectinatus* en los canales de desague del Valle Bonaerense del Río Colorado. In *II. Reunión sobre malezas subacuáticas en los canales de desague de CORFO, La Plata (Argentina)*, pp. 47–68. Publicaciones CIC.

Irving, N. S. and Beshir, M. O. (1982). Introduction of some natural enemies of water hyacinth to the White Nile, Sudan. *Tropical Pest Management*, **28** (1), 20–6.

Isensee, A. R., Kearney, P. C., Woolson, E. A., Jones, G. E., and Williams, V. P. (1973). Distribution of alkyl arsenicals in model ecosystems. *Environmental Science and Technology*, **7**, 841–5.

Ivens, G. W. (1967). Water weeds. In *Eastern African weeds and their control* (ed. G. W. Ivens), pp. 1–9. Oxford University Press, Nairobi.

Jackson, G. A. (1974). *A review of the literature on the use of copper sulfate in fisheries. Final report*. Bureau of Sport Fisheries & Wildlife, US Fish & Wildlife Service. NTIS PB-235 445/4GA. pp. 95.

Jackson, M. J. (1978). The changing status of aquatic macrophytes in the Norfolk Broads. *Transactions Norfolk Norwich Naturalists Society*, **24**, 137–52.

Jacot-Guillarmod, A. (1977*a*). *Myriophyllum*, an increasing water weed menace for South Africa. *South African Journal of Science*, **73**, 89–90.

Jacot-Guillarmod, A. (1977*b*). Some water weeds of the Eastern Cape Province. II. *Myriophyllum*. *Eastern Cape Naturalist*, **61**, 14–17.

Jacot-Guillarmod, A. (1979). Water weeds in Southern Africa. *Aquatic Botany*, **6**, 377–91.

Jähnichen, H. (ed.) (1973*a*) *Die biologische Krautung*. Internal Report, Institüt für Binnenfischerei, Berlin-Friedrichshagen. DDR. pp. 44.

Jähnichen, H. (1973*b*). Die Wirksamkeit von Amurkarpfen (*Ctenopharyngodon idella*) zur biologischen Wasserpflanzenbekämpfung in den Wasserlaufen der DDR. *Zeitschrift für die Binnenfischerei der DDR*, **20**, 14–28.

Jähnichen, H. (1973*c*). Weitere Erfolge beim Einsatz von Amurkarpfen (*C. idella*) zur biologischen Wasserpflanzenbekampfung in Wasserlaufen. *Zeitschrift für die Binnenfischerei der DDR*, **20**, 227–8.

James, W. F. (1984). *Effects of endothall treatment on phosphorus concentration and community metobolism of aquatic communities*. Miscellaneous Paper A-84-1. US Army Engineer Waterways Experiment Station, Vicksburg, Mississippi. pp. 83.

Jamil, K., Jamil, M. Z., Rao, P. V. R., and Thyagarajan, G. (1985). *Eichhornia crassipes* (Mart.) Solms. in relation to pH. *Indian Journal of Botany*, **8**, 156–8.

Jana, S. and Choudhuri, M. A. (1979). Photosynthetic, photorespiratory and respiratory behaviour of three submerged aquatic angiosperms. *Aquatic Botany*, **7**, 13–19.

Jana, S. and Choudhuri, M. A. (1982). Changes in the activities of ribulose 1,5-biphosphate and phosphoenolpyruvate carboxylases in submerged aquatic angiosperms during ageing. *Plant Physiology*, **70**, 1125–7.

Janauer, G. A. (1982). An estimation of submerged aquatic plants in the running waters of lower Austria. In *Studies on aquatic vascular plants* (ed. J. J. Symoens, S. S. Hooper, and P. Compère), pp. 272–6. Royal Botanical Society of Belgium, Brussels.

Janes, G. A. (1975). Controlled release copper herbicides. *Proceedings 1975 International Controlled Release Pesticide Symposium, Dayton, Ohio*, 326–33.

Jarman, M. L., Jarman, N. G., and Edwards, D. (1983). Remote sensing and vegetation mapping in South Africa. *Bothalia*, **14**, 271–82.

Jarvis, M. J. F., Mitchell, D. S., and Thornton, J. A. (1982). Aquatic macrophytes and *Eichhornia crassipes*. In *Lake McIlwaine: the eutrophication and recovery of a tropical African man-made lake*, (ed. J. A. Thornton), pp. 137–44. Junk, The Hague.

Jayanth, K. P. (1987). *Biological control of water hyacinth in India*. Technical Bulletin, **3**, pp. 28. Indian Institute of Horticultural Research, Bangalore.

Jensen, S. (1979). Classification of lakes in southern Sweden on the basis of their macrophyte composition, by means of multivariate methods. *Vegetatio*, **39**, 129–46.

Jewell, W. J. (1971). Aquatic weed decay: dissolved oxygen utilization and nitrogen and phosphorus regeneration. *Journal of Water Pollution Control Federation*, **43**, 1457–67.

Jhingaran, V. G. (1982). *Fish and fisheries of India* (2nd edn.). Hindustan Publishing Corporation, New Delhi.

Joglekar, V. R. and Sonar, V. G. (1987). Multiple applications for water hyacinth. *Biocycle*, **28**, 46–8.

Johannes, H. (1974). Mehrjährige Herbizidanwendung und ihr Einfluss auf das Ökosystem—Eine Gemeinschaftsarbeit. *Proceedings EWRC 4th Symposium on Aquatic Weeds* 1974, pp. 7–11.

Johannes, H., Heri, W., and Reynaert, J. (1975). Le triazine nella lotta contro le alghe e le piante vascolari sommerse. *Notizario sulle Malattie delle Piante*, 92/93, 39–59.

John, D. M. and Moore, J. A. (1985). Observations on the phytobenthos of the freshwater Thames. I. The environment, floristic composition and distribution of the macrophytes (principally macroalgae). *Archiv für Hydrobiologie*, **102**, 435–59.

Johnson, J. (1984). USAE Division/District Presentations: Aquatic Plant Problems—Operations Activities. Southwestern Division, Galveston District, *Proceedings 18th Annual Meeting Aquatic Plant Control Research Program 1983*, pp. 33–4. Miscellaneous Paper A-84-4. US Army Engineer Waterways Experiment Station, Vicksburg, Mississippi.

Johnson, M. G. (1965). Control of aquatic plants in farm ponds in Ontario. *US Fish & Wildlife Service, Progressive Fish-Culturist*, **27**, 23–30.

Johnson, R. E., and Bagwell, M. R. (1979). Effects of mechanical cutting on submerged vegetation in a Louisiana lake. *Journal of Aquatic Plant Management*, **17**, 54–7.

Johnson, T. (1978). Aquatic mosses and stream metabolism in a North Swedish river. *Verhandlungen der Internationalen Vereinigung für theoretische und angewandte Limnologie*, **20**, 1471–7.

Johnson, W. W. and Julin, A. M. (1974). *A review of the literature on the use of diuron in fisheries. Final report*. US Fish & Wildlife Service NTIS-PB-235 446/2GA. pp. 103.

Johnstone, I. M. (1982). Strategies for the control of macrophytes in hydroelectric impoundments. *Water Pollution Management Review 1982*, pp. 65–94.

Johnstone, I. (1986). Plant invasion windows: a time-based classification of invasion potential. *Biological Reviews*, **61**, 369–94.

Johnstone, I. M., Coffey, B. T., and Howard-Williams, C. (1985). The role of recreational boat traffic in interlake dispersal of macrophytes: a New Zealand case study. *Journal of Environmental Management*, **20**, 263–79.

Jones, T. W. and Winchell, L. (1984). Uptake and photosynthetic inhibition by atrazine and its degradation products in four species of submerged vascular plants. *Journal of Environmental Quality*, **13**, 243–6.

Jorga, W., Heym, W. D., and Weise, G. (1982). Shading as a measure to prevent mass development of submerged macrophytes. *Internationale Revue der gesamten Hydrobiologie*, **67**, 271–81.

Jorga, W., and Weise, G. (1977). [Submerged macrophyte biomass development in slow running waters in relation to oxygen balance]. In German. *Internationale Revue der gesamten Hydrobiologie*, **62**, 209–34.

Joy, P. J., Satheesan, N. V., Lyla, K. R., and Joseph, D. (1985). Successful biological control of the floating weed *Salvinia molesta* Mitchell using the weevil *Cyrtobagous salvinia* Calder & Sands in Kerala (India). *Proceedings 10th Asian–Pacific Weed Science Society Conference, Chaing-Mai, Thailand 1985*, pp. 622–6. Department of Agriculture, Thai Pesticides Association, Bangkok.

Joyce, J. C. and Haller, W. T. (1985). Effect of gibberellic acid and 2,4-D on water hyacinths. *Aquatic Botany*, **23**, 119–26.

Joyner, B. G. and Freeman, T. E. (1973). Pathogenicity of *Rhizoctonia solani* to aquatic plants. *Phytopathology*, **63**, 681–5.

Juge, R., Lodscrozet, B., Noetzlin, A., Perfetta, J., and Lachavanne, J. B. (1985). The macrophytic vegetation of a highly eutrophic lake on the Swiss Plateau—the Baldegger Lake. *Schweizerische Zeitschrift fuer Hydrologie*, **47**, 64–75.

Julien, M.H. (1981). Control of aquatic *Alternanthera philoxeroides* in Australia, another success for *Agasicles hygrophila*. *Proceedings 5th International Symposium on Biological Control of Weeds, Brisbane, Australia 1980*, (ed. E.S. Del Fosse), pp. 583–8. CSIRO, Australia.

Julien, M.H. (1987). *Biological control of weeds. A world catalogue of agents and their target weeds. Second edition.* Commonwealth Agricultural Bureau, Farnham Royal, UK.

Julien, M.H. and Bourne, A.S. (1986). Compensatory branching and changes in nitrogen content in the aquatic weed *Salvinia molesta* in response to disbudding. *Oecologia*, **70**, 250–7.

Julien, M.H. and Broadbent, J.E. (1980). The biology of Australian weeds. 3. *Alternanthera philoxeroides* (Mart.) Griseb. *Journal of the Australian Institute of Agricultural Science*, **46**, 150–5.

Jungwirth, M. (1980). Biologische Grabenentkrautung durch den Graskarpfen (*Ctenopharyngodon idella*). *Österreichische Wasserwirtschaft*, **32**, 47–52.

Junk, W.J. (1970). Investigation on the ecology and production biology of the 'floating meadows' on middle Amazon. I. The floating vegetation and its ecology. *Amazoniana*, **2**, 449–95.

Junk, W.J. (1977). Note on aquatic weeds in some reservoirs in Thailand. *Aquatic Botany*, **3**, 85–90.

Junk, W.J. (1980*a*). Areas inundaveis—un desafio para limnología. *Acta Amazonica*, **10**, 775–95.

Junk, W.J. (1980*b*). Aquatic macrophytes: ecology and use in Amazonian agriculture. In *Tropical ecology and development* (ed. J.I. Furtado), pp. 763–70. *Proceedings 5th International Symposium on Tropical Ecology 1979.* Kuala Lumpur, Malaysia.

Junk, W.J. (1984). Ecology of the várzea, floodplain of the Amazonian white water river. In *The Amazon: limnology and landscape ecology of a mighty tropical river and its basin*, (ed. H. Sioli), pp. 215–43. W. Junk Publishers, Boston and Dordrecht.

Junk, W.J. and Howard-Williams, C. (1984). Ecology of aquatic macrophytes in Amazonia. In *The Amazon: limnology and landscape ecology of a mighty tropical river and its basin* (ed. H. Sioli), pp. 269–93. W. Junk Publishers, Boston and Dordrecht.

Jupp, B.B. and Spence, D.H.N. (1977). Limitations of macrophytes in a eutrophic lake, Loch Leven. I: Effects of phytoplankton. *Journal of Ecology*, **65**, 175–86.

Jupp, B.P., Spence, D.H.N., and Britton, R.H. (1974). The distribution and production of submerged macrophytes in Loch Leven, Kinross. *Proceedings Royal Society Edinburgh, B*, **74**, 195–208.

Kachroo, P. (1959). Aquatic vegetation of Damodar valley. I. Phanerogamic flora of freshwater ponds and marshy lands with particular reference to Damodar–Eden canal area of West Bengal. *Journal of the Asiatic Society*, **1**, 271–98.

Kadono, Y. (1982*a*). Occurrence of aquatic macrophytes in relation to pH, alkalinity, Ca^{++}, Cl^{-} and conductivity. *Japanese Journal of Ecology*, **32**, 39–44.

Kadono, Y. (1982*b*). Germination of the turions of *Potamogeton crispus* L. *Physiological Ecology, Japan*, **19**, 1–5.

Kadono, Y. (1984). Comparative ecology of Japanese *Potamogeton*. *Japanese Journal of Ecology*, **34**, 161–72.

Kai, Y. and Kai-Yu, L. (1983). [Application of LANDSAT MSS digital image processing technique in investigation of aquatic plant distribution in Honghu Lake]. In Mandarin; English summary. *Acta Botanic Sinica*, **25**, (5), 472–81.

Kamal, I. A. and Little, E. C. S. (1970). The potential utilization of water hyacinth for horticulture in the Sudan. *Pest Articles and News Summaries*, **16** (3), 488–96.

Kamath, M. K. (1979). A review of biological control of insect pests and noxious weeds in Fiji (1969–1978). *Fiji Agricultural Journal*, **41**, 55–72.

Kammathy, R. V. (1967). *Salvinia auriculata* Aublet—a rapidly spreading exotic weed in Kerala. *Science & Culture*, **34**, 396.

Kangasniemi, B. J. (1983). Observations on herbivorous insects that feed on *Myriophyllum spicatum* in British Columbia. *Lake Restoration, Protection & Management: Proceedings 2nd Annual Conference North America. Lake Society US, Vancouver 1982*, pp. 214–18. Environmental Protection Agency, Washington DC.

Kanwisher, J. (1963). Effect of wind on CO_2 exchange across the sea surface. *Journal of Geophysical Research*, **68**, 3921–6.

Kappers, F. I. (1973). Giftige blauwwieren en de drinkwater voorziening. *H_2O*, **6**, 396–9.

Kappers, F. I., Leeuwang, P., Dekker, M., and Koerselman, W. (1981). Investigation of the presence of toxins produced by Cyanobacteria (blue-green algae) in the Netherlands. *Science of the Total Environment*, **18**, 359–61.

Kasahara, Y. (1961). *Studies on the control of a paddy weed, barnyard grass.* Publication Agricultural Technology Imp. 47, Agriculture, Forest and Fishery Research Council, Japan.

Kasno, K. A. A. and Soerjani, M. (1979). Prospects of biological control of weeds in Indonesia. *Proceedings 7th Asian–Pacific Weed Science Society Conference, Sydney*, Supplementary Volume, pp. 35–8.

Kasno, Aziz, K. A. and Soerjani, M. (1981). Prospects for biological control of weeds in Indonesia. *Biocontrol News Information*, **2**, 341.

Kasno, S. R. (1982). Recent research advances in biological control of water hyacinth (*Eichhornia crassipes*) in Indonesia. *Biocontrol News Information*, **3**, 247.

Katz, E. E. (1966). *Chemical destruction of aquatic fauna in the Withlacoochee River, Florida, through hyacinth control.* Report: Florida Game & Freshwater Fish Commissioners & Stetson University, Deland, Florida. pp. 57.

Kaul, V. and Zutshi, D. P. (1966). Some ecological considerations of floating islands in Srinagar lakes. *Proceedings National Academy of Science India* B36. pp. 273–80.

Kaul, V., Zutshi, D. P., and Vass, K. K. (1976). Aquatic weeds in Kashmir. In *Aquatic Weeds in south-east Asia* (ed. C. K. Varshney and J. Rzóska), pp. 79–84. W. Junk, The Hague.

Kavetskiy, V. N., Karnauknov, A. I., and Paliyenko, I. M. (1984). Content of heavy metals in water and some aquatic plants of the Danube and Dniester estuaries. *Hydrobiological Journal*, **20**, (2), 67–70.

Kay, S. H., Quimby, P. C., and Ouzts, J. D. (1982). H_2O_2: a potential algicide for aquaculture. *Proceedings Southern Weed Science Society 35th Annual Meeting: New Perspectives in Weed Science, Atlanta, Georgia, USA*, pp. 275–89.

Kay, S. H., Quimby, P. C., and Ouzts, J. D. (1983). Control of coontail with hydrogen peroxide and copper. *Journal of Aquatic Plant Management*, **21**, 38–40.

Kay, S. H., Quimby, P. C., and Ouzts, J. D. (1984). Photo-enhancement of hydrogen peroxide toxicity to submersed vascular plant and algae. *Journal of Aquatic Plant Management*, **22**, 25–34.

Keckemet, O. (1974). Endothall and the environment. *Proceedings EWRC 4th Symposium on Aquatic Weeds 1974*, pp. 53–7.

Keckemet, O. (1980). Endothall–potassium and environment. *Proceedings 1980 British Crop Protection Conference—Weeds*, **2**, pp. 715–22.

Keddy, P. A. (1983). Shoreline zonation in Axe Lake, Ontario: effects of exposure on zonation patterns. *Ecology*, **64**, 331–44.

Keddy, P. A. (1984). Plant zonation on lakeshores in Nova Scotia: a test of the resource specialization hypothesis. *Journal of Ecology*, **72**, 797–808.

Keeley, J. E. (1981). *Isoetes howellii*: submerged aquatic CAM plant? *American Journal of Botany*, **68**, 420–4.

Keeley, J. E. (1983). Crassulacean acid metabolism in the seasonally submerged aquatic *Isoetes howellii*. *Oecologia*, **58**, 57–62.

Keeley, J. E. and Bowes, G. (1982). Gas exchange characteristics of the submerged aquatic Crassulacean acid metabolism plant *Isoetes howellii*. *Plant Physiology*, **70**, 1455–8.

Keeves, D. R. (1983). The economics of spraying v. mechanical clearing. *Proceedings Aquatic Weeds Seminar, May 1983, Massey University, Palmerston North, New Zealand*, 89–94.

Kelcey, J. G. (1981). Weed control in amenity lakes. *Proceedings Association of Applied Biologists Conference: Aquatic Weeds and their Control 1981*, pp. 15–31. AAB, Wellesbourne, UK.

Kelsall, J. D. (1981). Weed problems in fisheries waters. *Proceedings Association of Applied Biologists Conference: Aquatic Weeds and their Control 1981*, pp. 1–4. AAB, Wellesbourne, UK.

Kemmerling, W. (1974). Die Bedeutung von Betriebshöfen für die Gewässerinstandhaltung. *Proceedings EWRC 4th Symposium on Aquatic Weeds 1974*, pp. 87–95.

Kemmerling, W. (1978). Mechanische Bekämpfung unerwünschter Pflanzen in und an Gewässern. *Proceedings EWRS 5th Symposium on Aquatic Weeds 1978*, pp. 27–34.

Kent, D. F. (1955). '*Egeria densa* Planch.' and '*Lagarosiphon major* (Ridley)' (in Plant Notes). *Proceedings Botanical Society of the British Isles*, **1**, 322–3.

Kern-Hansen, U. (1978). Drift of *Gammarus pulex* L. in relation to macrophyte cutting in four small Danish lowland streams. *Verhandlungen der Internationalen Vereinigung für theoretische und angewandte Limnologie*, **20**, 1440–5.

Kern-Hansen, U. and Dawson, F. H. (1978). The standing crop of aquatic plants of lowland streams in Denmark and the inter-relationship of nutrients in plant, sediment and water. *Proceedings EWRS 5th Symposium on Aquatic Weeds 1978*, pp. 143–50.

Kern-Hansen, U. and Holm, T. F. (1982). Aquatic plant management in Danish streams. *Proceedings EWRS 6th Symposium on Aquatic Weeds 1982*, pp. 122–31.

Kersting, K. (1975). Effects of diuron on the energy budget of a *Daphnia magna* population. In *Sublethal effects of toxic chemicals on aquatic animals* (ed. J. H. Koeman and J. J. T. W. A. Strik). Elsevier, Amsterdam.

Kersting, K. (1984). Development and use of an aquatic micro-ecosystem as a test system for toxic substances. Properties of an aquatic micro-ecosystem IV. *Internationale Revue der gesamten Hydrobiologie*, **69**, 567–607.

Khattab, A. F. and El Gharably, Z. (1984). The problem of aquatic weeds in Egypt and methods of management. *Proceedings EWRS 3rd Symposium on Weed Problems in the Mediterranean Area 1984*, **1**, pp. 335–44.

Khattab, A. F. and El Gharably, Z. (1986). Management of aquatic weeds in irrigation systems with special reference to the problem in Egypt. *Proceedings EWRS/AAB 7th Symposium on Aquatic Weeds 1986*, pp. 199–206.

Khobot'ev, V. G., Kapkov, V. I., Rukhadze, E. G., Turunina, N. V., and Shidlovskaya, N. A. (1975). [The toxic effects of copper complexes on algae]. In Russian. *Gidrobiologicheskii Zhurnal*, **11**, (5), 33–8.

Khogali, F. A. and El Moghraby, A. I. (1979). Observations on the toxicity of 2,4-D to the biota of the White Nile. *Proceedings Symposium on Weed Research in Sudan* 1, pp. 79–84. Berichte Fachgebeit Herbologie, Universität Hohenheim.

Kilambi, R. V. and Robison, W. R. (1979). Effects of temperature and stocking density of food consumption and growth of grass carp. *Ctenopharyngodon idella* Val. *Journal of Fish Biology*, **15**, 337–42.

Kilambi, R. V. and Zdinak, A. (1980). Food preference and growth of grass carp, *Ctenopharyngodon idella* and hybrid carp *Ctenopharyngodon idella* female × *Aristichthys nobilis* male. *Proceedings Vth International Symposium in Biological Control, Brisbane 1979*, pp. 281–6.

Kilgen, R. H. and Smitherman, R. O. (1971). Food habits of white amur stocked in ponds alone and in combination with other species. *Progressive Fish Culturist*, **33**, 123–7.

Killgore, K. J. (1984*a*). *Use of herbicide/adjuvant formulations for the control of* Myriophyllum spicatum L. Miscellaneous Paper A-84-8, US Army Engineer Waterways Experiment Station, Vicksburg, Mississippi.

Killgore, K. J. (1984*b*). *Field methods to measure aquatic plant treatment method efficacy*. Miscellaneous Paper A-84-3, US Army Engineer Waterways Experiment Station, Vicksburg, Mississipi, pp. 22.

Killgore, K. J. and Payne, B. S. (1984). Field methods to measure aquatic plant treatment method efficacy. Miscellaneous Paper A-84-3, pp. 22. US Army Engineer Waterways Experiment Station, Vicksburg, Mississippi.

Kimbel, J. C. (1982). Factors influencing potential intralake colonization by *Myriophyllum spicatum* L. *Aquatic Botany*, **14**, 295–307.

Kimbel, J. C. and Carpenter, S. R. (1981). Effects of mechanical harvesting on *Myriophyllum spicatum*. L. Regrowth and carbohydrate allocation to roots and shoots. *Aquatic Botany*, **15**, 121–7.

Kinori, B. Z. and Mevorach, J. (1984). *Manual of surface drainage engineering. II. Stream flow engineering and flood protection*. Elsevier, Amsterdam. pp. 523.

Kirby, G. E. and Shell, F. W. (1976). Karmex herbicide with fertilization and aeration to control filamentous algae in hatchery ponds. *Proceedings Annual Conference South East Game & Fish Commissioners 1975*, **19**, pp. 279–81.

Kirk, J. T. O. (1985). Effect of suspensoids (turbidity) on penetration of solar radiation in aquatic ecosystems. *Hydrobiologia*, **125**, 195–208.

Klaine, S. J. and Ward, C. H. (1984). Environmental and chemical control of vegetative dormant bud production in *Hydrilla verticillata*. *Annals of Botany*, **53**, 503–14.

Kliemand, G. (1974). Vorschläge auf Grund von Versuchserfahrungen zur Beseitigung schwer bekämpfbarer Wasserunkräuter. *Proceedings EWRC 4th Symposium on Aquatic Weeds 1974*, pp. 20–7.

Klötzli, F. (1971). Biogenous influence on aquatic macrophytes, especially *Phragmites communis*. *Hydrobiologia Bucuresti*, **12**, 107–11.

Klötzli, F. and Züst, S. (1973). Conservation of reed beds in Switzerland. *Polskie Archiwum Hydrobiologie*, **20**, 229–35.

Kluge, M. and Ting, I. P. (1978). *Crassulacean acid metabolism: Analysis of an ecological adaptation*. Springer-Verlag, Berlin.

Knipling, E. B., West, S. H., and Haller, W. T. (1970). Growth characteristics, yield potential, and nutritive content of water hyacinth. *Soil Crop Science Society Florida Proceedings*, **30**, pp. 51–63..

Kobylinski, G. J., Miley, W. W., Van Dyke, J. M., and Leslie, A. J. (1980). *The effects of grass carp* (Ctenopharyngodon idella *Val.*) *on vegetation, water quality, zooplankton and macroinvertebrates of Deer Point Lake, Bay County, Florida*. Final Report, Florida Department of National Resources Bureau of Aquatic Plant Research and Control, Tallahassee, Florida. pp. 114.

Koch, W., Harris, G., El Tigani, K. B., Hamza, F. R., Obeid, M., Akasha, M., Leffler, V., and Häfliger, T. (1978). Investigations on the chemical control of *Eichhornia crassipes* (Mart.) Solms in the Sudan. *Proceedings EWRS 5th Symposium on Aquatic Weeds 1978*, pp. 415–27.

Koegel, R. G, Livermore, D. F., and Bruhn, H. D. (1975). Harvesting aquatic plants. *Agricultural Engineering 1975*, pp. 20–1.

Koegel, R. G., Livermore, D. F., and Bruhn, H. D. (1977). Costs and productivity in harvesting of aquatic plants. *Journal of Aquatic Plant Management*, **15**, 12–17.

Koegel, R. G., Sy, S. H., Bruhn, H. D., and Livermore, D. F. (1973). Increasing efficiency of aquatic plant management through processing. *Hyacinth Control Journal*, **11**, 24–30.

Koeman, J. H., Horsmans, T. H., and van der Maas, H. L. (1969). Accumulation of diuron in fish. *Mededelingen van de Faculteit Landbouwwetenschappen, Rijksuniversiteit, Gent*, **34**, (3), 428.

Kohler, A., Wonnenberger, R., and Zeltner, G. (1973). Die Bedeutung chemischer und pflanzlicher 'Verschnutzungsindikatoren' im Fliessgewassersystem Moosach (Münchener Ebene). *Archiv für Hydrobiologie*, **72**, 533–49.

Kokordak, J. (1978). Einsatz des weissen Amurkarpfen (*Ctenopharyngodon idella*) in Bewasserungskanalen. *Acta Hydrochimica et Hydrobiologica*, **6**, 227–34.

Koopman, H. and Daams, J. (1960). 2,6-dichlorobenzonitrile: a new herbicide. *Nature* (*London*), **186**, 89–90.

Korelyakova, I. L. (1970). Chemical composition of the higher aquatic vegetation of Kiev Reservoir. *Hydrobiological Journal*, **6**, (5), 15–21.

Korsak, N. B. and Myakushko, V. K. (1980). The vegetation of the Tuzda Reservoir and its production. *Hydrobiological Journal*, **16 (1)**, 22–7.

Kotalawala, J. (1976). Noxious water vegetation in Sri Lanka: the extent and impact of existing infestations. In *Aquatic weeds in south-east Asia* (ed. C. K. Varshney and J. Rzoska), pp. 51–8. W. Junk, The Hague.

Koumpli-Sovantzi, L. (1983). [Studies on the tracheophytes in the lakes and adjacent hydrobiotopes of Aetoloakarnania, Greece]. In Greek; English summary. University of Athens, Athens, pp. 346.

Kouwen, N. and Li, R. M. (1980). Biomechanics of vegetative channel linings. *Proceedings Hydraulics Division American Society Civil Engineers*, **106**, 1085–1103.

Kovács, G. and Oláh, J. (1982). Silver carp (*Hypophthalmichthys molitrix* Val.) dominated domestic sewage oxidation fish pond technology. *Proceedings 2nd International Symposium on Herbivorous Fish 1982*, pp. 32–9. EWRS, Wageningen, The Netherlands.

Kovács, I. (1983). [Possibilities of chemical weed control in canals and other areas belonging to Water Conservancy]. In Hungarian. *Novenyvedelem*, **19**, 125–7.

Kovrizhnykh, A. I. (1978). Decomposition of green filamentous algae and microphytobenthos on various substrates of the Severskiy Donets–Donbas Canal. *Gidrobiologicheskii Zhurnal*, **14**, (6), 100.

Kramer, D., Schmaland, G., and Nanzke, E. (1974). [Recent results on chemical weed control]. In German. *Wasserwirtschaft und Wassertechnik*, **24**, (7), 239–43.

Krause, W. (1977). Die makrophytische Wasservegetation der südlichen Oberrheinaue—Die Aschenregion. *Archiv für Hydrobiologie* Supplement **37**, 337–45.

Krause, W. (1981). Characeen als Bioindikatoren für den Gewasserzustand. *Limnologica, Berlin*, **13**, 399–418.

Krinke, H. and Dyckova, I. (1974). Filme über Böschungsmähmachinen. *Proceedings EWRC 4th Symposium on Aquatic Weeds, 1974*, pp. 96–106.

Krishnamoorthi, K. P. (1976). Aquatic plants in relation to public health aspects in Nagpur district and elsewhere. In *Aquatic weeds in south-east Asia*, (ed. C. K. Varshney and J. Rzoska), pp. 162–66. W. Junk, The Hague.

Krotkevic, P. G. (1959). [On the effects of timing and height of reedcutting on its regeneration and growth]. In Russian. *Sbornik Trudov UKS Nauchno-Issledovatel'skogo inst. Tsellyul. Burnaz. Promsti*, **3**, 8–17.

Krsnik-Rasol, M. (1975). Effects of some herbicides on the growth of *Lemna gibba* and *Spirodela polyrrhiza*. *Acta Botanica Croatica*, **34**, 190.

Krupauer, V. (1971). Use of herbivorous fishes for ameliorative purposes in Central and Eastern Europe. *Proceedings EWRC 3rd Symposium on Aquatic Weeds 1971*, pp. 95–103.

Krykhtin, M. L. and Gorbach, E. I. (1982). Reproductive ecology of the grass carp *Ctenopharyngodon idella* and the silver carp *Hypophthalmichthys molitrix* in the Amur Basin. *Journal of Ichthyology*, **21**, 109–24.

Krzywosz, T., Krzywosz, W., and Radziej, J. (1980). The effect of grass carp (*Ctenopharyngodon idella* Val.) on aquatic vegetation and ichthyofauna of Lake Dgal Wielki. *Ekologiya Polska*, **28**, 433–50.

Kucklentz, V. (1985). Restoration of a small lake by combined mechanical and biological methods. *Verhandlungen der Internationalen Vereinigung für theoretische und angewandte Limnologie*, **22**, 2314–17.

Kulshreshta, M. and Gopal, B. (1983). Allelopathic influence of *Hydrilla verticillata* (L.f.). Royle on the distribution of *Ceratophyllum* species. *Aquatic Botany*, **16**, 207–9.

Kunii, H. (1984). Effects of light intensity on the growth and buoyancy of detached *Elodea nuttallii* (Planch.) St. John during winter. *Botanical Magazine, Tokyo*, **97**, 287–95.

Küthe, K. (1974). Zeitpunktwahl als Beitrag zur umweltfreundlichen Verminderung unerwünchster Wasserpflanzen. *Proceedings EWRC 4th Symposium on Aquatic Weeds 1974*, pp. 122–7.

Kṽet, J. and Hejny, S. (1986). Biology, ecology and identification of aquatic weeds in relation to control methods. *Proceedings EWRS/AAB 7th Symposium on Aquatic Weeds 1986*, p. 419–27.

Květ, J., Svoboda, J., and Fiala, K. (1969). Canopy development in stands of *Typha latifolia* L. and *Phragmites communis* Trin. in South Moravia. *Hidrobiologia Bucuresti*, **10**, 63–75.

Lachavanne, J. B. (1977). Contribution à l'étude des macrophytes du Léman. Thèse no. 1760, Université de Genève, Department de Biologie Végétale. pp. 71.

Lachavanne, J. B. (1979*a*). Le végétation macrophytique du Bergaschisee. *Bericht der Schweizerischen Botanischen Gesellschaft*, **89**, 92–104.

Lachavanne, J. B. (1977*b*). Les macrophytes du lac de Morat. *Bericht der Schweizerischen Botanischen Gesellschaft*, **89**, 114–32.

Lachavanne, J. B. (1977*c*). Les macrophytes du lac de Bienne. *Schweizerische Zeitschrift fuer Hydrologie*, **41**, 356–73.

Lachavanne, J. B. (1982). Influence de l'eutrophisation des eaux sur les macrophytes des lacs Suisses: résultats préliminaires. In *Studies on aquatic vascular plants*, (ed. J. J. Symoens, S. S. Hooper, and P. Compère), pp. 333–9. Royal Botanical Society of Belgium, Brussels.

Lachavanne, J. B. (1985). The influence of accelerated eutrophication on the macrophytes of Swiss lakes: abundance and distribution. *Verhandlungen der Internationalen Vereinigung für theoretische und angewandte Limnologie*, **22**, 2950–5.

Lachavanne, J. B. and Wattenhofer, R. (1975). Evolution du couvert végétale de la Rade de Gènève. *Saussurea*, **6**, 217–30.

Lachner, E. A., Robins, C. R., and Courteney, W. R. (1970). Exotic fishes and other aquatic organisms introduced into North America. *Smithsonian Contributions to Zoology* 59, Washington DC. pp. 29.

Laing, R. L. (1979). The use of multiple inversion and 'Clean-Flo Lake Cleanser' in controlling aquatic plants. *Journal of Aquatic Plant Management*, **17**, 33–8.

Laing, W. A. and Browse, J. (1985). A dynamic model for photosynthesis by an aquatic plant, *Egeria densa*. *Plant, Cell and Environment*, **8**, 639–49.

Lallana, V. H. (1980). Productividad de *Eichhornia crassipes* (Mart.) Solms en una laguna isleña del Río Paraná Medio. 2. Biomasa y dinámica de población. *Ecología*, **5**, 1–16.

Lallana, V. H. (1981). Productividad de *Eichhornia crassipes* (Pontederiaceae) en una laguna isleña del Río Paraná Medio. 1. Análisis de crecimiento. *Boletin de la Sociedad Argentina de Botanica*, **20**, 99–107.

Lallana, V. H. (1982). *Vegetación acuática de deriva en el Río Paraná Medio*. Santo Tomé (Santa Fé), INALI (Inst. Nac. Limnol.) Inf. Final, 2da parte. 198 pp. 198.

Lallana, V. H. (1984*a*). Aspectos reproductivos del repollito de agua *Pistia stratiotes* L. Chapter 9 in *Resúmenes 2. Jornadas Ciencias Naturales Litoral, Paraná* (*Argentina*). pp. 95.

Lallana, V. H. (1984*b*). Transporte y dispersión de diseminulos de plantas acuáticas en el Río Paraná. Chapter 10 in *Resúmenes 2. Jornadas Ciencias Naturales Litoral, Paraná (Argentina)*. pp. 95.

Lallana, V. H. (1987). Evaluación de la fructificación de *Eichhornia crassipes* (Mart.) Solms—Laubach ('Camalote'). *Buletin de la Asociacion Ciencias Naturales Litoral*, **7**, 5–10.

Lallana, V. H. and Marta, M. C. (1980). Biologiá floral de *Eichhornia crassipes* (Mart.) Solms en el Río Paraná Medio. *Revista Asociacion Ciencias Naturales Litoral*, **11**, 73–81.

Lallana, V. H. and Marta, M. C. (1981). Biologiá floral de *Eichhornia azurea* (SW.) Kunth (Pontederiaceae). *Revista Asociacion Ciencias Naturales Litoral* **12**, 128–35.

Lallana, V. H., Marta, M. C., and Sabattini, R. A. (1981). Macrófitas acuáticas del valle aluvial, del Río Paraná Medio. Revisión crítica. In *Estudio Ecológico del Río Paraná Medio*, pp. 81–135. INALI (Inst. Nac. Limnol.) Santo Tomé, (Santa Fé). pp. 135.

Lallana, V. H., Sabbatini, R. A., and Lallana, M. C. (1987). Evapotranspiration from *Eichhornia crassipes, Pista stratiotes, Salvinia herzogii* and *Azolla caroliniana*. *Journal of Aquatic Plant Management*, **25**, 48–50.

Lambers, H. (1985). Respiration in intact plants and tissues: its regulation and dependence on environmental factors, metabolism and invaded organisms. *Encyclopaedia of Plant Physiology*, **18**, 418–73.

Lambert, J. M. (1946). The distribution and status of *Glyceria maxima* (Hartm.) Holmb. in the region of Surlingham and Rockland Broads. *Norfolk Journal of Geology*, **33**, 230–67.

Lambert, J. M. (1947). Biological flora of the British Isles: *Glyceria maxima* (Hartm.) Holmb. *Journal of Ecology*, **34**, 310–44.

Lambert, J. M. (1949). The British species of *Glyceria*. In *British flowering plants and modern systematic methods*, (ed. A. J. Wilmott), pp. 86–9. Botanical Society of the British Isles, London.

Lammers, W. (1968). *Report on the isolation and characterization of micro-organisms from* Myriophyllum spicatum. Report: Tennessee Valley Authority, Tennessee, June 1968. pp. 5.

Landolt, E. (1957). Physiologische und ökologische untersuchungen an Lemnaceen. *Berichte der Schweizerischen Botanischen Gesellschaft*, **67**, 269–410.

Landolt, E. (1982). Distribution pattern and ecophysiological characteristics of the European species of the Lemnaceae. *Veröffentlichungen des Geobotanischen Institüts der ETH, Zürich*, **49**, 127–45.

Landolt, E. (1986). The family of Lemnaceae—a monographic study. 1. In Volume 2. Biosystematic investigations in the family of duckweeds (Lemnaceae). *Veröffentlichungen des Geobotanischen Institüts der ETH, Stiftung Rubel, Zürich*, **71**, 1–566.

Landolt, E. and Kandeler, R. (1987). The family of Lemnaceae, a monographic study, Vol. 2. Phytochemistry; physiology; application; bibliography. In *Volume 4. Biosystematic investigations in the family of duckweeds (Lemnaceae). Veröffentlichungen des Geobotanischen Institüts der ETH, Zürich*, **95**, 1–638.

Langeland, K. A., Haller, W. T., and Thayer, D. D. (1983). Phytotoxicity of DPX 5648 to water hyacinth. *Journal of Aquatic Plant Management*, **21**, 106–7.

Langeland, K. A. and Schiller, D. (1983). *Hydrilla* in North Carolina: a successful program begins. *Aquatics,* **5**, 8–14.

Langeland, K. A. and Smith, C. B. (1984). *Hydrilla* produces viable seed in North Carolina lakes—a mechanism for long distance dispersal. *Aquatics,* **6**, 20–1.

Langeland, K. A. and Sutton, D. L. (1980). Regrowth of *Hydrilla* from axillary buds. *Journal of Aquatic Plant Management,* **18**, 27–9.

Larigauderie, A., Roy, J., and Berger, A. (1986). Long-term effects of high CO_2 concentration on photosynthesis of water hyacinth (*Eichhornia crassipes* (Mart.) Solms). *Journal of Experimental Botany,* **37**, 1303–12.

Larsen, V. (1978). Undersøgelser over plantevoeksten nogle vestjyske vandløls. *Beretning Danske Hedelskab Forsogsvirksanneden,* Nr. 18. pp. 84.

Leach, G. J. and Osborne, P. L. (ed.) (1985). *Freshwater plants of Papua New Guinea.* University of Papua New Guinea Press, Port Moresby. pp. 254.

Lebrun, J. (1959). La lutte contre le développement de l'*Eichhornia crassipes. Bulletin Agricole du Congo Belge,* **50**, 251–2.

Leclerc, L. (1977). Végétation et caracteristiques physiochimiques de deux rivières de Haute Ardennes (Belgique): la Helle et la Roer supérieure. *Lejeunia N.S.*, **88**. pp 42.

le Cosquino de Bussy, I. J. (1971). Research on substituted phenylureas as a control of algae in open-air swimming pools and recreational ponds in the Netherlands. *Proceedings EWRC 3rd Symposium on Aquatic Weeds 1971,* pp. 285–93.

Lee, K. W. and Furtado, J. E. (1977). The chemical control of *Salvinia molesta* (Mitchell) and some related toxicological studies. *Hydrobiologia* **56**, 49–61.

Legner, E. F. and Fisher, T. W. (1980). Impact of *Tilapia zillii* on *Potamogeton pectinatus, Myriophyllum spicatum* var. *exalbescens*, and mosquito reproduction in lower Colorado Desert irrigation canals. *Acta Oecologica Applicata,* **1**, 3–14.

Lekić, M. (1971). *Establishing of insect species, causes of diseases and other parasitic organisms which reduce the numerical strength of the populations of* Myriophyllum spicatum. Report: Project No. E30-ENT-10, Grant No. FG-Yu-133, Institute Plant Protection, Beograd, Yugoslavia.

Leonard, J. M. (1982). Improved timing of control measures for Eurasian watermilfoil in Lake Seminole. *Proceedings 16th Annual Meeting Aquatic Plant Control Research Planning and Operations Review*, Miscellaneous Paper A-82-3, US Army Engineer Waterways Experiment Station, Vicksburg, Mississippi. pp. 176–86.

Lesel, R., Fromageot, C., and Lesel, M. (1986). Cellulose digestibility in grass carp, *Ctenopharyngodon idella*, and in goldfish, *Carassius auratus. Aquaculture,* **54**, 11–17.

Leslie, A. C. and Walters, S. M. (1983). The occurrence of *Lemna minuscula* Herter in the British Isles. *Watsonia*, **14**, 243–8.

Leslie, A. C. (1988). Literature review of drawdown for aquatic plant control. *Aquatics,* **10**, (1), 12, 16–18.

Leslie, A. J., Nall, L. E., and Van Dyke, J. M. (1983). Effects of vegetation control by grass carp on selected water quality variables in four Florida lakes. *Transactions American Fisheries Society,* **112**, 777–87.

Leung, T. S., Naqvi, S. M., and Leblanc, C. (1983). Toxicities of two herbicides (Basagran, Diquat) and an algicide (Cutrine-Plus) to mosquitofish (*Gambusia affinis*). *Environmental Pollution* (*Series A*), **30**, 153–60.

Leung, T. S., Naqvi, S. M., and Naqvi, N. Z. (1980). Paraquat toxicity to Louisiana crayfish (*Procambarus clarkii*). *Bulletin of Environmental Contamination Toxicology,* **25,** 465–9.

Leventer, H. (1981). Biological control in reservoirs by fish. *Bamidgeh,* **33,** 3–23.

Lewis, D. H., Wile, I., and Painter, D. S. (1983). Evaluation of terratrack and aquascreen for control of aquatic macrophytes. *Journal of Aquatic Plant Management,* **21,** 103–4.

Lewis, G. and Williams, G. (1984). *Rivers and wildlife handbook—a guide to practices which further the conservation of wildlife on rivers.* Royal Society for the Protection of Birds & Royal Society for Nature Conservation, Lincoln, UK.

Lhoste, J. and Roth, P. (1946). Sur l'action des solutions aqueeuses de 2,4-D sur l'évolution des œufs de *Rana temporaria. Compte Rendu des Séances de la Societé de Biologie,* **140,** 272–3.

Li, R. and Shen, H. W. (1973). Effect of tall vegetation on flow and sediment. *Proceedings Hydraulics Division, ASCE* (American Society of Civil Engineers), **99,** pp. 793–814.

Li, S. (1987). Energy structure and efficiency of a typical Chinese integrated fish farm. *Aquaculture,* **65,** 105–18.

Liederman, L. and Figueiredo, P. (1967). Exterminacão do aguapé de coroae *Eichhornia azurea* Kunth no Río Ribiera pelo aplicacão de herbicidas. *Biologico,* **33,** 121–5.

Liederman, L. and Grassi, N. (1970). Chemical control of water hyacinth in the Río Preto, Perniba, São Paulo. *Biologico,* **36,** 157–9.

Lieffers, V. J. (1983). Growth of *Typha latifolia* in boreal forest habitats, as measured by double sampling. *Aquatic Botany,* **15,** 335–48.

Light, J. J. (1975). Clear lakes and aquatic bryophytes in the mountains of Scotland. *Journal of Ecology,* **63,** 937–43.

Limon, J. G. (1984). Mexican agency studies aquatic weeds. *Aquaphyte*, Fall 1984, 3.

Linacre, E. T., Hicks, B. B., Sainty, G. R., and Graza, G. (1970). The evapotranspiration from a swamp. *Agricultural Meteorology* **7,** 375–86.

Lindner, C. (1978). Eine neue *Elodea* in Lunz. *Verhandlungen der Zoologisch–Botanischen Gesellschaft in Österreich,* **116/117,** 79–81.

Link, L. E. and Long, K. S. (1978). *Remote sensing: a rapid and cheap method for detecting and monitoring aquatic plant infestations.* APCRP Information Exchange Bulletin, A-78-2, Vicksburg, Missippi. pp. 6.

Lips, G. (1985). [*An investigation of growth of the aquatic macrophyte* Hydrilla verticillata *Royle under various environmental conditions*]. In Dutch. Report Limnol. Inst. Nieuwersluis/Oosterzee 1985/23. pp. 81.

Little, E. C. S. (1967). Progress report on transpiration of some tropical water weeds. *Pest Articles and News Summaries, Section C, Weed Control,* **13,** 127–32.

Little, E. C. S. (1969*a*). Weeds and man-made lakes. In *Man-made Lakes: the Accra symposium* (ed. L. E. Obeng), pp. 284–91. Ghana Academy of Science, Accra, Ghana.

Little, E. C. S. (1969*b*). The floating islands of Rawa Pening. *Pest Articles & News Summaries,* **15,** 146–53.

Little, E. C. S. (1979). *Handbook of utilization of aquatic plants. A review of world literature.* FAO Fisheries Technical Paper No. 187. pp. 176. Food and Agriculture Organization, Rome.

Little, E. C. S., Robson, T. O., and Johnstone, D. R. (1964). A report on a project for drift-free aerial spraying of water weeds. *Proceedings 7th British Weed Control Conference*, pp. 920–4.

Liu, L. C. and Cedeno-Maldonado, A. (1974). Effects of fluometuron, prometryne, ametryne, and diuron on growth of two *Lemna* species. *Journal of Agricultural University Puerto Rico*, **58**, 483–8.

Livermore, D. F. and Koegel, R. G. (1979). Mechanical harvesting of aquatic plants: an assessment of the state of the art. In *Aquatic plants, lake management and ecosystem consequences of lake harvesting* (ed. J. Breck, R. Prentki, and O. Loucks). Institute for Environmental Studies, University of Wisconsin, Madison, Wisconsin.

Livermore, D. F., Koegel, R. G., Bruhn, H. D., Sy, S. H., and Link, H. F. (1975). *Aquatic plant harvesting: development of high-speed harvesters and processing and utilization of harvested vegetation.* Technical Report WIS WRC 75–02. Water Resources Center, University of Wisconsin, Madison, Wisconsin.

Livermore, D. F. and Wunderlich, W. E. (1969). Mechanical removal of organic production from waterways. In *Eutrophication: causes, consequences, correctives*, National Academy of Sciences, Washington, DC. pp. 512.

Lloyd, N. D. H., Canvin, D. T., and Bristow, J. M. (1977). Photosynthesis and photorespiration in submerged aquatic vascular plants. *Canadian Journal of Botany*, **55**, 3001–5.

Lloyd, R. (1977). *Are short-term fish toxicity tests a dead end?* Paper presented at British Association for the Advancement of Science Meeting, Aston, Birmingham 1977.

Lockett, E. P. (1983*a*). Control de malezas en arroz de riego con mezclas de herbicidas. *Malezas*, **11**, 37–42.

Lockett, E. P. (1983*b*). Determinación de la competencia de las malezas con el cultivo de arroz. *Malezas*, **11**, 45–51.

Lohammar, G. (1965). The vegetation of Swedish lakes. *Acta Phytogeographica Suecica*, **50**, 28–47.

Lollar, A. Q., Coleman, D. C., and Boyd, C. E. (1971). Carnivorous pathway of phosphorus uptake by *Utricularia inflata*. *Archiv für Hydrobiologie*, **69**, 400–4.

Lombardo, A. (1970). *Las plantas acuáticas y las plantas florales.* Intend Municipale, Montevideo, Uruguay. pp. 293.

Long, K. S. (1979). *Remote sensing of aquatic plants.* Technical Report WES-TR-A-79-2, US Army Engineer Waterways Experiment Station, Vicksburg, Mississippi. pp. 103.

Long, K. S. and Smith, P. A. (1975). *Effects of CO_2 laser on water hyacinth growth.* Technical Report 11, US Army Engineer Waterways Experimental Station, Vicksburg, Mississippi. pp. 156.

Longstreth, D. J., Bolanos, J. A., and Goddard, R. H. (1985). Photosynthetic rate and mesophyll surface area in expanding leaves of *Althernanthera philoxeroides* growth at two light levels. *American Journal of Botany*, **72**, 14–9.

Longstreth, D. J., Bolanos, J. A., and Smith, J. E. (1984). Salinity effects on photosynthesis and growth in *Alternanthera philoxeroides* (Mart.) Griseb. *Plant Physiology*, **75**, 1044–7.

Longstreth, D. J. and Mason, C. B. (1984). The effect of light on growth and dry matter allocation patterns of *Alternanthera philoxeroides* (Mart.) Griseb. *Botanical Gazette*, **145**, 105–9.

Lopinot, A. C. (1963). *Aquatic weeds: their identification and methods of control.* State of Illinois Department of Conservation Fisheries Bulletin No. 4. pp. 47.

Lorenzi, H. (1982). *Plantas daninhas do Brasil—terrestres aquáticas, parasitas, tóxicas é medicinais.* Nova Odessa, São Paulo. pp. 425.

Lottausch, W., Buchloch, G., and Kohler, A. (1980). Vegetationskundliche Untersuchungen in kryptogammenreichen Gebirgsbächen. *Verhandlungen der Gesellschaft für Ökologie,* **8**, 351–6.

Lourteig, A. (1951). Ranunculáceas de Sudamérica templada. *Darwiniana,* **9**, 397–608.

Lourteig, A. (1956). Ranunculáceas de Sudamérica tropical. *Memorias de la Sociedad de Ciencias Naturales 'la Salle' (Caracas)*, **16**, 19–88 and 125–228.

Loyal, D. S. and Grewal, R. K. (1967). Some observations on the morphology and anatomy of *Salvinia* with particular reference to *S. auriculata* Aubl. and *S. natans* All. *Research Bulletin Punjab University,* **18**, 13–28.

Lubke, R. A., Raynham, G. L., and Reavell, P. E. (1981). Reassessment of plant succession on spoil heaps along the Boro River, Okavango Delta, Botswana. *South African Journal of Science,* **77**, 21–3.

Lubke, R. A., Reavell, P. E., and Dye, P. J. (1984). The effects of dredging on the macrophytic vegetation of the Boro River, Okavango Delta, Botswana. *Biological Conservation,* **33**, 211–36.

Luther, H. (1951*a*). Verbreitung und Ökologie der höheren Wasserpflanzen im Brackwasser der Ekenäsgegend in Südfinnland. I. *Acta Botanica Fennici,* **49**, 1–232.

Luther, H. (1951*b*). Verbreitung und Ökologie der höheren Wasserpflanzen im Brackwasser der Ekenäsgegend in Südfinnland. II. *Acta Botanica Fennici,* **50**, 1–370.

Lytle, R. W. and Hull, R. J. (1980). Photo-assimilate distribution in *Spartina alterniflora* Loisel. II. Autumn and winter storage and spring regrowth. *Agronomy Journal,* **72**, 938–42.

Maberley, S. C. and Spence, D. H. N. (1983). Photosynthetic inorganic carbon use by freshwater plants. *Journal of Ecology,* **71**, 705–24.

Macan, T. T. (1977). Changes in the vegetation of a moorland fishpond in 21 years. *Journal of Ecology,* **65**, 95–106.

MacArthur, R. H. and Wilson, E. O. (1967). *Theoretical island biogeography.* Princeton University Press, Princeton. pp. 203.

Maceina, M. J. and Shireman, J. V. (1980). The use of a recording fathometer for determination of distribution and biomass of hydrilla. *Journal of Aquatic Plant Management,* **18**, 34–9.

Maceina, M. J., Shireman, J. V., Langeland, K. A., and Canfield, D. E. (1984). Prediction of submersed plant biomass by use of a recording fathometer. *Journal of Aquatic Plant Management,* **22**, 35–8.

Macek, K. J., Buxton, K. S., Sauter, S., Gnilka, S., and Dean, J. W. (1976). *Chronic toxicity of atrazine to selected aquatic invertebrates and fishes.* Proc. Publication EPA-600/3-76-047. Environmental Protection Agency, Washington DC. pp. 50.

Mackenthun, K. M. (1950). Aquatic weed control with sodium arsenite. *Sewage—Industrial Wastes,* **22**, 1062–7.

Mackenthun, K. M. (1955). The control of submergent aquatic vegetation through the use of sodium arsenite. *Proceedings North Eastern Weed Control Conference,* **9**, pp. 545–55.

Mackenthun, K. M. (1960). Some limnological investigations on the long term use of sodium arsenite as an aquatic herbicide. *Proceedings Annual North Central Weed Control Conference,* **17**, pp. 30–1.

Mackenthun, K. M. and Ingram, W. M. (1967). *Biological associated problems in freshwater environments.* US Government Printing Office, Washington DC.

MacKenzie, C. J. G., Oldham, W. K., and Powrie, W. D. (1979). *Interim report of the advisory committee on the control of Eurasian water milfoil in the Okanagan Lake system of British Columbia. VI (1978).* Ministry of Environment, British Columbia, Canada. p. 8.

MacKenzie, D., Frank, R., and Sirons, G. J. (1983). Aquatic plant control and disappearance of terbutryn from treated waters in Ontario, Canada 1974–77. *Journal of Aquatic Plant Management,* **21**, 11–16.

MacKenzie, R. C. and Murphy, K. J. (1986). *Aquatic macrophytes of Lothian lochs: Linlithgow, Duddingston, Dunsapie and Beecraigs Lochs 1985.* Report to Nature Conservancy Council, Edinburgh. University of Glasgow, Glasgow, UK. pp. 47

Maddox, D. M., Andres, L. A., Hennessey, R. D., Blackburn, R. D., and Spencer, N. R. (1971). Insects to control alligator weed. *Bioscience,* **21**, 985–91.

Maembe, T. W. (1981). *Final report on the improvement of fish processing and transportation on Lake Chad.* Food and Agriculture Organization, Rome.

Magee, C. A. and Colmer, A. R. (1959). Decomposition of 2,2-dichloropropionic acid by soil bacteria. *Canadian Journal of Microbiology,* **5**, 255–60.

Mahler, M. J. (1979). Hydrilla the number one problem. *Aquatics,* **1**, 56.

Maier-Bode, H. (1972). Verhalten von Herbiziden in Wasser, Schlamm und Fischen nach Applikation in Fischtein. *Schriftenreiche des Vereins für Wasser-Boden-und Lufthygiene,* **37**, 67–75.

Maity, B. R. and Samaddar, K. R. (1977). A toxic metabolite from *Alternaria eichhorniae:* production and properties. *Phytopathologische Zeitschrift,* **88**, 78–84.

Majid, F. Z. (1986). *Aquatic weeds: utility and development.* Agro-Botanical Publishers (India), Bikanar. pp. 96.

Majid, F. Z. and Akhtar Jahan, M. A. (1984). Study on the effect of some aquatic weed extract solutions on seed germination and primary seedling vigour of some economic crops. Paper to *4th National Botanical Convention, Dhaka.* pp. 12.

Malhotra, S. P. (1976). Remedy for aquatic weeds in Bhakra canals. In *Aquatic weeds in south east Asia* (ed. C. K. Varshney and J. Rzóska), pp. 253–4. W. Junk, The Hague.

Malicky, G. (1984). [Long- and short-term vegetation changes in the NE Bay of the Lunzer Untersee (Austria)]. In German; English summary. *Archiv für Hydrobiologie,* **101**, 221–9.

Malik, H. C. (1961). It pays to grow singhara and bhen. *Indian Farming,* **11**, (8), 23–4.

Mangoendihardjo, S. and Nasroh, A. (1976). *Proxenus* spp. (Lepidoptera: Noctuidea), a promising natural enemy of water lettuce (*Pistia stratiotes* L.). *Proceedings 5th Asian Pacific Weed Science Society Conference, Tokyo,* pp. 444–6.

Mangoendihardjo, S. and Soerjani, M. (1978). Weed management through biological control in Indonesia. *Proceedings Plant Protection Conference Kuala Lumpur,* pp. 323–37.

Mangoendihardjo, S., Setyawati, O., Syed, R. A., and Sosromarsono, S. (1977). Insects and fungi associated with some aquatic weeds in Indonesia. In: *Proceedings 6th Asian–Pacific Weed Science Society Conference, Jakarta.*

Mangoendihardjo, S. and Syed, R. A. (1974). Studies of natural enemies on *Eichhornia crassipes* and *Pistia stratiotes.* Paper presented at South East Asian Workshop on Aquatic Weeds, Malang, 1974.

Manning, J. H. and Johnson, R. E. (1975). Water level fluctuation and herbicide application: an integrated control method for *Hydrilla* in a Louisiana reservoir. *Hyacinth Control Journal,* **13,** 11–17.

Manning, J. H. and Sanders, D. R. (1975). Effects of water fluctuation on vegetation in Black Lake, Louisiana. *Hyacinth Control Journal,* **13,** 17–21.

Manning, R. (1891). On the flow of water in open channels and pipes. *Transactions Institute of Civil Engineers of Ireland,* **20,** 161–207.

Mara, M. J. (1976*a*). Estimated values for selected waterhyacinth by-products. *Economic Botany,* **27,** 383–7.

Mara, M. J. (1976*b*). Estimated costs of mechanical control of water hyacinths. *Journal of Environmental Economics and Management,* **2,** 273–94.

Marchesini, E. (1985). *Avances en el control químico del capín* Echinocloa *spp.* Entre Ríos (Argentina) Estacione Experiment Agrop. C. del Uruguay. Boletino Técnico Ser. Prod. Vegetal. 29. pp. 13.

Markmann, P. N. (1984). Introduction of grass carp (*Ctenopharyngodon idella* Val.) into Denmark. In *Symposium on stock enhancement in the management of freshwater fisheries, Budapest,* pp. 325–334. EIFAC/82/Symposium 4/40A. EIFAC Technical papers 42, Supplement 2.

Marks, T. S. (1974). Trials with terbutryne for the control of aquatic weeds. *Proceedings 12th British Weed Control Conference,* **1,** pp. 233–8.

Marler, J. E. (1969). Studies on the germination of *Heteranthera limosa* seed. *Proceedings 5th Weed Science Society Conference,* **22,** pp. 337.

Marquis, L. Y., Comes, R. D., and Yang, C. P. (1981). Absorption and translocation of fluridone and glyphosate in submerged vascular plants. *Weed Science Society of America Journal,* **29,** 229–36.

Marshall, B. E. and Junor, F. J. R. (1981). The decline of *Salvinia molesta* in Lake Kariba. *Hydrobiologia,* **83,** 477–84.

Marshall, C. D. and Rutschky, C. W. (1974). Single herbicide treatment: effect on the diversity of aquatic insects in Stone Valley Lake, Huntingdon Co., Pennsylvania. *Proceedings Pennsylvania Academy of Sciences,* **48,** pp. 127–31.

Marshall, E. J. P. (1981). The ecology of a land drainage channel. 1. Oxygen balance. *Water Research,* **15,** 1075–85.

Marshall, E. J. P. (1984). The ecology of a land drainage channel. II. Biology, chemistry, and submerged weed control. *Water Research,* **18,** 817–25.

Marshall, E. J. P., Wade, P. M., and Clare, P. (1978). Land drainage channels in England and Wales. *The Geographical Journal,* **144 (2),** 254–63.

Marshall, E. J. P. and Westlake, D. F. (1989). Water velocities around water plants in chalk streams. *Proceedings 1st International Workshop on Aquatic Plants 1981.* In press.

Marta, M. C. (1983). *Plantas acuáticas del litoral Guía para su reconocimiento a campo y otros temas.* Santo Tomé (Santa Fé). INALI, (Instituto Nacionale Limnologico).

Martin, D. F. and Reid, G. A. (1976). Uptake of manganese in *Hydrilla verticillata* Royle. *Journal of Agricultural Food Chemistry,* **24**, 1161–5.

Martin, D. F., Victor, D. M., and Dooris, P. M. (1976). Effects of artificially introduced ground water on the chemical and biochemical characteristics of six Hillsborough County (Florida) lakes. *Water Research,* **10**, 65–9.

Martyn, R. D. (1985). Water hyacinth decline in Texas caused by *Cercospora piaropi. Journal of Aquatic Plant Management,* **23**, 29–32.

Martyn, R. D. and Freeman, T. E. (1978). Evaluation of *Acremonium zonatum* as a potential biocontrol agent of water hyacinth. *Plant Disease Reports,* **62**, 604–8.

Marzocca, A. (1976). *Manual de malezas.* Hemisferio Sur, Buenos Aires. pp. 564.

Mason, H. L. (1957). *A flora of the marshes of California.* University of California Press, Los Angeles. pp. 857.

Mason, R. (1970). Macroscopic water plants. In Part 1: Background statement. *Proceedings New Zealand Water Conference,* **9**, 1–9.11.

Mason, R. (1975). The macrophytes. In *New Zealand lakes* (ed. V. H. Jolly and J. M. A. Brown), pp. 231–43. University of Auckland Press, Auckland.

Massini, P. (1961). Movement of 2,6-dichlorobenzonitrile in soils and plants in relation to its physical properties. *Weed Research,* **1**, 142–6.

Masters, C. O. (1974). *Encyclopaedia of the water-lily.* pp. 512. TFH Publications Inc., Neptune City, New Jersey.

Matsunaka, S. (1983). Evolution of rice weed control practices and research: world perspective. *Proceedings Conference on Weed Control in Rice,* pp. 5–17. International Rice Research Institute/International Weed Science Society, Manila.

Matthews, L. J. (1967). Seedling establishment of water hyacinth. *Pest Articles and News Summaries,* **13**, 7-8.

Mauck, W. L. (1974). *A review of the literature on the use of simazine in fisheries.* Bureau of Sport Fisheries and Wildlife, Wisconsin, Report FWS CR-74-16. pp. 62.

Mauck, W. L., Mayer, F. L., and Holz, D. D. (1976). Simazine residue dynamics in small ponds. *Bulletin of Environmental Contamination Toxicology,* **16**, (1), 17.

Mayer, A. M., and Poljakoff-Mayber, A. (1982). *The germination of seeds,* (3rd edn.). Pergamon Press, Oxford. pp. 211.

Mayer, J. R. (1978). Aquatic weed management by benthic semi-barriers. *Journal of Aquatic Plant Management,* **16**, 31–3.

McBride, J. C., McBride, A. C., and Togasaki, R. K. (1977). Isolation of *Chlamydomonas reinhardi* mutants resist to the herbicide DCMU. *Photosynthetic Organelles: Special Issue, Plant & Cell Physiology,* pp. 239–41.

McBride, J. R., Dye, H. M., and Donaldson, E. M. (1981). Stress response of juvenile sockeye salmon (*Onchorhynhus nerka*) to the butoxyethanol ester of 2,4-dichlorophenoxyacetic acid. *Bulletin of Environmental Contamination Toxicology,* **27**, 877–84.

McComb, A. J. and Lukatelich, R. J. (1986). Nutrients and plant biomass in Australian estuaries, with particular reference to south-western Australia. In *Limnology in Australia* (ed. P. De Deckker and W. D. Williams), pp. 433–55. CSIRO, Australia, Melbourne.

McCorkle, F. M., Chambers, J. E., and Yarbrough, J. D. (1977). Acute toxicities of selected herbicides to fingerling channel catfish *Ictalurus punctatus*. *Bulletin of Environmental Contamination Toxicology*, **18**, (3), 267–70.

McCowen, M. C., Young, C. L., West, S. D., Parka, S. J., and Arnold, W. R. (1979). Fluridone, a new herbicide for aquatic plant management. *Journal of Aquatic Plant Management*, 17, 27–30.

McCraren, J. P., Cope, O. B., and Eller, L. (1969). Some chronic effects of diuron on bluegills. *Weed Science*, 17, 497–504.

McRoy, C. P., Barsdate, R. J., and Nebert, M. (1972). Phosphorus cycling in an eelgrass (*Zostera marina* L.) ecosystem. *Limnology and Oceanography*, **17**, 58–67.

McDonnell, A. J. (1971). Variations in oxygen consumption by aquatic macrophytes in a changing environment. *Proceedings 14th Conference on Great Lakes Research*, pp. 52–9.

McGahee, C. F. and Davis, G. J. (1971). Photosynthesis and respiration in *Myriophyllum spicatum* L. as related to salinity. *Limnology and Oceanography*, **16**, 826–9.

McGhee, J. T. (1979). Mechanical hydrilla control in Orange Lake, Florida. *Journal of Aquatic Plant Management*, **17**, 58–61.

McGuire, M. J., Jones, R. M., *et al.* (1984). Controlling attached blue-green algae with copper sulphate. *Journal of the American Water Works Association*, **76**, pp. 60–5.

McHendry, W. B. (1963). Soil sterilant evaluation trials for the control of submerged aquatic weeds. *Research Progress Report, Western Weed Control Conference 1963*, pp. 67–9.

McIntyre, S. (1987). Population studies of *Diplachne fusca* (L.) Beauv. in relation to the weed flora of rice in New South Wales. Ph.D. thesis, University of Melbourne.

McKee, J. G. and Wolf, J. W. (1963). *Water quality criteria*. Resources Agency, California State Water Quality Control Board. Publication No. 3-A. Sacramento, California. pp. 548.

McKnight, D. M., Chisholm, S. W., and Morel, F. M. M. (1981). *Copper sulfate treatment of lakes and reservoirs: chemical and biological considerations*. Report on INCRA Project No. 252, from Department of Civil Engineering, MIT, Cambridge, Massachusetts. pp. 70.

McLay, C. (1976). Effect of pH on population growth of 3 species of duckweed: *Spirodela, Lemna minor*, and *Wolffia arrhiza*. *Freshwater Biology*, **6**, 137–44.

McLintock, W. L., Frye, J. W., and Hogan, W. D. (1974). Development of the bottom placement technique for hydrilla and eelgrass control. *Hyacinth Control Journal*, **12**, 16–17.

McNabb, C. D. (1976). The potential of submerged vascular plants for reclamation of wastewater. In *Biological control of water pollution* (ed. J. Tourbier and R. W. Pearson). University Press, Philadelphia.

McNabb, T. and Anderson, L. W. J. (1985). Aquatic weed control. In *Principles of weed control in California: California Weed Conference*. Thomson Publications, California. pp. 474.

McNaughton, S. J. (1966). Ecotypic function in the *Typha* community-type. *Ecological Monographs*, **36**, 297–325.

McNaughton, S. J. (1968). Autotoxic feedback in relation to germination and seedling growth in *Typha latifolia*. *Ecology*, **49**, 367–9.

McNaughton, S. J. (1975). r- and K-selection in *Typha*. *American Midland Naturalist*, **109**, 251–61.

McNaughton, S. J. and Fullem, L. W. (1970). Photosynthesis and photorespiration in *Typha latifolia*. *Plant Physiology*, **45**, 703–7.

Mehta, I. (1979). Problem and control of *Typha* in the Chambal Command area. *Indian Journal of Weed Science*, **11**, (1–2), 36–46.

Mehta, I. and Sharma, R. K. (1976). Effects of weeds on the flow capacity of Chambal Irrigation System in Kota, Rajasthan. In *Aquatic weeds in south east Asia* (ed. C. K. Varshney and J. Rzóska), pp. 85–96. W. Junk, The Hague.

Mehta, I., Sharma, R. K., and Tuank, A. P. (1976). The aquatic weed problem in the Chambal Irrigated Area and its control by grass carp. In *Aquatic weeds in south-east Asia* (ed. C. K. Varshney and J. Rzóska), pp. 307–14. W. Junk, The Hague.

Melack, J. M. (1984). Amazon floodplains lakes: shape, fetch and stratification. *Verhandlungen der Internationalen Vereinigung für theoretische und angewandte Limnologie*, **22**, 1278–82.

Melack, J. M. (1985). Interactions of detrital particulates and plankton. In *Perspectives in Southern Hemisphere Limnology* (ed. B. R. Davies and R. D. Walmsley), pp. 209–20. W. Junk, The Hague.

Melzer, A. (1976). Makrophytische Wasserpflanzen als Indikatoren des Gewässerzustandes oberbayerischer seen. *Dissertationes Botanicae*, **34**, 1–195, Cramer, Vaduz.

Melzer, A., Haben, W., and Kohler, A. (1977). Floristische-Ökologische Charakterisierung und Gliederung der Ostersee (Oberbayern) mit Hilfe von submerse Makrophyten. *Mitteilungen der Florist-soziologischen Arbeitsgemeinschaft*, **19/20**, 139–51.

Mendez, E. (1984). Observaciones ecológicas sobre la vegetación adventicia de cauces de riego en Mendoza. *Parodiana*, **3**, 185–96.

Mercado-Noriel, L. R. and Mercado, B. T. (1978). Floral anatomy and seed morphology of water lettuce (*Pistia stratiotes*). *Philippine Agriculture*, **61**, 281–90.

Meriaux, J.-L. (1978). Étude analytique et comparative de la végétation d'étangs et marais du Nord de la France (Vallée de la Sensee et Bassin Homiller du Nord—Pas-de-Calais). *Documents Phytosociologiques*, **3**, 1–244. Cramer, Vaduz.

Mestrov, M. and Tavcar, V. (1973). Problems of the recovery of Tralcoscan Lake and application of grass carp (*C. idella* [Val.]). *Bulletin Scientifique Conseil des Académies de Sciences et Arts, RST, Yugoslavia* (*A*), **18**, (4/6), 80–1.

Meulemans, J. T. and Best, E. P. H. (1981). Research into the characteristics of C_3 and C_4 plants in *Ceratophyllum demersum* L. *Hydrobiological Bulletin*, **15**, 199–205.

Michaud, M. T., Atchinson, G. J., McIntosh, A. W., Mayes, R. A., and Nelson, D. W. (1979). Changes in phosphorus concentrations in a eutrophic lake as a result of macrophyte kill following herbicide application. *Hydrobiologia*, **66**, 105–11.

Micotz, L. (1981). Vegetación acuática superior. In *Calidad del agua en el Proyecto Paranó Medio. Inf. Tec. 9*, pp. 16–19. Agua y Energia Eléctrica, Sector Ecología, Salud y Desarrollo, Santa Fé (Argentina).

Mikol, G. F. (1985). Effects of harvesting on aquatic vegetation and juvenile fish populations at Saratoga Lake, New York. *Journal of Aquatic Plant Management*, **23**, 59–63.

Miles, W. D. (1976). Land drainage and weed control. *Proceedings Conference on Aquatic Herbicides, British Crop Protection Council Monograph No.* **16**, pp. 7–13.

Miley, W. W., Sutton, D. L., and Stanley, J. G. (1979). The role and impact of the introduced grass carp (*Ctenopharyngodon idella* Val.) in the Union of Soviet Socialist Republics and several other European countries. *Proceedings Grass Carp Conference 1978*, pp. 177–83.

Miller, J. L., Garrard, L. A., and Haller, W. T. (1976). Some characteristics of *Hydrilla* tubers taken from Lake Ocklawaha during drawdown. *Journal of Aquatic Plant Management*, **14**, 29–31.

Miller, R. A., Norris, L. A., and Hawkes, C. L. (1973). Toxicity of TCDD in aquatic organisms. *Environmental Health Perspectives*, **1973**, 177–86.

Mills, C. A. (1981). The spawning of roach *Rutilus rutilis* L. in a chalk stream. *Fisheries Management*, **12**, (2), 49–54.

Milnes, M. H. (1971). Formation of 2,3,7,8-tetrachloro-dibenzo-dioxin by thermal decomposition of sodium 2,4,5-trichlorophenate. *Nature* (*London*), **232**, 395–6.

Ministry of Agriculture, Fisheries & Food (1979). *Guidelines for the use of herbicides on weeds in or near watercourses and lakes*. Booklet 2078, MAFF (Publications), London.

Ministry of Agriculture, Fisheries & Food (1986). *Guidelines for the use of herbicides on weeds in or near watercourses and lakes*. MAFF (Publications), Booklet B2078, Alnwick, UK.

Ministry of Environment, British Columbia (1980). *Studies on aquatic macrophytes. 19. Eurasian watermilfoil treatments with 2,4-D in the Okanagan valley, 1977–1978, Volume 3. Herbicide application effects on Eurasian watermilfoil*. British Columbia, Ministry of Environment: Victoria, BC, Canada. pp. 69.

Ministry of Environment, British Columbia (1981). *A review of the use of 2,4-D for Eurasian water milfoil control in the Okanagan Valley Lakes, 1976–1980*, Volume X. Information Bulletin, Aquatic Plant Management Program, Ministry of Environment, British Columbia, Victoria, BC. pp. 10.

Ministry of Environment and Parks, British Columbia (1986). *Aquatic Plant Management Programme: Information Bulletin No. XII*. Ministry of Environment and Parks, British Columbia, Canada. pp. 28.

Misra, A., Patro, G. K., and Tosh, G. C. (1976). Studies on chemical control of *Chara*. in *Aquatic weeds in south east Asia*, (ed. C. K. Varshney and J. Rzoska), pp. 265–8. W. Junk, The Hague.

Misra, R. D. (1938). Edaphic factors in the distribution of aquatic plants in the English lakes. *Journal of Ecology*, **26**, 411–51.

Mitchell, C. P. (1977). Use of grass carp for submerged weed control. *Fisheries Research Publication New Zealand Ministry of Agriculture and Fisheries* No. 38, pp. 145–8.

Mitchell, C. P. (1980). Control of water weeds by grass carp in two small lakes. *New Zealand Journal of Marine and Freshwater Research*, **14**, 381–90.

Mitchell, C. P. (1981). Grass carp and waterweed. *Soil and Water*, **17**, 22–6.

Mitchell, C. P. (1986). Effects of introduced grass carp on populations of two species of small native fishes in a small lake. *New Zealand Journal of Marine and Freshwater Research*, **20**, 219–30.

Mitchell, C. P., Fish, G. R. and Burnet, A. M. R. (1984). Limnological changes in a small lake stocked with grass carp. *New Zealand Journal of Marine and Freshwater Research*, **18**, 103–14.

Mitchell, D. S. (1969). The ecology of vascular hydrophytes on Lake Kariba. *Hydrobiologia,* **34**, 448–64.

Mitchell, D. S. (1970). Autoecological studies of *Salvinia auriculata* Aubl. Ph.D. thesis, University of London. pp. 669.

Mitchell, D. S. (1972). The Kariba weed, *Salvinia molesta. British Fern Gazette,* **10**, 251–2.

Mitchell, D. S. (1973). Aquatic weeds in man-made lakes. In *Man-made lakes: their problems and environmental effects* (ed. W. C. Ackermann, G. F. White and E. B. Worthington), pp. 606–11. American Geophysical Union, Washington DC.

Mitchell, D. S. (ed.) (1974). *Aquatic vegetation and its use and control.* UNESCO, Paris. pp. 135.

Mitchell, D. S. (1976). The growth and management of *Eichhornia crassipes* and *Salvinia* spp. in their native environment and in alien situations. In *Aquatic weeds in south-east Asia* (ed. C. K. Varshney and J. Rzoska), pp. 167–76. W. Junk, The Hague.

Mitchell, D. S. (1977). Water weed problems in irrigation systems. In *Arid land irrigation in developing countries,* (ed. E. B. Worthington), pp. 317–28. Pergamon Press, Oxford.

Mitchell, D. S. (1978). *Aquatic weeds in Australian inland waters.* Australian Government Publishing Service, Canberra. pp. 189.

Mitchell, D. S. (1979*a*). Assessment of aquatic weed problems. *Journal of Aquatic Plant Management,* **17**, 19–21.

Mitchell, D. S. (1979*b*). Formulating aquatic weed management programs. *Journal of Aquatic Plant Management,* **17**, 22–4.

Mitchell, D. S. (1979*c*). *The incidence and management of* Salvinia molesta *in Papua New Guinea.* Office of Environment & Conservation, Waigani; Department of Primary Industries, Konedou, Papua New Guinea.

Mitchell, D. S. (1985*a*). African aquatic weeds and their management. In *The ecology and management of African wetland vegetation* (ed. P. Denny), pp. 177–202. W. Junk, The Hague.

Mitchell, D. S. (1985*b*). Surface-floating aquatic macrophytes. In *The ecology and management of African wetland vegetation* (ed. P. Denny), pp. 109–24. W. Junk, The Hague.

Mitchell, D. S. (1986). The impact of aquatic weed control on aquatic ecosystems. *Proceedings EWRS/AAB 7th Symposium on Aquatic Weeds 1986,* pp. 213–23.

Mitchell, D. S., Petr, T., and Viner, A. B. (1980). The water fern *Salvinia molesta* in the Sepik river, Papua New Guinea. *Environmental Conservation*, **7**, 115–22.

Mitchell, D. S. and Rose, D. J. W. (1979). Factors affecting fluctuations in extent of *Salvinia molesta* on Lake Kariba. *Pest Articles & News Summaries,* **25**, 171–7.

Mitchell, D. S. and Tur, N. M. (1975). The rate of growth of *Salvinia molesta* (*S. auriculata* Aubl.) in laboratory and natural conditions. *Journal of Applied Ecology,* **12**, 213–25.

Mitchell, J. (1957). *Karmex W for the control of higher aquatic plants and algae.* Inland Fisheries Administration Report No. 57–20, California Department of Fish & Game. pp. 2.

Mitra, E. (1960). Contributions to our knowledge of Indian freshwater plants. 3. Behaviour of *Hydrilla verticillata* Presl in nature and under experimental conditions. *Bulletin of the Botanical Society of Bengal,* **14**, 73–5.

Mitra, E. (1977). *A report on copper sulphate treatments of fishery ponds for the control of weeds, 1952–1970.* Bulletin No. 26, Central Inland Fisheries Research Institute, Barrochpore, W. Bengal, India. pp. 26.

Mitsch, W. J. (1975). Systems analysis of nutrient disposal in cypress wetlands and lake ecosystems in Florida. Dissertation, University of Florida, Gainesville, Florida. pp. 421.

Mitsch, W. J. (1976). Ecosystems modelling of water hyacinth management in Lake Alice, Florida. *Ecological Modelling,* **2**, 69–89.

Mitzer, L. (1978). Evaluation of biological control of nuisance aquatic vegetation by grass carp. *Transactions American Fisheries Society,* **107**, 135–45.

Mixon, W. W. (1974). Hydout, an improved formulation for aquatic weed control. *Proceedings 27th Annual Meeting Southern Weed Science Society,* pp. 307–9.

Miyazaki, S., Sikka, H. C., and Lynch, R. S. (1975). Metabolism of dichlobenil by microorganisms in the aquatic environment. *Journal of Agriculture and Food Chemistry,* **23**, 365.

Mohamed, B. F. and Bebawi, F. F. (1973*a*). Burning as a supportive treatment in controlling water hyacinth in the Sudan. 1. Routine burning. *Hyacinth Control Journal,* **11**, 31–4.

Mohamed, B. F. and Bebawi, F. F. (1973*b*). Burning as a supportive treatment in controlling water hyacinth in the Sudan. II. Backburning. *Hyacinth Control Journal,* **11**, 34–7.

Moody, K. (1981). *Major weeds of rice in south and southeast Asia.* International Rice Research Institute, Manila. pp. 86.

Moody, K. (1985). *Salvinia molesta* found in Philippine rice fields. *International Rice Research Newsletter*, **10**, 25–6.

Mook, J. H., and van der Toorn, J. (1982). The influence of environmental factors and management on stands of *Phragmites australis.* II. Effects on yield and its relationship with shoot density. *Journal of Applied Ecology,* **19**, 501–17.

Moore, A. W. (1969). *Azolla:* biology and agronomic significance. *Botanical Reviews,* **35**, 17–34.

Moore, C. A. M. and Spillett, P. B. (1982). The ecological effects of introducing grass carp in a small lake. *Proceedings 2nd International Symposium on Herbivorous Fish 1982,* pp. 165–75. EWRS, Wageningen, The Netherlands.

Moore, G. T. and Kellerman, K. F. (1904). *A method of destroying or preventing the growth of algae and certain pathogenic bacteria in water supplies.* Department of Agriculture, Bureau of Plant Industry Bulletin No. 64, pp. 15–44.

Moore, J. A. (1986). *Charophytes of Great Britain and Ireland.* Botanical Society of the British Isles, London. pp. 140.

Moore, P. D. (1983). Alien duckweed hiding quietly beside the Cam. *Nature* (*London*), **302**, (5907), 384.

Moreira, I., Fernandes, J. D., Vasconcelos, T., and Fernandes, E. (1983). Plantas vasculares em valas e canais. *Engenharia-Ciencia e Tecnica,* **2**, (4), 20–7.

Morinaga, T. (1926*a*). Effect of alternating temperatures upon the germination of seeds. *American Journal of Botany,* **13**, 141–58.

Morinaga, T. (1926*b*). The favourable effect of reduced O_2 supply upon the germination of certain seeds. *American Journal of Botany,* 13, 159–66.

Morrice, C. P. (1977). The effects of copper sulphate ($CuSO_4.5H_2O$) on the feeding of roach (*Rutilus rutilus* (L.)). *Fisheries Management,* **8**, (3), 82–5.

Morrison, W. G. and Courtney, A. D. (1981). A brief review of herbicide efficacy assessed during a two year experimental programme, in minor watercourses in N. Ireland. *Proceedings Association of Applied Biologists Conference: Aquatic Weeds and their Control, 1981,* pp. 131–9. AAB, Wellesbourne, UK.

Mortensen, E. (1977). Density-dependent mortality of trout fry (*Salmo trutta* L.) and its relationship to the management of small streams. *Journal of Fish Biology,* **11**, 613–17.

Morton, J. F. (1975). Cattails (*Typha* spp.) weed problem or potential crop? *Economic Botany,* **29**, 7–29.

Moss, B. (1982). *Ecology of fresh waters.* Blackwell Scientific Publications, Oxford. pp. 332.

Moss, B., Balls, H. Irvine, K., and Stansfield, J. (1986). Restoration of two lowland lakes by isolation from nutrient-rich water sources with and without removal of sediment. *Journal of Applied Ecology,* **23**, 391–414.

Mossier, J. N. (1968). *Methods for harvesting or control of aquatic plants.* University of Wisconsin Water Resources Center Report OWRR-B-019-W15, Madison, Wisconsin.

Mount, D. I. and Stephen, C. E. (1967). A method for establishing acceptable toxicant limits for fish—malathion and the butoxyethanol ester of 2,4-D. *Transactions American Fish Society,* **96**, 185–93.

Muehlberger, C. (1969). The resistance of aquatic plants to copper sulphate. Note. *Morba/Monatsschr. Ornithol. Vivarienkunde Angsg. B. Aquarien Terrarien,* **16**, (6), 208.

Mugridge, R. E. R., Buckley, B. R., Fowler, M. C., and Stallybrass, H. G. (1982). An evaluation of the use of grass carp (*Ctenopharyngodon idella* Val.) for controlling aquatic plant growth in a canal in Southern England. *Proceedings 2nd International Symposium on Herbivorous Fish 1982,* pp. 8–16. EWRS, Wageningen, The Netherlands.

Muhlberg, H. (1982). *The complete guide to water plants,* (trans. from German by I. Lindsay). EP Publishing, German Democratic Republic. pp. 392.

Muir, D. C. G. (1980). Determination of terbutryn and its degradation products in water, sediment, aquatic plants and fish. *Journal of Agriculture and Food Chemistry,* **28**, 714–18.

Muir, D. C. G. and Grift, N. P. (1982). Fate of fluridone in sediment and water in laboratory and field experiments. *Journal of Agriculture and Food Chemistry,* **30**, 238–44.

Muir, D. C. G., Grift, N. P., Blouw, A. P., and Lockhart, W. L. (1980). Persistence of fluridone in small ponds. *Journal of Environmental Quality,* **9**, (1), 151–6.

Muir, D. C. G., Pitze, M., Blouw, A. P., and Lockhart, W. L. (1981). Fate of terbutryn in macrophyte-free and macrophyte-containing farm ponds. *Weed Research,* **21**, 59–70.

Müller, R. (1978). Ein Versuch zur biologischen Entkrautung von Gewässern. *Österreichs Fischerei,* **31**, 151–6.

Mullison, W. R. (1970). Effects of herbicides on water and its inhabitants. *Weed Science,* **18**, 738–50.

Mulrennan, J. A. (1962). The relationship of mosquito breeding to aquatic plant production. *Hyacinth Control Journal* 1, 6–7.

Murfitt, R. F. A. and Haslam, S. M. (1981). Some unanswered questions relating to the mechanical control of weeds in water channels. *Proceedings Association of Applied Biologists Conference: Aquatic Weeds and their Control 1981*, pp. 87–91. AAB, Wellesbourne, UK.

Murphy, K. J. (1980). The ecology and management of aquatic macrophytes in canals. Ph.D. thesis, University of Liverpool, Liverpool, UK.

Murphy, K. J. (1982). The use of methylthio-triazine herbicides in freshwater systems: a review. *Proceedings EWRS 6th Symposium on Aquatic Weeds, 1982*, pp. 263–77.

Murphy, K. J. (1988). *Aquatic weed problems and their management in Scottish fresh waters 1988*. Report to Nature Conservancy Council, Edinburgh. University of Glasgow, Glasgow, UK. pp. 13.

Murphy, K. J. and Eaton, J. W. (1981*a*). Ecological effects of four herbicides and two mechanical clearance methods used for aquatic weed control in canals. *Proceedings Association of Applied Biologists Conference: Aquatic Weeds and their Control 1981*, pp. 201–17. AAB, Wellesbourne, UK.

Murphy, K. J. and Eaton, J. W. (1981*b*). Water plants, boat traffic and angling in navigable canals. *Proceedings 2nd British Freshwater Fisheries Conference 1981*, pp. 173–87. University of Liverpool, Liverpool, UK.

Murphy, K. J. and Eaton, J. W. (1983). Effects of pleasure-boat traffic on macrophyte growth in canals. *Journal of Applied Ecology*, **20**, 713–29.

Murphy, K. J. and Pearce, H. G. (1987). Habitat modification associated with freshwater angling. In *Angling & wildlife in fresh waters*: ITE Symposium No. 19, (ed. P. S. Maitland and A. K. Turner), pp. 31–46. Institute of Terrestrial Ecology, Grange-over-Sands, Cumbria, UK.

Murphy, K. J., Eaton, J. W., and Hyde, T. M. (1980). A survey of aquatic weed growth and control in the canals and river navigations of the British Waterways Board. *Proceedings 1980 British Crop Protection Conference—Weeds*, **2**, pp. 707–14.

Murphy, K. J., Eaton, J. W., and Hyde, T. M. (1982). The management of aquatic plants in a navigable canal system used for amenity and recreation. *Proceedings EWRS 6th Symposium on Aquatic Weeds 1982*, pp. 141–51.

Murphy, K. J., Fox, A. M., and Hanbury, R. G. (1987). A multivariate assessment of plant management impacts on macrophyte communities in a Scottish canal. *Journal of Applied Ecology*, **24**, 1063–79.

Murphy, K. J., Hanbury, R. G., and Eaton, J. W. (1981). The ecological effects of 2–methylthio triazine herbicides used for aquatic weed control in navigable canals. I. Effects on aquatic flora and water chemistry. *Archiv für Hydrobiologie*, **91**, 294–331.

Musil, C. F. and Breen, C. M. (1977). The applications of growth kinetics to the control of *Eichhornia crassipes* (Mart.) Solms through nutrient removal by mechanical harvesting. *Hydrobiologia*, **53**, 165–71.

Musil, C. F., Grunow, J. O., and Bornman, C. H. (1973). Classification and ordination of aquatic macrophytes in the Pongola River Pams, Natal. *Bothalia*, **11**, 181–90.

Mycock, E. (1979). *Investigation of possible acute effects of the aquatic herbicide diquat-alginate on the benthic invertebrate community downstream of a river treatment trial*. Report BN34 (4/79) Rivers Division, North West Water Authority, Carlisle, UK.

Mycock, E. (1983): *Report on a biological survey of watercourses on the Solway plain, with a view to management of aquatic and marginal weed growths.* Report BN 79–10–83, Rivers Division, North West Water Authority, Carlisle, UK. pp. 20.

Nag Raj, T. R. (1965). Thread blight of water hyacinth. *Current Science,* **43,** 618–19.

Nag Raj, T. R. and Ponnappa, K. M. (1967). Some interesting fungi of India. *Technical Bulletin* No. 9, pp. 31–43, Commonwealth Institute of Biological Control, Bangalore, India.

Nag Raj, T. R. and Ponnappa, K. M. (1970*a*). Some interesting fungi occurring on aquatic weeds and *Striga* spp. in India. *Journal of the Indian Botanical Society,* **49,** 64–72.

NagRaj, T. R., and Ponnappa, K. M. (1970*b*). Blight of water hyacinth caused by *Alternaria eichhorniae* sp. nov. *Transactions of the British Mycological Society,* **55,** 123–30.

Nakagawa, K. and Miyahara, M. (1967). The effect of foliage treatment of several herbicides on the control of *Cyperus serotinus* and *Eleocharis tuberosa* var. *kuroguwai. Weed Research, Japan,* **6,** 107–11.

Napompeth, B. (1982). Biological control, research and development in Thailand. In *Proceedings International Conference on Plant Protection in the Tropics.* pp. 23.

Napompeth, B. (1984). Biological control of water hyacinth in Thailand. In *Proceedings International Conference on Water Hyacinth* (ed. G. Thyagarajan), pp. 811–22. UN Environment Programme, Nairobi.

Napompeth, B. (1985). Biological methods of aquatic vegetation management in tropical Asia. Paper presented at *Workshop on Ecology & Management of Aquatic Vegetation in the Tropics,* Centre for Studies of the Environment and Human Resources, University of Indonesia, Jakarta.

Naqvi, S. M., Leung, T. S., and Naqvi, N. Z. (1980). Toxicities of paraquat and diquat herbicides to freshwater copepods (*Diaptomus* sp. and *Eucyclops* sp.). *Bulletin of Environmental Contamination Toxicology,* **25,** 918–20.

Narayanayya, D. V. (1928). The aquatic weeds in Deccan irrigation channels. *Journal of Ecology,* **16,** 123–33.

Narine, D. R. and Guy, R. D. (1982). Binding of diquat and paraquat to humic acid in aquatic environments. *Soil Science,* **33,** 356–63.

National Academy of Sciences (1976). *Making aquatic weeds useful: some perspectives for developing countries. Ad Hoc* Committee, National Academy of Sciences, Washington DC, USA. pp. 175.

National Committee on Management of Aquatic Weeds (1982). *Water weeds in Australia: a national approach to management.* Australian Government Publishing Service, Canberra; Department of National Development and Energy, Australia Water Resources Council.

Neel, J. K., Peterson, S. A., and Smith, W. L. (1973). *Weed harvest and lake nutrient dynamics.* Ecological Research Service, US Environmental Protection Agency, Washington DC. pp. 91.

Negonovskaya, I. I. (1980). On the results and prospects of the introduction of phytophagous fishes into waters of the USSR. *Journal of Ichthyology,* **20,** 101–11.

Neiff, J. J. (1975). Fluctuaciones anuales en la composición fitocenótica y biomasa de la hidrofitia en lagunas isleñas del Paraná Medio. *Ecosur,* **2,** 153–83.

Neiff, J. J. (1979). Fluctuaciones de la vegetación acuática en ambientes del valle de inundación del Paraná Medio. *Physis,* **38,** 41–53.

Neiff, J. J. (1981). Panorama ecológico de los cuerpos de agua del noroeste argentino. *Simposio IV Jornadas Argentinas de Zoologia*, pp. 115–51.

Neiff, J. J. (1986). Las grandes unidades de vegetación y ambiente insular del Río Paraná en el tramo Candelaria—Ita Ibaté. *Revista Asociacion Ciencias Naturales Litoral*, **17**, 7–30.

Neiff, J. J. and Bonetto, A. A. (1983). Aportes alóctonos de vegetación flotante de arrastre y sus posibilidades de desarrollo en el embabe. In *Estudios Ecológicos en el área de embalse del Paraná Medio* (*Cierre Norte*), pp. 155–75. CECOAL-A. E.E. Tomo 1. pp. 532.

Neiff, J. J. and Marchessi, E. (1978). Caracterización sinóptica de la vegetación acuática y anfibia en el área del futuro embalse de Salto Grande, estimación del riesgo potencial del desarrollo de hidrófitos en el embalse. *5a. Reunión sobre aspectos de desarrollo ambiental. Salto* (*Uruguay*)—*Concordia* (*Argentina*) *Centro de Ecologia Aplicada del Litoral 1978*. pp. 34.

Nelson, M. L., Gangstad, E. O., and Seaman, D. E. (1970). *Potential growth of aquatic plants of the lower Mekong River Basin, Laos-Thailand.* Report, US Agency for International Development. pp. 96.

Neogy, B. P., Kachroo, P., and Biswas, K. (1957). Aquatic vegetation of Bokaro reservoir in relation to *Anopheles* mosquito larvae and its control by application of herbicides. *Journal of Malariology*, **11**, 191–212.

Nesbitt, H. J. and Watson, J. R. (1980*a*). Degradation of the herbicide 2,4-D in river water. I. Description of study area and survey of rate-determining factors. *Water Research*, **14**, 1683–8.

Nesbitt, H. J. and Watson, J. R. (1980*b*). Degradation of the herbicide 2,4-D in river water. II. The role of suspended sediment, nutrients and water temperature. *Water Research*, **14**, 1689–94.

Neururer, H. (1974*a*). Stand der chemischen Bekämpfung von Wasserunkräutern in Österreich. *Proceedings EWRC 4th Symposium on Aquatic Weeds 1974*, pp. 152–7.

Neururer, H. (1974*b*). Lenkung und Überwachung des Herbizideinsatzes in Österreichischen Gewässern. *Proceedings EWRC 4th Symposium on Aquatic Weeds 1974*, pp. 267–74.

Neururer, H. (1978). Erfahrungen mit der Kombination von biologischen und chemischen Massnahmen zur langfristigen Kontrolle unerwünschter Wasserpflanzen. *Proceedings EWRS 5th Symposium on Aquatic Weeds 1978*, pp. 239–44.

Newbold, C. (1974). The ecological effects of the herbicide dichlobenil within pond ecosystems. *Proceedings EWRC 4th Symposium on Aquatic Weeds 1974*, 37–51.

Newbold, C. (1975). Herbicides in aquatic systems. *Biological Conservation*, **7**, 97–118.

Newbold, C. (1976). Environmental effects of aquatic herbicides. *Proceedings Symposium on Aquatic Herbicides, British Crop Protection Council Monograph No.* **16**, pp. 78–90.

Newbold, C. (1982). The management principles of nature conservation and land drainage—two antagonistic aims? *Proceedings EWRS 6th Symposium on Aquatic Weeds 1982*, pp. 86–95.

Newman, J. F. (1967). The ecological effects of bipyridyl herbicides used for aquatic weed control. *Proceedings EWRC 2nd Symposium on Aquatic Weeds 1967*, 168–74.

Newroth, P. R. (1985). A review of Eurasian watermilfoil impacts and management in British Columbia. In *1st International Symposium on Watermilfoil* (Myriophyllum spicatum) *and related Haloragaceae species* (ed. L. W. J. Anderson), 139–53. Aquatic Plant Management Society, Vicksburg, Mississippi.

New South Wales Water Resources Commission (1986). *Water and the natural resources of New South Wales.* Water Resources Commission, New South Wales, Sydney.

New Zealand Ministry of Agriculture and Fisheries and Agricultural Chemicals Board (1975). *A guide for the use of herbicides on weeds in and near watercourses, ponds and lakes.* Ministry of Agriculture and Fisheries/Agricultural Chemicals Board, New Zealand.

Nezodoliy, V. K. and Mitrofanov, V. P. (1975). Natural reproduction of the grass carp, *Ctenopharyngodon idella* in the Ili River. *Journal of Ichthyology,* **15**, 927–33.

Nguyen-van-Vuong (1973*a*). *Report on the aquatic weed problem of the Brantas River Multipurpose Project.* Document WR/73/054, Biotrop, Bogor.

Nguyen-van-Vuong (1973*b*). *Review of the genus* Salvinia *Adans.* Document WR/73/056, Biotrop, Bogor, Indonesia.

Nguyen-van-Vuong (1979). Weed flora in rice fields in South Vietnam. *Biotrop Bulletin,* **11**, 155–62.

Nguyen-van-Vuong and Sumartono, T. (1979). Some notes on the dispersal of *Salvinia* spp. in Java. *Biotrop Bulletin,* **11**, 227–38.

Nichols, D. S. and Keeney, D. R. (1973). Nitrogen and phosphorus release from decaying water milfoil. *Hydrobiologia,* **42**, 509–25.

Nichols, D. S. and Keeney, D. R. (1976). Nitrogen nutrition of *Myriophyllum spicatum.* II. Uptake and translocation of ^{15}N by shoots and roots. *Freshwater Biology,* **6**, 145–54.

Nichols, S. A. (1973). The effects of harvesting aquatic macrophytes on algae. *Transactions of the Wisconsin Academy of Sciences,* **61**, 165–72.

Nichols, S. A. (1974). *Mechanical and habitat manipulation for aquatic plant management. A review of techniques.* Department of Natural Resources Technical Bulletin 77, Madison, Wisconsin. pp. 34.

Nichols, S. A. (1975*a*). Identification and management of Eurasian water milfoil in Wisconsin. *Transactions of the Wisconsin Academy of Sciences,* **63**, 116–28.

Nichols, S. A. (1975*b*). The use of overwinter drawdown for aquatic vegetation management. *Water Research*, **11**, 1137–48.

Nichols, S. A. (1979). Macrophyte biology to management techniques: bridging the information gap. In *Aquatic plant, lake management and ecosystem consequences of lake harvesting: Proceedings Congress at Madison, Wisconsin* (ed. J. E. Breck, R. T. Prentki, and O. L. Loucks), pp. 63–78.

Nichols, S. A. (1984*a*). Qualitative methods for assessing macrophyte vegetation. *Special Technical Publication American Society for Testing Materials STP 843,* pp. 7–15.

Nichols, S. A. (1984*b*). Macrophyte community dynamics in a dredged Wisconsin lake. *Water Resources Bulletin,* **20**, 573–6.

Nichols, S. A. and Cottam, G. (1972). Harvesting as a control for aquatic plants. *Water Resources Bulletin,* **8**, 1205–10.

Niehuss, M. and Boerner, H. (1971). Persistence and distribution of diuron, dichlobenil, diquat, and paraquat following application in drainage ditches. *Pflanz und Pflanz,* **78**, 385.

Niemann, E. (1980). Zur Ansprache des 'Verkrautungszustandes' in Fliessgewässern. *Acta Hydrochimica et Hydrobiologica,* **8**, 47–58.

Nikolayevski, V. B. (1971). Research into the biology of the common reed in the USSR. *Folia Geobotanica et Phytotoxonomica Praha,* **6**, 221–30.

Nikolsky, G. V. and Aliyev, D. D. (1974). The role of far-eastern herbivorous fishes in the ecosystems of natural waters in which they are acclimatized. *Journal of Ichthyology,* **14**, 842–7.

Noda, K. (1969). Species of hazardous weeds and their control on paddy rice fields. *Proceedings Asian–Pacific Weed Science Society Conference,* **2**, 97–111.

Noda, K. (1979). Recent changes of weed population in paddy fields. *Biotrop Bulletin,* **11**, 93–8.

Noda, K. and Obayash, H. (1971). Ecology and control of knotgrass (*Paspalum distichum*). *Weed Research, Japan,* **11**, 35–9.

Nolan, W. J. and Kirmse, D. W. (1974). The papermaking properties of water hyacinth. *Hyacinth Control Journal,* **12**, 90–7.

Norton, J. and Ellis, J. (1976). Management of aquatic vegetation with simazine. *Proceedings Southern Weed Science Society,* **29**, 359–64.

Numata, M. and Shinozaki, H. (1967). Floristic composition and dynamics of weed communities in rice fields. *Bulletin Marine Laboratory, Chiba University,* **9**, 24–9.

Obeid, M. and Tag el Seed, M. (1976). Factors affecting dormancy and germination of *Eichhornia crassipes* (Mart.) Solms. from the Nile. *Weed Research,* **16**, 71–80.

Ochse, J. J. and Bakhuizen van den Brink, R. C. (1931). *Vegetables of the Dutch East Indies.* Department of Agriculture, Industry and Commerce, Netherlands East Indies, Buitenzorg, Java.

Ochyra, R. (1979). [Flora of the Karst sink-holes in the vicinity of Staszow (S. Poland) I. Vascular plants]. In Polish. *Fragmenta Floristica Geobotanica,* **25**, 209–36.

Ogg, A. G. (1972). Residues in ponds treated with two formulations of dichlobenil. *Pesticide Monitoring Journal,* **5**, 356–9.

Oglesby, R. T., Vogel, A., Peverly, J. H., and Johnson, R. (1976). Changes in submerged plants at the south end of Cayuga Lake following tropical storm Agnes. *Hydrobiologia,* **48**, 251–5.

Ogren, W. L. (1984). Photorespiration: pathways, regulation and modification. *Annual Review Plant Physiology,* **35**, 415–42.

Okafor, L. I. (1980). Weeds of Lake Chad and their potential problems in the South Chad Irrigation Project. In *10th Annual Conference, Weed Science Society Nigeria, Benin City, Nigeria.*

Okafor, L. I. (1982). A preliminary study of some major aquatic weeds in Lake Chad, Nigeria. *Proceedings EWRS 6th Symposium on Aquatic Weeds*, pp. 10–19.

Økland, K A. (1974). Macrovegetation and ecological factors in two Norwegian lakes. *Norwegian Journal of Botany,* **21**, 137–59.

Oksiyuk, O. P. (1976). Control of growth of autotrophic organisms in water-supplying canals. *Limnologica,* **10**, 393–7.

Olah, J., Kintzly, A. V., and Varadi, L. (1982). Liquid manure utilization and purification by silver carp (*Hypophthalmichthys molitrix* Val.). *Proceedings 2nd International Symposium on Herbivorous Fish 1982*, pp. 47–52. EWRS, Wageningen, The Netherlands.

O'Leary, P. J. and Ouzts, J. D. (1983). Torward integrated control of water hyacinth. *Proceedings 36th Annual Meeting Southern Weed Science Society 1983, Biloxi, Mississippi*, p. 340.

O'Loughlin, E. M. (1975. Predicting the concentration of aquatic herbicide residues in waterways. *Technical Conference Institute of Engineers Hydrology Symposium, Armidale 1975*, pp. 56–60.

O'Loughlin, E. M. and Bowmer, K. H. (1975). Dilution and decay of aquatic herbicides in flowing channels. *Journal of Hydrology* (*Amsterdam*), **26**, 217–35.

Olsen, S. (1944). Danish Charophyta. *Botanisk Tidsskrift*, **51**, 242–7.

Ondok, J. P. (1973). Photosynthetically active radiation in a stand of *Phragmites communis* Trin. I. Distribution of irradiance and foliage structure. *Photosynthetica*, **7**, 8–17.

Ondok, J. P. (1978). Radiation climate in fishpond littoral plant communities. In *Pond littoral ecosystems*, Ecological Studies No. 28 (ed. D. Dykyjova and J. Květ), pp. 113–25. Springer-Verlag, Berlin.

Ondok, J. P. and Gloser, J. (1978*a*). Net photosynthesis and dark respiration in a stand of *Phragmites communis* Trin. calculated by means of a model. I. Description of the model. *Photosynthetica*, **12**, 328–36.

Ondok, J. P. and Gloser, J. (1978*b*). Net photosynthesis and dark respiration in a stand of *Phragmites communis* Trin. calculated by means of a model. II. Results. *Photosynthetica*, **12**, 337–43.

Ondok, J. P. and Gloser, J. (1978*c*). Modeling of photosynthetic production in littoral helophyte stands. In *Pond littoral ecosystems*, Ecological Studies No. 28 (ed. D. Dykyjova and J. Květ), pp. 234–45. Springer-Verlag, Berlin.

Ondok, J. P. and Pokorny, J. (1982). Model of diurnal regime of O_2 and CO_2 in stands of submerged aquatic vegetation. *Ecologia* (*Czechoslovakia*), **1**, 381–94.

Ondok, J. P., Pokorny, J., and Květ, J. (1984. Model of diurnal changes in oxygen, carbon dioxide and bicarbonate concentrations in a stand of *Elodea canadensis* Michx. *Aquatic Botany*, **19**, 239–305.

O'Neal , S. N. and Lembi, C. A. (1983). Effect of simazine on photosynthesis and growth of filamentous algae. *Weed Science*, **31**, 899–903.

O'Neill, R. V., Goldstein, R. A., Shugart, H. H., and Mankin, J. B. (1972). *Terrestrial ecosystem energy model.* US-IBP-EDFB Memo Report 72–19.

Opuszynski, K. (1972). Use of phytophagous fish to control aquatic plants. *Aquaculture*, **1**, 61–74.

Opuszynski, K. (1979). Weed control and fish production. *Proceedings of the Grass Carp Conference, Gainesville*, pp. 103–38.

Opuszynski, K. (1982*a*). Rearing grass carp larvae in ponds fed with warm water discharged by power stations. *Proceedings 2nd International Symposium on Herbivorous Fish 1982*, pp. 140–8. EWRS, Wageningen, The Netherlands.

Opuszynski, K. (1982*b*). Method of phytophagous fish culture used in China and countries of Eastern Europe. *Proceedings 2nd International Symposium on Herbivorous Fish 1982*, pp. 121–32. EWRS, Wageningen, The Netherlands.

Orchard, A. E. (1981). A revision of South American *Myriophyllum* (Haloragaceae) and its repercussions on some Australian and North America species. *Brunonia*, **4**, 27–65.

Orloci, I. (1966). Geometric models in ecology. I. The theory and application of some ordination methods. *Journal of Ecology*, **54**, 193–215.

Ormeno, N. J. (1983). Prospección de las principales malezas asociadas al cultivo de arroz (*Oryza sativa* L.). *Agricultura Técnica* (*Santiago*), **43**, 285–7.

Ortscheit, A., Jaeger, P., Carbinier, R., and Kapp, E. (1982). Les modifications des eaux et de la végétation aquatique du Waldrhein consécutives à la mise en place de l'ouvrage hydroélectrique de Gombsheim, au nord de Strasbourg. In *Studies on aquatic vascular plants* (ed. I. J. Symoens, S. S. Hooper, and P. Compère), pp. 277–83. Royal Botanical Society of Belgium, Brussels.

Osborne, J. A. (1982). The potential of the hybrid grass carp as a weed control agent. *Journal of Freshwater Ecology*, **1**, 353–60.

Osborne, J. A. (1985). A preliminary study of the efficacy of hybrid grass carp for *Hydrilla* control. *Journal of Aquatic Plant Management*, **23**, 16–20.

Osborne, J. A. and Sassic, N. M. (1981). The size of grass carp as a factor in the control of *Hydrilla*. *Aquatic Botany*, **11**, 129–36.

Osgerby, J. M. (1975). The release of cyanatryn from a controlled-release formulation into flowing water. *Pesticide Science*, **6**, 675–85.

Osman, H. E., El Hag, G. A., and Osman, M. M. (1975). Studies on the nutritive value of water hyacinth (*Eichhornia crassipes* Mart. Solms.). In *Aquatic weeds in the Sudan with special reference to water hyacinth* (ed. M. Obeid), pp. 104–27. National Council for Research, Kartoum.

Otto, N. E. (1975). *Survey of irrigation canal ecological parameters influencing aquatic weed growth*. Report: US Bureau of Reclamation, REL-ERC-75-9. pp. 59.

Overdieck, D. (1978). CO_2—gaswechsel und transpiration von schilf bei abgestuften stickstoff—und phosphorgaben. *Verhandlungen der Gesellschaft für Ökologie*, 369–76.

Ozimek, T. and Kowalczewski, A. (1984). Long-term changes of the submerged macrophytes in eutrophic Lake Mikolajskie (North Poland). *Aquatic Botany*, **19**, 1–11.

Pain, S. (1987). Australian invader threatens Britain's waterways. *New Scientist*, 23 July **1987**, 26.

Painter, D. S., and Waltho, J. I. (1985). Short-term impact of harvesting on Eurasian Watermilfoil. In *1st International Symposium on Watermilfoil* (Myriophyllum spicatum) *and related Haloragaceae species* (ed. L. W. J. Anderson), pp. 187–201. Aquatic Plant Management Society Inc., Vicksburg, Mississippi.

Palazzo, A. J., Tice, A. R., Oliphant, J. L., and Graham, J. M. (1984). *Effects of low temperatures on the growth and unfrozen water content of an aquatic plant*. Cold Regions Research and Engineering Laboratory Report 84–14, US Army Corps of Engineers.

Pallis, M. (1916). The structure and history of plav: the floating fen of the delta of the Danube. *Journal of Linnaean Society, Botany*, **43**, 233–90.

Pancho, J. V. (1979). Key to the families of weed flora of southeast Asia. *Biotrop Bulletin*, **11**, 139–50.

Pancho, J. V. and Soerjani, M. (1978). *Aquatic weeds of southeast Asia*. National Publishing Corporation, Quezon City, Philippines. pp. 130.

Pancho, J. V., Vega, M. R., and Plucknett, D. L. (1969). *Some common weeds of the Philippines*. Weed Science Society, Philippines, Manila.

Parija, P. (1934). A note on the reappearance of water hyacinth seedlings in cleared tanks. *Indian Journal of Science,* **4,** 1049.

Park, R. A. *et al.* (1974). A generalized model for simulating lake ecosystems. *Simulation,* **24,** 33–50.

Parka, S. J., Arnold, M. C., McCowen, M. C., and Young, C. L. (1978). Fluridone: a new herbicide for use in aquatic weed control systems. *Proceedings EWRS 5th Symposium on Aquatic Weeds 1978*, pp. 179–87.

Parker, C. (1966). Water hardness and phytotoxicity of paraquat. *Nature,* (*London*), **212,** 1465.

Paterson, D. M. and Wright, S. J. L. (1983). Herbicide effects on communities of aquatic micro-algae. *Abstracts 3rd International Symposium on Microbial Ecology, Michigan State University*, 70.

Patnaik, S. and Ramachandran, V. (1976). Control of blue-green algal blooms with simazine in fish ponds. In *Aquatic weeds in south-east Asia* (ed. C. K. Varshney and J. Rzóska), pp. 285–91. W. Junk, The Hague.

Patten, B. C. (1976). *Systems analysis and simulation in ecology III.* Academic Press, New York. pp. 601.

Patterson, D. J. and Duke, S. O. (1979). Effect of growth irradiance on the maximum photosynthetic capacity of water hyacinth (*Eichhornia crassipes* (Mart.) Solms.). *Plant and Cell Physiology,* **20,** 177–84.

Payne, B. (1982). Problem of identification and assessment for aquatic plant management. *Proceedings 16th Annual Meeting Aquatic Plant Control Research Planning and Operations Review*. Miscellaneous Paper A-82-3, pp. 110–2: US Army Engineer Waterways Experiment Station, Vicksburg, Mississippi.

Payne, D. H. (1974). Aquatic weed control with the new herbicide WL 63611. *Proceedings EWRC 4th Symposium on Aquatic Weeds 1974*, pp. 210–17.

Pearce, H. G. and Eaton, J. W. (1983). Effects of recreational boating on freshwater ecosystems—an annotated bibliography. Appendix B in *Waterways ecology and the design of recreational craft*, pp. 13–68. Inland Waterways Amenity Advisory Council, London.

Pearcy, R. W., Berry, J. A., and Bartholomew, B. (1974). Field photosynthetic performance and leaf temperatures of *Phragmites communis* under summer conditions in Death Valley, California. *Photosynthetica,* **8,** 104–8.

Pearcy, R. W., Bjorkman, O., Caldwell, M. M., Keely, J. E., Monson, R. K., and Strain, B. R. (1987). Carbon gain by plants in natural environments. *Bioscience,* **37,** 21–9.

Pearlmutter, N. L. and Lembi, C. A. (1976). Effects of copper sulphate on the green alga *Pithophora. Proceedings North Central Weed Control Conference 1976,* **31,** pp. 103.

Pearsall, W. H. (1917). The aquatic and marsh vegetation of Esthwaite Water. *Journal of Ecology,* **5,** 180–202.

Pearsall, W. H. (1920). The aquatic vegetation of the English lakes. *Journal of Ecology,* **8,** 163–99.

Pearsall, W. H. (1921). The development of vegetation in the English lakes, considered in relation to the general evolution of glacial lakes and rock basins. *Proceedings of the Royal Society B,* **92,** 259–84.

Pearson, R. G. and Jones, N. V. (1975). The effects of dredging operations on the benthic community of a chalk stream. *Biological Conservation*, **8**, 273–8.

Pearson, R. G. and Jones, N. V. (1978). Effects of weedcutting on the macro-invertebrate fauna of a canalized section of the R. Hull, a North English chalk stream. *Journal of Environmental Management*, **7**, 91–9.

Peck, D. E., Corwin, D. L., and Farmer, W. J. (1980). Absorption-desorption of diuron by freshwater sediments. *Journal of Environmental Quality*, **9**, 101–6.

Pedralli, G., Irgang, B. E., and Pereira, C. P. (1985). Macrófitos aquáticos do municipio de Río Grande do Sul, Brasil. *Revista AGROS*, **20**, 45–52.

Penfound, W. T. (1940). The biology of *Achyranthes philoxeroides* (Mart.) Standley. *American Midland Naturalist*, **24**, 248–52.

Penfound, W. T. and Earle, T. T. (1948). The biology of the water hyacinth. *Ecological Monographs*, **18**, 449–72

Peng, S. Y. and Twu, L. T. (1979). Studies on the regenerative capacity of rhizomes of torpedo grass (*Panicum repens* L.). Part 1. Characteristics in sprouting of rhizomes and resistance to herbicides and environmental adversities. *Journal of the Agricultural Association, China*, **107**, 61–74.

Pennington, J. C. (1986). *Bioengineering technology meeting Sept 1983*. Miscellaneous Paper A-86-1, US Army Engineer Waterways Experiment Station, Vicksburg, Mississippi.

Peñuelas, J. (1983). Vegetación briofítica acuática del Río Muga y sus affluentes (verona). In *II Congreso Nacional de la Sociedad Espagnola di Lemnologia, Murcia*.

Pereyra-Ramos, E. (1982). The ecological role of Characeae in the lake littoral. *Ekologiya Polska*, **29**, 167–209.

Pérez del Viso, R., Tur, N. M., and Montovani, V. (1968). Estimación de la biomasa de hidrófitos en cuencas isleñas del Paraná Medio. *Physis*, **28**, 219–26

Pérez-Moreau, R. A. (1938). Revisión de las *Hydrocotyle* argentinas. *Lilloa*, **2**, 413–63.

Perkins, B. D. (1972). Potential for water hyacinth management with biological agents. *Proceedings Annual Tall Timbers Conference on Ecological Animal Control by Habitat Management, February 24–25, 1972*. pp. 53–64.

Perkins, B. D. (1973). Release in the United States of *Neochetina eichhorniae* Warner. *Proceedings of the 26th Annual Meeting of the Southern Weed Science Social*, (*USA*). p. 368.

Perkins, B. D. (1976). Enhancement of effect of *Neochetina eichhorniae* for biological control of water hyacinth. *Proceedings IV International Symposium on Biological Control of Weeds*, pp. 87–92.

Perkins, B. D. (1977). Preliminary results of integrating chemical and biological controls to combat water hyacinth. *Proceedings Research Planning Conference Aquatic Plant Control Program 1976*, pp. 230–5. Miscellaneous Paper A-77-3, US Army Engineer Waterways Experiment Station, Vicksburg, Mississippi.

Perkins, M. A., Boston, H. L., and Curren, E. F. (1980). The use of fiberglass screens for control of Eurasian watermilfoil. *Journal of Aquatic Plant Management*, **18**, 13–19.

Perkins, M. A. and Sytsma, M. D. (1981). Efficacy of mechanical harvesting and its influence upon carbohydrate accumulation in Eurasian watermilfoil. *Proceedings 15th Annual Meeting Aquatic Plant Control Research Planning and Operations Review, Miscellaneous Paper A-81-3*, pp. 464–79. US Army Engineer Waterways Experiment Station, Vicksburg, Mississippi.

Perkins, M. A. and Sytsma, M. D. (1987). Harvesting and carbohydrate accumulation in Eurasian watermilfoil. *Journal of Aquatic Plant Management,* **25**, 57–62.

Perrier, E. R. and Gibson, A. C. (1982). *Simulation for harvesting of aquatic plants.* Technical Report A-82-1, US Army Engineer Waterways Experiment Station, Vicksburg, Mississippi. pp. 49.

Perring, F. H. and Farrell, L. (1983). *British red data books. I. Vascular plants.* Royal Society for Nature Conservation, Lincoln. pp. 98.

Perry, F. (1961). *Water Gardening* (3rd edn.). Country Life, London.

Pesacreta, G. J., Hodson, R. G., and Langeland, K. A. (1984. Integrated management of hydrilla in N. Carolina: the first year. *Abstract in: Proceedings 24th Annual Meeting Aquatic Plant Management Society 1984, Richmond, USA.*

Peterson, S. A., Smith, W. L., and Malveg, K. W. (1974). Full scale harvest of aquatic plants: nutrient removal from a eutrophic lake. *Journal of the Water Pollution Control Federation,* **46**, 697–707.

Petetin, C. A. and Molinari, E. P. (1977). Plantas acuáticas y palustres. In *Clave Ilustrada para el reconocimiento de malezas en el campo al estado vegetativo*, pp. 13–30. Colección Científica. Inst. Nac. Tecnol. Agrop. (INTA), Buenos Aires. pp. 243.

Petryk, S. and Bosmajian, G. (1975). Analysis of flow through vegetation. *Proceedings of the Hydraulics Division American Society of Civil Engineers,* **101**, 871–84.

Peverly, J. H. and Johnson, R. L. (1979). Nutrient chemistry in herbicide-treated ponds of differing fertility. *Journal of Environmental Quality,* **8**, 294–300.

Peverly, J. H., Miller, G., Brown, W. H., and Johnson, K. L. (1974). *Aquatic weed mananagement in the Finger Lakes.* Cornell University Water Resources and Marine Science Center, Technical Report 90. pp. 26.

Pflieger, W. L. (1978). Distribution and status of grass carp (*Ctenopharyngodon idella*) in Missouri streams. *Transactions American Fisheries Society,* **107**, 113–18.

Pheang, C. T. and Muchsin, I. (1975). Aquatic weed control using grass carp (*Ctenopharyngodon idella* Val.). *Proceedings 3rd Indonesian Weed Science Conference, Bandung*, pp. 406–19.

Philipose, M. T. (1968). Present trends in the control of weeds in fish cultural waters of Asia and the Far East. *FAO Fisheries Report,* **44**, 26–52.

Philipose, M. T. (1976). Fifty years of aquatic weed control in India. In *Aquatic weeds in south-east Asia* (ed. C. K. Varshney and J. Rzoska), pp. 215–23. W. Junk, The Hague.

Philipose, M. T., Ramachandran, V., Singh, S. B., and Ramaprabhu, T. (1970). Some observations on the weeds of cultivable freshwaters in Orissa. *Journal of the Inland Fisheries Society, India,* **2**, 61–84.

Philipose, S. A. (1963). Making fishponds weed free. *Indian Livestock,* **1**, 20–1, 34–6.

Philipp, O., El Tayeb, A., and Hag Yousif, B. (1979). Some studies and aims of the utilization of water hyacinth, *Eichhornia crassipes* (Mart.) Solms. in Sudan. In *Weed Research in Sudan: Proceedings of a Symposium* (ed. M. E. Beshir and W. Koch), *Berichte Fachgebeit Herbologie, Universität Hohenheim, Heft 18*, pp. 106–15.

Philipp, O., Koch, W., and Köser, H. (1983). *Utilization and control of water hyacinth in Sudan.* Eschbom. German Agency for Technical Cooperation 1983. pp. 224.

Phillips, G. L., Eminson, D., and Moss, B. (1978). A mechanism to account for macrophyte decline in progressively eutrophicated freshwaters. *Aquatic Botany,* **4**, 103–26.

Piaget, J. and Schliemann, G. (1973). Weeds choking Western Cape rivers—appearance of *Myriophyllum* spp. in the Boland. *Deciduous Fruit Grower,* **23**, 176–9.

Piccoli, F. and Gerdol, R. (1981). Rice-field weed communities in Ferrara province (Northern Italy). *Aquatic Botany,* **10**, 317–28.

Pierce, B. A. (1983). Grass carp status in the United States: a review. *Environmental Management,* **7**, 151–60.

Pierce, N. D. (1970). *Inland dredging evaluation.* Department of Natural Resources Technical Bulletin 46, Madison, Wisconsin. pp. 68.

Pierce, P. C. and Opoku, A. (1971). Summary of aquatic weed survey and control data of Volta Lake during 1969. *Hyacinth Control Journal,* **9**, 49–54.

Pieters, A. J. and de Boer, F. G. (1971). Dichlobenil in the aquatic environment. *Proceedings EWRC 3rd Symposium on Aquatic Weeds 1971*, 183–91.

Pieterse, A. H. (1977*a*). *Control of tropical aquatic weeds.* Bulletin 300, Department of Agricultural Research. Koninklijk Instituut voor de Tropen, Amsterdam. pp. 20.

Pieterse, A. H. (1977*b*). De bestrijding van tropische wateronkruiden. *Natuur en Techniek,* **45**, 50–64.

Pieterse, A. H. (1978). The water hyacinth (*Eichhornia crassipes*)—a review. *Abstracts on Tropical Agriculture,* **4**, 9–42.

Pieterse, A. H. (1979). Aquatic weed control in tropical and sub-tropical regions. In *Weed Research in Sudan: Proceedings of a Symposium* (ed. M. E. Beshir and W. Koch). *Berichte Fachgebeit Herbologie, Universität Hohenheim,* **1**, 130–6.

Pieterse, A. H. (1981). *Hydrilla verticillata:* a review. *Abstracts on Tropical Agriculture,* **7**, 9–34.

Pieterse, A. H., de Lange, L., and van Vliet, J. P. (1977). A comparative study of *Azolla* in the Netherlands. *Acta Botanica Neerlandica,* **26**, 433–49.

Pieterse, A. H., de Lange, L., and Verhagen, L. (1981). A study of certain aspects of seed germination and growth of *Pistia stratiotes* L. *Acta Botanica Neerlandica,* **30**, 47–57.

Pieterse, A. H., Ebbers, A. E. H., and Verkleij, J. A. C. (1983). A comparative study on isoenzyme patterns in *Hydrilla verticillata* (L.f.) Royle from Ireland and North Eastern Poland. *Aquatic Botany,* **18**, 299–303.

Pieterse, A. H., de Boer, W., Hassing, P. A. G., Blom, P., Bus, G. P., and Vogel, L. C. (1987). *A resource planning data review on major African inland swamp and flood plain ecosystems.* Report to the World Bank. Royal Tropical Institute, Amsterdam. pp. 157.

Pieterse, A. H. and Roorda, F. A. (1982). Synergistic effect of gibberellic acid and chlorflurenol on 2,4-D with regard to water hyacinth control. *Aquatic Botany,* **13**, 69–72.

Pieterse, A. H., Roorda, F. A., and Verhagen, L. (1980). Ten-fold enhancement of 2,4-D effect on water hyacinth by addition of gibberellic acid. *Experientia,* **36**, 650–1.

Pieterse, A. H. and Schulten, G. G. M. (1976). *Final report of mission to Venezuela in May 1976 on aquatic weed problems.* Royal Tropical Institute, Amsterdam. pp. 26.

Pieterse, A. H., Siregar, H., and Soemarwoto, O. (1975). The spread of noxious aquatic weeds in the Citarum basin. *Proceedings 3rd Indonesian Weed Science Conference, Bandung*, pp. 458–62.

Pieterse, A. H., Staphorst, H. P. M., and Verkleij, J. A. C. (1984). Some effects of nitrogen and phosphorus concentration on the phenology of *Hydrilla verticillata* (L.f.) Royle. *Journal of Aquatic Plant Management,* **22**, 62–3.

Pieterse, A. H. and van Rijn, P. J. (1974). A preliminary study on the response of *Eichhornia crassipes, Salvinia auriculata* and *Pistia stratiotes* to glyphosate. *Mededelingen Fakulteit Landbouwwetenschappen Gent* (*Belgium*), **39**, (2), 422–7.

Pieterse, A. H. and van Zon, J. C. J. (1983). Geintegreerde bestrijding van wateronkruiden in Egypte. In *Geintegreerde bestrijding in de derde wereld*, Wageningen, The Netherlands, pp. 111–19. Werkgroep Pesticiden en Ontwikkelingslanden van de Vereniging van Wetenschappelijk Werkers.

Pieterse, A. H., Verkleij, J. A. C., and Staphorst, H. P. M. (1985). A comparative study of isoenzyme patterns, morphology, and chromosome number of *Hydrilla verticillata* (L.f.) Royle in Africa. *Journal of Aquatic Plant Management,* **23**, 72–6.

Pimental, D. (1971). *Ecological effects of pesticides on non-target species.* Office of Science and Technology, Executive Office of the Resident. Government Printing Office, Washington DC. pp. 220.

Pinter, K. (1980). Exotic fishes in Hungarian waters, their importance in fishery utilization of natural water bodies and fish farming. *Fisheries Management,* **11**, 163–7.

Pirie, N. W. (1960). Water hyacinth: a curse or a crop. *Nature* (*London*), **185**, 116–17.

Pirie, N. W. (1966). Leaf protein as human food. *Science, Washington,* **152**, 1701–5.

Pisano, V. E. (1976). Cormófitos acuáticos de Magellanes. *Anales del Instituto Patagonia,* **7**, 115–36.

Pitlo, R. H. (1978). Regulation of aquatic vegetation by interception of daylight. *Proceedings EWRS 5th Symposium on Aquatic Weeds 1978*, pp. 91–9.

Pitlo, R. H. (1982). Flow resistance of aquatic vegetation. *Proceedings EWRS 6th Symposium on Aquatic Weeds 1982*, pp. 225–34.

Pitlo, R. H. (1986). Towards a larger flow capacity of vegetated channels. *Proceedings EWRS/AAB 7th Symposium on Aquatic Weeds 1986*, pp. 245–50.

Pligin, Y. V. (1983). Macrobenthos of shoreline macrophytes of Kremenchug Reservoir. *Hydrobiological Journal,* **19 (5)**, 15–21.

Poi de Neiff, A. (1977). Estructura de la fauna asociada a tres hidrófitos flotantes en ambientes leníticos del noroeste argentino. *Commun. Cient. CECOAL,* **6**, 1–16.

Poi de Neiff, A. and Neiff, J. J. (1977). El pleuston de *Pistia stratiotes* de la laguna Barranqueras (Chaco, Argentina). *Ecosur,* **4**, 69–101.

Poi de Neiff, A., Neiff, J. J., and Bonetto, A. A. (1977). Enemigos naturales de *Eichhornia crassipes* en el nordeste argentino y possibilidades de su aplicaión al control biológico. *Ecosur,* **4**, 137–56.

Pokorny, J., Květ, J., Ondok, J. P., Toul, Z., and Ostry, I. (1984). Production—ecological analysis of a plant community dominated by *Elodea canadensis* Michx. *Aquatic Botany,* **9**, 263–92.

Pokorny, J., Lhotsky, O., Denny, P., and Turner, E. G. (ed.) (1987). Waterplants and wetland processes. *Archiv für Hydrobiologie Beiheft Ergebnisse der Limnologie,* **27**, p. 1–265.

Pokorny, J., Mentberger, J., Losos, B., Hartman, P., and Hetesa, J. (1971). Changes in hydrochemical and hydrobiological relations occurring when *Elodea* was controlled with paraquat. *Proceedings EWRC 3rd Symposium on Aquatic Weeds*, pp. 217–29.

Polak, B. (1951). *Contributions from General Agricultural Research Station, Bogor, Indonesia, No. 121.* Bogor, Indonesia.

Polprasert, C., Edwards, P., Rajput, V. S., and Pachararrakiti, C. (1986). Integrated biogas technology in the tropics. I. Performance of small-scale digesters. *Solid Waste Management Research,* **4,** 197–213.

Ponnappa, K. M. (1970). On the pathogenicity of *Myrothecium roridum—Eichhornia crassipes* isolate. *Hyacinth Control Journal,* **8,** 18–20.

Ponnappa, K. M. (1971). A highly virulent strain of *Myrothecium roridum* on *Trapa bispinosa. Indian Journal of Mycology and Plant Pathology,* **1,** 90–4.

Ponnappa, K. M. (1977). Records of plant pathogens associated with eight aquatic weeds in India. *Technical Bulletin Commonwealth Institute of Biological Control,* **18,** 65–74.

Pons, T. L. (1982). Factors affecting weed seed germination and seedling growth in lowland rice in Indonesia. *Weed Research,* **22,** 155–61.

Portier, J. (1971). Le problème des algues dans les canaux bétonnes d'irrigation du sud-est de la France: récents essais de traitements. *Proceedings EWRS 3rd Symposium on Aquatic Weeds 1971.* pp. 265–71.

Pot, R. (1986). *De relatie graskarper-plankton.* Internal Report. Gemeenterwaterleidingen Amsterdam. pp. 33.

Potvin, C. (1986). Biomass allocation and phenological differences among southern and northern populations of the C_4 grass *Echinochloa crus-galli. Journal of Ecology,* **74,** 915–23.

Powell, K. E. C. (1972). *The roughness characteristics of the reach of the River Bain at Fulsby Lock.* Report, Hydrological Section, Lincolnshire River Authority, UK 1972. pp. 25.

Powell, K. E. C. (1978). Weed growth—a factor of channel roughness. In *Hydrometry: principles and practices* (ed. R. W. Herschey), pp. 327–52. Wiley, Chichester.

Praeger, R. L. (1913). On the buoyancy of the seeds of some Britannic plants. *Scientific Proceedings of the Royal Dublin Society,* **14,** 13–62.

Praeger, R. L. (1934). Fifteen miles of *Elatine hydropiper. Irish Naturalist Journal,* **5,** 101–4.

Prandtl, L. (1904). Über Flussigkeits Bewegung bei sehr kleiner Reibung. In *Verhandlungen III Internationalen Mathematiker Kongresses Heidelberg 1904,* pp. 484–91.

Prescott, G. W. (1948). Objectionable algae with reference to the killing of fish and other animals. *Hydrobiologia,* **1,** 1–13.

Présing, M. (1981). On the effects of Dikonirt (sodium salt of 7,4-dichlorophenoxyacetic acid) on the mortality and reproduction on *Daphnia magna. Hydrobiologia,* **83,** 511–16.

Priban, K. (1973). Microclimatic measurements of temperatures in pure reed stands. In *Ecosystem study on wetland biom in Czechoslovakia.* (ed. S. Hejny). PT–PP/IBP Report 3, pp. 65–70.

Price, D. R. H. (1976). Problems arising from the practical use of aquatic herbicides. *Proceedings of a Symposium on Aquatic Herbicides, British Crop Protection Council Monograph No.* **16,** pp. 51–62.

Price, H. (1981). A review of current mechanical methods. *Proceedings Association of Applied Biologists Conference: Aquatic Weeds and their Control 1981,* pp. 77–86. AAB, Wellesbourne, UK.

Prins, H. B. A., Snel, J. F. H., Zanstra, P. E., and Helder, R. J. (1982). The mechanism of bicarbonate assimilation by the polar leaves of *Potamogeton* and *Elodea.* CO_2 concentrations at the leaf surface. *Plant, Cell and Environment,* **5,** 207–14.

Prinsloo, J. F. and Schoonbee, H. J. (1987). Investigations into the feasibility of a duck–fish–vegetable integrated agriculture–aquaculture system for developing areas in South Africa. *Water SA*, **13**, 109–18.

Prishchepov, G. P. (1974). Selectivity in feeding of grass carp—nutritive value to the fish of some plant species of Belorussian lakes. *Trudy Belorusskkogo Nauchno Issledovatel'skogo Institut Rybnogo Khozyaistva*, **10**, 196.

Probatova, N. S. and Buch, T. G. (1981). [*Hydrilla verticillata* (Hydrocharitaceae) in the Soviet Far East]. In Russian, English summary. *Botanicheskii Zhurnal*, **66**, (2), 208–14.

Proctor, J. (1980). The macrophyte vegetation of Bangula Lagoon, Malawi. *Kirkia (Zimbabwe)*, **12**, 141–9.

Proctor, V. W. (1968). Long-distance dispersal of seeds by retention in the digestive tract of birds. *Science*, **160**, 321–2.

Provoost, K. J. (1979). Graskarpers en waterplantenbeheersing. *Waterschapsbelangen*, **64**, 460–2.

Provoost, K. J. (1981). Beheersaspecten van graskarpers bij waterplantenbeheersing. *Verslag graskarpercontactdag, Noordwijkerhout*, pp. 27–33.

Pullman, D. (1981). Pond bottom liner is more than a cover up. *Weeds, Trees and Turf*, **20**, (6), 36.

Que Hee, S. S. and Sutherland, R. G. (1981). *The phenoxyalkanoic herbicides*, Vol. I & II. CRC Press, Boca Raton, Florida, USA.

Quimby, P. C. (1981). Preliminary evaluation of hydrogen peroxide as a potential herbicide for aquatic weeds. *Journal of Aquatic Plant Management*, **19**, 53–5.

Quimby, P. C., Potter, J. R., and Duke, S. O. (1978). Photosystem II and hypoxic quiescence in alligator weed. *Physiologia Plantarum*, **44**, 246–50.

Rabe, R., Schuster, H., and Kohler, A. (1982). Effects of copper chelate on photosynthesis and some enzyme activities of *Elodea canadensis*. *Aquatic Botany*, **14**, 167–75.

Raghavendra, A. S. and Das, V. S. R. (1976). Distribution of the C_4 dicarboxylic acid pathway of photosynthesis in local monocotyledonous plants and its taxonomic significance. *New Phytologist*, **76**, 301–5.

Raitala, J. and Lämpinen, J. (1985). A LANDSAT study of the aquatic vegetation of the Lake Luodonjärvi reservoir, Western Finland. *Aquatic Botany*, **21**, 325–46.

Rakvidyasastra, V., Iemwimangsa, M., and Petcharat, V. (1981). Host range of fungi pathogenic to water hyacinth (*Eichhornia crassipes* (Mart.) Solms.). *Biocontrol News Information*, **2**, 154.

Ramachandran, V. and Ramaprabhu, T. (1976). Use of ammonia for aquatic weed control—a review. In *Aquatic weeds in south-east Asia* (ed. C. K. Varshney and J. Rzoska), pp. 293–8. W. Junk, The Hague.

Ramaprabhu, T., Kumaraiah, P., Parameswaran, S., Sukumarian, P. K., and Raghavan, S. L. (1987). Water hyacinth control by natural water level fluctuations in Byramangala Reservoir, India. *Journal of Aquatic Plant Management*, **25**, 63–4.

Ramaprabhu, T., Ramachandran, V., and Reddy, P. V. G. K. (1982). Some aspects of the economics of aquatic weed control in fish culture. *Journal of Aquatic Plant Management*, **20**, 41–5.

Ramarakoto, M. R. (1965). *Water hyacinth in Madagascar*. Working paper, FAO Conference on Quelea and Water Hyacinth Control, Douala.

Ramey, V. (1982). Mechanical control of aquatic plants. *Aquaphyte*, **2**, 1, 3–6.

Ramírez, C., Godoy, R., and Hauenstein, E. (1981). Las especies de 'Luchecillos' (Hydrocharitaceae) que prosperan en Chile. *Anales del Musco Historia Natural de Valparaiso* (*Chile*), **14**, 47–55.

Ramírez, G. C. and Beck, S. (1981). Makrophytische Vegetation und Flora in Gewässern der Umgebung von La Paz, Bolivien. *Archiv für Hydrobiologie*, **91**, 82–100.

Ramírez, G. C., Romero, A. M., and Riveros, G. M. (1980). Lista de cormófitos palustres de la region valdiviana. *Boletin Museo Nacional de Historia Natural* (*Santiago de Chile*), **37**, 153–77.

Ramírez, G. C. and Stegmeier, W. E. (1982). Formas de vida de hidrófitos chilenos. *Ambientes Acuática*, **6**, 43–54.

Ramírez, M. C., Romero, M., and Riveros, M. (1979). Habit, habitat, origin and geographical distribution of Chilean vascular hydrophytes. *Aquatic Botany*, **7**, 241–53.

Ramsay, A. J. and Fry, J. C. (1976). Response of epiphytic bacteria to the treatment of two macrophytes with the herbicide paraquat. *Water Research*, **10**, 453–9.

Ranade, S. B. and Burns, W. (1925). The eradication of *Cyperus rotundus* L. (a study in pure and applied botany). In *Memoranda of the Department of Agriculture, India: Botany Series*, 13, pp. 99–192. Calcutta: Government of India Central Publications Branch.

Rangel-C. O. and Jaime Aguirre, C. (1983). Communidades acuáticas alto andinas. I. Vegetación sumergida y de ribera en el lago de Tota, Boyaca, Colombia. *Caldasia*, **13**, 719–42.

Rani, V. U. and Bhambie, S. (1983). A study on the growth of *Salvinia molesta* in relation to light and temperature. *Aquatic Botany*, **17**, 119–24.

Ransom, J. K., Oelke, E. A., and Wyse, D. L. (1983). Behaviour of 2,4-D in common waterplantain (*Alisma triviale*). *Weed Science*, **31**, 766–70.

Rantzien, H. H. (1952). Notes on some tropical African species of *Najas* in Kew Herbarium. *Kew Bulletin*, **7**, 29–40.

Rao, V. P. (1970). *Evaluation of natural enemies associated with witchweed, nutsedge and several other aquatic weeds occurring in India*. Commonwealth Institute of Biological Control Report, pp. 18.

Rao, K. S., Moorthy, K. S., Naidu, M. D., Chetty, C. S., and Swami, K. S. (1983). Changes in nitrogen metabolism in tissues of fish (*Sarotherodon mossambicus*) exposed to benthiocarb. *Bulletin of Environmental Contamination Toxicology*, **30**, 473–8.

Raschke, R. L. (1984). Mapping—surface and ground survey. *Special Technical Publications American Society for Testing Materials STP 843*, pp. 88–91. American Society for Testing Materials.

Raschke, R. L. and Rusanowski, P. C. (1984). Aquatic macrophyton field collection methods and laboratory analyses. *Special Technical Publications American Society for Testing Materials STP 843*, pp. 16–27. American Society for Testing Materials.

Rataj, K. (1970). Las Alismataceae de la República Argentina. *Darwiniana*, **16**, 9–39.

Rataj, K. (1972). Revision of the genus *Sagittaria*. The species of West Indies, Central and South America. *Annot. Zoolog. Bot. Bratislava*, **78**, 1–61.

Rataj, K. (1975). *Revision of the genus* Echinodorus *Ric*. Studie CSAV, Cislo 2, Praha (Czechoslovakia). pp. 156.

Rataj, K. (1978). Alismataceae of Brazil. *Acta Amazonica*, **8** (Suplemento 1). pp. 53.

Rautava, E. (1972). Amphiphytic and aquatic mass vegetation in the rivers Vaskojoki and Kettijoki in Finnish Lapland. *Annales Universitatis Turkuensis, ser. A,* **11**, (49), 99–107.

Raven, J. A. (1984). *Energetics and transport in aquatic plants.* MBL Lectures in Biology, Vol. 4. Alan R. Liss, New York. pp. 587.

Raven, P. J. (1985). *The use of aquatic macrophytes to assess water quality changes in some Galloway lochs: an exploratory study.* Working Paper No. 9. Palaeoecology Research Unit, Department of Geography, University College, London. pp. 76.

Raven, P. J. (1986). Changes of in-channel vegetation following two-stage channel construction on a small rural clay river. *Journal of Applied Ecology,* **23**, 333–45.

Ravera, O., Garavaglia, G., and Stella, M. (1984). The importance of the macrophytes in two lakes with different trophic degree: Lake Comabbio and Lake Monate (Province of Varese, Northern Italy). *Verhandlungen der Internationalen Vereinigung für theoretische und angewandte Limnologie,* **22**, 1119–30.

Rawls, C. K. (1975). Mechanical control of Eurasian watermilfoil in Maryland with and without 2,4-D application. *Chesapeake Science,* **16**, 266–81.

Rawson, R. M. (1985). History of the spread of Eurasian water milfoil through the Okanogan and Columbia river systems (1978–1984). In *1st International Symposium on Watermilfoil* (Myriophyllum spicatum) *and related Haloragaceae species* (ed. L. W. J. Anderson), pp. 35–8. Aquatic Plant Management Society, Vicksburg, Mississippi.

Redding-Coates, T. A. and Coates, D. (1981). On the introduction of phytophagous fishes into gravity flow irrigation systems in the Sudan. *Fisheries Management,* **12**, 89–99.

Reddy, K. R. (1984). Water hyacinth *Eichhornia crassipes* biomass production in Florida. *Biomass,* **6**, 167–81.

Reddy, K. R. and DeBusk, W. F. (1984). Growth characteristics of aquatic macrophytes cultured in nutrient-enriched water. I. Water hyacinth, water lettuce, and pennywort. *Economic Botany,* **38**, 229–39.

Reddy, K. R. and Portier, K. M. (1987). Nitrogen utilization by *Typha latifolia* L. as affected by temperature and rate of nitrogen application. *Aquatic Botany,* **27**, 127–38.

Reddy, K. R. and Smith, W. H. (ed.) (1987). *Aquatic plants for water treatment and resource recovery.* Magnolia Publishers, Orlando, Florida. pp. 1032.

Reddy, K. R., Sutton, D. L., and Bowes, B. (1983). Freshwater aquatic plant biomass production in Florida. *Soil Crop Science Florida Proceedings,* **42**, 28–40.

Ree, W. O. and Crow, F. R. (1977). *Friction factors for vegetated waterways of small slope.* US Department of Agriculture, Agricultural Research Service, Publication S-151. pp. 56.

Reed, C. F. (1965). Distribution of *Salvinia* and *Azolla* in South America and Africa, in connection with studies for control of insects. *Phytologia,* **12**, 121–30.

Reed, C. F. (1977). History and distribution of Eurasian watermilfoil in the United States and Canada. *Phytologia,* **36**, 417–36.

Reeders, H. H., van Schoubroek, M. E. P., van Vierssen, W., Gopal, B., and Pieterse, A. H. (1986). Aquatic weeds and their implications for agriculture in the Chambal irrigated area, Kota (Rajasthan), India. *Proceedings EWRS/AAB 7th Symposium on Aquatic Weeds 1986*, pp. 251–5.

Reimold, R. J. (1972). The movement of P through the salt marsh cord grass, *Spartina alterniflora*. *Limnology and Oceanography*, **17**, 606–11.

Reinert, K. H. and Rodgers, J. H. (1984). Influence of sediment types on the sorption of endothall. *Bulletin of Environmental Contamination Toxicology*, **32**, 557–64.

Rejmankova, E. (1975). Biology of duckweeds in a Pannonian fishpond. *Symposia Biologica Hungarica*, **15**, 125–31.

Rekas, A. M. B. and Bailey, P. D. (1981). *An adaptation of existing instrumentation technology to aquatic plant monitoring*. Miscellaneous Paper A-81-3. Planning & Operations Review, pp. 85–100. US Army Engineer Waterways Experiment Station, Vicksburg, Mississippi.

Reschke, M. (1978). Untersuchungen zur Entwicklung von Unkrautbekämpfungssystemen in Entwässerungsgraben im Weser–Ems–Gebiet. *Proceedings EWRS 5th Symposium on Aquatic Weeds 1978*, pp. 231–8.

Reschke, M. and Blaszyk, P. (1974). Erfolge met der chemischen Unkrautbekämpfung an und in Entwässerungsgräben im nordwestdeutschen Küstengebiet. *Proceedings EWRC 4th Symposium on Aquatic Weeds 1974*, pp. 240–3.

Rezk, M. R. and Edany, T. Y. (1981). Ecology of *Phragmites australis* (Cav.) Trin. ex Steud., Shatt-Al-Arab, Iraq. II. Reed growth as affected by the chemical composition of its beds. *Polskie Archiwum Hydrobiologie*, **28**, 19–31.

Rhodes, R. C. (1980). Studies with ^{14}C-labelled hexazinone in water and bluegill sunfish. *Journal of Agricultural and Food Chemistry*, **28**, 306–10.

Richard, D. J. and Small, J. W. (1984). Phytoplankton responses to reduction and elimination of submerged vegetation by herbicides and grass carp in four Florida lakes. *Aquatic Botany*, **20**, 307–19.

Richard, D. I., Small, J. W., and Osborne, J. A. (1985). Response of zooplankton to the reduction and elimination of submerged vegetation by grass carp and herbicide in four Florida lakes. *Hydrobiologia*, **133**, 97–108.

Richards, A. J. and Blakemore, J. (1975). Factors affecting the germination of turions in *Hydrocharis morsus-ranae* L. *Watsonia*, **10**, 273–5.

Richardson, L. V. (1975). Water level manipulation: a tool for aquatic weed control. *Hyacinth Control Journal*, **13**, 8–11.

Ridings, W. H. and Zettler, F. W. (1973). *Aphanomyces* blight of Amazon sword plants. *Phytopathology*, **63**, 289–95.

Ridley, H. N. (1930). *The dispersal of plants throughout the world*. Reeve, Ashford (UK). pp. 744.

Riemens, R. G. (1978). Grasscarp (*Ctenopharyngodon idella* Val.) as a tool in fishery management. *Proceedings EWRS 5th Symposium on Aquatic Weeds 1978*, pp. 351–8.

Riemens, R. G. (1981). Vier jaar praktijkervaring met graskarper. *Verslag graskarpercontactdag, Noordwijkerhout*, pp. 9–12.

Riemens, R. G. (1982*a*). The result of grass carp stocking for weed control in the Netherlands. *Proceedings 2nd International Symposium on Herbivorous Fish 1982*, pp. 1–7. EWRS, Wageningen, The Netherlands.

Riemens, R. G. (1982*b*). The survival of grass carp stocked in a prey/predator population. *Proceedings 2nd International Symposium on Herbivorous Fish 1982*, pp. 149–57. EWRS, Wageningen, The Netherlands.

Riemer, D. N. (1976). Long-term effects of glyphosate applications to *Phragmites*. *Journal of Aquatic Plant Management*, **14**, 39–43.

Riemer, D. N. (1984). *Introduction to freshwater vegetation*. AVI Publishing, Connecticut.

Riemer, D. N. and MacMillan, W. W. (1967). Effect of defoliation on spatterdock (*Nuphar advena*). *Abstracts Proceedings Northeast Weed Conference*, **21**, 556.

Rijks, D. A. (1969). Evaporation from a papyrus swamp. *QJR Metereological Society*, **95**, 643–9.

Rintanen, T. (1976). Lake studies in eastern Finnish Lappland. I. Aquatic flora: phanerogams and charales. *Annales Botanicae Fennici*, **13**, 137–48.

Rintz, R. E. (1973). A zonal leaf spot of water hyacinth caused by *Cephalosporium zonatum*. *Hyacinth Control Journal*, **11**, 41–4.

Robel, R. J. (1962). Changes in submerged vegetation following a change in water level. *Journal of Wildlife Management*, **26**, 221–4.

Roberts, H. A. (ed.) (1982). *Weed control handbook: principles* (7th edn). British Crop Protection Council and Blackwell, London. pp. 533.

Roberts, L. I. N., Winks, C. J., and Sutherland, D. R. W. (1984). *Biological control of alligator weed*. Report for the period July 1980–April 1984. DSIR, Auckland, New Zealand. pp. 29.

Robinson, E. W. (1969). The use of herbicides in the maintenance of land drainage channels on Romney Marsh. *Journal of the Institution of Water Engineers*, **23**, 159–76.

Robinson, G. W. (1971). Practical aspects of chemical control of weeds in land drainage channels in UK. *Proceedings EWRC 3rd Symposium on Aquatic Weeds 1971*, pp. 297–302.

Robinson, G. W. (1986). Weed control in land drainage channels in England and Wales with reference to conservation. *Proceedings EWRS/AAB 7th Symposium on Aquatic Weeds 1986*, pp. 257–62.

Robinson, G. W. and Leeming, J. B. (1969). The experimental treatment of some waters in Kent with diuron to control aquatic weed growth. *Association of River Authorities Yearbook 1969*, pp. 58–63.

Robinson, M. C. and Morley, R. L. (1980). *Eurasian water milfoil studies, Volume II. A monitoring study of the effects of 2,4-D on fish and waterfowl as applied in lakes of the Okanagan Valley, 1977*. Ministry of Environment, British Columbia, File No. 0316533. pp. 35.

Robson, T. O. (1966*a*). Further progress with the aerial application of aqueous solutions of dalapon to drainage ditches. *Proceedings 8th British Weed Control Conference*, **2**, pp. 598–9.

Robson, T. O. (1966*b*). Studies of persistence of 2,4-D in natural surface waters of Britain. *Proceedings 8th British Weed Control Conference*, **2**, p. 594.

Robson, T. O. (1968). Persistence of 2,4-D in natural surface waters. *Proceedings 9th British Weed Control Conference*, **1**, p. 404.

Robson, T. O. (1970). Water weed control: the problem in Britain. *Chemistry and Industry*, **1970**, 1419–23.

Robson, T. O. (1974). Mechanical control. In *Aquatic vegetation and its use and control* (ed. D. S. Mitchell), pp. 72–84. UNESCO, Paris.

Robson, T. O. (1975). *A survey of aquatic weed control methods used in Internal Drainage Boards, 1973*. Technical Report Agricultural Research Council Weed Research Organisation, No. 35 Oxford (UK). pp. 31.

Robson, T. O. (1976*a*). Water weeds: current trends in their control. *Span*, **19**, (2), 78–9.

Robson, T. O. (1976*b*). A review of the distribution of aquatic weeds in the tropics and subtropics. In *Aquatic weeds in south-east Asia* (ed. C. K. Varshney and J. Rzóska), pp. 25–30. W. Junk, The Hague.

Robson, T. O. (1978). The present status of chemical aquatic weed control. *Proceedings EWRS 5th Symposium on Aquatic Weeds 1978*, pp. 17–25.

Robson, T. O. (1986). Aquatic plant management in Europe. *Proceedings EWRS/AAB 7th Symposium on Aquatic Weeds 1986*, pp. 263–9.

Robson, T. O. and Barrett, P. R. F. (1977). Review of effects of aquatic herbicides. In *Ecological effects of pesticides* (ed. F. M. Perring and K. Mellanby), pp. 111–18. Academic Press, London.

Robson, T. O., Barrett, P. R. F., and Spencer-Jones, D. H. (1982). Chapter 18: *The control of aquatic weeds*. In *Weed control handbook: principles* (7th edn), (ed. H. A. Roberts), pp. 438–48. British Crop Protection Council and Blackwell, London.

Robson, T. O. and Fearon, J. H. (ed.) (1976). *Aquatic herbicides*. Proceedings of a Symposium: British Crop Protection Council, London, Monograph No. **16**. pp. 119.

Robson, T. O. and Fillenham, L. F. (1976). *Weed control in land drainage channels in Britain*. FAO. European Commission on Agriculture. Report ECA: WR/76/3(h).

Robson, T. O., Fowler, M. C., and Barrett, P. R. F. (1976). Effect of some herbicides on freshwater algae. *Pesticide Science*, **7**, 391–402.

Robson, T. O., Fowler, M. C., and Hanley, S. (1978). Observations on a lake treated with terbutryne in 3 alternate years. *Proceedings EWRS 5th Symposium on Aquatic Weeds 1978*, pp. 303–13.

Rodewald-Rudescu, L. (1974). *Das Schilfrohr*. Schweizbart'sche Verlag Stuttgart (Die Binnengewässer, Band 27).

Rodgers, C. A. (1970). Uptake and elimination of simazine by green sunfish. *Weed Science*, **18**, 134–6.

Roelofs, J. G. M. (1983). Impact of acidification and eutrophication on macrophyte communities in soft waters in the Netherlands. I. Field observations. *Aquatic Botany*, **17**, 139–55.

Rogers, H. H. and Davis, D. E. (1972). Nutrient removal by water hyacinth. *Weed Science*, **20**, 423–8.

Rogers, K. H. and Breen, C. M. (1980). Growth and reproduction of *Potamogeton crispus* in a South African lake. *Journal of Ecology*, **68**, 561–71.

Room, P. M. (1983). 'Falling apart' as a lifestyle: the rhizome architecture and population growth of *Salvinia molesta*. *Journal of Ecology*, **70**, (2), 349–65.

Room, P. M. (1986*a*). Equations relating growth and uptake of nitrogen by *Salvinia molesta* to temperature and the availability of nitrogen. *Aquatic Botany*, **24**, 43–59.

Room, P. M. (1986*b*). Biological control is solving the world's *Salvinia molesta* problems. *Proceedings EWRS/AAB 7th Symposium on Aquatic Weeds 1986*, pp. 271–6.

Room, P. M., Forno, I. W., and Taylor, M. F. J. (1984). Establishment in Australia of two insects for biological control of the floating weed *Salvinia molesta*. *Bulletin of Entomological Research*, **74**, 505–16.

Room, P. M., Harley, K. L. S., Forno, I. W., and Sands, D. P. A. (1981). Successful biological control of the floating weed *Salvinia*. *Nature* (*London*), **294**, 78–80.

Room, P. M. and Thomas, P. A. (1985). Nitrogen and establishment of a beetle for biological control of the floating weed *Salvinia* in Papua New Guinea. *Journal of Applied Ecology*, **22**, 139–56.

Room, P. M. and Thomas, P. A. (1986). Nitrogen, phosphorus and potassium in *Salvinia molesta* Mitchell in the field: effects of weather, insect damage, fertilizers and age. *Aquatic Botany*, **24**, 213–32.

Roorda, F. A., Schulten, G. G. M., and Pieterse, A. H. (1978). The susceptibility of *Orthogalumna terebrantis* Wallwork (Acarina: Galumnidae) to various pesticides. *Proceedings EWRS 5th Symposium on Aquatic Weeds*, pp. 375–81.

Rørslett, B. (1969). On the dispersal of *Elodea canadensis* Michx. in the lowlands of south-east Norway 1961–68. *Blyttia*, **27**, 185–93.

Rørslett, B. (1977). *Elodea canadensis* in southern Norway until 1976. *Blyttia*, **35**, 61–6.

Rørslett, B. (1984). Environmental factors and aquatic macrophyte response in regulated lakes—a statistical approach. *Aquatic Botany*, **9**, 199–220.

Rørslett, B. (1987). A generalized spatial niche model for aquatic macrophytes. *Aquatic Botany*, **29**, 63–81.

Rørslett, B. and Agami, M. (1987). Downslope limits of aquatic macrophytes; a test of the transient niche hypothesis. *Aquatic Botany*, **29**, 83–95.

Rørslett, B., Berge, D., and Johansen, S. W. (1986). Lake enrichment by submersed macrophytes: a Norwegian whole-lake experience with *Elodea canadensis*. *Aquatic Botany*, **26**, 325–40.

Rørslett, B., Green, N. W., and Kvalvågnaes, K. (1978). Stereophotography as a tool in aquatic biology. *Aquatic Botany*, **4**, 73–81.

Rosenberg, A. (1984). 2,3,6-trichlorophenylacetic acid (fenac) degradation in aqueous and soil systems. *Bulletin of Environmental Contamination Toxicology*, **32**, 383–90.

Ross, S. L., Doughty, C. R, and Murphy, K. J. (1986). Cause, effects and environmental management of a *Lemna* problem in a Scottish canal. *Proceedings EWRS/AAB 7th Symposium on Aquatic Weeds, 1986*, pp. 277–83.

Rossi, J. B. and Tur, N. M. (1976). Autoecologia de *Scirpus californicus*. II. Decarollo del rizoma. *Boletin de la Sociedad Argentina de Botanica*, **17**, 280–8.

Rowe, D. K. (1984). Some effects of eutrophication and the removal of aquatic plants by grass carp (*Ctenopharyngodon idella*) on rainbow trout (*Salmo gairdnerii*) in Lake Parkinson, New Zealand. *New Zealand Journal of Marine and Freshwater Research*, **18**, 115–27.

Rubin, J. L., Gaines, C. G., and Jensen, R. A. (1982). Enzymological basis for the herbicidal action of glyphosate. *Plant Physiology*, **70**, 833–9.

Rueppel, M. L., Brightwell, B. B., Schaefer, J., and Marvel, J. T. (1977). The metabolism and degradation of *N*-phosphonomethylglycine in soil and water. *Journal of Agricultural and Food Chemistry*, **25**, 517–28.

Rumpho, M. E. and Kennedy, R. A. (1983). Activity of the pentose phosphate and glycolytic pathways during anaerobic germination of *Echinochloa crus-galli* (barnyard grass) seeds. *Journal of Experimental Botany*, **34**, 893–902.

Rushing, W. N. (1976). Integrated control of water hyacinths with four biological agents. *Proceedings Research Planning Conference Aquatic Plant Control Program*, pp. 106–9. Miscellaneous Paper A-76-1, US Army Engineer Waterways Experiment Station, Vicksburg, Mississippi.

Rutherford, J. C. (1975). *Simulation of water quality in the Waikato and Tarawera Rivers*. Report 119, School of Engineering, University of Auckland, Auckland, New Zealand.

Rutherford, J. C. (1977). Modelling effects of aquatic plants in rivers. *Journal of Environmental Engineering Division American Society of Civil Engineers*, **103**, 575–91.

Rybicki, N. B., Carter, V., Anderson, R. T., and Trombley, T. J. (1985). The Potomac River *Hydrilla* project. *Hydrilla verticillata* in the tidal Potomac River 1983 and 1984. In *Proceedings 19th Annual Meeting Aquatic Plant Control Research Program*, pp. 170–4. Miscellaneous Paper A-85-4, US Army Engineer Waterways Experiment Station, Vicksburg, Mississippi.

Rzóska, J. (1974). The Upper Nile swamps. *Freshwater Biology*, **4**, 1–30.

Rzóska, J. (ed.) (1976*a*). *The Nile, biology of an ancient river*. Dr W. Junk, The Hague.

Rzóska, J. (1976*b*). The invasion of *Eichhornia crassipes* in the Sudanese White Nile. In *The Nile, biology of an ancient river* (ed. J. Rzóska). W. Junk, The Hague.

Sabbatini, M. R., Argüello, J. A., Fernández, O. A., and Bottini, R. A. (1983). Dormancy and growth inhibitor levels in oospores of *Chara contraria* A. Braun ex Kütz (Charophyta). *Aquatic Botany*, **28**, 189–94.

Sabbatini, M. R., Fernández, O. A., and Argüello, J. A. (1986*b*). Estudio sobre la germinación y conservación de oosporas de *Chara contraria* A. Braun ex Kütz (Charophyta). *Phyton*, **46**, 69–76.

Sabbatini, R. A., Irigoyen, J. H., and Fernández, O. A. (1986*a*). Phenology and biomass dynamics of *Chara contraria* A. Braun ex Kütz. in drainage channels of a temperate irrigation area in Argentina. *Proceedings EWRS/AAB 7th Symposium on Aquatic Weeds 1986*, pp. 285–9.

Sabbatini, R.AA., Lallana, V. H., and Marta, M. C. (1983). Inventario y biomasa de plantas acuáticas en un tramo del valle aluvial del Río Paraná Medio. *Revista Asociacion Ciencias Naturales Litoral*, **14**, 179–91.

Sabol, B. M. (1983). Simulation modelling to evaluate aquatic plant mechanical control in Buffalo Lake, Wisconsin. *Proceedings Southern Weed Science Society 36th Annual Meeting*, pp. 322–35.

Sabol, B. M. (1984*a*). Conceptual developments of methods for determining effectiveness of control techniques. *Proceedings 18th Annual Meeting Aquatic Plant Control Research Program 1983*, pp. 43–50. Miscellaneous Paper A-84-4, US Army Engineer Waterways Experiment Station, Vicksburg, Mississippi.

Sabol, B. M. (1984*b*). Development and use of the Waterways Experiment Station's hydraulically-operated submerged aquatic plant sampler. *Special Technical Publications American Society for Testing Materials STP 843*, pp. 46–57. American Society for Testing Materials.

Sabol, B. M. (1987). Environmental effects of aquatic disposal of chopped hydrilla. *Journal of Aquatic Plant Management*, **25**, 19–23.

Sacher, R. M. (1978). Safety of Roundup[R] in the aquatic environment. *Proceedings EWRS 5th Symposium on Aquatic Weeds 1978*, pp. 315–22.

Saha, S. (1968). The genus *Typha* in India—its distribution and uses. *Bulletin Botanical Society Bengal*, **22**, 11–18.

Sahni, B. (1927). A note on the floating islands and vegetation of Khajia near Chamba in the N.W. Himalaya. *Journal of the Indian Botanical Society*, **6**, 1–7.

Sainty, G. R. (1973). *Aquatic plants identification guide*. Water Resources Commission, New South Wales, Sydney.

Sainty, G. R. and Jacobs, S. W. L. (1981). *Water plants of New South Wales*. Water Resources Commission, New South Wales, Sydney. pp. 550.

Saitoh, M., Narita, K., and Iskawa, S. (1970). Photosynthetic nature of some aquatic plants in relation to temperature. *Botanical Magazine, Tokyo*, **83**, 10–12.

Sala, S. E. and Inartagio, C. (1985). Estudio taxonómico de fitoplancton del embalse Paso Piedras, Buenos Aires, Argentina. *Lilloa*, **36**, 249–63.

Sale, P. J. M. and Orr, P. T. (1981). Growth and photosynthesis of *Salvinia* and *Eichhornia*. *Proceedings 13th International Botanical Congress*, Sydney, Australia (Abstract 10.15.57).

Sale, P. J. M. and Orr, P. T. (1986). Gas exchange of *Typha orientalis* Presl. communities in artificial ponds. *Aquatic Botany*, **23**, 329–39.

Sale, P. J. M., Orr, P. T., Shell, G. S., and Erskine, D. J. C. (1985). Photosynthesis and growth rates in *Salvinia molesta* and *Eichhornia crassipes*. *Journal of Applied Ecology*, **22**, 125–37.

Sale, P. J. M. and Wetzel, R. G. (1983). Growth and metabolism of *Typha* species in relation to cutting treatments. *Aquatic Botany*, **15**, 321–34.

Salvucci, M. E. and Bowes, G. (1981). Induction of reduced photorespiratory activity in submerged and amphibious aquatic macrophytes. *Plant Physiology*, **67**, 335–40.

Salvucci, M. E. and Bowes, G. (1982). Photosynthetic and photorespiratory responses of the aerial and submerged leaves of *Myriophyllum brasiliense*. *Aquatic Botany*, **13**, 147–64.

Salvucci, M. E. and Bowes, G. (1983*a*). Two photosynthetic mechanisms mediating the low photo-respiratory state in submerged aquatic angiosperms. *Plant Physiology*, **73**, 488–96.

Salvucci, M. E. and Bowes, G. (1983*b*). Ethoxyzolamide repression of the low photo-respiration state in two submerged angiosperms. *Planta*, **158**, 27–34.

Samuelsson, G. (1934). Die Verbreitung der hoheren Wasserpflanzen in Nord-Europa. *Acta Phytogeographica Suecica*, **6**, 1–211.

Sanchez, L. and Vasquez, E. (1986). Notes on aquatic macrophytes in the lower section of the Orinoco flood-plain system, Venezuela. Contribution II in *Conference on Research and Application of Aquatic Plants for Water Treatment and Resource Recovery 1986*. Orlando, Florida.

Sandberg, C. L. and Burkhalter, A. P. (1983). Alligator weed control with glyphosate. *Proceedings Southern Weed Science Society 36th Annual Meeting*, pp. 336–9.

Sanders, D. R. and Theriot, R. F. (1979). *Evaluation of two fluridone formulations for the control of hydrilla in Gatun Lake, Panama Canal Zone*. Technical Report A-79-3, US Army Engineer Waterways Experiment Station, Vicksburg, Mississippi. pp. 29.

Sanders, H. L. (1986). Marine benthic diversity: a comparative study. *American Naturalist*, **102**, 243–82.

Sanders, H. O. (1969). Toxicity of pesticides to the crustacean *Gammarus lacustris*. *Technical Paper Bureau of Sport Fisheries & Wildlife (USA) No. 25*, pp. 18.

Sanders, H. O. (1970). Pesticide toxicities to tadpoles of the western chorus frog and Fowler's toad. *Copeia*, **1970**, (2), 246–51.

Sanders, H. O. and Cope, O. B. (1966). Toxicities of several pesticides to two species of cladocerans. *Transactions American Fisheries Society*, **95**, 165–9.

Sanders, H. O. and Cope, O. B. (1968). The relative toxicities of several pesticides to naiads of three species of stoneflies. *Limnology and Oceanography*, **13**, 112–17.

Sand-Jensen, K. (1983). Photosynthetic carbon sources of stream macrophytes. *Journal of Experimental Botany*, **34**, 198–210.

Sand-Jensen, K. and Gordon, D. M. (1984). Differential ability of marine and freshwater macrophytes to utilize HCO_3^- and CO_2. *Marine Biology*, **80**, 247–53.

Sand-Jensen, K. and Søndegaard, M. (1979). Distribution and quantitative development of aquatic macrophytes in relation to sediment characteristics in oligotrophic Lake Kalgaard, Denmark. *Freshwater Biology*, **9**, 1–12.

Sand-Jensen, K. and Søndegaard, M. (ed.) (1986). Submerged macrophytes: carbon metabolism, growth regulation and role in macrophyte-dominated ecosystems. *Aquatic Botany*, **26**, 200–394.

Sands, D. P. A., Kassulke, R. C., and Harley, K. L. S. (1982). Host specificity of *Disonycha argentinensis* (Coleoptera: Chrysomelidae), an agent for the biological control of *Alternanthera philoxeroides* (alligator weed) in Australia. *Entomophaga*, **27**, 163–72.

Sands, D. P. A., Schotz, M., and Bourne, A. S. (1983). The feeding characteristics and development of larvae of a salvinia weevil *Cyrtobagous* sp. (Coleoptera: Curculionidae). *Entomological Experimental Applications*, **34**, 291–6.

Sankaran, T. and Krishna, K. (1967). The biology of *Nanophyes* spp. (Coleoptera: Curculionidae) infesting *Jussiaea repens* in India. *Bulletin Entomological Research*, **57**, 337–41.

Sankaran, T. and Rao, V. P. (1972). An annotated list of insects attacking some terrestrial and aquatic weeds in India, with records of some parasites of the phytophagous insects. *Technical Bulletin Commonwealth Institute of Biological Control*, **15**, 131–57.

Sankaran, T., Srinath, D., and Krishna, K. (1966). Studies on *Gesonula punctifrons* Stal. (Orthoptera: Acrididae: Cyrtacanthacridinae) attacking water hyacinth in India. *Entomophaga*, **11**, 433–40.

San Martín, A. J. M. and Ramírez, G. C. (1983). Flora de malezas en arrozales de Chile central. *Ciencia e Investigacion Agraria*, **10**, 207–22.

Sarkany, S. (1982). [Experiments for chemical weed control in drainage canals]. In Hungarian. *Novenyvedelem*, **18**, (6), 258–60.

Sarpe, N., Balan, C., Melachrinos, A., and Florea, A. (1974). Chemical control of aquatic weeds in irrigation ditches and rice plantations using paraquat, amitrol, dalapon, propanil and bentason and their combinations. *Proceedings EWRC 4th Symposium on Aquatic Weeds 1974*, pp. 164–72.

Sastroutomo, S. S. (1980). Dormancy and germination in axillary turions of *Hydrilla verticillata*. *Botanical Magazine Tokyo*, **93**, 265–73.

Sastroutomo, S. S., Ikusima, I., Numata, M., and Iizumi, S. (1979). The importance of turions in the propagation of pondweed (*Potamogeton crispus L.*) *Ecological Reviews*, **19**, 75–88.

Sathal, H. (1979). Weed problems on lowland rice fields in Khmer. *Biotrop Bulletin*, **11**, 163–6.

Satpathy, B. (1964). Kalami sag (*Ipomoea reptans*)—a new addition to our greens. *Indian Farming*, **14**, (8), 12, 16.

Saunders, D. G. and Mosier, J. W. (1983). Photolysis of the aquatic herbicide fluridone in aqueous solution. *Journal of Agricultural and Food Chemistry*, **31**, 237–41.

Savage, K. E., Truelove, B., and Wiese, A. F. (1978). *Herbicide movement from application sites and effects on non-target species*. Southern Cooperative Bulletin 234. pp. 16.

Sazima, I. and Zamprogno, C. (1985). Use of water hyacinth as shelter, foraging place and transport by young piranhas *Serresalmus spilopheura*. *Environmental Biology Fishes*, **12**, 237–40.

Schardt, J. D. (1983). *1983 Aquatic flora of Florida survey report*. Florida Department of Natural Resources, Bureau of Aquatic Plant Research and Control. pp. 143.

Schardt, J. (1986). The status of water hyacinth and hydrilla in Florida waters. *Aquatics*, **8**, 13–17.

Schardt, J. D. (1987). *1987 Florida aquatic plant survey*, Report, Florida Department of Natural Resources, Bureau of Aquatic Plant Management. Tallahassee, Florida. pp. 49.

Schardt, J. D. and Nall, L. E. (1982). *1982 Aquatic flora of Florida survey report*. Florida Department of Natural Resources, Bureau of Aquatic Plant Research and Control. pp. 116.

Schelpe, E. A. C. (1961). The ecology of *Salvinia auriculata* and associated vegetation on Kariba Lake. *Journal of South African Botany*, **27**, 181–7.

Scheppe, W. A. (1986). Effects of human activities on Zambia's Kafue Flats ecosystems. *Environmental Conservation*, **12**, 49–54.

Schiele, S. (1986). Introduction and spread of three mudplantains (Pontederiaceae) in the rice fields of Europe. *Proceedings EWRS/AAB 7th Symposium on Aquatic Weeds 1986*, pp. 297.

Schiemer, F. and Prosser, M. (1976). Distribution and biomass of submerged macrophytes in Neusiedlersee. *Aquatic Botany*, **2**, 289–307.

Schindler, D. W., Brunskill, G. J., Emerson, S., Broecker, W. S., and Peng, T. H. (1972). Atmospheric carbon dioxide: its role in maintaining phytoplankton standing crops. *Science*, **177**, 1102–94.

Schlettwein, C. H. G. (1985). Distribution and densities of *Cyrtobagous singularis* Hustache (Coleoptera: Curculionidae) on *Salvinia molesta* Mitchell in the Eastern Caprivi Zipfel. *MADOQUA* **14**, 291–3.

Schlott, G. and Malicky, G. (1984). [Biomass and phosphorus-content of the macrophytes of the NE bay of the Lunzer Untersee (Austria) in relation to nutrient-rich inflows and to the sediment]. In German; English summary. *Archiv für Hydrobiologie*, **101**, 265–77.

Schneller, J. J. (1980). Cytotaxonomic investigations of *Salvinia herzogii* de la Sota. *Aquatic Botany*, **9**, 279–83.

Schneller, J. J. (1981). Chromosome number and spores of *Salvinia auriculata* Aublet s. str. *Aquatic Botany*, **10**, 181–4.

Schoen, L. (1983). Gevoeligheid van verschillende stammen van het wateronkruid *Hydrilla verticillata* (L.f.) Royle voor de schimmel *Fusarium roseum* 'Culmorum'. M.Sc thesis, University of Amsterdam and Royal Tropical Institute, Amsterdam, The Netherlands. pp. 39.

Schoonbee, H. J., Vermaak, J., and Swanepoel, J. H. (1985). Use of the Chinese grass carp, *Ctenopharyngodon idella*, in the control of the submerged water weed *Potamogeton pectinatus* in an inland lake in the Transvaal, South Africa. *Proceedings VIth International Symposium on the Biological Control of Weeds, Vancouver*, pp. 557–65. Agriculture Canada, Canada.

Schotsman, H. D. (1967). *Les Callitriches—espèces de France et taxa nouveaux d'Europe*. Flore de France 1, Le Chevalier, Paris. pp. 152.

Schotsman, H. D. (1977). Callitriches de la région Mediterranéenne. *Bulletin du Centre d'Études Recherches Scientifique, Biarritz*, **11**, 241–312.

Schröder, R. and Schröder, H. (1982). Changes in the composition of the submerged macrophyte community in Lake Constance. A multi-parameter analysis with various environmental factors. *Memorie dell'Istituto Italiano di Idrobiologie*, **40**, 25–52.

Schryer, F. and Ebert, V. W. (1972). *Determination of the effects of fertilizer induced phytoplankton turbidity, supplemented by herbicides, on submerged aquatic plants*. Report, Kansas Forestry, Fish and Game Commission. pp. 11.

Schultz, D. P. and Harman, P. D. (1974). *Review of literature on use of 2,4-D in fisheries*. Final Report No. PB-235 451/9GA. Bureau of Sport, Fisheries & Wildlife, South East Fisheries Control Laboratory, Warm Springs, Georgia, USA. pp. 94.

Schulz, A. G. (1961). Notas sobre la vegetación acuática chaqueña 'Esteros y Embalsados'. *Boletin de la Sociedad Argentina*, **9**, 141–50.

Schulz, D. (1972). Proliferative endocarditis in the heart of the carp after exposure to the herbicide Karmex. *Zentralblatt für Veterinaermedizin*, **19**, (5), 390.

Schulz, D. (1981). Herbizide im Fisch. Wirkungen und Rueckstaende. In *Beitraege zür Fischtoxicologie und – Parasitologie* (ed. H. H. Reichenbach-Klinke and W. Ahne), pp. 1–17. Gustav Fisher Verlag, Stuttgart.

Schumaker, R. W. and Bontwell, J. E. (1975). *Glyphosate dissipation in static water and influence of treatment on selected water chemistry and ecological parameters*. USDA—Agricultural Research Service Aquatic Weed Research Report. US Department of Agriculture, Denver, Colorado.

Schwimmer, D. and Schwimmer, M. (1968). Medical aspects of phycology. In *Algae, man and the environment* (ed D. F. Jackson). Proceedings of an International Symposium, Syracuse 1967, pp. 279–359. Syracuse University Press, Syracuse, New York.

Scorgie, H. R. A. (1980). Ecological effects of the aquatic herbicide cyanatryn on a drainage channel. *Journal of Applied Ecology*, **17**, 207–25.

Scorgie, H. R. A. and Cooke, A. S. (1979). Effects of the triazine herbicide cyanatryn on aquatic animals. *Bulletin of Environmental Contamination Toxicology*, **22**, 135–42.

Scott, T. G., Schultz, H. C., and Eschmeyer, P. H. (eds) (1978). 2,4-D DMA influences pond community interrelations. In *Sport Fishery and Wildlife Research 1975–76*, pp. 17–18. US Fish & Wildlife Service, Denver, Colorado.

Scott, W. E., Ashton, P. J., and Steyn, D. J. (1979). *Chemical control of the water hyacinth on Hartebeespoort Dam*. Water Research Commission, National Institute for Water Research and Hydrological Institute. Pretoria, South Africa.

Scotter, C. N. G., Wade, P. M., Marshall, E. J. P., and Edwards, R. W. (1977). The Monmouthshire levels drainage system: its ecology and relation to agriculture. *Journal of Environmental Management*, **5**, 75–86.

Sculthorpe, C. D. (1967). *The biology of aquatic vascular plants.* Edward Arnold, London. [Reprinted by Koeltz Scientific, Königstein, W. Germany, 1985].

Seaman, D. E. and Porterfield, W. A. (1964). Control of aquatic weeds by the snail *Marisa cornuarietis. Weeds*, **12**, 87–92.

Sebacher, D. I., Harriss, R. C., and Bartlett, K. B. (1985). Methane emissions to the atmosphere through aquatic plants. *Journal of Environmental Quality*, **14**, 40–6.

Secer, S. (1976). Morphologische Untersuchungen am Verdauungstrakt des Grasfisches (*Ctenopharyndogon idella* Val.). Thesis, Ludwig Maximilian Universität, München. pp. 78.

Seddon, B. (1965). The macrophytic vegetation of Welsh lakes in relation to factors of the lacustrine environment. *Journal of Ecology*, **53**, 828–9.

Seddon, J. C. (1981). The control of aquatic weeds with the isopropylamine salt of *N*-phosphonomethyl glycine. *Proceedings Association of Applied Biologists Conference: Aquatic Weeds and their control 1981*, pp. 141–8. AAB, Wellesbourne, UK.

Seidel, K. (1971). Macrophytes as functional elements in the environment of man. *Hidrobiologica (Bucuresti)*, **12**, 121–30.

Senaratna, J. E. (1943). *Salvinia auriculata* Aublet, a recently introduced free-floating water weed. *Tropical Agricultural Magazine, Ceylon*, **99**, 146–9.

Serns, S. L. (1975). The effects of dipotassium endothal on the zooplankton and water quality of a small pond. *Water Resources Bulletin* (*American Water Resources Association*), **11**, (6), 1221–31.

Serns, S. L. (1977). The effects of dipotassium endothall on rooted aquatic plants and adult and first year generation bluegill. *Water Resources Bulletin* (*American Water Resources Association*), **13**, (1), 71–80.

Seshavatharam, V. and Venu P. (1981). Observations on the ecology of Kolleru lake. *International Journal of Ecology and Environmental Science*,**7**, 33–44.

Seth, A. K., Fua, J. M., and Yusoff, Y. B. M. (1973). The use of paraquat and 2,4-D for the control of water hyacinth (*Eichhornia crassipes*). *Proceedings 4th Asian-Pacific Weed Science Society Conference*, pp. 322–8.

Setiadarma, D., Soerjani, M., Widyanto, L. S., and Soewardi, K. (1976). The effect of some herbicides on phytoplankton populations. *Biotrop Newsletter*, **15**, 13.

Shahjahan, M., Husne Ara Khan, Nilofar Akhtar, Aminur Rahman, A. S. M., and Majid, F. Z. (1981). Use of aquatic weeds and algae as poultry feed. *Proceedings 2nd Annual Conference Maximum Livestock on Minimum Land*, 271–8. Bangladesh Agricultural University, Mymensingh.

Shankar, V. (1966). Weeds of paddy fields of Varanasi district. *Journal of Scientific Research BHU*, **16**, (2), 139–45.

Shapovalova, I. M. and Vologdin, M. P. (1979). Procedure for quantitative estimation of submerged vegetation and the phytophilous fauna. *Hydrobiological Journal*, **16**, 89–91.

Sharif El Din, H. and Jones, K. W. (1954). Copper sulphate as an aquatic herbicide. *Nature (London)*, **174**, (4421), 187.

Sharma, K. P. (1977). Effect of cutting on the growth and flowering behaviour of *Typha elephantina* Roxb. *Current Science*, **47**, 275–6.

Sharma, K. P. (1978). Effect of submergence on the growth of *Typha* species. *Current Science*, **47**, 349.

Sharma, K. P. and Gopal, B. (1979*a*). Effect of water regime on the growth and establishment of *Typha angustata* seedlings. *International Journal of Ecology and Environmental Science*, **5**, 69–74.

Sharma, K. P. and Gopal, B. (1979*b*). Effect of light intensity on seedling establishment and growth in *Typha angustata* Bory and Chaub. *Polskie Archiwum Hydrobiologie*, **26**, 495–500.

Shay, J. M. (1983). *Post-fire performance of* Phragmites australis. University of Manitoba, Delta Marsh Field Station Annual Report 18. pp. 7.

Shay, J. M., Thompson, D. J., and Shay, C. T. (1987). Post-fire performance of *Phragmites australis* (Cav.) Trin. in the Delta Marsh, Manitoba, Canada. *Archiv für Hydrobiologie: Beiheft Ergebnisse der Limnologie*, **27**, 95–103.

Sheffield, C. W. (1967). Water hyacinth for nutrient removal. *Hyacinth Control Journal*, **6**, 27–30.

Shekhov, A. G. (1974). Effect of cutting time on renewal of stands of reed and cattail. *Hydrobiological Journal*, **10**, (3) 45–8.

Sheldon, R. B. and Boylen, C. W. (1978). An underwater survey method for estimating submerged macrophyte population density and biomass. *Aquatic Botany*, **4**, 65–72.

Shelton, W. L., Boney, S. P., and Rosenblatt, E. M. (1982). Monosexing grass carp by sex reversal and breeding. *Proceedings 2nd International Symposium on Herbivorous Fish 1982*, pp. 184–94. EWRS, Wageningen, The Netherlands.

Shelyag-Sosonko, Yu R., and Semenikhina, K. A. (1984). [Plants of Lakes Vaden and Trubin (the Desua Floodplain).] In Ukrainian; English summary. *Ukr. bot. Zhurnal*, **41**, 28–33.

Shibayama, H. (1981). Aquatic weeds in creeks and their control in Japan. *Weeds & Weed Control in Asia*, **20**, 240–56.

Shibayama, H. and Miyahara, M. (1977). Seasonal changes of growth of aquatic weeds and their control. *Proceedings 6th Asian–Pacific Weed Science Conference*, *Jakarta*, **1**, pp. 258–63.

Shiralipour, A., Haller, W. T., and Garrard, L. A. (1981). Effect of nitrogen sprays on biomass production and phosphorus uptake in water hyacinth. *Journal of Aquatic Plant Management*, **19**, 44–7.

Shireman, J. V., Colle, D. E., and Martin, R. G. (1979). Ecological study of Lake Wales, Florida after introduction of grass carp. *Proceedings Grass Carp Conference 1978*, pp. 49–89.

Shireman, J. V., Haller, W. T., Colle, D. E., Watkins, C. E., Durant, D. F., and Canfield, D. E. (1983). *Ecological impact of integrated chemical and biological aquatic weed control.* Report: Center for Aquatic Weeds, Gainesville, Florida for US Environmental Protection Agency, Florida: EPA-600/3-83-098. pp. 333.

Shireman, J. V. and Maceina, M. J. (1983). *Recording fathometer techniques for* Hydrilla *distribution and biomass studies.* Miscellaneous Paper A-83-1, prepared by the University of Florida, Gainesville, Florida, for US Army Engineer Waterways Experiment Station, Vicksburg, Mississippi. pp. 43.

Shireman, J. V. and Smith, C. R. (1983). Synopsis of biological data on the grasscarp *Ctenopharyngodon idella* (Cuvier et Valenciennes 1844). *FAO Fisheries Synopsis No 135.* FAO Rome. pp. 86.

Shiyan, P. N. and Merezhko, A. I. (1972). Effect of hydrogen ion concentration on photosynthesis and radiocarbon metabolism in aquatic plants. *Hydrobiological Journal*, **8**, 23–9.

Shugart, H. H., Goldstein, R. A., and O'Neill, R. V. (1974). TEEM: a terrestrial energy model for forests. *Oecologia Plantarum*, **25**, 251–84.

Shultz, D. P. and Harman, P. D. (1974). Residues of 2,4-D in pond waters, mud and fish, 1971. *Pesticide Monitoring Journal*, **8**, 179–9.

Siddall, J. D. (1885). The American water weed, *Anacharis alsinastrum* Bab.: its structure and habit; with some notes on its introduction in Great Britain, and the causes affecting its rapid spread at first and apparent present diminution. *Proceedings of the Chester Society Natural Science*, **3**, 125–33.

Siemelink, M., Zonneveld, N., De Blocq van Scheltinga, F., and El Enbabay, O. A. M. (1982). Controlled reproduction and cage culture of grass carp (*Ctenopharyngodon idella* Val.) in Egypt. *Proceedings 2nd International Symposium on Herbivorous Fish 1982*, pp. 76–82. EWRS, Wageningen, The Netherlands.

Sierra, F. J., Vera, H. A., Fullerton, T. M., and Cardenas, J. (1970). *Problemas de malezas en sistemas de riego.* Documento 690, ICA-INCORA, Bogota. pp. 32.

Sifton, H. B. (1959). The germination of light sensitive seed of *Typha latifolia*. *Canadian Journal of Botany*, **37**, 719–39.

Sikka, H. C., Ford, D., and Lynch, R. S. (1975). Uptake, distribution and metabolism of endothall in fish. *Journal of Agricultural and Food Chemistry*, **23**, 849–51.

Sikka, H. C., Lynch, R., and Lindenberger, S. (1974). Uptake and metabolism of dichlobenil by emersed aquatic plants. *Journal of Agricultural and Food Chemistry*, **22**, 230–4.

Sikka, H. C., Pack, E. J., Appleton, H. T., Hsu, R., and Cunningham, D. (1982). *Environmental fate, effects and health hazards of fenac.* Technical Report A-82-2, US Army Engineer Waterways Experiment Station, Vicksburg, Mississippi. pp. 120.

Sikka, H. C. and Rice, C. P. (1973). Persistence of endothall in aquatic environments as determined by gas–liquid chromatography. *Journal of Agricultural and Food Chemistry*, **25**, 842–6.

Sills, J. B. (1964). A report on the use of Karmex to control filamentous algae in fish ponds. *Proceedings 18th Annual Meeting South East Association of Game and Fisheries Commissioners*, pp. 474.

Silveira-Guido, A. (1971). Datos preliminaires de biologia y especifidad de *Acigona ignitalis* Hamps. (Lepidoptera, Puralidae) sobre el hospedero *Eichhornia crassipes* (Mart.) Solms—Laubach (Pontederiaceae). *Revista de la Sociedad Entomológica Argentina*, **33**, 137–45.

Silveira-Guida, A. and Perkins, D. (1975). Biology and host-specificity of *Cornops aquaticum* (Burner) (Orthoptera: Acrididae), a potential biological control agent for water hyacinth. *Environmental Entomology*, **4**, 400–5.

Simon, J. P. (1987). Differential effects of chilling on the activity of C_4 enzymes in two ecotypes of *Echinochloa crus-galli* from sites of contrasting climates. *Physiologia Plantarum*, **69**, 205–10.

Simpson, D. A. (1984). A short history of the introduction and spread of *Elodea* Michx. in the British Isles. *Watsonia*, **15**, 1–9.

Simpson, N. D. (1932). *A report on the weed flora of the irrigation channels in Egypt*. Ministry of Public Works, Government Press, Cairo.

Simsiman, G. V. (1974). I. Chemical control of aquatic weeds and its effect on the nutrient and redox status of water and sediment. II. Persistence of diquat and endothall in the aquatic environment. Ph.D thesis, University of Wisconsin. Madison, Wisconsin.

Simsiman, G. V. and Chesters, G. (1975). Persistence of endothall in the aquatic environment. *Water, Air and Soil Pollution*, **4**, 399–413.

Simsiman, G. V., Chesters, G., and Daniel, T. C. (1972). Chemical control of aquatic weeds and its effect on the nutrient and redox status of water and sediment. *Proceedings 15th Conference Great Lakes Research 1972*, pp. 166–70.

Simsiman, G. V., Daniel, T. C., and Chesters, G. (1976). Diquat and endothall: their fates in the environment. *Residue Reviews*, **62**, 131–74.

Singh, P. K. (1974). Algicidal effect of 2,4-dichlorophenoxy acetic acid on blue-green alga *Cylindrospermum* sp. *Archives of Microbiology*, **97**, 69–72.

Singh, S. P. and Moolani, M. K. (1973). Control of cattail (*Typha angustata*) in relation to period of stubble submergence. *Proceedings 4th Asian-Pacific Weed Science Society Conference*, pp. 329–38.

Singh, S. P., Sukamaran, K. K., Piliai, K. K., and Charabarti, P. C. (1967). Observations on the efficacy of grass carp (*Ctenopharyngodon idella* Val.) in controlling and utilizing aquatic weeds in ponds in India. *Proceedings Indo–Pacific Fisheries Council*, **12**, 220–35.

Singh, S. P., Pahuja, S. S., and Moolani, M. K. (1976). Cultural control of *Typha angustata* at different stages of growth. In *Aquatic weeds in south-east Asia* (ed C. K. Varshney and J. Rzóska), pp. 245–52. W. Junk, The Hague.

Sinha, V. R. P. and Gupta, M. V. (1975). On the growth of grass carp, *Ctenopharyngodon idella*, in composite fish culture at Kalyani, West Bengal (India). *Aquaculture*, **5**, 283–90.

Siregar, H. and Soemarwoto, O. (1976). Studies of *Panicum repens* L. in West Java. In *Aquatic weeds in south-east Asia*, (ed. C. K. Varshney and J. Rzóska), pp. 211–14. W. Junk, The Hague.

Sirjola, E. (1969). Aquatic vegetation of the River Teuronjoki, South Finland and its relation to water velocity. *Annales Botanicae Fennici*, **6**, 68–75.

Sisounthone, C. and Sisombat, L. (1979). Brief information on weeds in rice fields in Laos. *Biotrop Bulletin*, **11**, 151–4.

Ska, B. and vander Borght, P. (1986). The problem of *Ranunculus* development in the River Semois. *Proceedings EWRS/AAB 7th Symposium on Aquatic Weeds 1986*, pp. 307–14.

Skender, A., Filicić, N., and Steuk, I. (1974). Total vegetation control on banks of irrigation and drainage canals effected by combining Hyvar X and Karmex. *Fragmenta Herbologica Jugoslavia*, **48**, 1–11.

Skender, A., Stević, I., Radovanović, A., Tabacki, M., Ljubicić, Z., Hristov, I., and Milosević, S. (1982). Aquatic flora of the hydroameliorative systems in the regions of Banat, Baranja, Eastern Slavonia and Istria. *Proceedings EWRS 6th Symposium on Aquatic Weeds 1982*, pp. 1–8.

Slade, P. and Smith, A. E. (1967). Photochemical degradation of diquat. *Nature (London)*, **213**, 919–20.

Slamet, S. and Sukowati, S. (1975). Interaction between light intensities and nutrient concentration on the growth of water hyacinth. (*Eichhornia crassipes*). *Proceedings 3rd Indonesian Weed Science Conference, Bandung, West Java*, 377–91.

Small, J. W., Richard, D. I., and Osborne, J. A. (1985). The effects of vegetation removal by grass carp and herbicides on the water chemistry of four Florida lakes. *Freshwater Biology*, **15**, 587–96.

Smid, P. (1973). Microclimatological characteristics of reedswamps at the Nesyt Fishpond. In *Littoral of the Nesyt Fishpond*, (ed. J. Květ), *Studie CSAV*, **15**, 29–38.

Smith, D. W. (1985). Biological control of excessive phytoplankton growth and enhancement of aquacultural production. *Canadian Journal of Fisheries and Aquatic Sciences*, **42**, 1940–5.

Smith, E. V. and Swingle, H. S. (1941). The use of fertilizer for controlling pondweed *Najas guadalupensis*. *Transactions of the North American Wildlife Conference*, **6**, 245–51.

Smith, G. E. (1971). Résumé of studies and control of Eurasian water milfoil (*Myriophyllum spicatum* L.) in the Tennessee Valley from 1960 through 1969. *Hyacinth Control Journal*, **9**, 23–5.

Smith, L. M. and Kadlec, J. A. (1983). Seedbanks and their role during drawdown of a North American marsh. *Journal of Applied Ecology*, **20**, 673–84.

Smith, S. G. (1987). *Typha*: its taxonomy and the ecological significance of hybrids. *Archiv für Hydrobologie: Beiheft Ergebnisse der Limnologie*, **27**, 129–38.

Smits, A. J. M. and Wetzels, A. M. M. (1986). Germination studies on three nymphaeid species (*Nymphaea alba* L., *Nuphar lutea* (L.) Sm. and *Nymphoides peltata* (Gmel.) O. Kuntze). *Proceedings EWRS/AAB 7th Symposium on Aquatic Weeds 1986*, pp. 315–20.

Snyder, J. H. (1983). Water and agriculture. In *A guidebook to California agriculture* (ed. A. F. Scheuring). University of California Press. pp. 413.

Snyder, R. L. and Boyd, C. E. (1987). Evapotranspiration by *Eichhornia crassipes* (Mart.) Solms and *Typha latifolia* L. *Aquatic Botany*, **27**, 217–27.

Sobolev, I. A. (1970). Food interrelationships of young grass carp, silver carp and carp reared jointly in ponds in Belorussia. *Journal of Ichthyology*, **10**, (4), 528–33.

Soemartono, T. (1979). Competition between *Salvinia molesta* D. S. Mitchell and rice. *Biotrop Bulletin*, **11**, 189–92.

Soerjani, M. (ed.) (1971). Tropical weeds: some problems, biology and control. *Proceedings 1st Indonesian Weed Science Conference, Bogor*. SEAMEO-Biotrop, Bogor, Indonesia.

Soerjani, M. (1976). Aquatic weed problems in Indonesia, with special reference to the construction of man-made lakes. In *Aquatic weeds in South East Asia* (ed. C. K. Varshney and J. Rzóska, pp. 63–78. W. Junk, The Hague.

Soerjani, M. (1977). Integrated control of weeds in aquatic areas. In *Integrated control of weeds* (ed J. D. Fryer and S. Matsunaka), pp. 121–151. University Press, Tokyo.

Soerjani, M. (1978). Aquatic weed problems and their control. *Philippines Journal of Weed Science*, **5**, 44–53.

Soerjani, M. (1980). Aquatic plant management in Indonesia. In *Tropical ecology and development* (ed. J. I. Furtado), pp. 725–37. *Proceedings 5th International Symposium on Tropical Ecology 1979.* Kuala Lumpur, Malaysia.

Soerjani, M., Combar, J. B., and Tjitrosoepomo, G. (eds) (1979). Weed problems in south-east Asia. Weed Science Conference, Yogyakarta. *Biotrop Bulletin*, **11**, pp. 343.

Soerjani, M., Kostermans, A. J. G. H., and Tjitrosoepomo, G. (1987). *Weeds of rice in Indonesia.* Balai Pustaka Jakarta, Indonesia. pp. 716.

Soerjani, M. and Pancho, J. V. (1974). *Aquatic weed problem, control, research needs and coordination in southeast Asia.* Paper 1.9, South East Asian Workshop on Aquatic Weeds, Malong Indonesia. Mimeo.

Soerjani, M., Parker, C., Tjitrosemito, S., Allen, G. E., Varshney, C. K., Mitchell, D. S., and Pancho, J. V. (eds) (1976). *Proceedings south-east Asian Workshop on Aquatic Weeds. Biotrop Special Publication*, **1**, pp. 45. Biotrop, Bogor, Indonesia.

Soewardi, K. (1979). Some ecological impacts of the introduction of grass carp (*Ctenopharyngodon idella* Val.) for aquatic weed control. *Proceedings 6th Asian-Pacific Weed Science Society Conference, Jakarta*, **2**, pp. 451–8.

Solander, D. (1982). Production of macrophytes in two small subarctic lakes in northern Sweden. In *Studies on aquatic vascular plants* (eds J. J. Symoens, S. S. Hooper, and P. Compère), pp. 181–6. Royal Botanical Society of Belgium, Brussels.

Søndergaard, M. and Sand-Jensen, K. (1979). Carbon uptake by leaves and roots of *Littorella uniflora* (L) Aschers. *Aquatic Botany*, **6**, 1–12.

Sorsa, K. K., Nordheim, E. V., and Andrews, J. H. (1988). Integrated control of Eurasian watermilfoil, *Myriophyllum spicatum*, by a fungal pathogen and a herbicide. *Journal of Aquatic Plant Management*, **26**, 12–17.

Sosa Gonzalez, R. (1956). Control de la vegetación sumergida en canales de riego. *Agronomia Tropical*, **5**, 235–8.

Soulsby, P. G. (1974). The effect of a heavy cut on the subsequent growth of aquatic plants in a Hampshire chalk stream. *Journal of the Institute of Fisheries Management*, **5**, 49–53.

South, E. L. and Robinson, M. C. (1982). *Monitoring the effects of 2,4-D on fish and waterfowl, and the effect of Eurasian watermilfoil* (Myriophyllum spicatum L.) *on fish and waterfowl in the Okanagan valley, 1978.* Eurasian Watermilfoil Studies, Volume III. Ministry of Environment, British Columbia, Water Management Branch. pp. 51.

Spagnoli, R. (1973). [Chemical weed control in canals]. In Italian. *Lotta Antiparassitaria*, **25**, (8), 3–5.

Speirs, J. M. (1948). Summary of the literature on aquatic weed control. *Canadian Fish Culturist*, **3**, 20–32.

Spence, D. H. N. (1964). The macrophytic vegetation of freshwater lochs, swamps and associated fens. In *The vegetation of Scotland* (ed. J. H. Burnett), pp. 306–425. Oliver & Boyd, Edinburgh.

Spence, D. H. N. (1981). Light quality and plant response underwater. In *Plants and the daylight spectrum* (ed. H. Smith), pp. 245–75. Academic Press, London.

Spence, D. H. N. (1982). The zonation of plants in freshwater lakes. *Advances in Ecological Research*, **12**, 37–125.

Spence, D. H. N. and Allen, E. D. (1979). The macrophytic vegetation of Loch Urigill and other lochs of the Ullapool area. *Transactions Botanical Society of Edinburgh*, **43**, 131–44.

Spence, D. H. N. and Maberley, S. C. (1985). Occurrence and ecological importance of HCO_3^- use among aquatic higher plants. In *Inorganic carbon uptake by aquatic photosynthetic organisms* (ed. W. J. Lucas and J. A. Berry). American Society Plant Physiology, Rockville, Maryland.

Spence, D. H. N., Milburn, T. R., Ndawula-Denyimba, M., and Roberts, E. (1971). Fruit biology and germination of two typical *Potamogeton* species. *New Phytologist*, **70**, 197–212.

Spencer, D. F. (1986). Tuber demography and its consequences for *Potamogeton pectinatus* L. *Proceedings EWRS/AAB 7th Symposium on Aquatic Weeds 1986*, pp. 321–5.

Spencer, N. R. and Coulson, J. R. (1976). The biological control of alligatorweed, *Alternanthera philoxeroides* in the United States of America. *Aquatic Botany*, **2**, 177–90.

Spencer, W. and Bowes, G. (1985). *Limnophila* and *Hygrophila*: a review and physiological assessment of their weed potential in Florida. *Journal of Aquatic Plant Management*, **23**, 7–16.

Spencer, W. and Bowes, G. (1986). Photosynthesis and growth of water hyacinth under CO_2 enrichment. *Plant Physiology*, **82**, 528–33.

Spencer-Jones, D. H. (1971). Trials to evaluate the environmental and herbicidal effects of dichlobenil granules applied to water for the control of aquatic weeds. *Proceedings EWRC 3rd Symposium on Aquatic Weeds 1971*, pp. 173–82.

Spencer-Jones, D. H. and Wade, P. M. (1986). *Aquatic plants: a guide to recognition.* ICI Professional Products, Farnham (UK).

Spillett, P. B. (1981). The use of groynes and deflectors in river management. *Proceedings Association of Applied Biologists Conference; Aquatic Weeds & their Control 1981*, pp. 189–98. AAB, Wellesbourne, UK.

Spira, W. M., Huq, A., Ahmed, Q. S., and Saeed, Y. A. (1981). Uptake of *Vibrio cholerae* biotype eltor from contaminated water by water hyacinth (*Eichhornia crassipes*). *Applied Environmental Microbiology*, **42**, 550–3.

Stalling, C. L. and Huckins, J. N. (1978). Metabolism of 2,4-dichlorophenoxyacetic acid (2,4-D) in bluegills and water. *Journal of Agricultural and Food Chemistry*, **26**, 447–52.

Standing Committee of Analysts (1987). *Methods for the use of aquatic macrophytes for assessing water quality 1985–6*. London: HMSO. Methods for the examination of waters and associated materials. pp. 176.

Stanković, A., Arsenović, M., Dimitrijević, M., and Konstantinović, B. (1982). A study of the possibility of the elimination of emerged vegetation in irrigation systems by new herbicides. *Proceedings EWRS 6th Symposium on Aquatic Weeds 1982*, pp. 185–92.

Stanley, J. G. (1976*a*). Reproduction of the grass carp (*Ctenopharyngodon idella*) outside its native range. *Fisheries Bulletin of the American Fisheries Society*, **1**, (3), 7–10.

Stanley, J. G. (1976*b*). Production of hybrid, androgenetic and gynogenetic grass carp and carps. *Transactions of the American Fisheries Society*, **105**, 10–16.

Stanley, J. G., Miley, W. W., and Sutton, D. L. (1978). Reproductive requirements and likelihood for naturalization of escaped grass carp in the USA. *Transactions of the American Fisheries Society*, **107**, 119–28.

Stanley, R. A. (1975). Response of Eurasian watermilfoil to heat. *Hyacinth Control Journal*, **13**, 62.

Stanley, R. A. (1976). Response of Eurasian watermilfoil to subfreezing temperature. *Journal of Aquatic Plant Management*, **14**, 36–9.

Stanley, R. A. and Naylor, A. W. (1972). Photosynthesis in Eurasian watermilfoil (*Myriophyllum spicatum* L.). *Plant Physiology*, **50**, 149–51.

Stanley, R. A., Shackelford, E., Wade, D., and Warren, C. (1976). Effects of season and water depth on Eurasian watermilfoil. *Journal of Aquatic Plant Management*, **14**, 32–6.

Statzner, B. and Stechmann, D. H. (1977). Der Einfluss einer mechanischen Entkrautungsmassnahme auf der Driftraten der Makroinvertebraten im unteren Schierenseebach. *Faunistisch-Ökologische Mitteilungen*, **5**, 93–109.

Steele, J. H. (1962). Environmental control of photosynthesis in the sea. *Limnology and Oceanography*, **7**, 137–50.

Steemann-Nielsen, E. (1947). Photosynthesis of aquatic plants with special reference to the carbon sources. *Dansk Botanisk Arkiv*, **12**, 3–71.

Steenis, J. H. (1968). The phases of the Eurasian watermilfoil problem in the Chesapeake Bay. *Abstracts Weed Science Society of America 1968*, 56–7.

Steeves, T. A. (1952). Wild rice—Indian food and a modern delicacy. *Economic Botany*, **6**, 107–42.

Stein, H. and Heri, O. (1986). *Pseudorasbora parva*, eine neue Art der mitteleuropaischen Fischfauna. *Der Fischwirt*, **36**, 1–2.

Steinberger, N. and Goschel, H. (1979). *Handbuch UdSSR*. VEB Bibliografisches Institüt, Leipzig. pp. 611.

Stent, C. J. and Hanley, S. (1985). A recording echo-sounder for assessing submerged aquatic plant populations in shallow lakes. *Aquatic Botany*, **21**, 377–94.

Stepanaviciene, V. (1985). [New find-places of 15 rare macrophytes in lakes of the Lithuanian SSR]. In Russian. *Lietuvos TSR Mokslu Akademija Darbai B Serija*, **1**, (89), 34–8.

Stephenson, M., Turner, G., Pope, P., Colt, C., Knight, A., and Tchobanoglous, G. (1980). *The use and potential of aquatic species for wastewater treatment: the environmental requirements of aquatic plants*. Report No. 65: University of California/California State Water Resources Board, Sacramento, California.

Stevens, K. L., Badar-ud-Din, A. A., and Ahmad, M. (1979). The antibiotic bostrycin from *Alternaria eichhorniae*. *Phytochemistry*, **18**, 1579–80.

Steward, K. K. (1970). Nutrient removal potentials of various aquatic plants. *Hyacinth Control Journal*, **8**, 34–5.

Steward, K. K. (1984). Growth of hydrilla (*Hydrilla verticillata*) in hydrosoils of different composition. *Weed Science*, **32**, 371–5.

Steward, K. K. and Nelson, L. L. (1972). Evaluations of controlled release PVC and Attaclay formulations of 2,4-D on Eurasian watermilfoil. *Hyacinth Control Journal*, **10**, 35–8.

Steward, K. K. and Ornes, W. H. (1975). The autecology of sawgrass in the Florida Everglades. *Ecology*, **56**, 162–71.

Steward, K. K., Van, T. K., Carter, V., and Pieterse, A. H. (1984). *Hydrilla* invades Washington, D.C. and the Potomac. *American Journal of Botany*, **71**, 162–3.

Steward, K. K., Van, T. K., and Jones, A. O. (1982). Controlled delivery of chemicals for managing aquatic plant growths. *Proceedings EWRS 6th Symposium on Aquatic Weeds 1982*, 212–24.

Stewart, W. D. P., Tuckwell, S. B., and May, E. (1975). Eutrophication and algal growth in Scottish freshwater lochs. In *The ecology of resource degradation and renewal*, (ed. M. J. Chadwick and G. T. Goodman), pp. 57–80. Blackwell, Oxford.

Steyn, D. G. (1945). Poisoning of animals and human beings by algae. *South African Journal of Science*, **41**, 243–4.

St. John, H. (1963). Monograph of the genus *Elodea* (Hydrocharitaceae). Part 3. The species found in northern and eastern South America. *Darwiniana*, **12**, 639–52.

St. John, H. (1964). Monograph of the genus *Elodea* (Hydrocharitaceae). Part 2. The species found in the Andes and western South America. *Caldasia*, **9**, 95–113.

St. John, H. (1965). Monograph of the genus *Elodea*: Summary. *Rhodora*, **67**, 155–80.

Stocker, R. (1987). *Quarterly report to the Hydrilla Technical Advisory Committee*. Report: Imperial Irrigation District, Imperial, California.

Stodola, J. (1967). *Encyclopedia of water plants*, pp. 368. TFH Publications Inc., Neptune City.

Stokoe, R. (1983). *Aquatic macrophytes in the tarns and lakes of Cumbria*. Occasional Publication No. 18, Freshwater Biological Association, Cumbria (UK).

Stott, B. (1979). Grass carp research and public policy in England. *Proceedings Grass Carp Conference 1978*, pp. 147–58.

Stott, B. (1981). Progress towards the practical use of grass carp for water weed control. *Proceedings Association of Applied Biologists Conference: Aquatic Weeds and their Control 1981*, pp. 117–23. AAB, Wellesbourne, UK.

Stovell, F. R. (1966). The use of herbicides as invert emulsions for aquatic weed control. *Proceedings 8th British Weed Control Conference*, pp. 600–6.

Strange, R. J. (1976). Nutrient release and community metabolism following the application of herbicide to macrophytes in microcosms. *Journal of Applied Ecology*, **13**, 889–97.

Strange, R. J. and Schreck, C. B. (1976). Response of aerobic community metabolism to chemical treatment of aquatic macrophytes. *Journal of Aquatic Plant Management*, **14**, 45–50.

Straskraba, M. (1963). The share of the littoral region in the productivity of two ponds in Southern Bohemia. *Rozpr. csl. Akad. Ved.*, **73**, (13), 1–63.

Streit, B. (1979). Uptake, accumulation and release of organic pesticides by benthic invertebrates. 2. Reversible accumulation of lindane, paraquat and 2,4-D from aqueous solution by invertebrates and detritus. *Archiv für Hydrobiologie/Supplement*, **55**, 349–72.

Streit, B. and Peter, H. M. (1978). Long-term effects of atrazine to selected freshwater invertebrates. *Archiv für Hydrobologie/Supplement*, **55**, 62–77.

Stroband, H. W. J. (1977). Growth and diet dependent structural adaptations of the digestive tract in juvenile grass carp (*Ctenopharyngodon idella* Val.). *Journal of Fish Biology*, **11**, 167–74.

Suasa-Ard, W. (1976). Ecological investigation on *Namangana pectinicornis* Hampson (Lepidoptera: Noctuidae), as a potential biological control agent of the water lettuce, *Pistia stratiotes* L. (Arales: Araceae). M.Sc. thesis, Department of Entomology, Kasetrart University, Bangkok, Thailand.

Suasa-Ard, W. and Napompeth, B. (1982). *Investigation on* Episammia pectinicornis *(Hampson) (Lepidoptera: Noctuidae) for biological control of the water lettuce in Thailand.* Technical Bulletin No 3, National Biological Control Research Centre, Kasetrart University. National Research Council, Bangkok, Thailand.

Subagyo, T. (1975). Some aspects of the biology of *Nymphula responsalis* Wlk. attacking *Salvinia sp. Biotrop Bulletin*, **12**, 14.

Subramanyan, K. (1962). *Aquatic angiosperms.* Botanical Monograph No. 3, Council of Scientific and Industrial Research, New Delhi, India, pp. 190.

Sukopp, H. (1971). Effects of man, especially recreational activities, on littoral macrophytes. *Hidrobiologia (Bucuresti)*, **12**, 331–40.

Sullivan, D. S., Sullivan, T. P., and Bisalputra, T. (1981). Effects of Roundup^R herbicide on diatom populations in the aquatic environment of a coastal forest. *Bulletin of Environmental Contamination Toxicology*, **26**, 91–6.

Sundaresan, A. and Reddy, N. (1979). *Salvinia molesta* (Mitchell)—a serious water weed in Fiji. *Fiji Agricultural Journal*, **41**, 103–7.

Sutanto, L., Widyanto, L. S., and Soerjani, M. (1976). The effect of ametryne and cyanatryn on water hyacinth control. *Biotrop Newsletter*, **15**, 13.

Sutcliffe, J. F. (1962). *Mineral salt absorption in plants.* Pergamon Press, London.

Sutton, D. L. (1968). Uptake and translocation of three growth regulators in parrot feather (*Myriophyllum brasiliense* Camp.). Ph.D. thesis, Virginia Polytechnic Institute. pp. 95.

Sutton, D. L. (1974). Utilization of hydrilla by the white amur. *Hyacinth Control Journal*, **12**, 66–70.

Sutton, D. L. (1985). Biology and ecology of *Myriophyllum aquaticum. Proceedings 1st International Symposium on water milfoil* (Myriophyllum spicatum) *and related Haloragaceae species* (ed L. W. J. Anderson), pp. 59–71. Aquatic Plant Management Society, Vicksburg, Mississippi.

Sutton, D. L. and Bingham, S. W. (1968). Translocation patterns of simazine in *Potamogeton crispus* L. *Proceedings 22nd Northeast Weed Control Conference*, pp. 357–61.

Sutton, D. L. and Bingham, S. W. (1970). Uptake and translocation of 2,4-D-1-^{14}C in parrot feather. *Weed Science*, **18**, 193–6.

Sutton, D. L. and Blackburn, R. D. (1971*a*). Uptake of copper in hydrilla. *Weed Research*, **11**, 47–53.

Sutton, D. L. and Blackburn, R. D. (1971*b*). Uptake of copper by parrot feather. *Weed Science*, **19**, 282–5.

Sutton, D. L., Blackburn, R. D., and Barlowe, W. C. (1971). Response of aquatic plants to combinations of endothall and copper. *Weed Science*, **19**, 643–6.

Sutton, D. L., Durham, D. A., Bingham, S. W., and Foy, C. L. (1969). Influence of simazine on apparent photosynthesis of aquatic plants and herbicide residue removal from water. *Weed Science*, **17**, 56–9.

Sutton, D. L., Haller, W. T., Steward, K. K., and Blackburn, R. D. (1972). Effect of copper on uptake of diquat-^{14}C by hydrilla. *Weed Science*, **20**, 581–3.

Sutton, D. L., Littell, R. C., and Langeland, K. A. (1980). Intraspecific competition of *Hydrilla verticillata*. *Weed Science*, **28**, 425–8.

Suvatabandhu, K. (1950). *Weeds in paddy fields in Thailand*. Report: Department of Agriculture, Bangkok, Thailand.

Svachka, O., Irigoyen, J. H., Sabbatini, M. R., and Fernández, O. A. (1982). Uso de acroleina para el control de malezas submergidas en canales de desague. *IX Reunion Argentina sobre la maleza y su control Santa Fé, Agosto 1982. Resumenes*, **7**, 120 (Abstract).

Svenson, H. K. (1929). Monographic studies in the genus *Eleocharis*. *Rhodora*, **31**, 43–77, 123–35, 152–63, 199–219.

Svenson, H. K. (1939). Monographic studies in the genus *Eleocharis*. *Rhodora*, **41**, 1–19.

Svobodova, Z., Faina, R., Máchová, J., and Prausova, J. (1984). Persistence of Midstream herbicide in pond environment. *Pracé Vyskumneho Ustavu Ryb, Hydrobiol. Vodnany*, **13**, 34–40.

Swales, S. (1982). Impacts of weed cutting on fisheries: an experimental study in a small lowland river. *Fisheries Management*, **13**, (4), 125–38.

Swarbrick, J. T., Finlayson, C. M., and Cauldwell, A. J. (1981). The biology of Australian weeds. 7. *Hydrilla verticillata* (L.f.) Royle. *Journal of the Australian Institute of Agricultural Science*, **7**, 183–90.

Swift, J. (1976). *Ducks, ponds and people. A guide to the management of small lakes and ponds for wildfowl*. Wildfowlers Association of Great Britain and Ireland, Rossett.

Swindale, D. N. and Curtis, J. T. (1957). Phytosociology of the larger submerged plants in Wisconsin lakes. *Ecology*, **38**, 397–407.

Swindells, P. (1983). *Water-lilies*. Croom Helm, Beckenham. pp. 159.

Swingle, H. S. and Smith, E. V. (1947). Pond weeds and their control. *Alabama Agricultural Experiment Station Bulletin*, **254**, 24–8.

Syed, R. A. (1979). Some aspects of biological control of weeds in south-east Asia. *Biotrop Bulletin*, **11**, 311–18.

Symoens, J. J., Hooper S. S., and Compère, P. (ed.) (1982). *Studies on aquatic vascular plants*. Royal Botanical Society of Belgium, Brussels. pp. 424.

Szczepanska, W. (1971). Allelopathy among the aquatic plants. *Polskie Archiwum Hydrobologie*, **18**, 17–30.

Szmeja, J. (1979). [The abundance of indicator species in *Lobelia* lakes in the southern part of the Kashubian Lake District (Northern Poland)]. In Polish; English summary. *Fragmenta Floristica Geobotanica*, **25**, 123–43.

Szumiec, J. (1963). Wplyw rosliunosci wynurzonej is sposobu jej koszenia na faune denna stawow rybnych. *Acta Hydrobiologica*, **5**, 315–35.

Taekema, L. (1980). De herontdekking van veerkoloniale waterwegen. *Recreatie*, **18**, (2), 25–31.

Takle, J. C. C., Beitinger, T. L., and Dickson, K. L. (1983). Effect of the aquatic herbicide endothall on the critical thermal maximum of red shiner *Notripos lutrensis*. *Bulletin of Environmental Contamination Toxicology*, **31**, (5), 512–17.

Talling, J. F. (1957). Photosynthetic characteristics of some freshwater plankton diatoms in relation to underwater radiation. *New Phytologist*, **56**, 29–50.

Tang, Y. (1960). Reproduction of chinese carps *Ctenopharyngodon idella* and *Hypophthalmichthys molitrix* in a reservoir in Taiwan. *Japanese Journal of Ichthyology*, **8**, 1–2.

Tansley, A. G. (1949). *The British islands and their vegetation*. Cambridge University Press, Cambridge. pp. 930.

Tarver, D. P. (1980). Water fluctuation and the aquatic flora of Lake Miccosukee. *Journal of Aquatic Plant Management*, **18**, 19–23.

Task Force on Friction Factors in Open Channels (1963). Friction factors in open channels. *Journal of the Hydraulics Division, Proceedings of the American Society of Civil Engineers*, **89**, 97–143.

Tatum, W. M. and Blackburn, R. D. (1962). Preliminary study of the effects of diquat on the natural bottom fauna and plankton in two subtropical ponds. *Proceedings of the South East Association of Game and Fisheries Commissioners 1962*, pp. 13.

Taubayev, T. T. (1958). [Weed control in infested canals]. In Russian. *Minist. Sel'sk. Khoz. Uzb. SSR.*, **7**, 50–2.

Taussig, J. K. (1969). The development cycle of a hydraulic aquatic weed control device. *Proceedings North Eastern Weed Control Conference*, **23**, pp. 367–74.

Tazik, P. P., Kodrich, W. R., and Moore, J. R. (1982). Effects of overwinter drawdown on bushy pondweed. *Journal of Aquatic Plant Management*, **20**, 19–21.

Teles, A. N. and Pinto da Silva, A. R. (1975). A 'Pinheirinha' (*Myriophyllum aquaticum* (Vell.) Verd.), uma agressiva infestante aquática. *Agronomia lusitania*, **36**, (3), 307–23.

Tell, G. (1985). *Catálogo de las algas de agua dulce de la República Argentina*. Biblioteca Phycológica, Vaduz. pp. 283.

Temby, F. E. (1973). Dimethyltridecylamine oxide salts of endothall as aquatic herbicides. *Proceedings of North Central Weed Control Conference*, **28**, pp. 107.

Templeton, G. E. (1982). Status of weed control with plant pathogens. In *Biological control of weeds with plant pathogens* (ed. R. Charudattan and H. L. Walker), pp. 22–44. John Wiley, New York.

Tenhunen, J. D., Yocum, C. S., and Gates, D. M. (1976). Development of a photosynthesis model with an emphasis on ecological applications. I. Theory. *Oecologia* (*Berlin*), **26**, 89–100.

Terrell, J. W. and Fox, A. C. (1975). Food habits, growth and catchability of grass carp in the absence of aquatic vegetation. *Proceedings of the South Eastern Association of Game & Fish Commissioners*, **18**, 251–9.

Terrell, J. W. and Terrell, T. T. (1975). Macrophyte control and food habits of the grass carp in Georgia ponds. *Verhandlung der Internationalen Vereinigung für theoretische und angewandte Limnologie*, **19**, 2515–20.

Terry, P. J. (1981). Weeds and their control in the Gambia. *Tropical Pest Management*, **27**, 44–52.

Terry, P. J., Robson, T. O., and Hanley, S. (1981). Localized control of aquatic weeds with dichlobenil. *Proceedings Association of Applied Biologists Conference: Aquatic Weeds & their Control 1981*, pp. 165–76. AAB, Wellesbourne, UK.

Tevyashova, L. Ye. and Tevyashova, O. Ye. (1973). Optimum plant growth in fish-culture ponds of Don River breeding-nursery farms. *Gidrobiologicheskii Zhurnal*, **6**, 45–50.

Thakurta, S. C. and Mitra, G. (1977). Absorption of ions from the environment for the destruction of aquatic plants. I. Treatment with copper sulphate. *Science & Culture*, **43**, 402–5.

Thamasara, S. (ed.) (1982). *Proceedings of the First Tropical Weed Science Conference, Thailand*. Bangkok, Thailand. pp. 161.

Tharp, B. C. (1917). Texas parasitic fungi. *Mycologia*, **9**, 105–24.

Thayer, D. D. and Haller, W. T. (1982). Effects of surfactants on the penetration of 2,4-D in water hyacinth. *Aquatics*, **4**, (4), 16, 18.

Thayer, D. and Ramey, V. (1986). *Mechanical harvesting of aquatic weeds*. Florida Department of Natural Resources Technical Publication. pp. 39. Tallahassee, Florida.

Thiegs, B. J. (1955). The stability of dalapon in soils. *Down to Earth*, **11**, 1–4.

Thirumalachar, M. J. and Govindu, H. C. (1954). Notes on some Indian Cercosporae. *Sydowia*, **8**, 343–8.

Thomas, B. (1981). The operation of the Pesticide Safety Precautions Scheme. *Proceedings Association of Applied Biologists Conference: Aquatic Weeds and their Control 1981*, pp. 275–85. AAB, Wellesbourne, UK.

Thomas, E. A. (1978). Mass growth of algae and macrophytes in streams: method, cause and prevention. *Verhandlungen der Internationalen Vereinigung für theoretische und angewandte Limnologie*, **20**, 1796–9.

Thomas, G. J., Allen, D. S., and Grose, M. P. B. (1980). Aquatic plants at the Ouse Washes in 1978. *Nature in Cambridgeshire*, **28**, 29–39.

Thomas, G. L., Marino, D. A., Thorne, R. E., and Pauley, G. B. (1985). An evaluation of fisheries sonar techniques as a tool for measuring aquatic macrophyte biomass. In *Proceedings 19th Annual Meeting Aquatic Plant Control Research Program*, pp. 153–7. Miscellaneous Paper A-85-4, US Army Engineer Waterways Experiment Station, Vicksburg, Mississippi.

Thomas, K. J. (1976). Observations on the aquatic vegetation of Trivandrum, Kerala. In *Aquatic weeds in south east Asia* (ed. C. K. Varshney and J. Rzóska), pp. 99–102. W. Junk, The Hague.

Thomas, L. and Anderson, L. W. J. (1984). Water hyacinth control in California. *Aquatics*, **6**, 11–20.

Thomas, P. A. (1985). The management of *Salvinia molesta* in Papua New Guinea. *Plant Protection Bulletin, FAO*, **32**, 50–6.

Thomas, P. A. and Room, P. M. (1986). Taxonomy and control of *Salvinia molesta*. *Nature (London)*, **320**, 581–4.

Thomas, T. M. and Seaman, D. E. (1968). Translocation studies with endothall-^{14}C in *Potamogeton nodosus* Poir. *Weed Research*, **8**, 321–6.

Thommen, G. H. and Westlake, D. F. (1981). Factors affecting the distribution of populations of *Apium nodiflorum* and *Nasturtium officinale* in small chalk streams. *Aquatic Botany*, **11**, (1), 21–36.

Thompson, G. T. and Roberson, J. A. (1976). A theory of flow resistance for vegetated channels. *Transactions of the American Society of Agricultural Engineers*, **19**, 288–93.

Thompson, K., Shewry, P. R., and Woolhouse, H. W. (1979). Papyrus swamp development in the Upemba Basin, Zaire: studies of population in *Cyperus papyrus* stands. *Botanical Journal of the Linnaean Society*, **78**, 299–316.

Thurston, R. V. *et al.* (1984). Chronic toxicity of ammonia to rainbow trout. *Transactions of the American Fisheries Society,* **113**, 56–73.

Thyagarajan, E. (ed.) (1984). *Proceedings of the International Conference on Water Hyacinth.* UN Environment Programme, Nairobi.

Thyssen, N. (1982). Aspects of the oxygen dynamics of a macrophyte dominated lowland stream. In *Studies on aquatic vascular plants* (ed. J. J. Symoens, S. S. Hooper, and P. Compère), pp. 202–13. Royal Botanical Society of Belgium, Brussels.

Timmer, C. E. and Weldon, L. W. (1967). Evapotranspiration and pollution by water hyacinth. *Hyacinth Control Journal,* **6**, 34–7.

Timmons, F. L. (1960). *Weed control in western irrigation and drainage systems.* Crops Research: a joint report, Agricultural Research Service and Bureau of Reclamation, USA. ARS 34–14. pp. 22.

Timmons, F. L. (1966). Control of weeds harmful to water uses in the west. *Journal of the Waterways and Harbors Division, American Society of Civil Engineers,* **92**, 47–58. No. WW1, Paper 4645.

Tischler, N., Bates, J. C., and Gorgonio, P. Q. (1951). A new group of defoliant-herbicidal chemicals. *Proceedings of the Northeastern Weed Control Conference 1951,* **4**, pp. 51–84.

Titus, J. E. and Adams, M. S. (1979). Co-existence and the comparative light relations of the submerged macrophytes *Myriophyllum spicatum* L. and *Vallisneria americana* Michx. *Oecologia,* **40**, 273–86.

Titus, J., Goldstein, R. A., Adams, M. S., Mankin, J. B., O'Neill, R. V., Weiler, P. R., Shueart, H. H., and Booth, R. S. (1975). A production model for *Myriophyllum spicatum* L. *Ecology,* **56**, 1129–38.

Tivy, J. (1980). *The effect of recreation on freshwater lochs and reservoirs in Scotland.* Countryside Commission for Scotland, Perth, UK. pp. 202.

Tjitrosemito, S., Soerjani, M., and Mercado, B. L. (1979). Photosynthesis of and the effect of photosynthetic inhibitors on water hyacinth *Eichhonia crassipes* (Mart.) Solms. *Proceedings 6th Asian–Pacific Weed Science Society Conference, Jakarta 1977,* **2**, 548–54.

Toerien, D. F., Cary, P. R., Finlayson, C. M., Mitchell, D. S., and Weerts, P. G. J. (1983). Growth models for *Salvinia molesta. Aquatic Botany,* **16**, 173–9.

Toetz, D. W. (1974). Uptake and translocation of ammonia by freshwater macrophytes. *Ecology,* **55**, 199–201.

Toivonen, H. (1983). Changes in dominant macrophytes of 54 small Finnish lakes in 30 years. *Proceedings International Symposium on Aquatic Macrophytes, Nijmegen 1983,* pp. 220–4.

Toivonen, H. and Lappalainen, T. (1980). Ecology and production of aquatic macrophytes in the oligotrophic mesohumic lake Suomunjärvi, eastern Finland. *Annales Botanicae Fennicí,* **17**, 69–85.

Tölg, L. (1967). Die limnologische Bedeutung der Östasiatischen pflanzenfressenden Fische im Europäischen Fischbestand. *Acta Zoologica Academiae Scientarum Hungaricae,* **8**, 445–8.

Tomaszewicz, H. (1969). The water and swamps vegetation of closed meanders of River Bug in Warsaw region. *Acta Societas Botanica Poloniae,* **38**, (2), 217–45.

Toner, E. D., O'Riordan, A., and Twomey, E. (1965). The effects of arterial drainage works on the salmon stock of a tributary of the River Moy. *Irish Fisheries Investigations, Series A (Freshwater),* **1**, 36–55.

Tooby, T. E. (1971). The toxicity of aquatic herbicides to freshwater organisms—a brief review. *Proceedings EWRS 3rd Symposium on Aquatic Weeds,* pp. 129–37.

Tooby, T. E. (1976). Effects of aquatic herbicides on fisheries. *Proceedings Symposium on Aquatic Herbicides: British Crop Protection Council Monograph No.* **16**, pp. 62–77.

Tooby, T. E. (1978). A scheme for the evaluation of hazards to non-target aquatic organisms from the use of chemicals. *Proceedings EWRS 5th Symposium on Aquatic Weeds 1978,* pp. 287–94.

Tooby, T. E. (1981). Predicting the direct toxic effect of aquatic herbicides to non-target organisms. *Proceedings Association of Applied Biologists Conference: Aquatic Weeds and their Control 1981,* pp. 265–74. AAB, Wellesbourne, UK.

Tooby, T. E., Durbin, F. J., and Rycroft, R. J. (1974). Accumulation and elimination of residues of the aquatic herbicide dichlobenil in two species of British freshwater fish. *Proceedings EWRS 4th Symposium on Aquatic Weeds 1974,* pp. 202–9.

Tooby, T. E., Hursey, P. A., and Alabaster, J. S. (1975). Acute toxicity of 102 pesticides and miscellaneous substances to fish. *Chemistry & Industry,* (21 June 1975), 523–6.

Tooby, T. E., Lucey, J., and Stott, B. (1980). The tolerance of grass carp, *Ctenopharyngodon idella* Val., to aquatic herbicides. *Journal of Fisheries Biology,* **16**, 591–7.

Tooby, T. E. and Macey, D. J. (1977). Absence of pigmentation in corixid bugs (Hemiptera) after the use of the aquatic herbicide dichlobenil. *Freshwater Biology,* **7**, 519–25.

Tooby, T. E. and Spencer-Jones, D. H. (1978). The fate of the aquatic herbicide dichlobenil in hydrosoil, water and roach (*Rutilus rutilus* L.) following treatment of areas of a lake. *Proceedings EWRS 5th Symposium on Aquatic Weeds 1978,* pp. 323–31.

Toro, G. J., Briones, V. J., and Pinoargote, Ch. M (1982). *Controle la 'totora' con herbicidas.* Estacione Experimentale 'Portoviejo' INIAP Bol. Divulg. 125. pp. 6.

Toscani, H. A. (1981). Información sobre ensayos de control de malezas en canales de desagüe y taludes de diques. *Malezas,* **9**, 8–11.

Toscani, H. A. (1983). Progresos en el control de malezas acuáticas sumergidas en embalses. *Malezas,* **11**, 232–60.

Toscani, H. A. (1984). Control de malezas acuáticas en un canal del Delta del Paraná con métodos mecánicos y químicos. *Malezas,* **12**, 11–19.

Toscani, H. A., Pizzolo, G., and Maradei, G. (1983). Avances en el control químico de malezas acuáticas en canales y zanjas de desagüe y drenaje de la región Delta del Paraná. *Malezas,* **11**, 145–75.

Triest, L. (1982). *Lagarosiphon major* (Hydrocharitaceae) een Zuidafrikaans wateronkruid in Nieuw Zeeland en Europa. *Ver. Onderwijs. Biol.,* **5**, 113–21.

Triest, L. (1986). *Najas* L. species (Najadaceae) as rice field weeds. *Proceedings EWRS/AAB 7th Symposium on Aquatic Weeds 1986,* pp. 357–62.

Trivedy, R. K., Sharma, K. P., Goel, P. K., and Gopal, B. (1978). Some ecological observations on floating islands. *Hydrobiologia,* **60,** 187–90.

Trudeau, P. N. (1982). *Nuisance aquatic plants and aquatic plant management programs in the United States. Volume 3. North-eastern and north central region.* Report: The Mitre Corporation, Metrek Division, McLean, Virginia.

Tsuchiya, M. (1979*a*). Natural reproduction of grass carp in the Tone River and their pond spawning. *Proceedings of the Grass Carp Conference, Gainesville,* pp. 185–200.

Tsuchiya, M. (1979*b*). Control of aquatic weeds by grass carp (*Ctenopharyngodon idella*). *Japan Agricultural Research Quarterly,* **13,** 200–3.

Tu, C. C. and Kimbrough, J. W. (1978). Systematics and phylogeny of fungi in the *Rhizoctonia* complex. *Botanical Gazette,* **139,** 454–66.

Tubea, B., Hawxby, K., and Mehta, R. (1981). The effects of nutrient, pH and herbicide levels on algal growth. *Hydrobiologia,* **79,** 221–7.

Tucker, C. S. and Boyd, C. E. (1981). Relationships between pond sediments and simazine loss from waters of laboratory systems. *Journal of Aquatic Plant Management,* **19,** 55–7.

Tucker, C. S., Busch, R. L., and Lloyd, S. W. (1983). Effects of simazine treatment on channel catfish production and water quality in ponds. *Journal of Aquatic Plant Management,* **21,** 7–11.

Tur, N. M. (1972). Embalsados y camalotes de la region isleña del Paraná Medio. *Darwiniana,* **17,** 397–407.

Tur, N. M. (1977). Plantas vasculares. In *Biota Acuática de Sudamérica Austral* (ed. S. H. Hulbert), pp. 37–45. San Diego State University Press, San Diego. pp. 342.

Tur, N. M. (1982). Revisión del género *Potamogeton* L. en la Argentina. *Darwiniana,* **24,** 217–65.

Tur, N. M. and Rossi, J. B. (1976). Autoecología de *Scirpus californicus.* I. Crecimiento y desarollo de la parte aérea. *Boletin de la Sociedad Argentina de Botanica,* **17,** 73–82.

Turner, D. J. (1977). *Safety of herbicides 2,4-D and 2,4,5-T.* Forestry Commission Bulletin No. 57. H.M.S.O., London, pp. 56.

Tyndall, R. W. (1982). *Nuisance aquatic plants and aquatic plant management programmes in the United States, Volume 1. South-western Region.* Report: The Mitre Corporation, No. MTR-82W47-01, McLean, Virginia. pp. 142.

Tyndall, R. W., Trudeau, P. N., Aurand, D. V., and Smith, D. L. (1982). *Integrated aquatic weed management—principles and applications for reservoirs in the southeastern United States.* Report from Mitre Corporation to US EPA Office of Environmental Processes and Effects Research. pp. 189.

'Typha' (1983). Let us spray. *ADA Gazette* 1986, 27–8. Association of Drainage Authorities, UK.

Tyson, D. (1974). The fate of terbutryne in the aquatic ecosystem and its effects on non-target organisms. In *Report of a Technical Symposium: use of terbutryne as an aquatic herbicide,* held at Royal Commonwealth Society, London, pp. 21–32.

Ueki, K. and Oki, Y. (1979). Seed production and germination of *Eichhornia crassipes* in Japan. *Proceedings 7th Asian–Pacific Weed Science Society Conference,* pp. 257–60.

Uhler, F. M. (1944). Control of undesirable plants in waterfowl habitats. *Transactions of the North American Wildlife Conference,* **9,** 295–303.

Ulrich, K. E. and Burton, T. M. (1985). The effects of nitrate, phosphate and potassium fertilization on growth and nutrient uptake patterns of *Phragmites australis* (Cav.) Trin. ex Steudel. *Aquatic Botany,* **21**, 53–62.

Ultsch, G. R. and Anthony, D. S. (1973). The role of the aquatic exchange of carbon dioxide in the ecology of the water hyacinth. *Florida Scientist,* **36,** (1), 16–22.

Uotila, P. (1971). Distribution and ecological features of hydrophytes in the polluted Lake Vanajavesi, South Finland. *Annales Botanicae Fennici,* **8,** 257–95.

USBR (1949). *Control of weeds on irrigation systems.* US Bureau of Reclamation. pp. 140.

USBR (1985). *1985 Summary Statistics, Vol. 1. Water, land and related data.* US Bureau of Reclamation, Division of Water and Land Technical Services. pp. 319.

USDA (1963). *Chemical control of submerged waterweeds in western irrigation and drainage canals.* US Department of Agriculture Agricultural Research Service and US Department of the Interior: Reclamation Joint Report ARS 34–57. pp.14.

USDA (1987). *Weekly Rice Market News,* **68,** (15), 5. US Department of Agriculture.

USDI, Fish and Wildlife Service (1981). Food of Everglade kite threatened. *US Department of the Interior Fish & Wildlife Research Report 1980,* pp. 47–8.

USDI, Fish and Wildlife Service (1982). Fenatrol: a potential herbicide for fish ponds. Chronic toxicity of Aquazine. *US Department of the Interior Fish and Wildlife Research Report 1981,* p. 79.

US Department of the Navy, Army and the Air Force (1971). *Military entomology operational handbook.* Superintendent of Documents, US Government Printing Office, Washington DC.

Vaas, K. F. (1951). Notes on water hyacinth in Indonesia and its eradication by spraying with 2,4-D. *Contributions General Agricultural Research Station, Bogor,* **120,** 1–61. Bogor, Indonesia.

Van, T. K., Haller, W. T., and Bowes, G. (1976). Comparison of the photosynthetic characteristics of three submerged aquatic macrophytes. *Plant Physiology,* **58,** 761–8.

Van, T. K., Haller, W. T., and Bowes, G. (1978). Some aspects of the competitive biology of *Hydrilla. Proceedings EWRS 5th Symposium on Aquatic Weeds 1978,* pp. 117–26.

Van, T. K., Haller, W. T., Bowes, G., and Garrard, L. A. (1977). Effects of light quality on growth and chlorophyll composition in *Hydrilla. Journal of Aquatic Plant Management,* **15,** 29–31.

Van, T. K., Haller, W. T., and Garrard, L. A. (1978). The effect of daylength and temperature on *Hydrilla* growth and tuber production. *Journal of Aquatic Plant Management,* **16,** 57–9.

Van, T. K. and Steward, K. K. (1985). The use of controlled-release fluridone fibres for control of hydrilla (*Hydrilla verticillata*). *Weed Science,* **34,** 70–6.

van Aart, R. (1985). Irrigation research and development in Malawi. In *IILRI annual report 1984,* pp. 12–29. International Institute for Land Reclamation and Improvement, Wageningen, The Netherlands.

van Busschbach, E. J. and Elings, H. (1967). The use of dichlobenil against aquatic weeds in ditches and ponds. *Proceedings EWRC 2nd Symposium on Aquatic Weeds 1967,* pp. 130–7.

van Dam, H., and Kooyman van Blokland, H. (1978). Man-made changes in some Dutch moorland pools, as reflected by historical and recent data about diatoms and macrophytes. *Internationale Revue der Gesamten Hydrobiologie,* **63,** 587–607.

van den Bosch, R., Leigh, T. F., Falcon, L. A., Stern, V. M., Gonzales, D., and Hagen, K. S. (1971). The development program of integrated control of cotton pests in California. In *Biological control* (ed. C. B. Huffaker), pp. 377–94. Plenum, New York.

van den Linden, J. H. A. (1980). Nitrogen economy of reed vegetation in the Zuidelijk Flevoland polder. I. Distribution of nitrogen among shoots and rhizomes during the growing season, and loss of nitrogen due to line management. *Oecologia Plantarum,* **1,** (3), 219–30.

van der Bliek, A. M., El Gharably, Z., Pieterse, A. H., and Scheepens, M. H. M. (1982). Observations of the phenology of *Potamogeton pectinatus* L. and other submerged weeds in irrigation systems in Egypt. *Proceedings EWRS 6th Symposium on Aquatic Weeds 1982,* pp. 37–44.

vander Borght, P., Ska, B., Schmitz, A., and Wollast, R. (1982). Eutrophisation de la rivière Semois: le développement de *Ranunculus* et ses conséquences sur l'écosystème aquatique. In *Studies on aquatic vascular plants* (ed. J. J. Symoens, S. S. Hooper, and P. Compère). pp. 340–5. Royal Botanical Society of Belgium, Brussels.

van der Eijk, M. (1978). Notes on the experimental introduction of grass carp (*C. idella*) into the Netherlands. *Proceedings EWRS 5th Symposium on Aquatic Weeds 1978,* pp. 245–51.

van der Ploeg, D. T. E. (1966). *Elodea nuttalli* (Planch.) St. John in Friesland. *Gorteria,* **3,** 76.

van der Toorn, J. (1978). Effect of cutting on reed performance. *Progress Report 1978 Royal Netherlands Academy of Arts & Science,* pp. 335–9.

van der Toorn, J. and Mook, J. H. (1982). The influence of environmental factors and management on stands of *Phragmites australis.* I. Effects of burning, frost and insect damage on short density and shoot size. *Journal of Applied Ecology,* **19,** 477–99.

van der Valk, A. G. and Davies, C. B. (1978). The role of seed bank in the vegetation dynamics of prairie glacial marshes. *Ecology,* **59,** 322–35.

van der Velde, G. (1978). Structure and function of a nymphaeid-dominated system. *Proceedings EWRS 5th Symposium on Aquatic Weeds 1978,* pp. 127–33.

van der Velde, G., Custers, C. P. C., and de Lyon, M. J. H. (1986). The distribution of four nymphaeid species in the Netherlands in relation to selected abiotic factors. *Proceedings EWRS/AAB 7th Symposium on Aquatic Weeds 1986,* pp. 363–8.

van der Velde, G. and Peelen-Bexkens, P. M. M. (1983). Production and biomass of floating leaves of three species of Nymphaeaceae in two Dutch waters. *Proceedings International Symposium on Aquatic Macrophytes, Nijmegen 1983,* pp. 230–5.

van der Weert, R. and Kamerling, G. E. (1974). Evapotranspiration by water hyacinth (*Eichhornia crassipes*). *Journal of Hydrology,* **22,** 201–12.

van der Weij, H. G. (1966). Recent developments and investigations in chemical control of aquatic weeds in the Netherlands. *Proceedings 8th British Weed Control Conference 1966,* **3,** pp. 835–41.

van der Weij, H. G., Hoogers, B. J., and Blok, E. (1971). Mercaptotriazines compared with diuron for aquatic weed control in stagnant waters. *Proceedings EWRS 3rd Symposium on Aquatic Weeds 1971*, pp. 149–60.

van der Werf, A. (1986). [*Effect of the reserve substances on the respiration rate of the submerged macrophyte* Ceratophyllum demersum *L.: a simulation model for growth of* Ceratophyllum demersum *L.*]. In Dutch. Report Limnologisch Instituut Nieuwersluis/Oosterzee 1986/8. pp. 61.

van der Zweerde, W. (1980). *Graskarper en macrofauna, verslag over onderzoek in een zestal proefsloten*. Internal Report, Centrum voor Agrobiologisch Onderzoek, Wageningen. pp. 127.

van der Zweerde, W. (1982*a*). The influence of grass carp (*Ctenopharyngodon idella* Val.) on macro-invertebrates in some experimental ditches in the Netherlands. *Proceedings 2nd International Symposium on Herbivorous Fish, 1982*, pp. 158–64. EWRS, Wageningen, The Netherlands.

van der Zweerde, W. (1982*b*). Some introductory experiments on the influence of day-length, light intensity and temperature on turion formation and flowering in two strains of *Hydrilla verticillata* (L.f.) Royle. *Proceedings EWRS 6th Symposium on Aquatic Weeds 1982*, pp. 71–6.

van der Zweerde, W. (1983). The use of grass carp (*Ctenopharyngodon idella* Val.) in the management of watercourses in the Netherlands: effects and side-effects. *Proceedings International Symposium on Aquatic Macrophytes, Nijmegen 1983*, pp. 332–6.

van der Zweerde, W., Hoogers, B. J., and van Zon, J. C. J. (1978). Effects of grass carp on microflora and -fauna, macroflora, macro-invertebrates and chemical properties in the water. *Proceedings EWRS 5th Symposium on Aquatic Weeds 1978*, pp. 343–50.

van Dord, J. C. J., Hoogers, B. J., and van Zon, J. C. J. (1974). Studies on the side effects of herbicides used in the aquatic environment. *Proceedings EWRS 4th Symposium on Aquatic Weeds 1974*, pp. 173–9.

Van Dyke, J. M., Leslie, A. J., and Nall, L. E. (1984). The effect of grass carp on the aquatic macrophytes in four Florida lakes. *Journal of Aquatic Plant Management*, **22**, 87–95.

van Himme, M. Stryckers, J., and Bulcke, R. (1977). *Bespreking van de resultaten bereikt door het Centrum voor Onkruidonderzoek tijdens de proefjaren 1975–1976–1977. Biologie, ekologie en bestrijding van draadwieren,* Vaucheria *spp. Biotoetsen voor het opsporen van herbiciden en algiciden in water*. Report, Rijksuniversiteit Gent 1977, pp. 121–5.

van Ieperen, H., and Herfst, M. S. (1986). Laboratory experiments on the flow resistance of aquatic weeds. In *Proceedings 2nd International Conference on Hydraulic Design in Water Resources Engineering: Land Drainage April 1986*.

van Leeuwen, B. H. (1979). Grass carp: a threat to our amphibia? *Environmental Conservation*, **6**, 264.

van Overbeek, J., Hughes, W. J., and Blondeau, R. (1959). Acrolein for the control of water weeds and disease-carrying water snails. *Science* (*New York*), **129**, 335–6.

van Schayck, C. P. (1985). Laboratory studies on the relation between aquatic vegetation and the presence of two bilharzia-bearing snail species. *Journal of Aquatic Plant Management*, **23**, 87–91.

van Schayck, C. P. (1986). The effect of several methods of aquatic weed control on two bilharzia-bearing snail species. *Aquatic Botany,* **24,** 303–9.

van Vierssen, W. (1982*a*). Some notes on the germination of seeds of *Najas marina* L. *Aquatic Botany,* **12,** 201–3.

van Vierssen, W. (1982*b*). The ecology of communities dominated by *Zannichellia* taxa in western Europe. I. Characterization and autecology of the *Zannichellia* taxa. *Aquatic Botany,* **12,** 103–55.

van Vierssen, W., Breukelaar, A. W., and Peppelenbos, H. W. (1986). A comparison of some morphological characteristics of four *Hydrilla* strains under different environmental conditions. *Proceedings EWRS/AAB 7th Symposium on Aquatic Weeds 1986,* 369–74.

van Vierssen, W., van Dijk, G. M., and Breukelaar, A. W. (1985). Grenzen aan de groei? Enkele notities betreffende een waterplantengroeipotentietoets. *H_2O* 21, 440–3.

van Vierssen, W., van der Zee, J. R., and van Kessel, C. M. (1984). On the germination of *Ruppia* taxa in Western Europe. *Aquatic Botany,* **19,** 381–93.

van Vierssen, W. and van Wijk, R. J. (1982). On the identity and autoecology of *Zannichellia peltata* Bertol. in western Europe. *Aquatic Botany,* **13,** 367–83.

van Weerd, J. H. (1985). Growth and survival in drainage channels of grass carp, *Ctenopharyngodon idella* Val., fry and their potential for weed control. *Aquaculture and Fisheries Management,* **1,** 7–23.

van Wijk, R. J. (1983). Life cycles and reproductive strategies of *Potamogeton pectinatus* L. in the Netherlands and the Camargue (France). *Proceedings International Symposium on Aquatic Macrophytes, Nijmegen 1983,* pp. 317–21.

van Wijk, R. J. (1986). Life-cycle characteristics of *Potamogeton pectinatus* L. in relation to control. *Proceedings EWRS/AAB 7th Symposium on Aquatic Weeds 1986,* pp. 375–80.

van Zon, J. C. J. (1974). The grasscarp in Holland. *Proceedings EWRS 4th Symposium on Aquatic Weeds 1974,* pp. 128–33.

van Zon, J. C. J. (1976). Status of biotic agents other than insects or pathogens as biocontrols. *Proceedings 4th International Symposium on Biological Control of Weeds, Gainesville, Florida 1976,* pp. 245–50.

van Zon, J. C. J. (1977). Grass carp (*Ctenopharyngodon idella* Val.) in Europe. *Aquatic Botany,* **3,** 143–55.

van Zon, J. C. J. (1979). The use of grass carp in comparison with other aquatic weed control methods. *Proceedings Grass Carp Conference, Gainesville, Florida 1976,* pp. 15–26.

van Zon, J. C. J. (1981). Status of the use of grass carp (*Ctenopharyngodon idella* Val.). *Proceedings 5th International Symposium on Biological Control of weeds, Brisbane,* pp. 249–60.

van Zon, J. C. J. (1982). *Aquatic weeds.* Chapter 38 in *Biology and ecology of weeds* (eds W. Holzner and N. Numata). W. Junk, The Hague. pp. 449–56.

van Zon, J. C. J. (1985). Conflicts of interest in the use of grass carp. *Proceedings VIth International Symposium of Biological Control of Weeds, Vancouver,* pp. 399–403.

van Zon, J. C. J., van der Zweerde, W. I., and Hoogers, B. J. (1978). The grass carp, its effect and side effects. *Proceedings 4th International Symposium on Biological Control of Weeds, Gainesville 1976,* pp. 251–6.

Vardia, H. K. and Durve, V. S. (1981). The toxicity of 2,4-D to *Cyprinus carpio* var. *communis* in relation to the seasonal variation in temperature. *Hydrobiologia,* **77**, 155–9.

Vardia, H. K., Rao, P. S., and Durve, V. S. (1984). Sensitivity of toad larvae to 2,4-D and endosulfan pesticides. *Archiv für Hydrobiologie,* **100,** 395–400.

Varshney, C. K. and Rzóska, J. (ed.) (1976). *Aquatic weeds in south-east Asia.* W. Junk, The Hague. pp. 396.

Varshney, C. K. and Singh, K. P. (1976). A survey of the aquatic weed problem in India. In *Aquatic weeds in south east Asia* (ed. C. K. Varshney and J. Rzóska), pp. 31–42. W. Junk, The Hague.

Veber, K., Zahradnik, J., Breyl, I., and Kredl, F. (1981). Toxic effect and accumulation of atrazine in algae. *Bulletin of Environmental Contamination Toxicology,* **27**, 872–8.

Velu, M. (1976). Development of equipment for eradication of aquatic weeds. In *Aquatic weeds in south-east Asia* (ed. C. K. Varshney and J. Rzóska), pp. 233–40. W. Junk, The Hague.

Vera, A. (1970). Evaluación de las malezas en vias acuáticas en algunos distritos de riego en Columbia. *Agricultura Tropical, Bogota,* **26**, 649–56.

Vercruysse, J. (1985). The epidemiology of human and animal schistosomiasis in the Senegal River Basin. *Acta Tropica,* **42**, 249–59.

Verhoeven, J. T. A. (1979). The ecology of *Ruppia*-dominated communities in Western Europe. I. Distribution of *Ruppia* representatives in relation to their autoecology. *Aquatic Botany,* **6**, 197–268.

Verigin, B. V., Makeyeva, A. P., and Zaki Mokhamed, M. I. (1979). Natural spawning of the silver carp (*Hypophthalmichthys molitrix*, the bighead carp (*Aristichthys nobilis*) and the grass carp (*Ctenopharyngodon idella*) in the Syr-Dar'ya River. *Journal of Ichthyology,* **18**, 143–6.

Verkleij, J. A. C. and Pieterse, A. H. (1986). Identification of *Hydrilla verticillata* (L.f.) Royle strains by means of isoenzyme patterns. *Proceedings EWRS/AAB 7th Symposium on Aquatic Weeds 1986,* pp. 381–8.

Verkleij, J. A. C., Pieterse, A. H., Horneman, G. J. T., and Torenbeek, M. (1983*a*). A comparative study of the morphology and isoenzyme patterns of *Hydrilla verticillata* (L.f.) Royle. *Aquatic Botany,* **17**, 43–59.

Verkleij, J. A. C., Pieterse, A. H., Staphorst, H. P. M., and Steward, K. K. (1983*b*). Identification of two different genotypes of *Hydrilla verticillata* (L.f.) Royle in the USA by isoenzyme studies. *Proceedings International Symposium on Aquatic Macrophytes, Nijmegen 1982,* 251–61.

Verloop, A. and Nimmo, W. B. (1970). Transport and metabolism of dichlobenil in wheat and rice seedlings. *Weed Research,* **10**, 65–70.

Verloop, A., Nimmo, W. B., and de Wilde (1974). The fate of the herbicide dichlobenil in fresh water fish. *Proceedings EWRC 4th Symposium on Aquatic Weeds 1974,* pp. 186–91.

Victoria Filho, R. and Carvalho, J. B. (1981). Controle de plantas daninhas na cultura do arroz de sequeiro (*Oryza sativa* L.). *Planta Daninha,* **4**, 11–16.

Vietmeyer, N. (1974). The endangered but useful manatee. *Smithsonian Magazine,* **5**, 60–5.

Vinogradov, V. K. and Zolotova, Z. K. (1974). [Effect of grass carp on ecosystems of reservoirs]. In Russian. *Gidrobiologicheskii Zhurnal,* **10,** 90–6.

Vollenweider, R. A. (1968). *Scientific fundamentals of the eutrophication of lake and flowing waters, with particular reference to nitrogen and phosphorus as factors in eutrophication.* Technical Report, Water Management Research OECD. Paris. pp. 159.

von Karman, T. (1934). Turbulence and skin friction. *Journal of Aeronautical Science,* **1,** 1–20.

von Lukowicz, M. (1976). Pflanzenfresser als Nebenfische in der Teichwirtschaft. *Bayerisches Landwirtschaftliches Jahrbuch,* **10,** 72–8.

von Menzel, A. (1974). Gewässerreinhaltung durch pflanzenfressende Fische. *Proceedings EWRC 4th Symposium on Aquatic Weeds 1974,* pp. 139–43.

Vooren, C. M. (1972). Ecological aspects of the introduction of fish species into natural habitats in Europe, with special reference to the Netherlands: a literature survey. *Journal of Fish Biology,* **4,** 565–83.

Vovk, P. S. (1976). [The use of white amur as a biocontrol agent for canals and industrial reservoirs]. In Russian. *Biologiya dal'nevostochnykh rastitel'noyadnykh rgbi ikh khozyaistvennoe ispol'zovanie v vodoemalch Ukrainy,* pp. 183–206.

Voyer, R. A. and Heltsche, J. F. (1984). Factor interactions and aquatic toxicity testing. *Water Research,* **18,** (4), 441–8.

Vrhovsek, D., Martincić, A., and Kralj, M. (1981). Evaluation of the polluted river Savinja with the help of macrophytes. *Hydrobiologia,* **80,** 97–110.

Wade, P. M. (1978). The effect of mechanical excavators on the drainage channel habitat. *Proceedings EWRS 5th Symposium on Aquatic Weeds 1978,* pp. 333–42.

Wade, P. M. (1981). The long-term effects of aquatic herbicides on the macrophyte flora of freshwater habitats—a review. *Proceedings Association of Applied Biologists Conference: Aquatic Weeds and their Control 1981,* pp. 233–40. AAB, Wellesbourne, UK.

Wade, P. M. (1982). The long-term effects of herbicide treatment on aquatic weed communities. *Proceedings EWRS 6th Symposium on Aquatic Weeds 1982,* pp. 278–85.

Wade, P. M. (1987). A review of the provision made for the identification of wetland macrophytes as an aid to the study and management of wetlands. *Archiv für Hydrobiologie: Beiheft Ergebnisse der Limnologie,* **27,** 105–13.

Wade, P. M. and Bowles, F. (1981). A comparison of the efficiency of freshwater macrophyte surveys carried out from underwater with those from the shore or a boat. *Progress in Underwater Science* (*Report of 14th Symposium of the Underwater Association*), **6,** 7–11.

Wade, P. M. and Edwards, R. W. (1980). The effect of channel maintenance on the aquatic macrophytes of the drainage channels of the Monnouthshire levels, South Wales, 1840–1976. *Aquatic Botany,* **8,** 307–22.

Wade, P. M., Vanhecke, L., and Barry, R. (1986). The importance of habitat creation, weed management and other habitat disturbance to the conservation of the rare aquatic plant *Callitriche truncata* Guss. *Proceedings EWRS/AAB 7th Symposium on Aquatic Weeds 1986,* pp. 389–94.

Wahlquist, H. (1972). Production of water hyacinths and resulting water quality in earthen ponds. *Hyacinth Control Journal,* **10,** 9–11.

Waldrep, T. W. and Taylor, H. M. (1976). 1-methyl-3 phenyl [3-(trifluoromethyl) phenyl]-4 (1*H*)-pyridinone, a new herbicide. *Agricultural Food and Chemistry,* **24,** 1250–1.

Walker, A. O. (1912). The distribution of *Elodea canadensis* Michaux in the British Isles in 1909. *Proceedings of the Linnean Society, London,* **124,** 71–7.

Walker, C. R. (1960). Herbicide toxicity and ecology in Missouri fish ponds with reference to aquatic weed control. *Proceedings North Central Weed Control Conference,* **17,** pp. 30.

Walker, C. R. (1963). Endothall derivatives as aquatic herbicides in fishery habitats. *Weeds,* **8,** 226–32.

Walker, C. R. (1965). Diuron, fenuron, monuron, neburon, and TCA mixtures as aquatic herbicides in fish herbicides in fish habitats. *Weeds,* **13,** 297.

Walker, C. R. (1971). Toxicological effects of herbicides and weed control on fish and other organisms in the aquatic ecosystem. *Proceedings EWRC 3rd Symposium on Aquatic Weeds 1971,* pp. 119–27.

Wallsten, M. (1981). Changes of lakes in Uppland, Central Sweden during 40 years. *Acta Universitatis Upsaliensis, Symbolae Botanicae Upsalienses,* **23,** (3), pp. 84.

Wallsten, M. (1982). Changes in aquatic vegetation of eutrophic Lake Tämnaren. In *Studies on aquatic vascular plants* (ed. J. J. Symoens, S. S. Hooper, and P. Compère), pp. 293. Royal Botanical Society of Belgium, Brussels.

Wallsten, M. (1983). Starch content in roots in relation to vegetative growth in natural and treated aquatic macrophyte areas. *Proceedings International Symposium on Aquatic Macrophytes, 1983, Nijmegen,* pp. 292–7.

Walsh, G. E., Miller, C. W., and Heitmuller, P. T. (1971). Uptake and effects of dichlobenil in a small pond. *Bulletin of Environmental Contamination Toxicology,* **6,** 279–38.

Wapshere, A. J. (1982). Biological control of weeds. In *Biology and Ecology of Weeds* (ed. W. Holzner and N. Numala). pp.47–56.

Ward, J. and Talbot, J. (1984). Distribution of aquatic macrophytes in Lake Alexandrina, New Zealand. *New Zealand Journal of Marine and Freshwater Research,* **18,** 211–20.

Ware, F. J., Gasaway, R. D., Marz, R. A., and Drda, T. F. (1975). Investigations of herbivorous fishes in Florida. *Proceedings Symposium on Water Quality Management through Biological Control, Gainesville 1975,* 79–84.

Ware, G. W. (1983). *Pesticides: theory and application.* W. H. Freeman & Co., San Francisco. pp. 308.

Warrington, P. D. (1980). *Studies on aquatic macrophytes. XXXIII. Aquatic plants of British Columbia.* Ministry of Environment, British Columbia.

Warrington, P. D. (1985). Factors associated with the distribution of *Myriophyllum* in British Columbia. In *1st International Symposium on water milfoil* (Myriophyllum spicatum) *and related Haloragaceae species* (ed. L. W. J. Anderson), pp. 79–94. Aquatic Plant Management Society, Vicksburg, Mississippi.

Watkins, C. E., Thayer, D. D., and Haller, W. T. (1985). Toxicity of adjuvants to bluegill. *Bulletin of Environmental Contamination Toxicology,* **34,** 138–42.

Way, J. M., Newman, J. F., Moore, N. W., and Knaggs, F. W. (1971). Some ecological effects of the use of paraquat for the control of weeds in small lakes. *Journal of Applied Ecology,* **8,** 509–32.

Weber, E. (1974). Bekämpfung unerwünschter Wasserpflanzen durch den Weissen Amur. *Proceedings EWRC 4th Symposium on Aquatic Weeds 1974,* pp. 134–8.

Weber, E. (1984). Die Ausbreitung der Pseudokeilfleckbarben im Donauraum. *Österreichs Fischerei,* **37**, 63–5.

Weber, J. A. (1979). The ecology of *Ruppia*-dominated communities in western Europe. I. Distribution of *Ruppia* representatives in relation to their autecology. *Aquatic Botany,* **6**, 192–267.

Weber, J. A. and Noodén, L. D. (1974). Turion formation and germination in *Myriophyllum verticillatum:* phenology and its interpretation. *Michigan Botanist,* **13**, 151–8.

Weber, J. A. and Noodén, L. D. (1976*a*). Environmental and hormonal control of turion formation in *Myriophyllum verticillatum. Plant and Cell Physiology,* **17**, 721–31.

Weber, J. A. and Noodén, L. D. (1976*b*). Environmental and hormonal control of turion germination in *Myriophyllum verticillatum. American Journal of Botany,* **63**, 936–44.

Weber, J. A., Tenhunen, J. D., Westrin, S. S., Yocum, C. S., and Gates, D. M. (1981). An analytical model of photosynthetic response of aquatic plants to inorganic carbon and pH. *Ecology,* **62**, 697–705.

Weber, J. B. (1972). Interaction of organic pesticides with particulate matter in aquatic and soil systems. In *Fate of organic pesticides in the aquatic environment, Advances in chemistry series III,* pp. 55. American Chemistry Society.

Weber-Oldecop, D. W. (1970). Water plant communities in Eastern Lower Saxony I. *Internationale Revue der Gesamten Hydrobiologie,* **55**, 913–67.

Weber-Oldecop, D. W. (1971). Water plant communities in Eastern Lower Saxony II. *Internationale Revue der Gesamten Hydrobiologie,* **56**, 79–122.

Weber-Oldecop, D. W. (1977). *Elodea nuttalli* (Planch.) St. John, Hydrocharitaceae, a new limnic phanerogam of German flora. *Archiv für Hydrobiologie,* **79**, 397–403.

Weidenbacher, W. D. and Willenbring, P. R. (1984). Limiting nutrient flux into an urban lake by natural treatment and diversion. *Lake & Reservoir Management: Proceedings 3rd Annual Conference, North American Lake Management Society,* pp. 525–6. Environmental Protection Agency, Washington DC, USA.

Welker, W. V. and Riemer, D. N. (1982). Fragrant waterlily (*Nymphaea odorata*) control with multiple applications of glyphosate. *Weed Science,* **30**, 145–6.

Werkgroep Afvoerberekeningen (1979). *Richtlijnen voor het berekenen van afwateringsstelsels in landelijke gebieden*. Sectie en Studiekring voor Cultuurtechniek. pp. 155.

Werkgroep Graskarper (1984). *Graskarpers in Nederland 1984*. Werkgroep Graskarper, NRLO, Wageningen. pp. 144.

West, G. (1905). A comparative study of the dominant phanerogamic and higher cryptogamic flora of aquatic habit in three lake areas in Scotland. *Proceedings Royal Society of Edinburgh,* **25**, 967–1023.

West, S. D., Burger, R. O., Poole, G. M., and Mowrey, D. H. (1983). Bioconcentration and field dissipation of the aquatic herbicide fluridone and its degradation products in aquatic environments. *Journal of Agricultural and Food Chemistry,* **31**, 579–85.

West, S. D., Day, E. W., and Burger, R. O. (1979). Dissipation of the experimental aquatic herbicide fluridone from lakes and ponds. *Journal of Agricultural and Food Chemistry,* **27**, 1067–72.

West, S. D. and Parka, S. J. (1981). Determination of the aquatic herbicide fluridone in water and hydrosoil: effect of application method on dissipation. *Journal of Agricultural and Food Chemistry*, **29**, 223–6.

Westerdahl, H. E. (1983*a*). Effect of Hydout and Aquathol K on hydrilla in Gatun Lake, Panama. *Journal of Aquatic Plant Management*, **21**, 17–21.

Westerdahl, H. E. (1983*b*). *Field evaluation of two endothall formulations for managing hydrilla in Gatun Lake, Panama*. Technical Report A-83-3. US Army Engineer Waterways Experiment Station, Vicksburg, Mississippi, pp. 83.

Westerdahl, H. E. (1986). Chemical control technology development, a review. In *Proceedings 20th Annual Meeting, Aquatic Plant Control Research Program, 1985*, pp. 26–30. Miscellaneous Paper A-86-2, US Army Engineer Waterways Experiment Station, Vicksburg.

Westerdahl, H. E. and Getsinger, K. D. (ed.) (1988). *Aquatic plant identification and herbicide use guide. Volume II. Aquatic plants and susceptibility to herbicides*. pp. 104. Technical Report A-88-9, US Army Engineer Waterways Experiment Station, Vicksburg, Mississippi.

Westerdahl, H. E. and Hall, J. F. (1983). Threshold 2,4-D concentrations for control of Eurasian watermilfoil and sago pondweed. *Journal of Aquatic Plant Management*, **21**, 22–5.

Westerdahl, H. E. and Hoeppel, R. E. (1982). 2,4-D residue dissipation studies to support expansion of the Federal level. *Proceedings 16th Annual Meeting Aquatic Meeting Aquatic Plant Control Research Planning & Operations Review*, pp. 87–92. Miscellaneous Paper A-82-3. US Army Engineer Waterways Experiment Station, Vicksburg, Mississippi.

Westlake, D. F. (1961). Aquatic macrophytes and the oxygen balance of running water. *Verhandlungen der Internationalen Vereinigung für theoretische und angewandte Limnologie*, **14**, 499–504.

Westlake, D. F. (1963). Comparisons of plant productivity. *Biological Reviews*, **38**, 385–425.

Westlake, D. F. (1964). Light extinction, standing crop and photosynthesis within weed beds. *Verhandlungen der Internationalen Vereinigung für theoretische und angewandte Limnologie*, **15**, 415–25.

Westlake, D. F. (1966*a*). A model for quantitative studies of photosynthesis by higher plants in streams. *Air and Water Pollution International Journal*, **10**, 883–96.

Westlake, D. F. (1966*b*). The biomass and productivity of *Glyceria maxima*. I. Seasonal changes in biomass. *Journal of Ecology*, **54**, 745–53.

Westlake, D. F. (1967). Some effects of low velocity currents on the metabolism of aquatic macrophytes. *Journal of Experimental Botany*, **18**, 187–205.

Westlake, D. F. (1968). The weight of water weed in the River Frome. *Yearbook of the Association of River Authorities 1968*, pp. 3–12.

Westlake, D. F. (1975). Primary productivity of freshwater macrophytes. In *Photosynthesis and productivity in different environments*, International Biological Programme, No. 3 (ed. J. P. Cooper), pp. 189–206. Cambridge University Press, London & New York.

Westlake, D. F. (1981). The development and structure of aquatic weed populations. *Proceedings Association of Applied Biologists Conference: Aquatic Weeds and their Control 1981*, pp. 33–48. AAB, Wellesbourne, UK.

Westlake, D. F. and Dawson, F. H. (1982). Thirty years of weed cutting on a chalk stream. *Proceedings EWRS 6th Symposium on Aquatic Weeds 1982,* pp. 132–40.

Westlake, D. F. and Dawson, F. H. (1986). The management of *Ranunculus calcareus* by pre-emptive cutting in Southern England. *Proceedings EWRS/AAB 7th Symposium on Aquatic Weeds 1986,* pp. 395–400.

Westlake, D. F., Spence, D. H. N., and Clarke, R. T. (1986). *The direct determination of biomass of aquatic macrophytes and measurement of underwater light 1985.* Methods for the Examination of Waters and Associated Materials. HMSO, London. pp. 45.

Wetzel, R. G. (1975). *Limnology*, W. Saunders & Co., Philadelphia. pp. 743.

Wharfe, J. R., Taylor, K. S., and Montgomery, H. A. C. (1984). The growth of *Cladophora glomerata* in a river receiving sewage effluent. *Water Research,* **18,** 971–9.

Whigham, D. F. and Simpson, R. L. (1982). Germination and dormancy studies of *Pontederia cordata* L. *Bulletin of the Torrey Botanical Club,* **109,** 524–8.

White, M. P. and Lembi, C. A. (1976). The persistence and efficacy of a water-soluble blue dye in aquatic weed control. *Proceedings North Central Weed Control Conference 1976,* **31,** 103.

Whitton, B. A. (1971). Filamentous algae as weeds. *Proceedings EWRC 3rd Symposium on Aquatic Weeds 1971,* pp. 249–61.

Widyanto, L. S., Burhan, M. A., Susilo, H., and Nur, M. (1977). New combinations of herbicides to control water hyacinth (*Eichhornia crassipes* (Mart.) Solms). *Proceedings 6th Asian-Pacific Weed Science Society Conference 1977,* pp. 285–93.

Wiegleb, G. (1984). A study of habitat conditions of the macrophytic vegetation in selected river systems in Western Lower Saxony (Federal Republic of Germany). *Aquatic Botany,* **18,** 313–52.

Wiersma-Roem, W. J., Vischer, L. W. A., Frederix-Wolters, E. M. H., and Harmsen, E. G. M. (1978). Sub-acute toxicity of the herbicide dichlobenil (2,6 dichlorobenzonitrile) in rainbow trout (*Salmo gairdnerii*). *Proceedings EWRS 5th Symposium on Aquatic Weeds 1978,* pp. 261–8.

Wild, H. (1961). Harmful aquatic plants in Africa and Madagascar. *Kirkia,* **2,** 1–66.

Wile, I. (1978). Environmental effects of mechanical harvesting. *Journal of Aquatic Plant Management,* **16,** 14–20.

Willems, P. J. J. (1981). Graskarperkering bij Salland. *Waterschapsbelangen,* **66,** 429.

Willemsen, J., Schaap, L. A., Rolloos, H. M., and van den Ham, C. J. (1978). *De ontwikkeling van de visstand in Gasuniesloten na introductie van graskarper.* Internal Report, Rijksinstituut voor Visserijonderzoek, Ijmuiden. pp. 18.

Williams, E. H., Mather, E. L., and Carter, S. M. (1984), Toxicity of the herbicides endothall and diquat to benthic crustacea. *Bulletin of Environmental Contamination Toxicity,* **33,** 418–22.

Williams, R. H. (1956). *Salvinia auriculata* Aubl. The chemical eradication of a noxious aquatic weed in Ceylon. *Tropical Agriculture, Trinidad,* **33,** 145–57.

Williams, W. T. and Lambert, J. M. (1959). Multivariate methods in plant ecology. 1. Association-analysis in plant communities. *Journal of Ecology,* **47,** 83–101.

Williams, W. T. and Lambert, J. M. (1966). Multivariate methods in plant ecology. V. Similarity analysis and information analysis. *Journal of Ecology,* **54,** 427–45.

Wilson, D. C. and Bond, C. E. (1969). The effects of the herbicides diquat and dichlobenil (casoron) on pond invertebrates. I. Acute toxicity. *Transactions of the American Fisheries Society*, **98**, 438–42.

Wilson, S. D. and Keddy, P. A. (1985). Plant zonation along a lake shore gradient: physiological response curves of component species. *Journal of Ecology*, **73**, 851–60.

Wingfield, G. J. and Johnson, J. M. (1981). Deoxygenation of water following use of the herbicide terbutryn simulated in a batch culture system. *Bulletin of Environmental Contamination Toxicology*, **26**, 65–72.

Wingfield, G. I. and Bebb, J. M. (1982). Simulation of deoxygenation of water containing vascular plants following treatment with terbutryne. *Proceedings EWRS 6th Symposium on Aquatic Weeds, 1982*, pp. 255–62.

Winter, K. (1978). Short-term fixation of ^{14}carbon by the submerged aquatic angiosperm *Potamogeton pectinatus*. *Journal of Experimental Botany*, **29**, 1169–72.

Wium-Andersen, S. (1971). Photosynthetic uptake of free CO_2 by the roots of *Lobelia dortmanna*. *Physiologia Plantarum*, **25**, 245–8.

Wium-Andersen, S. and Andersen, J. M. (1972). The influence of vegetation on the redox profile of the sediment of Grane Langsø, a Danish 'Lobelia' lake. *Limnology and Oceanography*, **17**, 948–52.

Wlosinski, J. H. (1981). *Workshop on modelling of aquatic macrophytes*. Miscellaneous Paper A-81-1. US Army Engineers, Washington DC. pp. 17.

Wolek, J. (1974). [Critical survey of Polish floating plant communities]. In German. *Fragmenta Floristica Geobotanica*, **20**, 365–79.

Wolff, P. (1980). Die Hydrillae (Hydrocharitaceae) in Europa. *Gött. Flor. Rundbr.*, **14**, 33–56.

Wolseley, P. A. (1986). The aquatic macrophyte communities of the ditches and dykes of the Somerset levels and their relation to management. *Proceedings EWRS/AAB 7th Symposium on Aquatic Weeds 1986*, pp. 407–18.

Wolverton, B. C. (1987). Aquatic plants for wastewater treatment: an overview. In *Aquatic plants for water treatment and resource recovery* (ed. K. R. Reddy and W. H. Smith), pp. 3–15. Magnolia Publications, Orlando, Florida.

Wolverton, B. and McDonald, R. C. (1976). Don't waste waterweeds. *New Scientist*, **71**, (1013), 318–20.

Wolverton, B. C., McDonald, R. C., and Gordon, J. (1975). *Bioconversion of water hyacinths into methane gas. Part 1*. NASA Technical Memorandum TM-X-72725.

Wong, S. L. and Clark, B. (1979). The determination of desirable and nuisance plant levels in streams. *Hydrobiologia*, **63**, 223–30.

Wood, R. D. (1963). Adapting SCUBA to aquatic plant ecology. *Ecology*, **44**, 416–19.

Wood, R. D. (1972). Characeae of Australia. *Nova Hedwigia*, **22**, 1–120.

Wood, R. D., and Imahori, K. (1959). Geographical distribution of Characeae. *Bulletin Torrey Botanical Club*, **86**, 172–83.

Wood, R. D. and Imahori, K. (1965). *A revision of Characeae. 1. Monograph of the Characeae*. J. Cramer Verlag, Mannheim. pp. 904.

Wood, R. D. and Mason, R. (1977). Characeae of New Zealand. *New Zealand Journal of Botany*, **15**, 87–180.

Woolson, E. A. (1975). Bioaccumulation of arsenicals. In *Arsenical Pesticides*. American Chemical Society Symposium Series 7, pp. 97–107.

Worthing, C. R., Richardson, W. G., and Taylor, W. A. (1982). Chapter 5: *Herbicides and their properties*. In *Weed control handbook: principles* (7th edn.), (ed. H. A. Roberts), pp. 206–57, British Crop Protection Council and Blackwell, London.

Worthing, C. R. and Walker, S. B. (1983). *The Pesticide Manual* (7th edn.). British Crop Protection Council, Croydon. pp. 695.

Wright, A. D. (1981). Biological control of water hyacinth in Australia. *Proceedings 5th International Symposium on Biological Control of Weeds* (ed. E. S. Del Fosse), pp. 529–36. CSIRO, Australia. Melbourne.

Wright, A. D. (1984). Effect of biological control agents on water hyacinth in Australia. *Proceedings International Conference on Water Hyacinth* (ed. G. Thyagarajan), pp. 823–33. UNEP, Nairobi.

Wright, D. M. (1973). *The fly-fisher's plants: their value in trout waters*. David & Charles, Newton Abbot (UK).

Wright, J. F., Cameron, A. C., Hiley, P. D., and Berrie, A. D. (1982). Seasonal changes in biomass of macrophytes on shaded and unshaded sections of the River Lambourn, England. *Freshwater Biology*, **12**, 271–83.

Wright, R. M. and McDonnell, A. J. (1986*a*). Macrophyte growth in shallow streams: biomass model. *Journal of Environmental Engineering, American Society of Civil Engineers*, **112**, 967–84.

Wright, J. F., Hiley, P. D., Ham, S. F., and Berrie, A. D. (1981). Comparison of three mapping procedures developed for river macrophytes. *Freshwater Biology*, **11**, 369–79.

Wright, R. M. and McDonnell, A. J. (1986*b*). Macrophyte growth in shallow streams: field investigations. *Journal of Environmental Engineering, American Society of Civil Engineers*, **112**, 953–66.

Wunderlich, W. E. (1967). The use of machinery in the control of aquatic vegetation. *Hyacinth Control Journal*, **6**, 22–5.

Yabuno, T. (1983). Biology of *Echinochloa* species. In *Weed control in rice*, pp. 307–18. International Rice Research Institute, Manila.

Yadav, P. S. and Varshney, C. K. (1981). Notes on the ecological and socio-economic importance of wetlands of north-eastern India. *International Journal of Ecology and Environmental Science*, **7**, 149–50.

Yamanaka, T. (1983). Effect of paraquat on growth of *Nitrosomonas europaea* and *Nitrobacter agilis*. *Plant and Cell Physiology*, **24**, 1349–52.

Yana, A. (1964). Essais de lutte contre les typhas en Tunisie. *Proceedings EWRC Symposium, Moyens de Destruction des Plantes Aquatiques, La Rochelle 1964*, pp. 111–25.

Yaschenko, P. T. (1984). [Biomorphological spectrum of the flora from Shatsk Lakes region]. In Ukrainian; English summary. *Ukr. Bot. Zhurnal*, **41**, 73–6.

Yefimova, T. A. and Nikanorov, Yu. I. (1977). Prospects for the introduction of phytophagous fishes into Ivankovskoye Reservoir. *Journal of Ichthyology*, **17**, 634–44.

Yeo, R. R. (1959). Germination of gigantic sago pondweed (*Potamogeton pectinatus* var. *interruptus*). *Proceedings Western Weed Control Conference*, pp. 70.

Yeo, R. R. (1964). Influence of water quality on toxicity of aquatic herbicides to certain plants and fish. *Abstracts 1964 Meeting, Weed Society of America*, pp. 107.

Yeo, R. R. (1965). Life history of sago pondweed. *Weeds*, **13**, 314–21.

Yeo, R. R. (1966). Yields of propagules of certain aquatic plants. *Weeds,* **14**, 110–13.
Yeo, R. R. (1967). Dissipation of diquat and paraquat and effects on aquatic weeds and fish. *Weeds,* **15**, 42–6.
Yeo, R. R. (1970). Dissipation of endothall and effects on aquatic weeds and fish. *Weed Science,* **18**, 282–4.
Yeo, R. R. (1980*a*). Spikerush may help control waterweeds. *Californian Agriculture,* April 80, 12–13.
Yeo, R. R. (1980*b*). Life history and ecology of dwarf spikerush (*Eleocharis coloradoensis*). *Weed Science,* **28**, 263–72.
Yeo, R. R. and Dechoretz, N. (1976). Diquat and copper-ion residues in salmon-spawning channels. *Weed Science,* **24**, 405–9.
Yeo, R. R. and Dechoretz, N. (1977). Acute toxicity of a herbicidal combination of diquat + copper ion to eggs, alevins and fry of rainbow trout and two aquatic macro-invertebrates. *Journal of Aquatic Plant Management,* **15**, 57–60.
Yeo, R. R. and Fisher, T. W. (1970). Progress and potential for biological control with fish, pathogens, competitive plants and snails. In *Technical Papers, First FAO International Conference on Weed Control,* pp. 450–63.
Yeo, R. R. and Thurston, J. R. (1984). The effect of dwarf spikerush (*Eleocharis coloradoensis*) on several submerged aquatic weeds. *Journal of Aquatic Plant Management,* **22**, 52–6.
Zambrano, C. J. O. (1974). Las malezas acuáticas. *Revista de la Facultad de Agronomia Universitad Zulía* (*Venezuela*), **2**, 87–94.
Zambrano, C. J. O. (1978). Aporte al estudio sistemático de las malezas acuáticas del Estado Zulía. *Revista de la Facultad de Agronomia Universitad Zulía* (*Venezuela*), **4**, 138–48.
Zanveld, J. S. (1940). The Charophyta of Malaysia and adjacent countries. *Blumea,* **4**, 1–223.
Zelitch, I. (1971). *Photosynthesis, photorespiration and plant productivity.* Academic Press, New York.
Zerov, K. K. (1976). *Formirovaniye rastitel nosti i zarastinye vodokhranilischch Dneprovskogo kaskada.* Naukova Dumka Press, Kiev.
Zettler, F. W. and Freeman, T. E. (1972). Plant pathogens as biocontrols of aquatic weeds. *Annual Review of Phytopathology,* **10**, 455–70.
Zonderwijk, P. and van Zon, J. C. J. (1974). A Dutch vision on use of herbicides in waterways. *Proceedings EWRS 4th Symposium on Aquatic Weeds,* pp. 158–63.
Zonderwijk, P. and van Zon, J. C. J. (1976). Waterplanten als bondgenoten bij het onderhoud. *Waterschapsbelangen,* **61**, (2), 21–3.
Zonderwijk, P. and van Zon, J. C. J. (1978). Aquatic weeds in the Netherlands: a case of management. *Proceedings EWRS 5th Symposium on Aquatic Weeds 1978,* pp. 101–16.
Zonneveld, N. and van Zon, J. C. J. (1985). The biology and culture of grass carp (*Ctenopharyngodon idella*), with special reference to their utilization for weed control. In *Recent advances on aquaculture,* Volume 2 (ed. J. F. Muir and J. K. Roberts), pp. 119–91. Croom Helm, London.
Zweig, R. D. (1985). Freshwater aquaculture in China: ecosystem management for survival. *Ambio,* **14**, 66–74.

INDEX OF PLANT NAMES

SUBJECT INDEX

Includes organisms other than aquatic plants and aquatic plants used as biological control agents.